OXFORD GEOGRAPHIC
ENVIRONMENTAL ST

Editors: Gordon Clark, Andrew Goudie, and Ceri Peach

CULTIVATED LANDSCAPES OF NATIVE NORTH AMERICA

ALSO PUBLISHED BY
OXFORD UNIVERSITY PRESS
IN THE OXFORD GEOGRAPHICAL AND
ENVIRONMENTAL STUDIES SERIES

The New Middle Class and the Remaking of the Central City
David Ley

Culture and the City in East Asia
Won Bae Kim, Mike Douglass, Sang-Chuel Choe, and Kong Chong Ho (eds.)

Energy Structures and Environmental Futures in Europe
Torleif Haugland, Helge Ole Bergesen, and Kjell Roland

Homelessness, Aids, and Stigmatization
Lois Takahashi

Island Epidemics
Andrew Cliff, Peter Haggett, and Matthew Smallman-Raynor

Pension Fund Capitalism
Gordon L. Clark

Class, Ethnicity, and Community in Southern Mexico
Oaxaca's Peasantries
Colin Clarke

Cultivated Landscapes of Native Amazonia and the Andes
William M. Denevan

Indigenous Land Management in West Africa
An Environmental Balancing Act
Kathleen Baker

Cultivated Landscapes of Native North America

William E. Doolittle

OXFORD
UNIVERSITY PRESS

OXFORD
UNIVERSITY PRESS
Great Clarendon Street, Oxford OX2 6DP
Oxford University Press is a department of the University of Oxford.
It furthers the University's objective of excellence in research, scholarship,
and education by publishing worldwide in
Oxford New York
Athens Auckland Bangkok Bogotá Buenos Aires Cape Town
Chennai Dar es Salaam Delhi Florence Hong Kong Istanbul Karachi
Kolkata Kuala Lumpur Madrid Melbourne Mexico City Mumbai Nairobi
Paris São Paulo Shanghai Singapore Taipei Tokyo Toronto Warsaw
with associated companies in Berlin Ibadan
Oxford is a registered trade mark of Oxford University Press
in the UK and certain other countries
Published in the United States
by Oxford University Press Inc., New York

First published 2000
First published in Paperback 2001

British Library Cataloguing in Publication Data
Data available
Library of Congress Cataloging-in-Publication Data
Doolittle, William Emery.
Cultivated landscapes of native North America / William E. Doolittle.
p. cm.
Includes bibliographical references.
1. Indians of North America–Agriculture. 2. Nature–Effect of human beings on–North America.
3. Landscape changes–North America. I. Title.
E98.A3 D66 2000 631'.089'97–dc21 00-022129
ISBN 0-19-925071-5
1 3 5 7 9 10 8 6 4 2
Typeset by Best-set Typesetter Ltd., Hong Kong
Printed in Great Britain
on acid-free paper by
Biddles Ltd
Guildford & Kings Lynn

Again,
for my brothers- and sisters-in-arms
who didn't make it home

FOREWORD

Soon after Columbus made landfall in the Bahamas in 1492 his hopes of reaching Asia by sailing west were disappointed. Instead he encountered a 'New World'. Although it was new to the inhabitants of Europe, who in the sixteenth century reacted with amazement and curiosity to its strange people, plants, and animals, we now know that this 'new' world had been occupied by people for at least 15,000 years. By 10,000 years ago some of them had begun to cultivate useful plants, and by AD 1500 many different and complex forms of agriculture were being practised in large parts of South, Middle, and North America. It is the evidence of this indigenous agricultural endeavour that concerns the authors of this trilogy of books: *Cultivated Landscapes of Native Amazonia and the Andes* by William M. Denevan, *Cultivated Landscapes of Native North America* by William E. Doolittle, and *Cultivated Landscapes of Middle America on the Eve of Conquest* by Thomas M. Whitmore and B. L. Turner II.

Of what does this evidence consist? Before answering that question, the authors' approach to the subject needs to be emphasized. As they correctly point out, research on indigenous American agriculture has been concerned more with the origins, domestication, and dispersal of crops than with the nature of agricultural landscapes and field systems. The strength and novelty of this trio of books is that they all focus on the spatial characteristics of agriculture—fields and landscapes of cultivation. It is no accident that all four authors are geographers and that they are all experienced first-hand observers of indigenous agricultural systems 'in the field'. And although they emphasize agricultural landscapes, they do not disregard the processes of agricultural production: what crops were grown, how they were planted, tended and harvested, what tools and techniques were used, as well as more general questions such as the demographic capacity and long-term sustainability of different systems of production.

To attempt a survey of indigenous agriculture throughout the Americas is a herculean task, even with four authors shouldering and sharing the burden. The range of potentially relevant evidence is prodigious. It includes descriptions by early explorers and later travellers, conquistadores, administrators, missionaries, and settlers; histories and other narratives; archaeological and biological evidence; drawings and photographs; and of course published and unpublished accounts of field projects, including the four authors' own first-hand field observations. But, despite its quantity and variety, much of this evidence frustratingly lacks detailed spatial information about the forms and functions of fields and the processes of production. It is also necessary to approach the historical sources, especially early European accounts, sceptically, prone as many of them are to exaggeration and sensationalism. These limitations have added greatly to the difficulties faced by the authors in attempting to review and synthesize knowledge of the indigenous agricultural landscapes of the Americas. That they have succeeded remarkably well will be clear to readers of the three volumes, who will also discover that the authors not only share a common aim and outlook but that their approaches differ interestingly in scope and emphasis.

The intellectual bond that links the authors in pursuit of their common goal derives from an academic genealogy that can be traced back to the influence of a group of scholars at the University of California, Berkeley, in the 1950s, particularly the geographer Carl O. Sauer. His early methodological insistence that the morphology of landscape was central to geographical enquiry, and his later explorations of New World culture history, including agricultural origins and dispersals, are seeds from which the three books can be said to have germinated. The senior author, Bill Denevan, was a graduate student in the Berkeley Geography Department in the 1950s, where he came under the influence not only of Carl Sauer but also of other scholars of Latin American geography (James Parsons), anthropology (John Rowe) and history (Woodrow Borah). I, too, was a Berkeley grad student at that time, and I remember Bill departing in 1956 to travel in South America, only to return many months later committed to what became a lifetime of research on the cultural ecology and history of Amazonia and the Andes, much of which he has distilled for us in his contribution to this trilogy. The genealogy then moves to the next generation, with Denevan supervising Bill Turner's doctoral research at the University of Wisconsin-Madison, and in due time to the next, with Turner supervising Bill Doolittle's PhD while at the University of Oklahoma and Tom Whitmore's later at Clark University, Massachusetts—a line of descent that is one of the sturdiest branches of the 'Sauer tree' (1999).[1]

It might be supposed that such direct academic descent would produce a stultifying intellectual uniformity in the four authors' approach to their joint task, but not so. Beyond their common focus on agricultural landscapes and field systems, individual differences of emphasis are allowed free rein. This adds greatly to the insights afforded by the three books, the format of which is sufficiently similar to allow easy comparison, but also flexible enough to bring out inherent contrasts in the nature and evidence of indigenous agriculture in North, Middle and South America. Thus, all the authors distinguish conceptually between rainfed, wetland, and dryland systems of cultivation, and describe the evidence of extensive pre-European landscape modelling by means of terracing, ditching, mounding, and canalization. They also give welcome attention to both horticulture and arboriculture. Those of us who have worked extensively in the tropics, whether in the New World or the Old World, know that house gardens are a widespread and important element in the domestic economy of rural communities, but because they are relatively inconspicuous and less visually impressive than extensive field systems, they tend to be overlooked and their contribution to subsistence underrated.

Likewise, forest cultivation, whether by swiddening or the more subtle exploitation of wild and semi-wild forest products, has attracted less attention than have terrace, raised-field, and canal irrigation systems. The emphasis on horticulture and arboriculture in the three volumes varies, with Whitmore and Turner devoting a chapter to them, Doolittle one to horticulture alone, and Denevan focusing on their role in Amazonian cultivation, but all the authors acknowledge the importance of these modes of production in indigenous American agriculture.

The unifying theme of this remarkable trilogy is its focus on the agricultural landscapes of the Americas 'on the eve of conquest', and the scale and comprehensiveness of the undertaking is enormously impressive. Although AD 1492 is taken as the temporal fulcrum, the authors do not allow this to inhibit them from delving deeper into the past as well as addressing the relevance of their theme to the present. Pre-European, historical, and contemporary Native American cultivation are all within their remit, although they differ in their emphasis on these three aspects of indigenous agriculture and in the extent to which they make use of archaeological in addition to historical and ethnographic evidence. They are not explicitly concerned with the origins of agriculture in the Americas, and so archaeobotanical evidence is not central to their theme, but the story they tell has fascinating implications for students of agricultural history and prehistory, and should stimulate new agro-archaeological research in the Old as well as the New World (such as the investigation of relict ditch-and-mound landscapes in lowland New Guinea and other parts of the Asian tropics). There are important implications too for agricultural development, especially among traditional communities still closely linked to the land, because many of the indigenous systems described were (and in some cases still are) both highly productive and sustainable over the long term.

Denevan disarmingly describes his contribution to the trilogy as 'to a large extent a reference work'. This is doubtless true, of his and the other two volumes (which will long be quarried by students as a guide to the vast literature), but collectively the trilogy amounts to much more than that. The authors' labours in the fields and the libraries over many years, and their determination to persevere with what must often have seemed a daunting and endless task, have resulted in a magnificent summary and synthesis of what is known, at the end of the twentieth century, about the agricultural landscapes that the Europeans encountered when they entered the 'New World' five centuries ago. It is also a testament to how far understanding of the scale and complexity of indigenous American agriculture has been transformed since the 1960s when Parsons's and Denevan's pioneering papers on ridged and drained fields in South America were published.

In 1964 Golomb and Eder (quoted by Denevan) could justly assert that 'among the surfaces deliberately fashioned by man, agricultural landforms are the most venerable, the most extensive and in many ways the least known and most ignored'.[2] Venerable and extensive they may be, but now, thanks largely to the three Bills and Tom, the cultivated landscapes of Native North, Middle, and South America can no longer be said to be either ignored or least known. With the publication of these three meticulously researched, engagingly written, and profusely illustrated books the study of indigenous American agriculture has come of age.

David R. Harris

University College London
April 2000

References

1 Brown, Scott S. and Kent Mathewson. 1999. 'Sauer's descent? Or Berkeley roots forever?' *Yearbook of the Association of Pacific Coast Geographers* **61**: 137–57.

2 Golomb, Berl and Herbert M. Eder. 1964. 'Landforms made by Man', *Landscape* 14/1: 1–7.

PREFACE

I don't remember the exact date but I do remember the time and place. It was a spring evening in 1986. William M. Denevan and I were enjoying before-dinner drinks on the patio of México Tipico Restaurante in east Austin and discussing topics of mutual interest. It was then that Bill hit me with his idea.

With retirement approaching, but still a few years away, Bill had begun to give some thought to writing a *magnum opus* that would be the capstone of his career. Having, at that time, spent more than twenty years studying prehistoric agriculture in South America, it seemed only appropriate for him to consider writing a definitive synthesis of works on the topic.

'A good idea,' I said, 'Tell me more.'

I ordered another round.

As it turned out, shortly before coming to Austin, Bill had attended a meeting in Wisconsin that involved scholars interested in topics related to the year AD 1492. The quincentennary of the so-called 'discovery' of the 'New World' was only a few years away and people were beginning to make big plans for the event which, as we all are now painfully aware, proved to be a non-event. At that meeting Bill learned that several public and private agencies were earmarking funds for projects and activities pertaining to the 500th anniversary.

'Why not,' Bill pondered, 'make the focus of his project indigenous American agriculture at the time of contact? And, why not cover the entire continent?'

'Surely,' I said, 'you're not thinking of writing a book on native agriculture in the whole of the New World in 1492!'

'Well,' Bill replied, 'I did think of that, but soon recognized the monumental size, and hence, impossible nature, of such a task.'

I should have changed the topic right then as it soon became clear that this discussion was about to have a profound impact on my future. Realizing that he had neither the time nor the expertise to write such a tome, Bill was thinking about a three volume set, each book written by an authority on native agriculture in a particular part of the hemisphere. Furthermore, he was thinking about keeping things in the family, so to speak, and restricting authorship to academic descendants of the late Carl O. Sauer. Given that Bill worked in South America, that one of his doctoral advisees was B. L. Turner II who knew more about Maya agriculture than anyone else, and that Turner directed my dissertation which launched my studies into indigenous agriculture in the American Southwest, the picture was rapidly coming into focus. Three books, three areas, three students (all named Bill); Denevan on South America, Turner on Mesoamerica, Doolittle on North America. I'm not sure who bought the next round of drinks, but the evening was beginning to get awfully expensive for me, emotionally, if not monetarily.

At the time, academic tenure seemed secure and I had begun working on another book which I had been thinking about for some time. I was on a roll and didn't want anything to impede progress on that book. Nothing did, and *Canal Irrigation in Prehistoric Mexico: The Sequence of Technological Change* was published by the University of Texas Press in 1990. If I was to accept Bill's proposal, I could only begin serious work on my 'North America' volume no earlier than the fall of 1987. That was fine by Bill.

'That will give us time to obtain some grant money,' he said. 'Let's eat.'

I could see the proverbial wheels turning in Bill's mind; one down, one to go. With Doolittle in, Turner can't refuse.

The idea of the three volumes was good at the time of inception and it remained a good idea throughout the course of my involvement. I accepted Bill's offer to participate for several reasons, the most important one being the significance I perceived of the set. Right from the start it was evident that these books would be like no others. They would have a decidedly geographical bent, nature-society and spatial relationships with an emphasis on the cultural landscapes especially *fields*. The temporal focus would be 1492, more or less. Trying to determine what was going on everywhere in the New World at the time of contact was immediately seen to be a near impossibility. Maintaining a common format for the volumes was one goal, but it was clear that the vast expanses of territory, different biophysical environments, and mosaic of native cultures would require each volume to have its own individual qualities. Problems were many from the start, and they remained numerous until the end. I am sure some readers will think that many problems remain unresolved. So be it. The job is now complete, or nearly so. Denevan's book will be out directly. Turner finally accepted the task but, like me, had other projects he wanted to finish first. Unlike me, however, Turner also took on other projects, and he recruited another of his students, Thomas M. Whitmore, as a co-author. Their book should appear in the near future.

Work on my book progressed in fits and starts. It took much longer than I had ever hoped. Two things slowed me down. First, as a southwesternist, there was a great deal of North America of which I possessed only a passing knowledge. I had much to learn. Second, tenure and promotion brought with it certain unexpected administrative duties. After accepting Denevan's offer, I was appointed graduate advisor for four years and then spent four years as department chair. I've now done my career's worth of bureaucratic paper shuffling.

Much as was the case with Michelangelo in the Sistine Chapel, work on this book has been at once 'agony and ecstasy'. Never in my wildest dreams had I anticipated such a project. It was at times an albatros around my neck. At other times, it was a most rewarding experience. Indeed, it may be the best work I've produced to date. By no means, however, do I have any delusions about this book being as great as the work of the Italian artist.

W. E. D.

ACKNOWLEDGEMENTS

Whatever success this book achieves is due in large part to the input and assistance I received from numerous people. Thanks go first to a great bunch of students who participated in either of the two graduate seminars I taught, or more accurately, conducted, in 1987 and 1990 on the topic of indigenous agriculture in North America. These students included Andrea Abel, Jennifer Cohen, Kyle Curry, Weston G. Hult, Juana Ibañez, Patrick J. Kroschewsky, Lance Lambdin, Dean P. Lambert, Michael L. Melvin, John Miller Morris, Cindy M. Mueller, Robert A. Ricklis, Kathleen A. Schroeder, Kathryn D. Stratos, and Christine G. Ward. Three other students participated in one of these seminars and provided extra service that proved to be invaluable as well. Deserving special thanks are Michael D. Myers, B. J. Perkins, and Michael D. Pool. Other students who helped along the way, in what might be called informal capacities, and have my gratitude include Clair E. Burleson, Carlos E. Córdova F. de A., Charles D. Frederick, Eric P. Perramond, and Andrew Sluyter.

Thanks are certainly due several of my colleagues at the University of Texas and in the Austin area. These include Karl W. Butzer, Darrell Creel, G. Lain Ellis, Stephen A. Hall, Richard A. Hubbard, Terry G. Jordan, Gregory W. Knapp, Campbell W. Pennington, and, especially, James A. Neely, and my wife, Diane Doolittle, who also drew some of the illustrations. And, speaking of illustrations, thanks go to Narda Lebo for drawing one, and to Leticia Ozuna, who used GIS to produce the maps.

In addition to students and colleagues in Texas, a number of individuals from around the country and the world provided assistance and input either when called upon or voluntarily. Others showed me around the field sites at which they were working, and some graciously provided illustrations and permission to publish them. Some of these folks might not remember the help they provided, and some might think their input was trivial and not worthy of mention. I remember, though, and I appreciate *all* the help I received. Thank you, Bradley Adams, Karen R. Adams, Jeffrey A. Altschul, Kurt F. Anschuetz, Roger Anyon, Anne Baldwin, Elizabeth D. Benchley, Jeffery W. Bentley, Robert Birmingham, Robert F. Boszhardt, Cory Dale Breternitz, R. B. Brown, Robert A. Bye, Jr., John S. Cable, Richard Ciolek-Torrello, Jim Clark, Jr., Linda S. Cordell, Patricia L. Crown, Allen Dart, Jeffrey S. Dean, Robin A. Donkin, Christian E. Downum, David E. Doyel, Douglas Deur, Richard W. Effland, Jr., Paul R. Fish, Suzanne K. Fish, Richard I. Ford, Melvin L. Fowler, Robert E. Gasser, William Gustav Gartner, Margerie Green, William Green, David H. Greenwald, Sally T. Greiser, George J. Gumerman, Leigh Jenkins, T. Kathleen Henderson, LaVerne C. Herrington, Barbara

Holmes, Jeffrey A. Homburg, Gary B. Huckleberry, R. Douglas Hurt, J. Charles Kelley, Keith W. Kintigh, Jonathan B. Mabry, W. Bruce Masse, Minnabell Laughlin, Kent Mathewson, Randall H. McGuire, Linda Mick, Charles H. Miksicek, Paul E. Minnis, David A. Phillips, Adrianne G. Rankin, John C. Ravesloot, J. Jefferson Reid, David Rhode, Caroll L. Riley, Christine K. Robinson, Scott C. Russell, Herman Saneyah, Jonathan A. Sandor, Robert Schiowitz, Harry J. Shafer, Steven D. Shelley, Bruce D. Smith, Jack E. Smith, Alan P. Sullivan, Scott C. Travis, Sharon F. Urban, R. Gwinn Vivian, John R. Welch, Stephanie M. Whittlesey, David R. Wilcox, W. H. Wills, J. Scott Wood, Anne I. Woosley, and Frank E. Wozniak.

A project with the scope of this one could not have been successful without the critical review of several persons more knowledgeable than myself in regards to specific native practices in particular parts of the continent. A number of distinguished and highly reputable authorities not only helped in the same way as those mentioned above, but they also read specific substantive chapters and provided special insight that only they possess: Vorsila L. Bohrer, Chapters 2 and 3; Clarissa T. Kimber, Chapter 4; Conrad E. Heidenreich, Chapter 5; William I. Woods, Chapters 5, 6, and 12; Charles R. Moffat, Chapter 6; Dale R. Lightfoot and Timothy D. Maxwell, Chapter 7; Laurance Herold, Chapter 8; Neal W. Ackerly, and Jerry B. Howard, Chapters 9 and 10; and James P. Gallagher, Chapter 12.

This project could not have been carried out without substantial financial support. The largest grant came from the National Endowment for the Humanities (No. RO-21458-87). Supplementary funds were provided by the Andrew W. Mellon Foundation through the Institute of Latin American Studies, and the Erich W. Zimmermann Professorship in the Department of Geography, both at the University of Texas at Austin. A fellowship for a different project allowed me to use the resources of the John Carter Brown Library to check many of my secondary sources against first edition sources. An early phase of the research, and a later phase of the writing were facilitated with time off from teaching through Faculty Research Assignments. I can think of no better university with which to be affiliated, and have always felt a deep sense of appreciation to the University of Texas for giving this Fort Worth boy a job. Mammas, *do* to let your babies grow up to be cowboys.

CONTENTS

LIST OF FIGURES

LIST OF TABLES

Part I

Introduction

1

Landscapes of Cultivation

That North America was a different place five hundred years ago than it is today should go without saying. However, many of the changes that have taken place during the past five centuries may well be more of nature than degree or kind. As we enter a new millennium, North America holds the distinction of being the breadbasket of the world; more food is produced there than on any other continent. But, this is really nothing new. Although everyone knows that North America has been an agricultural leader since at least the 1870s, few recognize that it may have been a global leader in food production for several centuries, well before people even began to think in global terms.

North America has been a rich, productive, and abundant land for a very long time. To be sure, agricultural technology has changed greatly, and the landscape has been altered accordingly. Fossil fuel-powered implements, and chemical fertilizers and pesticides allow a single farmer to cultivate tracts so large today as to be unimaginable to farmers only a few generations removed. The Plains of the Midwest are, as a result, characterized by 'amber waves of grain', corn in Iowa, and wheat in Kansas. Florida is famous for its oranges, California for its grapes, and, of course, New Jersey is the Garden State. As for non-food crops, cotton remains king and tobacco is still smoking (pun well-intended) in the South.

Many of the crops produced in North America today were not domesticated there, but rather on other continents. Indeed, most, but not all, began to arrive not long after the so-called 'New World' was 'discovered' by Europeans. The introduction of new crops as well as the development of new technologies certainly transformed the landscape of North America. However, the landscape that existed before this time was impressive in its own right. The aboriginal Americas have been described by some writers (e.g., Sale 1990) as having been a 'paradise', with the biophysical environment having been little disturbed by people who lived in harmony with nature, who took only what they needed, and who had developed ways of extracting produce such that the system was eternally sustainable. This romantic image is more myth than reality (Thompson and Smith 1971), and in not being truthful probably does more harm than good (C. Martin 1981).

Viewing aboriginal North America in such glowing terms connotes, either explicitly or implicitly, that non-native people and their descendants have defiled if not destroyed the landscape, and that present-day farmers are greedy and destructive. Rightly or wrongly, many people, and especially farmers in developing countries around the world today, look at North America and its agricultural landscape not as *a* paradise but as *Paradise*. Also, painting aboriginal North America as pristine belies evidence that the continent was widely and densely populated (Denevan 1992*a*; Thornton 1987; Verano and Ubelaker 1993), that these numerous people did transform the natural landscapes into cultural landscapes (Butzer 1990*a*; Krech 1994), that part of this transformation, in fact a very large part of it, involved agriculture (Doolittle 1992*c*), and that ancient farmers developed some ingenious production techniques.

The agricultural landscape created by indigenous North Americans was most impressive, far more so than many people may realize or appreciate. In contrast to the pristine image that many writers would like us to believe, huge tracts of forests were cleared and farmed, rivers were re-routed to irrigate desert lands that nature never intended for the raising of crops, mountain slopes were terraced and planted, and wetlands were drained for cultivation. As the pristine image would have us believe, however, many, indeed most, and perhaps all, of the farming practices employed by aboriginal North American farmers were sustainable. The natural environment was transformed, but what replaced it was ecologically sound. This book is about those cultivated landscapes.

Previous Works

This book is not the first written about agriculture as practiced by the aboriginal people of North America, and it most certainly will not be the last. Indeed, in recent years there has been a rather steady stream of books published on the topic, either for the continent as a whole (Ford 1985*a*; Gremillion 1997*a*; Hurt 1987) or for large portions of it (Green 1994; Keegan 1987*a*; Scarry 1993; B. Smith 1992), and the pace seems to be increasing. To one degree or another, each of these works have paid some attention to the methods of cultivation. Overwhelmingly, however, the focus of them all is on crops, the plants themselves. Three phenomena probably account for this emphasis. One may be simply a function of common perception. When most people think of agriculture, they tend to think of crops. A second factor may be the discipline from which most writers come—anthropological archaeology. With an overriding interest in prehistoric cultures, this group clearly recognizes the importance of subsistence practices and hence plant productivity. A third reason may be that those scholars conducting such research in North America are keeping up with, if not leading, a hemispherical (Johannessen and Hastorf 1994) and global (Cowan and Watson

1992; Gebauer and Price 1992; Harris and Hillman 1989; Price and Gebauer 1995*a*) trend.

Specific crop plants, be they defined by genus, species, or variety, require particular environmental conditions. Given that North America possesses a wide variety of biophysical environments, it should come as no surprise that a very old (e.g., Wissler 1922) and continuous (e.g., Weatherwax 1954) tradition that remains popular today (e.g., Hammett 1997) in research on the continent's aboriginal agriculture is to delineate interregional patterns of cultivation. As well recognized as the regional component may be in such studies, however, there is one geographical dimension of native agriculture that has not been addressed systematically until now—fields.

Geography and Fields

This book is unlike any that has been written before. It is written from a decidedly geographical, not an anthropological, perspective. Furthermore, the focus is on fields, not crops. Now, to those not familiar with geography as a profession, it should be pointed out that the academic discipline involves more than simply the study of countries, their capitals, and products, and it involves more than simply making maps. Geography can be defined as 'the study of the surface of the earth, with emphasis on the shaping processes and combinations of elements such as soils and human activities that result in distinctive regions. It involves understanding systemic relationships between certain elements rather than synthesizing diverse data' (Doolittle 1990*c*). Accepting that farmers operate on the earth's surface, that both natural and human forces constantly affect farming practices, that cultivation involves numerous elements, including plants, soil, moisture, and temperature, and that spatial or regional patterns which can be mapped are created as a result, agriculture, especially that of native North America, is as much a geographical as it is an anthropological topic.

The concept of *landscape* is paramount here. Developed out of German geography early in the 20th century (Sauer 1925), the study of landscapes is a hallmark characteristic of the so-called 'Berkeley School' of geography (Spencer 1976). Emphasis is placed on material phenomena, those characteristics that are visually perceptible, and which encompass the agricultural ecosystem in its entirety (Gulick 1986). Although plants, and by extension crops, certainly fit this description (see e.g., Gremillion 1997*a*), so do geomorphic features, and by extension agricultural landforms (Golomb and Eder 1965), perhaps even more so. To illustrate, most crops are annuals and therefore visible for only a few months each year, during the growing season. In contrast, agricultural features such as terraces, irrigation canals, and garden borders remain in place and identifiable year round. Indeed, such features are easier to see and study during the several months they are not in use, that is, after the harvest and before planting, than they are while crops are actually growing. Furthermore, vestiges of ancient fields often remain conspicuous centuries after abandonment; the same cannot be said of the crops cultivated. In a very real sense, this book follows the lead of European archaeological investigations of agriculture (see Gallagher 1992: 95); it is a study in anthropogenic geomorphology. The term 'patterned ground' (Krantz, Gleason, and Caine 1988; Siemens 1989) is particularly relevant here as fields and the features within them, that are essential to the production of crops (see also Martin and Plog 1973: 285), are the focus.

Minimal attention is paid here to the cultural, that is, the demographic, social, economic, and political implications of fields. Whereas previous studies of crops emphasized their anthropological significance, this book looks at fields simply because of their geographical importance. Various cultural issues are raised from time to time but only in so far as they contribute to a better understanding of agricultural landforms themselves. Could a book be written on the anthropological importance of fields? Yes, and one is probably needed. Arguably, however, that book might best be written by an anthropologist, not a geographer. And, what about a book on native North American agriculture by a botanist or an agronomist? None currently exists, principally because botany as a discipline is generally oriented away from plants of direct value to humans. Although there are some distinguished economic botanists, they are precious few and such a book is unlikely to appear in the near future. As for agronomy, the focus there has long been on plants of present-day and future value. Looking back, especially a half millennium or more, is not within the normal bounds of that discipline. To be sure, books about aboriginal agriculture on this continent written by botanists and agronomists would be most welcome and valuable. Some solace, however, can be taken in knowing that the anthropologists who have written on the topic are well-trained in plant sciences and have, as a result, done a most admirable job. Lacking, at least until now, is a book that looks at native fields, especially in their geographical contexts.

In a very real sense, agricultural fields are modified environments. The objective of modification, of course, is to optimize crop growth and productivity (Muenchrath and Salvador 1995: 318). Given that humans create fields at least implicitly for the purpose of taming nature, the term 'domestication' (Axtell 1992: 58; Blackburn and Anderson 1993; Yen 1989) has more than a little merit in discussing agricultural landscapes. There are, however, two problems with using the term here. First, there are more connotations to domestication than simply those of an agricultural nature. Domesticated landscapes can also include such things as house lots, cities, mines, parks, and reservoirs. Second, domestication is a process to which there are degrees. The degree to which certain environments were actually domesticated can be debated. Such discussion would likely produce more heat than light and, therefore, not result in increased understanding. In order to be more specific, and to avoid debate, the term 'cultivation' is preferred here for discussing native North American agricultural landscapes. Understanding the nature of, and the extent to which, the continent's landscape was domesticated by indigenous people is beyond the present goal. A geography of cultivated fields is hereby offered up.

Data

This book is based entirely on secondary source materials. It is, in essence, a synthesis of the literature available through mid 1999. An attempt was made to be as comprehensive as possible. Library catalogs were gleaned, and books and articles on the topic and related topics were not only read, but also had their bibliographies searched for yet additional materials. As a result, an extensive bibliography has been created (see also Edwards and Rasmussen 1942; Sturtevant 1965). Seemingly countless books and articles were perused. Many proved to be dead ends; they had promising titles but contained little in the way of information about fields, *per se*. Principal among these are some excellent publications that focus on specific native North American crops such as maize or corn (e.g., Carter 1945; Kempton 1936; Murphy 1971; Neuman 1961; Vogel and Van Der Merwe 1977; Weatherwax 1936), squash (e.g., Cutler and Murphy 1965; Whitaker and Cutler 1965), the sunflower (e.g., Heiser 1951, 1955), cotton (e.g., Jones 1936*a*; Lewton 1912; L. White 1941), and tobacco (e.g., Gilmore 1922; Haberman 1984; Jones and Morris 1960; Robert 1966; Setchell 1921; Spinden 1950; L. White 1943). Although a thorough counting was not undertaken, it would not be wrong to estimate that fewer than half of the works pertaining to aboriginal agriculture in North America are cited in this book. The exclusion of the other half should in no way be taken to mean these works were not considered or that they are not important, only that they contain nothing in direct regards to fields, the cultural landforms in which the crops were grown.

In terms of data collection and use, three types of publications were given special attention, European explorers' accounts, later ethnographies, and the findings of archaeological investigations.

Documentary Evidence

Primary among the materials used here are the first-hand descriptions by early explorers and settlers up to the end of the 18th century. Explorers such as Giovanni da Verrazzano (1910, 1970), missionaries such as the Jesuit Andrés Pérez de Ribas (1944, 1968), and early naturalists such as the Swedish botanist Pehr Kalm (1812, 1935, 1972; E. Larson 1935) provided some very graphic descriptions of the land and native agricultural practices on it. With very few exceptions (e.g., Le Page du Pratz 1763, which is actually the first English translation of a work that appeared initially in French), books other than the original publications were relied upon.

It is recognized that first edition books are typically more accurate sources than second, third, and even later editions, and that primary documents are better yet. However, at least four factors compensate for any deficiencies that use of republications might entail. These are abundance through availability, replication, comparative checking, and the limited need for translation.

Few scholars have the time and the resources to spend years scouring archives around the world. Similarly, few are fortunate enough to devote a year or more in a facility, for example the John Carter Brown Library (JCB), which specializes in collecting early books. Fortunately, all of the important explorers' accounts have been republished numerous times and several editions are available in every major library or through interlibrary loan. Obtaining copies of early European accounts was a relatively quick and easy task. Data therefore were, and are, abundant. Also, many sources were double-checked against first editions in the JCB.

One of the tenants of science is that studies should be repeatable. Although it is theoretically possible to repeat studies based on primary data, doing so is often time- and cost-prohibitive. If it is time-consuming and expensive to conduct a study using primary materials once, it is even more so to do it a second time, probably to the point of being counterproductive. It is highly doubtful that any funding agency would pay for such an endeavor. One of the side benefits of using widely available republished materials is that anyone, anywhere, anytime can check any aspect of the present work, including the entire opus. And, in order to facilitate checking, as well as to be as accurate and as unbiased as possible, descriptions of aboriginal agriculture are, with a few explained exceptions, presented as direct quotes, and not paraphrased. Data are presented in their raw form prior to interpretations.

Whereas rechecking is one benefit that accrues to readers and future scholars, comparative checking was one used in the actual research. Many of the sources used here have been published more than once. For instance, the Gentleman of Elvas' account of the 1538–43 expedition of Hernando de Soto has been republished at least twice (Elvas 1907, 1993). Various editions sometimes differ slightly as a result of the unique imprint of editors. Regardless of how well intended and slight they may be, editorial alterations can, although they need not, change the original meaning of the author and hence the integrity of her or his work. Multiple editions were relied upon in the writing of this book. A deliberate and careful attempt was made to clarify inconsistencies between individual passages and entire editions in an attempt to better understand the author's intent. In those cases where only one description is quoted, references to other editions are also cited, as was done above in regard to the writings of Elvas.

Research in historical geography often involves the use of materials written in languages different from that of the mother tongue of the author and the intended audience. Translations, be they done by the author of the work in which they appear, or by someone else and simply adopted by the author, always pose problems. Words and phrases do not always translate word for word. Furthermore, different translators can arrive at different interpretations.

Given that the subject region of this book is North America, that the principal audience is English-speaking, and that most of the principal explorers' accounts are in the English language, few translations were necessary. To be sure, some early descriptions were written in Spanish and a few in French, but overwhelmingly they are in English. Changes in meaning due to translations are, therefore, not a major problem. Furthermore, wherever a passage in a language other than English is presented, it is followed by a translation. In the way of an illustration, passages from the 1645 writings of Pérez de Ribas are presented as they first appeared in Spanish (1944) followed by how they appeared in English (1968). Readers who question translations, are given original transcriptions to translate themselves. Scholars will never stop debating the relative merits of various translations, for example, George P. Hammond's and Agapito Rey's (1940) as opposed to George Parker Winship's (1904) translation of documents pertaining to the journey of Francisco Vázquez de Coronado. They will also continue to argue about the ability of particular observers to accurately portray what they saw, and their underlying motives (see e.g., Greenblatt 1993).

The accounts of early Europeans should always be treated with much caution, and they were in the writing of this book. To begin with, early observer–writers had difficulty in knowing and hence conveying their precise locations. As a result, scholars have debated the locations of specific sites where observations were made and recorded, when in reality many will never be known (e.g., Elliott 1997: 253, 254). In this book, only general locations are noted, and then mapped with dots sufficiently large that no one should infer precision.

A second problem the early observer–writers encountered was in describing certain things. They were quite literally in a world new to them, and, accordingly, many were prone to errors. In terms of things agricultural, however, descriptions from the 16th through 18th centuries seem to be accurate (Peacock 1998). Certainly, the later ones are better than their earlier counterparts, but the early writers appear to have strove for accuracy as well. Their accounts of native crops and practices, often include parallels to features with which they were familiar in Europe (Axtell 1992: 59). For example, in describing maize or corn (*Zea mays*), as bizarre a plant as there has ever been, early writers typically equated it with wheat, the most common and widespread of Old World domesticated grains. Indeed, use of the term 'corn' for maize comes from references to it being the natives' principal crop. In the English-speaking world outside of the United States, 'corn' was, and to some extent still is, used principally as a generic reference to the staple grain of a country or people, regardless of the genus.

A third problem with explorers' accounts, and one not unlike those related to differences in translations and editions, is one involving the intent of the original author. Some scholars are skeptical of early European accounts because of what appear today as exaggerations. To a certain degree exaggerations are explainable. How else could one describe something as unfamiliar and as peculiar as a bison or a rattlesnake without using some creative superlatives? Be they true or not, one need not look further than recent accounts of extraterrestrial or bigfoot sightings for an answer.

Of greater concern than what might be considered unintentional exaggerations are the deliberate inaccuracies. Most, if not all, the explorers were advance men for governments or religious institutions with well-defined agendas. They knew their accounts would be read by influential people, and in many cases these texts were reports to such individuals (see e.g., Mathes 1997: 37). Now, one might think that explorers, especially those from religious orders, were noble and honest, but there is more than a little evidence that some stretched the truth, to put it mildly. Without pointing fingers at specific individuals, suffice it to say that the famed '*El Dorado*' was reportedly seen by some early explorers but never found by later ones.

Do deliberate exaggerations appear in early descriptions of agriculture? Perhaps. In more than a few places various explorers described what are almost unbelievably extensive fields, involving nearly unrealistic technology, and feeding huge populations. Such accounts could have been written to substantiate the need for advancing mission or political frontiers. They could, however, also have been true and accurate. All first-hand accounts were scrutinized in the writing of this book. One of the techniques employed here is the use of descriptions by multiple writers; several places were visited by more than one explorer who rendered his impressions in print. A second technique incorporated here involves the use of different types of data from different eras, ethnographic and archaeological evidence. In some cases different accounts and different types of data support each other, in other cases they do not. Interpretations are based accordingly.

Two final notes about reliance on early explorers' accounts are warranted. First, they are of sufficient quality and quantity that a number of scholars have relied on them in previous studies of various aspects of the continent and its inhabitants five centuries ago (e.g., Axtell 1992; W. Bray 1993; Doggett, Hulvey, and Ainsworth 1992; Greenblatt 1993; Josephy 1993; McManis 1972; Office of Indian Affairs 1921; Williams and Lewis 1993). In addition to being widely accepted, the early writers have been proven to be reliable because of their diligence (Franklin 1979).

Finally, consideration needs to be, and was, given to the times and places various descriptions were made. How early is early? And, early in one place is late in another. The early Europeans to describe North America did not describe a place totally unaffected by people from their home continent. Verrazzano (1910, 1970) set foot on northeastern shores in 1510, 18 years after the first Europeans arrived on a Caribbean island. And, Jacques Cartier (1906*a*) followed in 1534, 19 years after Hernán Cortés landed on the east coast of Mexico. As early as we might think of him, John Smith (Arber 1967) was a rather late arrival in 1606.

It is well-recognized that indigenous Americans had extensive trade routes and communications networks between themselves (e.g., Riley 1976), some early writers such as Alvar Nuñez Cabeza de Vaca (1946) in 1528 even described them. As a result of these connections, Old World diseases, to which Americans had no immunity, spread rapidly, in many cases wiping out entire villages before a European even saw them (Thornton 1987: 47). Population declines of this magnitude would have obviously had a major impact on agriculture, the principal one being that fields would have been abandoned. If anything, this means that grandiose descriptions may well be deliberate exaggerations. Also, in some cases early chroniclers reported seeing what were obviously Old World crops growing side by side with strange New World crops (e.g., Bartram 1940: 96; Tonti 1844: 64). Clearly, seeds and cuttings introduced in one place spread quickly to other places where they were immediately adopted (Deagan 1985). Descriptions of 'native' agriculture, therefore, have to be, and were here, screened or filtered carefully.

Were problems with the timing of the earliest descriptions not great enough, they are even greater for the later of the early descriptions. In this book, first-hand descriptions as late as *c.* AD 1800 are considered first. This does not, however, mean that descriptions of places, such as New England, in the late 1700s are accepted as accurate reflections of unadulterated native conditions (Bernstein 1990: 97). Far from it. By 1800 much, if not all, of the area east of the Appalachian Mountains was heavily influenced by settlers from Europe. Agriculture, especially that practiced by aboriginal people, was not the same as it was in 1500. However, just because native farmers had changed many of their practices, they continued to use some of those that were developed by their ancestors. Accounts were gleaned for evidence of these native practices.

In contrast to the east coast, where descriptions made in the 1700s can only be considered as relatively late, in other places, such as the Great Plains of what is now the Midwest, they were early. Europeans got to the interior of the continent much later than they got to the coastal areas. For example, Pierre Gaultier de Varennes (1980), the Sieur de la Vérendrye, explored the Missouri Valley in 1738. By the time such places were visited and described, aboriginal conditions, including agriculture, had changed from what they had been prior to the arrival of Europeans on the continent. As in the case of accounts from the east coast in the 18th century, these early descriptions should not be, and were not here, considered to be perfect. They are taken here only for their face value. These accounts are for the most part accurate descriptions of what writers saw, and they hint at what might have been.

Although the documentary evidence was scrutinized carefully, it is recognized that even after such consideration these data remain deficient. Two other types of evidence were, as a result, also relied on.

Ethnographic Evidence

In many cases early European explorers knew what they saw and were describing. In other cases, however, they did not have the foggiest idea. To be sure, they were trying, but some things were just too foreign. By 1800 this situation had changed. Indeed, throughout the 19th and 20th centuries, writings about aboriginals and their agriculture were done principally by trained professional academics, principally anthropologists. These scholars recognized that native cultures were not only changing at an increasing pace, but also that most were in real danger of disappearing. In attempts to preserve for posterity knowledge of these people, ethnographers spread to every corner of the continent. It has been said of New Guinea that every village has its anthropologist. Well, for the past two centuries one need not go as far as the South Pacific to find such a professional in every native settlement. There have long been plenty of them in North America.

Ethnographies of native North Americans were relied on here in order to bring into clearer perspective what the early Europeans were describing. These works can be dichotomized as those of a general nature in which agriculture constitutes only one part, or those which focus specifically on food production. To illustrate, cultivation makes up a small, but very important portion of William H. Kelly's (1977) study of the Cocopa, while it is the entirety of Edward F. Castetter's and Willis H. Bell's (1942) work on the Akimel O'odham (Upper Pima) and the Tohono O'odham (Papago). Numerous articles (e.g., Balmer 1926; Dodge 1957; Herndon 1967; Lewandowski 1987) and chapters (e.g., Denig 1930; S. Silver 1978*a*; K. Stewart 1977), as well as books (e.g., Denig 1961; Kroeber 1953) were gleaned for ethnographic portrayals of native agriculture.

Some ethnographies are earlier than others. While it might be convenient to conclude that later ones are not as accurate as their earlier counterparts in terms of understanding aboriginal farming practices, such might not be entirely correct. To be sure, native agriculture was not as heavily impacted by introduced practices in the early 1800s as was in the late 20th century. However, although the principal strength of early ethnographies lies in their first-hand descriptions, an added benefit of the later ones is their frequent reliance on unpublished accounts by earlier ethnographers. Many of the early professionals, be they anthropologists or practitioners from other fields, wrote ethnographic reports that were either never published or were published in forums with extremely limited distribution. Recent scholars who specialize in a particular group or on a certain area of the continent are typically more familiar with such obscure sources of information than those whose work is of a more general nature. In many cases, regional and group specialists have been able to use such data to draw conclusions about the concordance of conditions between different time periods (e.g., A. Williams 1983: 101). Their work, therefore, is considered here to be important, not so much for its complementarity, but for its supplementarity. Ethnographic evidence must be (Minnis 1985: 109), and is in this case, used with extreme caution. It is far from perfect, but it does add insight.

Archaeological Evidence

Whereas ethnographic evidence can aid in understanding the documentary evidence, neither type of evidence is suitable for understanding what was happening in earlier times. In order to better understand agriculture in North America prior to the arrival of the Europeans, archaeological evidence is needed (K. Lightfoot 1995), and is used here.

Archaeological evidence is particularly important as remnants of fields and field features often persist on the landscape for centuries. It is, however, far less than perfect for at least three reasons. For one thing, vestiges of prehistoric agricultural features deteriorate with age. In many cases, the remaining evidence is a mere fragment of what once existed. For another, by themselves archaeological data can be grossly misinterpreted. As one recent study (Butzer 1990*b*) has shown, conclusions based solely on archaeological evidence, without the benefit of either documentary or ethnographic evidence, are more often than not erroneous. The key here is not so much with the material evidence as it is with the inferences drawn from it (see Beaudry 1988 for several examples). This is particularly true of prehistoric groups who are typically characterized by their remains. To illustrate, one should not assume that just because something unusual was done at one time in the past that it was done by a different people. To do so, without positive evidence is:

> as great an absurdity as it would be to maintain, supposing all record of the fact to be lost, that Westminster Abbey was built by a people belonging to a different [group] from that which is known, formerly, to have lived in London, and for no better reason than because the English of today have ceased to build such abbeys (Carr 1893: 506).

A third problem with archaeological data of agricultural landscapes involves chronology. The ages of prehistoric agricultural fields and features are notoriously difficult to determine with accuracy as ancient farmers brought little to their places of work that can be dated. Some canals, for instance, may date to *c.* AD 500 while identical ones may date to AD 1500. Collaborating evidence in the form of ceramic artifacts found on the site or habitation sites nearby can be suggestive of ages, but they are not definitive. Ceramics may be from earlier or later occupations and therefore might represent contamination. Proximal sites might be contemporaneous, but they also might not. For this study, chronology is not a real problem principally because the archeological data are used primarily for the purpose of getting an idea of what was going on in the way of native cultivation prior to the arrival of the Europeans in the very broadest sense both temporally and regionally.

Despite its inherent shortcomings, archaeological evidence is important as it represents physical or material manifestations of past activities. Indeed, it is the only type of evidence that has this characteristic. When used in concert with documentary and ethnographic evidence, the archaeological remains provide a rich dimension to understanding native agriculture (Peacock 1998: 8–13).

A Temporal and Spatial Disclaimer

As discussed in the Preface, this book has undergone many transformations during the research and writing. One is the temporal focus. 'Prehistoric' first gave way to '1492' and then this date was relaxed because of data limitations. There is simply no definitive evidence of what agriculture was like for any place in North America when the first Europeans arrived in the New World. The best that can be hoped for, therefore, is a broad, and very general synthesis pulling together numerous shreds of evidence, temporally and regionally. No single type of evidence is seen to be perfect in every regard. Conversely, no type of evidence is seen to be so flawed as to be useless. In a very general sense, therefore, this book has much in common with an old tapestry that has been recently rewoven or reconstructed with only a few, faded, and well-worn threads. The picture is nowhere near complete and it is far less vibrant than might be desired. It is, however, the best that can be fabricated at the moment, and it is an attempt to cover and decorate an otherwise cold, bare wall.

Non-data

Although many references from the literature are used in this synthesis, many are not, and for good reason. At least four factors are not used here. First are references to stored foods. The documentary, ethnographic, and archaeological literature is replete with accounts and descriptions of food caches (e.g., Hudson 1990: 212, 214, 260, 262; see also DePratter, Hudson, and

Smith 1983: 129). Such references mean only that certain people in certain places at certain times had crops. This information is certainly nothing new, and it reveals absolutely nothing about fields or how the crops were grown. Second, and not unlike the first, are references to crops. Many early explorers and later ethnographers made comments such as 'They live on Indian corn' (Allouez 1903: 78). Passages of this nature can, of course, be interpreted to mean that the people mentioned actually grew the crop themselves. Prudence, however, suggests that there is a possibility that they did not. Others could have raised the crop and then traded it away. Possession is not the same as production. Given the abundance of production data, there was no reason to use flimsy possession data.

The third factor not considered here are vague references to productive country. As common and as frequent as references to stored and possessed crops are, references to bountiful countryside are equally abundant. If there exists one quote similar to 'This is very good land' (e.g., Hudson 1990: 306–7, 311–12), there are hundreds (see also E. Hall 1910: 342; Rosier 1905: 137, 140). Although such passages appear to say a great deal, in reality, they say very little. It is unclear if such references indicate high yields per hectare, numerous hectares, or both. Furthermore, they say nothing about production techniques. The final factor not given attention here are nondescript references to cultivators. As in the case of productive lands, there are seemingly innumerable passages in the literature such as 'These people were farmers'. Saying they were farmers, however, does not say enough. It certainly says nothing about the nature of the farming these people were engaged in. As will be seen in the following chapters, nearly every native group to one degree or another in North America were farmers who produced an abundance of food, possessed it, and stored it.

Quasi-data

Two types of information are relied on only in a limited manner in this work, ethnographic descriptions of certain people and unpublished papers. Some groups, particularly the Diné (Navajo) (e.g., Aberle 1983; Billy 1981; Brugge 1983; Hill 1938; Kluckholm and Leighton 1974; S. Russell 1983) and the Indé (Western Apache) (e.g., Basso 1983; Bingham 1979; Buskirk 1949, 1986; Goodwin 1942; Opler 1941, 1983*a*; Pool 1985, 1994) have been the subjects of detailed ethnographic research which included agriculture. These people, however, were rather late immigrants to the Southwest, the area with which they are traditionally identified, having arrived sometime around AD 1500 from the Great Plains. Prior to their arrival in northern and eastern Arizona, respectively, they were non-agricultural peoples. Much debate currently swirls in the anthropological literature about the precise dates of their arrival, and where and how they learned farming. Complicating matters is that historically the Diné are known principally as shepherds and the Indé were dependent in large part on raiding by means of horses. Sheep and horses, of course, were both Spanish introductions. The degree to which the known agricultural practices of the Diné and the Indé are actually 'native' therefore is quite speculative. Accordingly, information about these group's agricultural practices is used only sparingly, and typically in the context of suggesting what might have been going on in the areas which are now identified with them, prior to their arrival (see e.g., Culley *et al.* 1982).

As for unpublished papers, these sources of information are quite common in archaeology. They are also troublesome. There probably exists hundreds if not thousands of professional papers on native North American agriculture that have been written but never published. Many have been delivered orally at meetings of learned societies. Identifying these papers, much less using them, is next to impossible. The onus of disseminating knowledge falls on authors, and the most important outlet is the printed page. Written but unpublished papers are not easily accessible to a broad audience. Spoken words are nice, but they lack permanence. For all practical purposes, therefore, if a paper is not in print, it does not exist. With very few significant exceptions, for example, a series of papers presented in a symposium on prehistoric southwestern water control at the 1970 Pecos Conference (e.g., Ellis n.d.*a*; Plog n.d.), unpublished papers are not relied upon here.

Names and Numbers

The term 'native' is used in this book, but only after much deliberation. Thought was given to 'Indian' but that was soon abandoned principally because the people whose agriculture is being discussed come from neither the South Asian subcontinent nor from either set of islands known as the Indies. 'Native' is used in reference to the original inhabitants of North America, the people who were already living on the continent when the Europeans arrived. Given that the writings of these later people constitute a significant portion of the information employed here, and that these interlopers often referred the people who occupied the continent on their arrival as 'natives', it is most acceptable to use the term in the current context.

The term 'Native Americans', despite its current political correctness, is not used here. Millions of people born in North America cannot trace their ancestry to the original inhabitants. These people are by definition 'native Americans', but few would confuse them with descendants of people who lived on the continent prior to the 1500s. The terms 'aboriginal' and 'indigenous' are also used here interchangeably with 'native'. The reason for this is simply because the three terms are pretty much synonymous. Of course, some might argue that this debate is semantic and political (R. Wright 1992: ix–x). So be it.

What an author or an outsider calls a people is one thing. What a people call themselves is another. Many of the more familiar native names are not what these people called themselves, but rather what other people, typically their less than friendly neighbors, called them. More times than not, these names are less than flattering if not downright insulting and denigrating. To correct this situation, many groups have been reidentifying themselves recently. Following these peoples' wishes, an attempt is made here to identify them by the names they prefer. As was done above with the Diné and the Indé, the first time a group is referred to, its preferred name is stated and followed by its former, perhaps more recognizable, name in parentheses. Subsequent mention of these people includes only their preferred name. Although an attempt has been made to use native peoples' preferred names for themselves, there undoubtedly exist oversights. Some of the materials relied upon here predate the use of preferred names. Also, given the large number of native groups, some of which are quite small and known well only to scholars who specialize on them, it is likely that some preferred names eluded detection and inadvertently are not used in this book.

Any discussion of agriculture, especially one that focuses on fields, will contain references to units of measure. Inches, feet, miles, and acres are certainly familiar units in North America, and especially in the United States. They are also rapidly becoming antiquated. Two of the other countries in North America, Mexico and Canada, are on the metric system, as is most of the remainder of the world and the scientific community. In an attempt to be up with the time and trends, if not ahead of them, the metric system is used here.

Organization of the Text

This book is many things. In some respects it is a primer. In others, it is a catalog. While in yet others it is a handbook, or an encyclopaedia. It is at once definitive and speculative. It is thorough, yet some things have probably been missed. An effort was made to review all the literature that deals with native agricultural fields in North America and make some sense out of it. Coming at the topic with a geographical bent, the format is to examine how indigenous farmers modified habitats in order to make them productive. Given that perhaps the most basic of agricultural requirements is water, habitats were divided into three broad groups, those with sufficient moisture, those with moisture deficits, and those with moisture surpluses. A fourth group, one dealing with the incipient forms of agriculture, was also created as it sheds insight into the evolution of cultivated landscapes.

These groups constitute the four substantive parts of this six part book; Parts I and VI are this Introduction and the Conclusion, respectively. Part II covers Horticulture and contains three chapters, one each on the husbandry of small herbaceous plants, the husbandry of large, and woody plants, and gardens. Part III deals with Rainfed Systems. Its two chapters discuss plain fields in warm environments and ridged fields in cool environments. Part IV is the longest section and contains four chapters. Focusing on dry lands, it has chapters on dry farming, terracing, water harvesting, and canal irrigation. Part V is devoted to wetlands with chapters on flood recession farming, and draining and ridging.

Now, some readers will undoubtedly find this division arbitrary and some of the subsequent discussion cumbersome; there are, as we all know, *splitters* and *lumpers*. When it comes to classifying agricultural systems according to environments, the task of how to split and how to lump is certainly arduous. This book offers no exception. Take, for example, certain gardens, terraces, and canals. Some fields which might seem to be appropriate for Chapter 8—Terracing—are actually discussed in Chapter 4—Gardens—because they are small, located adjacent to houses, and facilitated the growth of specialty crops. Similarly, some fields which might seem to be appropriate for—Chapter 10—Canal Irrigation—are discussed in Chapter 8 because they are terraced. Were this not confusing enough, some gardens are both terraced and irrigated by means of canals! Clearly, they could be covered in any one of three chapters. The decision concerning which chapter was the most appropriate for discussing a particular field was based mainly on that field's principal landscape modification function. Readers interested in a particular type of field, and who might not wish to read this entire book, are advised, therefore, to read ancillary chapters.

Each of the substantive chapters opens with a general introduction. The intent here is to place the particular practices in larger contexts. For example, Chapter 12, which deals with draining and ridging wetlands, opens by discussing the nature of field raising in general, using examples from Lake Titicaca in South America, and the Maya Lowlands of Mesoamerica, to mention just two locales. North America is not an island, physically, literally, intellectually, or conceptually. Looking at its native agriculture without framing it in the larger context of studies done on similar agricultural landforms elsewhere would be anti-geographical at best, and intellectually myopic, provincial, and parochial at worst.

Following the introduction, each chapter is divided into discussions of the documentary, ethnographic, and archaeological evidence. Interpretations are made where needed. In a few places entire sections are devoted to specific themes. For example, in Chapter 5, which is concerned with rainfed agriculture on nondescript fields, much attention is given over to the debate about slash-and-burn cultivation and using fish as fertilizer. Tables are used to present data which are numerous but lacking detail.

Being a work in historical geography, this book includes two heuristic devices that characterize the discipline—illustrations and maps. Landscape studies by their very nature involve material features. Visual perception is critical. Accordingly, it is only fitting that a book that deals with such phenomena should be richly illustrated. Included here are many photographs, some published for the first time, others being reprinted. Also included are some sketches and drawings. In addition to the illustrations, maps, the principal tool of geography, are plentiful. In general, there are at least three maps for each chapter, one each showing the locations from which there is documentary, ethnographic, and archaeological evidence for the specific type of agricultural landscape discussed. There are also maps showing the locations of certain agriculturally related materials.

People and projects are mentioned frequently in the text. These are not, however, necessarily native peoples and the projects which resulted in the creation of their agricultural landscapes. Instead, they are people who were responsible for generating information about aboriginal agriculture and the projects in which they were a part. In no small way, this is a function on the heavy reliance on early European explorers' accounts. Discussing fields of the Huron, for example, without recognizing the information provided by Gabriel Sagard-Théodat (1964) and without recognizing his missionary activities among them in the early 1600s would be academically unacceptable. Citing the source of the observation and then noting the context in which the observer worked are paramount for lending veracity to the study. Ethnographers and archaeologists, and their projects, are recognized for the same reason. Furthermore, the mention of people and projects is intended to lend a personal and human touch to this work. The cultivated landscapes of native North America are fascinating topics in their own right, to be sure. A book written by a popular author and intended for a purely general audience could certainly do without the references to people and projects. This book, however, is written under the tenets of scholarship and for a more demanding audience.

Part II

Horticulture

2

Husbandry of Small Herbaceous Plants

One problem that has long perplexed scholars interested in prehistoric and early indigenous agriculture involves delineating the point at which gathering wild plant resources ended and agriculture began, or determining if specific plants were collected from naturally occurring stands, planted, or some variant of both (Keegan 1986; Bohrer 1991: 227). The vast body of literature on the process of domestication (see e.g., Darwin 1958; Rindos 1984; and B. Smith 1984: 165) is based in large part on this controversy. For the purposes here, of course, the details of that particular academic debate need not be of concern. However, discussion of various types of husbandry (Gremillion 1997*b*: 9; Scarry 1993: 6), or plant modifications and alterations of the medium in which plants were grown, is needed (see Asch 1994: 25; Harlan 1992: 22–3). To be sure, at contact as well as for several centuries earlier and a few centuries since, people indigenous to North America played an active role, to one degree or another, in manipulating wild plants and their habitats, the physical environments in which they were grown (Weatherford 1988: 86). Although not often considered an important aspect of aboriginal food production, because few if any associated field features or environmental modifications of the landscape remain evident, such activities were economically significant (Cocannouer 1964: 61–7).

A number of different terms have been used in discussing plant-soil modifications that lie in that broad transitional area between gathering and agriculture. Perhaps the first person to identify explicitly the activity in terms of the subsistence practices of indigenous Americans was Alfred F. Whiting (1937: 16, 1939: 17) who, in his studies of the Hopi of northern Arizona, labeled it 'embryonic agriculture', because it involved 'semi-cultivated wild plants'. According to him, this activity involved two general practices, one of which had what can be considered three sub-types: (1) certain weeds were allowed to grow unmolested in otherwise carefully hoed fields; (2a) some wild seed plants were sowed, while (2b) others were planted, and (2c) yet other species were transplanted (Whiting 1939: 16–17).

Perhaps fortunately, the somewhat archaic sounding and clearly value-laden term 'embryonic agriculture' never caught on. A bevy of equally confusing terms, cloaked in the context of domestication, have been used, however. For example, one scholar (Downs 1966: 45) used the phrases 'protoagricultural manipulations', 'incipient agriculture', and 'incipient domestication'. Others (Bean 1978: 578; Bean and Lawton 1973: xxxiii) used the term 'proto-agriculture', especially in regard to plants that were 'semi-domesticated' (see also Lopinot 1992: 46). Most recently others have employed the term 'quasi-domestication' (Keegan and Butler 1987: 119), and 'aboriginal agronomy' (Yen 1989).

In a more general sense, and one not encumbered with the conceptual trappings inherent in the domestication literature, are a number of additionally ambiguous terms. For example, the term 'semi-agricultural' has been used to define people who had certain subsistence practices (Heizer 1958: 23) while 'semi-agriculture' (Aller 1954: 66) and 'quasi-agriculture' (Ziegler 1968) have been applied to the practices themselves. Although most studies of plant manipulation have shown little concern for the medium in which the plants grew, the term 'pseudo-cultivation' (Yarnell 1972: 310) has been used in regard to human-modified plants and environments. This term is clearly akin to Whiting's semi-cultivation, which was picked up on initially by Castetter and Bell (1942: 108, 109), mentioned in their study of Akimel O'odham and Tohono O'odham subsistence in southern Arizona, and discussed nearly a decade later in their treatise on Yuman food production along the lower course of the Colorado River (Castetter and Bell 1951: 167–78).

Although Castetter and Bell used the term 'semi-cultivation', they actually did very little in the way of defining it and refining the way it had been used. Indeed, they did little more than talk about it as though everyone knew exactly what they meant. About as detailed as they got was to mention that some plants grow in 'waste places', while others are 'scattered', and yet others 'planted' (Castetter and Bell 1951: 170, 171). The scholar who has done the most to date in elaborating the term and the concept is Joseph C. Winter. According to him, semi-cultivation lies on 'a spectrum of increasing intervention' between 'encouraged wild species' and 'cultivated domesticates' (Winter 1974: 3, 109). Whereas agriculture includes domesticated species, soil preparation, the construction of various types of permanent or semi-permanent field features (e.g., terraces and windbreaks), planting, weeding, and an assortment of other possible cultivation activities (e.g., fertilizing), semi-cultivation involves wild species, the scattering or sowing of seeds, and relatively little if any extra care (Winter 1974: 109). Although Winter does not specifically provide for either planting or transplanting, as did Whiting, his recognition of the transplanting of cattails (*Typha angustifolia*) by the Hopi implies that both this practice and planting can be associated with semi-cultivation. Similarly, although Winter says that semi-cultivated plants are typically afforded little care, he recognizes that there exist some activities that involve additional attention. For example, in order to produce larger leaves, some aboriginal Californians pruned wild tobacco (*Nicotiana* sp.) plants, the seeds of which were sown (C. Fowler 1986: 93–5), and other groups in the Great Basin burned natural vegetation to clear the fields in preparation for planting (Downs 1966: 46).

Although Winter clearly defines semi-cultivation, he, and no one else for that matter, distinguishes it from 'cultivation'. If, as he suggests, agriculture involves cultivated domesticates, and semi-cultivation involves non-domesticated plants, then there probably is no difference between semi-cultivation and cultivation, other than the plants themselves. The latter term is, therefore, clearly superior to the former in that it is less confusing (see Harlan 1992: 64; Price and Gebauer 1995*b*: 6).

In addition to elaborating others' ideas on semi-cultivation, or more appropriately cultivation, Winter also developed further the concept of encouraging or enhancing the growth of certain wild plants, a notion first entertained by A. L. Kroeber (1941). Encouraging, sometimes thought of as nurturing (K. Anderson 1993: 158), lies on the spectrum of increasing intervention below or prior to cultivation. By definition (Winter 1974: 109), encouraging involves neither sowing nor planting, but instead involves disturbing wild plants and their habitats such that growth and reproduction is increased. Often, this involves burning (Shipek 1993: 382–4). For example, setting fire to patches of seed- and berry-bearing plants after the harvest stimulates rapid regeneration. By retarding the growth and encroachment of woody species (see e.g., Daubenmire 1968), and by rapidly returning plant nutrients back into the soil (see e.g., Baldanzi 1960), burning helps to maintain and even increase the production of a stand (Winter and Hogan 1986: 123), to expand the area of natural populations (Cowen 1985: 222, 231), to prolong the bearing period (Downs 1966: 44), and produce a juicier crop (Nabokov and Snow 1992: 124–5), all of which results in an increase in the supply of plant food.

Although burning in and of itself can and does result in harvests greater than those that could be obtained otherwise, it is not the only activity that can be considered to encourage or facilitate the growth of desired plants. Also important are fallowing and even irrigation, among other things. For example, indigenous Californians who lived near Palm Springs would often not gather seed from a plot one year in order 'to let the ground have more seed for plants another year' (Patencio 1943: 69), and those in the Owens Valley actually excavated canals to transport water from springs to stands of certain wild seed plants (Steward 1929).

In his general discussions of plant encouragement, Winter (1974: 109), picking up on the first of the practices identified by Whiting, also mentioned that protection, too, is an activity that affects the production of natural or wild plants. Although he provided no specific examples, and developed the concept no further, a number of other scholars (e.g., M. Fowler 1971; Wissler 1940: 20, 1967: 573) have addressed the issue of protection in a systematic fashion. For example, in a recent study Bruce D. Smith (1987: 28–34) discussed the nature of volunteer, pioneer, invader, ruderal, or spontaneous species in what were identified as 'domestilocalities'. Essentially, he maintained that continually disturbed places, 'anthropogenic habitats', such as areas between planted crops within fields, or trash middens near dwellings, are places where weeds, plants that quickly colonize recently opened areas (de Wet and Harlan 1975; Gasser and Kwaitkowski 1991*a*: 437), proliferate. Such places benefit from having abundant sunlight, soils that are aerated by regular disturbance, and fertilized by such things as increased earthworm activity and deposition of human waste, and are subjected to the continual introduction of new plants (B. Smith 1987: 28–9, 1995). Often, as seems to have been the case with goosefoot or lamb's quarter (*Chenopodium* sp.) in the Eastern Woodlands, weedy species were introduced accidentally or inadvertently (Asch and Asch 1978: 6–7, 24–5). Once their value was recognized these plants were protected. Later they were cultivated. Throughout this process there was a great deal of coexistence and interaction between weed and crop populations (Gremillion 1993, 1994).

In sum, the husbandry of plants is multifaceted and involves at least three different forms between the gathering of wild plant foods and what is normally thought of as agriculture (Table 2.1). Agriculture is distinguished by its involvement of domesticated plants, those that either cannot reproduce themselves or cannot compete successfully with other plants in the wild; they are dependent on human assistance for their survival (P. Smith 1976: 9–10). Cultivation is distinct from encouragement in that people take an active role in planting certain species in the former (Price and Gebauer 1995*b*: 6; Watson

Table 2.1. Traits associated with various forms or stages of plant husbandry

Agriculture
1. Domesticated species
2. Planting
3. Tending
4. Harvesting
Cultivation
1. Wild species
2. Planting
3. Tending
4. Harvesting
Encouragement
1. Wild species
2. Tending/Planting
3. Harvesting
Protection
1. Wild species
2. Tending
3. Harvesting
Gathering
1. Wild species
2. Harvesting

1995: 30–3), whereas they take what might best be described as a passive role in planting that is difficult to distinguish from tending plants in the latter. Protection is distinguished from gathering by virtue of desired species being given special attention and care.

Throughout this scheme, going from gathering to agriculture, are progressive increases in the amounts of labor, the degree of technology, and the extent of environmental modifications involved (K. Anderson 1991*a*: 22–3; Blackburn and Anderson 1993: 15–16). Stated simply, protection and, especially, gathering involve little work, a limited number of tools and skills, and minimal environmental alterations. Cultivation and, particularly, agriculture, on the other hand, require relatively greater amounts of human effort, additional tools and knowledge, and they substantially affect soils, slopes, water, flora, and fauna (Price and Gebauer 1995*b*).

Finally, both the husbandry of wild plants and agriculture itself, are different from gathering in that they generally, although not always, involve greater degrees of exclusive usership by individuals or small groups (Cowan 1985: 225). Even at the low end of Winter's spectrum of increasing intervention, there exists a need for people to control specific plots of land, however small they may be, in order to tend plants. Protection, by its very nature, involves dedication and this, in turn, axiomatically results in some degree of personal attachment. To illustrate, indigenous people in the Puget Sound area owned patches of root plants that they encouraged (Suttles 1951: 279, 281), while those in Wisconsin marked with stakes their private 'wild rice' plots that they clearly cultivated (Jenks 1900: 1057–9).

Data on the husbandry of small herbaceous plants *c.* AD 1500 are limited not only in a quantitative sense, as with virtually all other types of agroecosystems, but they are limited in a qualitative sense as well. Direct observations that have been recorded by explorers are few, and in many instances, perhaps most, those responsible for the documentation may not have had a full understanding of what they observed. For example, some ethnohistorical or documentary accounts refer to small herbs growing among what was clearly the principal subsistence crop, corn (*Zea mays*). Could this be suggestive of protection? Perhaps. There are also reports of yet other wild plants, typically berries in the case of New England, growing quite profusely in burned-over areas. Is this an indication of encouragement? Possibly. And, there are even more reports of herbs growing better under cultivation than in nature. Are these evidence of cultivation? Probably.

Most of the ethnographic data on the husbandry of small herbaceous plants are very precise, largely because professional researchers of the day were astute observers. Even if they lacked a thorough understanding of environmental alterations, they recorded in some detail what they saw in regard to plants and where and how they grew.

The archaeological data are fairly numerous but highly inferential. For the most part, the evidence consists of seeds that show degrees of inter- and intraspecific variation. For example, in most cases small seeds of a species found in excavated cave deposits are viewed as having been wild, while somewhat larger seeds are seen as being from plants that were manipulated, had their growing medium altered, or some combination of the two (Yarnell 1976: 268). Archaeological evidence of fields, gardens, or plots where plants were encouraged or cultivated is typically lacking. The macro remains of plants, however, clearly indicate that people were modifying the flora they used and the locales where it grew for several centuries, and even a few millennia before contact.

In this chapter, evidence for the husbandry of small herbaceous plants is reviewed and discussed. Herbs that were simply gathered or harvested and not otherwise manipulated by aboriginals, and large, woody plants that were manipulated are not included here.

Documentary Evidence

A mere handful of first-hand accounts, dating to the 1500s, 1600s, and 1700s, indicate that native North Americans manipulated small herbaceous plants and their habitats at the time of European contact. Most of these, perhaps not surprisingly, deal with areas along the Atlantic coast, or along major rivers that allowed explorers relatively easy access to interior regions, although a few are from California (Fig. 2.1). Unfortunately, most of these accounts are less than precise, and, therefore, should be accepted only with great caution.

Protection

Only four accounts exist to suggest that natives protected small herbaceous plants at contact; one may be somewhat suspect, as it comes from a rather late date and an area that was already undergoing European settlement. Writing of the Delaware Valley in 1749, Kalm (1972: 266) reported that a species of onion (*Allium* sp.) 'grows at present on almost all corn-fields' and 'on some fields it grew in great abundance'.

The second account of plant protection, and one that is probably more accurate given that it came from an earlier date and a more remote region, was made by Antoine Simon Le Page du Pratz (1763: 225) who noted of the Natchez along the Mississippi River in the early 1700s that 'near all their habitations, they have fields of maiz, and another nourishing grain called *Choupichoul*, which grows without cultivation'. The exact species of this plant has not been determined but at least two scholars (Cowan 1978: 281; Waselkov 1977: 516) have suggested that it was *Chenopodium*.

Dealing with conditions in the early 17th century, the third and fourth accounts of protection are not only the oldest, but also from the Atlantic

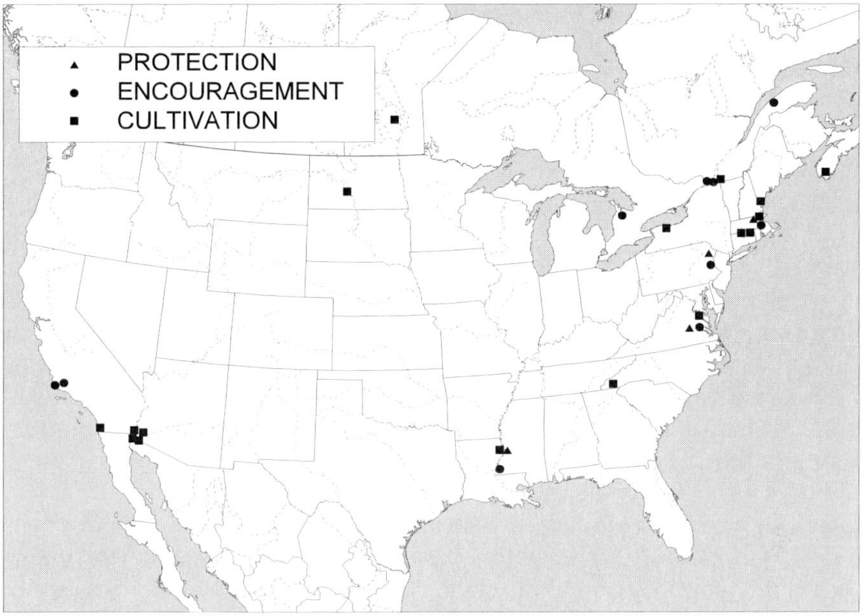

Fig. 2.1. Locations cited in the text for which there is documentary evidence of the husbandry of small herbaceous plants.

coastal region. In 1605, Samuel de Champlain (1907: 67) noted that near Boston Bay indigenous people 'brought also some purslane, which grows in large quantities among the Indian corn, and of which they make no more account than of weeds'. In his discussion of agricultural practices of the natives of Virginia in 1612, William Strachey (1953: 79, 120–1) wrote that a wild plant called '*Maracock* . . . groweth generally lower and creepeth in a manner amongst the Corne'. With an edible fruit described as being like a pomegranate, this plant was probably maypop (*Passiflora incarnata*).

That the onions reported by Kalm, the *Chenopodium* reported by Le Page du Pratz, the purslane reported by Champlain, and the maypop reported by Strachey all grew in fields with corn, but without being cultivated, indicates that they were protected. Were they not, and if they had posed a threat to the corn, then in all likelihood they would have been weeded out.

Encouragement

Ten reports of the encouragement of small herbaceous plants exist. Of these, three deal with hemp, four deal with berries growing in areas with disturbed soils, while three mention seeds growing in burned over areas.

In 1541 Cartier (1906*c*: 98) noted that in one place along the St. Lawrence River 'the land groweth full of Hempe which groweth of it selfe'. Marc Lescarbot (1977: 276), who never actually visited the Huron, but probably relied on discussions with Champlain who had (Conrad Heidenreich, personal communication 1996), reiterated Cartier's claim when he wrote in 1607 that in regards to hemp 'the ground produceth of itselfe'. About the same time Champlain (1907: 68) observed that people living north of what is today Boston Harbor made fishing line out of hemp (*Apocynum* sp.) which they gathered 'without being obliged to cultivate it'. Although these passages are vague and certainly not conclusive evidence of encouragement, Champlain's numerous references to both fishing and agriculture from the Atlantic Ocean to Lake Huron has led at least one writer (G. Hunt 1940: 43, 56, 59) to infer that hemp was both cultivated and widespread while another (H. Russell 1980: 131) has suggested that cultivation was only a possibility. More conservatively, a third researcher (Yarnell 1964: 92) has concluded that because it grew in disturbed areas hemp was only an encouraged plant.

Although no concrete evidence was provided in support of this third conclusion, a previously unrecognized passage in a well-used documentary source lends confirmation to the claim. Kalm (1972: 265) stated most explicitly that early Swedish settlers in Pennsylvania and New Jersey recalled that hemp once 'grew plentifully in old corn-grounds' of the aboriginals who inhabited the region. Kalm went on to note that during his own travels through the country of the Iroquois, he frequently saw filaments drawn from the stalks of hemp plants constituting the only materials used in the manufacture of fishing tackle. Both the reference to hemp growing in abandoned or fallowed fields and the exclusive use of the plant for catching fish in the St. Lawrence area confirm that hemp was indeed encouraged.

In addition to reporting one incident of indigenous protection of certain plants in the Louisiana Territory, Le Page du Pratz (1763: 42) commented that the seeds of a common, yet still unidentified, cane were used for food and that the canebrakes were burned periodically. In one treatise on early American agriculture, it was claimed that this burning was intended to make the cane plants 'bear grain more regularly than they do naturally' (Carrier 1923: 36).

Two similar accounts come from California. There in the mid 1770s military governor Fernándo Rivera y Moncada noted that the Chumash burned 'so that new weeds may grow to produce more seeds' (Timbrook, Johnson, and Earle 1993: 127). Fifteen years later naturalist José Longinos Martínez reported that these same people 'have the custom of burning the brush . . . so that with the first light rain or dew the shoots will come up which they call peillo and upon which they feed' (Timbrook, Johnson, and Earle 1993: 128). Fray Juan Crespí, who was one of the first missionaries in California reported widespread deliberate burning of grass and brush lands, but never explicitly stated a cause. In all likelihood, however, most southern California natives burned to encourage plant growth (Timbrook, Johnson, and Earle 1993: 147). The four early reports of berries being encouraged involve the northeastern part of the continent. Champlain (1907: 284) observed that the Huron of western Ontario not only cultivated a 'great quantity of corn . . . squashes, and sunflowers', but they also had 'excellent raspberries and strawberries' growing on land that was 'very extensively cleared up'. Writing in the early 1600s, Nicolas Denys (1908: 396) noted that if a field in Acadia was 'left for a year without cultivation, all of it is filled with raspberry bushes', while in Virginia George Percy (1969: 134–5) noted suggestively 'that the Sauages had beene there burning down the grasse . . . We past through excellent ground full of Flowers of diurse Kinds and . . . going a little further we came into a little plot of ground full of fine and beautiful Strawberries'. Robert Beverly (1947: 131) confirmed Percy's allusion to encouragement a century later when he noted that wild strawberries were plentiful, especially 'in the deserted old fields'.

One could, of course, argue that all of these accounts are highly speculative, and that, at best, they describe protection rather than encouragement. Evidence of cultivation of seeds and berries exists, however, and this indicates quite clearly that these plants were more than simply tended.

Cultivation

In addition to noting that some berries grew better in abandoned or fallow fields than in natural environments, Beverly (1947: 123) reported that at least a few of the aboriginal Virginians 'transplanted' these fruit-bearing plants. Although he was not specific about the propagation process, Roger Williams (1973: 168) reported in the 1640s that the indigenous inhabitants of New England 'planted' strawberries (see also Bidwell and Falconer 1925: 5; Carrier 1923: 30; H. Russell 1980: 127).

At least four accounts exist to demonstrate that small herbs valued for their root parts were both gathered and cultivated. The first of these is indirect and involves the onion (*Allium* sp.). In an assessment of a number of early explorers' reports, one scholar (Holmes 1912: 25) claimed that this plant was 'cultivated' and that the bulbs were eaten. Although the specific explorer to which the report can be attributed was not identified, onions are documented as formerly having been used (Parker 1910: 105, 107) and 'cultivated' (Hedrick 1933: 31) by the Iroquois.

Three accounts that indicate tubers were cultivated are more direct than that for onions. Lescarbot (1977: 279) observed that groundnuts (*Apios americana*) were 'planted' along the coast of northern New England. Other reports of groundnut use, for example those of Thomas Heriot (1812: 598) who wrote of Virginia in the late 1500s, and Kalm (1972: 248, 257) who described the Delaware Valley in the mid 1700s, refer only to the plant being collected in the wild. It has been argued that the species may have been 'cultivated more by protection than by extensive manipulation' (Cowan 1985: 217). Apparently, however, Lescarbot's description was not taken into account.

Champlain (1907: 90) observed 'Some very good roots, which the savages cultivate' around Gloucester Harbor. Thought to be Jerusalem artichoke (*Helianthus tuberosus*), this nutritious plant (Kaldy, Johnston, and Wilson 1980) is neither from Jerusalem nor an artichoke (Salaman 1940), and is sometimes inappropriately considered a domesticate (e.g., Grigg 1974: 288). In addition to having been cultivated, the plant is claimed to have been 'harvested without cultivation' at the time of European contact (Holmes 1912: 25, 26). Unfortunately, however, the person who purportedly made this observation remains unidentified. As in the case of the onions, the Iroquois might well have been the ones who both collected and cultivated the Jerusalem artichoke at contact, as they did both in more recent times (Hedrick 1933: 28).

That it grows in the wild is, in and of itself, evidence that the Jerusalem artichoke was not a true domesticate, and therefore, not an agricultural crop in the truest sense of the term (Heiser 1985: 61–2). The plant 'spreads into old fields and is easy both to maintain and to propagate through simple harvesting methods and the elimination of competitors' (Ford 1985*c*: 362; see also Yarnell 1976: 269, 1977: 871–2). Indeed, because it thrives in disturbed soils, Jerusalem artichokes may have been commensal with other crops rather than having been planted in fields (Sauer 1980: 235). Early reports of it having been cultivated are, however, supported by the ecological and biogeographical evidence. The Ohio and Mississippi River valleys have long been identified as the region in which the plant was native (Havard 1895: 101). Accordingly, the plant's presence far afield in both the Tidewater area and New England can only be explained as a function of the natives in the latter areas having 'planted' it (Heiser 1985: 62).

Cultivation of small herbs was not limited strictly to plants valued for either their berries or root parts. Also important were plants whose leaves were smoked and grasses whose seeds were eaten. The former, of course, was tobacco (*Nicotiana* sp.). Although not domesticated, this plant was cultivated widely at contact, and a number of documents verify such. In 1535 Cartier (1906*b*: 68) noted in his discussion of cultivated plants in Nova Scotia that '[t]here groweth also a certain kind of herbe', that after having been 'dried in the Sunne' was put in 'a pipe' and used by the natives to 'fill their bodies full of smoke'. Later in the 16th century Heriot (1812: 597) related that some natives in Virginia had 'an herb which is sowed apart by itself . . . the Spaniards generally call it TABACCO'. It was reported about this same time that aboriginals who lived along the St. Lawrence River had 'tobacco which they likewise cultivate' (Champlain 1907: 62), and 'plant great store of *Tobacco*' (Lescarbot 1977: 277). Elsewhere, it is reported that Roger Williams saw tobacco 'cultivated', 'planted', and 'tended' in New England in the 1620s

(Temple and Sheldon 1875: 46–7), William Bartram (1940: 96), who traveled through western Carolina in the late 18th century, reported that indigenous people there 'cleared land about the village' and planted a number of crops including 'tobacco (Nicotiana)', and, finally, Nicholas Biddle (1978: 529), one of Lewis and Clark's chroniclers, noted that along the Missouri River in present-day North Dakota, the Arikara and Mandan 'cultivate tobacco'.

Among the small herbaceous plants cultivated for their seeds, several are clearly identifiable. The first of these, and the one that may well have been the most widespread in North America at the time was *Chenopodium* (Holmes 1912: 25). Perhaps not surprisingly, it was Le Page du Pratz (1763: 239, 290) who provided the best early description of the manner in which this plant was treated. According to him, the Natchez not only protected *Chenopodium*, but in some places along the lower Mississippi River actually sowed it, the women and children using their feet to cover the seeds with soil. Given its physical similarity to *belle-dame*, the French *Chenopodium*, and because it was harvested at the same time and prepared in the same manner as its Old World counterpart (Gilmore 1931: 97–8), Le Page du Pratz named the plant '*Belle Dame Sauvage*'. He clearly identified it as cultivated *Chenopodium*. Writing of native Virginians, Heriot (1812: 596) described the sowing of a grain called Melden which he identified as 'a kind of Drage', a species of *Atriplex* of the *Chenopodiacea* (Asch and Asch 1977: 8), probably *Chenopodium berlandieri* (Lopinot 1992: 52). Cultivation of *Chenopodium*, therefore, cannot be doubted.

Other cultivated small herbs include two panic grasses (*Panicum hirticaule* and *P. sonorum*), grown by Yumans along and near the confluence of the Gila and Colorado rivers and an unidentified wild grass grown by the Tipai-Ipai (Diegueño) of southern California. In 1540, Hernando de Alcaron, a member of the Coronado expedition, reported of the Yumans that 'These people, in addition to maize, have some pumpkins and some grain like millet' (Hammond and Rey 1940: 130). In the mid and late 1700s, two priests who were members of the expedition of Juan Bautista de Anza to settle California, observed, independently of each other, that along with the standard domesticates these natives raised a grass the seeds of which were ground into flour. Pedro Font (1966: 99) noted that 'crops raised by the Indians are maize . . . beans . . . calabashes . . . and seeds of grasses'. Thomas Eixarch (1966: 372), provided confirmation and greater detail by stating 'that these Indians harvest another grain which they call *quies* . . . which they sow . . . and looks like the seeds of the fig, although somewhat smaller'. According to two authorities (Castetter and Bell 1951: 167), the multiple reference to a millet-like grain along with crops such as maize and pumpkins suggests that they were cultivated panic grasses (see also Nabhan and de Wet 1984).

Finally, in 1772 Father Francisco Palóu reported that the natives near what was to become Mission San Diego subsisted on seeds of a grass which they made into sheaves (Pourade 1960: 17). Sheaving has been determined to indicate that the grain must have been sown prior to the arrival of the Spaniards, as harvesting methods directly influence the species ability to reproduce itself (Bean and Lawton 1973: xxxiii–iv). With sheaving, the inflorescences remain intact and, over generations, harvesting will select in favor of seeds that do not dehisce. The species, therefore, needed to be sown if sheaving was practiced for more than a few years. This grass was clearly cultivated.

Of all the cultivated grasses, the one that was both widespread and abundant in natural stands but manipulated in only one area, was so-called 'wild' rice (*Zizania aquatica*). Found both growing and collected in places as distant from each other as the present-day states of Virginia, the Carolinas, Louisiana, and southwestern Missouri (Carr 1895: 168; Carrier 1923: 34–5), this plant is most closely identified with the western Great Lakes area, and the upper Mississippi Valley, particularly what is now Wisconsin.

The earliest reports of rice in this region come from French explorers. In a Jesuit relation of 1671, Father Claude D'Ablon (1899*a*: 169) said that natives living around the head of Lake Superior divided among themselves the marshes in which grew what he identified as 'wild oats'. As for its actual manipulation, Nicholas Perrot (1911: 103) noted that the Assiniboin living west of Lake Winnipeg in Canada 'sow in their marshes some wild oats, which they harvest' (see also Blair 1912: 207).

That rice was widely distributed and abundant in the wild begs two questions. Why would some people cultivate it? What precipitated either the desire or the need to plant? The answers may lie in the method of harvesting which was first described by Father James Marquette (1903: 9–10) in 1673. He referred to this species of rice as 'oats', as had his predecessors, reiterated that it grew in small bodies of shallow, slow-moving water with muddy bottoms, and also reported that harvesting involved shaking ripe grain heads over canoes that were paddled through stands of the grass (Fig. 2.2).

Father Louis Hennepin (1880: 201) contributed additional insight when in the late 1600s he said that rice harvesting involved tying several stalks together into bunches a few weeks earlier. This practice, much as was the case with the sheaving practiced by the Tipai-Ipai, prevented the seeds from being scattered by both wind and waterfowl and rendered the grain easier to collect than would have been possible otherwise (Jenks 1900: 1057). In effect, tying the rice plants together as well as harvesting in canoes allowed for nearly all of the grain to be collected (Jenness 1931: 478); too little fell back into the water to allow the stand to reseed itself for the following year. In some studies of wild grain harvesting (e.g., Harlan, de Wet, and Stemler 1976: 6) it was found that, only about 50 percent of the seed can be collected, the other 50 percent falls to the ground and contributes to the continued existence of the plant community. Given these conditions, aboriginals dependent on rice in the upper Midwest at contact probably had to sow seeds in order to compensate for their so thorough method of harvesting.

Fig. 2.2. Natives of the Great Lakes region harvesting so-called 'wild' rice that in some cases was cultivated (Schoolcraft 1851*b*). (Provided courtesy, and reprinted by permission, of the Ohio Historical Society, Neg. No. OHS 2888b.)

Ethnographic Evidence

Unlike the early explorer accounts which are few and sketchy for the most part, evidence of historic-era husbandry of small herbaceous plants by indigenous North Americans is abundant. Ethnographies written during the 1800s and 1900s, especially those written from about 1850 through the early 1900s, contain some information on protection, a great deal on encouragement, and a veritable wealth of insight on cultivation. They also provide coverage of the entire continent, especially the far west and Great Lakes regions (Fig. 2.3).

Protection

Although protection of small herbaceous plants has been recorded in relatively few cases, the ethnographic evidence does suggest that the practice was widespread, involved a variety of plants, each with different economically valuable parts and with different growth requirements, and was practiced to a much greater extent at contact than the explorers' reports alone suggest.

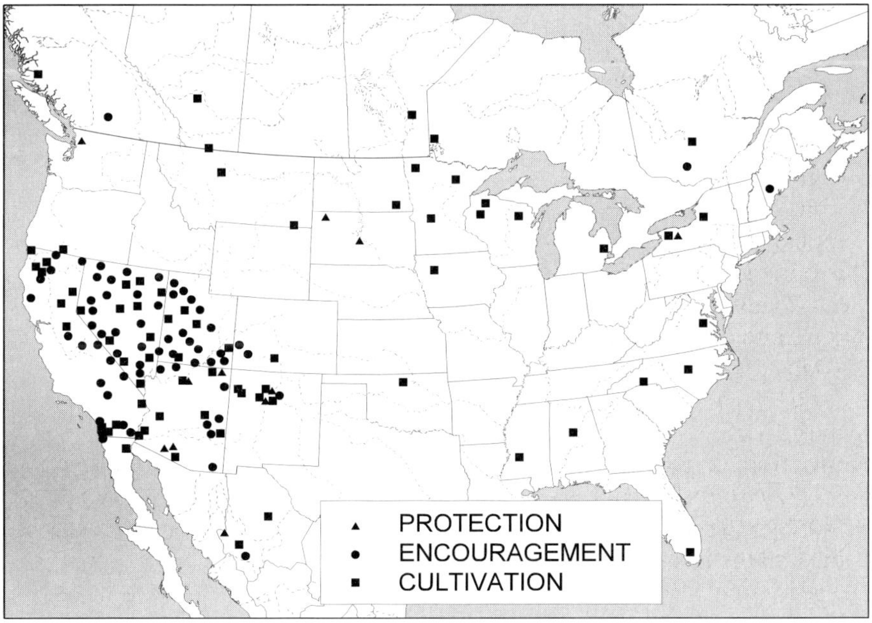

Fig. 2.3. Locations cited in the text and Table 2.2 for which there is ethnographic evidence of the husbandry of small herbaceous plants.

Protection was extended to plants valued for their tubers, seeds, berries, and greens.

In contrast to the earlier claims of Champlain, one writer (Parker 1910: 105–6) noted of the Jerusalem artichoke that the Iroquois 'never cultivated the plant but it frequently grew in their cornfields'. The firmness of this statement clearly indicates that the plant was not observed being cultivated at the turn of the century. Furthermore, use of the term 'never', although probably too definitive, suggests that cultivation was also not practiced during the previous century. Finally, the point that it grew in fields, among corn, sufficiently illustrates that the plant was tolerated and, therefore, protected.

Coast Salish people, who lived around Puget Sound, were found to protect 'potatoes' (species not identified) by keeping 'the ground loosened up so as to make digging easier, and one informant spoke of burning off the bed after digging' (Suttles 1951: 281). In this case, burning was probably not used to encourage growth of these potatoes so much as it was intended to destroy other plants whose roots made gathering of the subject plant difficult (Deur 1999: 140). Burning after digging clearly would have resulted in roots, and not just the above-ground parts of other plants, being destroyed. In separate studies carried out years apart (Whiting 1937: 16, 1939: 17, 90; Winter 1974: 22, 117) it was discovered that wild potatoes (*Solanum jamesii*) were occasional weeds that were allowed to grow in the corn fields of both the Hopi, and the Zuni of western New Mexico (Ford 1985*d*: 113). Although no more details have been provided, these observations clearly document protection.

As for plants protected because of their value as greens (leaves and/or stalks), Rocky Mountain beeweed (*Cleome serrulata*) is known to have been collected from unplanted areas between crops in the fields of San Juan Pueblo, New Mexico (Ford 1968: 191). This plant was harvested in a manner identical to that for mushrooms, by breaking off the stalks of young plants above the ground surface. The Hopi also collected beeweed from disturbed areas (Winter 1974: 25, 123–4). More specifically they allowed the plants to mature and disperse their seeds in corn fields providing an abundant supply of young plants to be consumed the following spring (Whiting 1937: 16, 1939: 16, 77).

In the mountains of northwestern Mexico, the Rarámuri (Tarahumar) are known to have protected at least two plants valued for their greens: pepper grass (*Lepidium* sp.), and a mustard (*Brassica* sp.) (Bennett and Zingg 1935: 39, 40; Bye 1979, 1981). Both are specifically said to be found growing in corn fields, and the last was 'carefully picked or cut from the roots so the plant will not be injured' (Bennett and Zingg 1935: 39).

Finally, the Tohono O'odham are reported to have protected a number of *quelites* or greens including goosefoot, Wright's saltbush (*Atriplex wrightii*), pigweed (*Amaranthus palmeri*), and purslane (*Portulaca* sp.) in the fields planted in other crops (Meals for Millions Foundation 1980).

Devil's claw (*Proboscidea parviflora* ssp. *parviflora*), a plant valued for its seed pods, the fibers of which were used to weave baskets, and berries were also protected by the Tohono O'odham (Castetter and Underhill 1935: 57). Reports indicate that long ago women guarded fertile patches of the plant (Castetter and Bell 1942: 113), and that it 'is never weeded out of a field' of corn planted by the Hopi (Whiting 1939: 92). The Zuni protect amaranth (Bohrer 1960, 1983; Ford 1985*d*: 60). *Chamaesaracha coronopus* and tomatoes (*Solanum triflorum*), plants also protected for their berries, have been found growing among other crops at Santo Domingo Pueblo, New Mexico (Winter 1974: 29, 118). And, Pawnee have been noted to not remove all the 'weeds' from their native squash fields, preferring only to thin them, as the weeds are later eaten and their presence improves the growth of crop plants (Cocannouer 1964: 63).

Encouragement

Data on the historic encouragement of small herbaceous plants by indigenous people is abundant, involves mainly plants valued for their seeds, burning, and comes principally from the arid western and southwestern part of the continent. The encouragement of herbs by means of burning was a practice reported as having been widespread in California (O. Stewart 1941), the

Great Basin (Steward 1941), and British Columbia (N. Turner 1991). According to one scholar (Baumhoff 1978: 23) the natives of the West burned after the harvest—which means it could have taken place any time from July to October, depending on the local environmental conditions and the species encouraged. Others (Bean and Shipek 1978: 552) have pointed out that the annual return from certain plants was maintained by burning at least once every third year.

Other details about burning in the West and Southwest are few. In most of the ethnographies, statements about the practice are simple and brief. For example, 'Indians purposely burned certain communities to favor plants such as wild tobacco and seed crops' (Harper 1986: 63). For the Great Basin, which includes the states of Nevada and Utah as well as parts of neighboring states, no fewer than thirty-nine groups, comprising the Ute (Callaway, Janetski, and Stewart 1986: 343), the Paiute (Downs 1966: 47; Kelly and Fowler 1986: 371), and the Shoshone (C. Fowler 1986: 93), have been identified as having encouraged the growth of small herbaceous plants (Winter 1974: 79). Of these groups, eleven burned only to promote the growth of tobacco, while two burned only to increase the production of seeds, including shadescale (*Atriplex argentea*), goosefoot, and sunu (*Artemisia argentea*). The other twenty-six groups used fire to encourage the growth of both tobacco and seed plants (Winter 1974: 80).

A total of ten California and five Arizona groups also encouraged small herbaceous plants by burning. The habitats of plants valued solely for their seeds (species not identified) were so manipulated by the Tipai-Ipai (Luomala 1978: 600), the Chiamariko (S. Silver 1978*a*: 208), the Luiseño (Bean and Shipek 1978: 552), the Cahuilla (Lawton and Bean 1968: 23; Patencio 1943: 69), the Diné, the Northern and Southern Tonto and White Mountain and Chiricahua branches of the Indé (Winter 1974: 65, 70, 72). The Northern Valley Yokuts (Cook 1960: 242, 248; Wallace 1978: 464), and the Karok (Baumhoff 1978: 22; Harrington 1932: 64) burned in order to stimulate the growth of a rice-like grass, while the Serrano encouraged the growth of chia (*Salvia columbariae*) through burning (Bean and Smith 1978: 571). Tobacco was encouraged through burning by the Kawaiisu (Zigmond 1986: 399, 403), the Tubatulabal (C. Smith 1978: 444), and the Shasta (S. Silver 1978*b*: 217).

The last two groups mentioned are interesting in terms of their encouragement of tobacco because, in addition to burning, they also manipulated plants as they were growing. Specifically, those Shasta who lived in the Shasta River Valley also thinned their plants while those who resided along the Klamath River practiced pruning. For this later practice, additional details come from the Kern River Valley of central California. According to their ethnographer (C. Smith 1978: 444), the Tubatulabal pruned their tobacco at least three times each year. In doing so, the women stripped off side shoots thereby ensuring larger growth of the remaining leaves. Although details are lacking, pruning of tobacco plants seems to have been widespread in the Great Basin (see e.g., C. Fowler 1986: 94; Steward 1939).

At least one western group, the Achumawi of northeastern California, not only encouraged the growth of tobacco and plants valued for their seed, but they also stimulated the production of some unidentified berries by burning high altitude meadows and grasslands (Olmstead and Stewart 1978: 225, 228); burning to encourage berry production in meadows was particularly common in British Columbia (Deur 1999: 141; N. Turner 1991). The Southern Sierra Miwok of the Yosemite region also encourage, by burning, the growth of bracken fern (*Pteridium aquilinum var. pubescens*) for medicinal purposes and deer grass (*Muhlenbergia rigens*) for use in weaving baskets (K. Anderson 1991*a*: 18, 20).

Burning, of course, was not the only activity employed to encourage plant growth. The Pomo of California treated stands of sedge (*Carex* sp.) almost like private property, and valued the plant for its rhizomes used in the manufacture of baskets (Peri and Patterson 1993). Collecting the rhizomes involved digging a 'bed' of these plants, and cutting the roots. Under natural conditions the rhizomes grow in a 'kinky' manner (Peri and Patterson 1993: 181). But, if cut regularly, the rhizomes become straight and better for baskets. In the process, the soil is disturbed thereby contributing to faster growth, and sometimes weavers even replant short pieces of root which are accidentally removed (Peri and Patterson 1993: 182; see also B. Ortiz 1993: 202–3).

During his study of puebloan ethnobotany, Richard I. Ford (1968) found yet other evidence of curious plant encouragement. The San Juan people collected spearmint (species not identified), a medicinal plant that grows on the banks of irrigation canals. Ford argued that encouragement rather than protection was the ongoing process of husbandry in this case because the annual disturbance resulting from deposition of sediment cleared from the canals helped maintain a variegated community.

In a similar vein, Robert A. Bye, Jr. (1985: 379–80) found that Rarámuri of the Sierra Madre encourage the growth of onions (*Allium* sp.) by disturbance. It has long been known that these people use this plant (Bennett and Zingg 1935: 37), but it has been only recently that the ecology of encouragement has been understood. According to Bye, adults prefer to collect the larger bulbs and in doing so, deliberately release the lateral daughter bulbs into the disturbed soil. The roots of other plants are broken, thus reducing severe competition from other species for a few years. Comparisons between collected locales and uncollected locales found significantly more onions present in disturbed areas. The Rarámuri also harvest and consume greens that grow prolifically in fields just prior to the planting of maize. They also rely on wild greens growing elsewhere but are cognizant that these greens grow better in disturbed environments (Graham 1994: 11–12).

Without doubt, one of the most fascinating forms of encouragement was that first reported by Julian Steward (1929) for the Ningwi (Southern Paiute) of the Owens Valley. There, people excavated canals which carried water from creeks, via gravity, to irrigate stands of certain 'wild' plants. Although a number of plants had their yields enhanced by this artificial application of water, two were more important than the rest—yellow nut-grass (*Cyperus esculentus*) and wild hyacinth or blue dicks (*Dichelostemma pulchella*) (Lawton *et al.* 1976: 33–6). The former was valued for its seed, the latter for its onion-like bulb.

Steward (1933) believed this practice to be aboriginal, and his writings on it lead one to believe it was limited largely to one locale. Later studies (Elston 1986: 148; Liljeblad and Fowler 1986: 417; Patch 1951; Treganza 1947) indicate, however, that it is widespread in eastern and southern California and confirm its existence in pre-European times. One (Lawton *et al.* 1976) even expands greatly on Steward's idea concerning what he called 'irrigation without agriculture'.

According to Steward (1929, 1933) two canals carried water out of Bishop Creek near the town of the same name. One canal was 3 km long and watered an area north of the river; the other was 5 km long and irrigated an area on the south side. The northern plot measured 6.5 km by 2 km, the southern plot measured approximately 3.3 km square (Steward 1929: 15). An interesting feature of these plots is that they were alternated for irrigation annually. Steward (1933: 47) was informed by one Ningwi that alternate irrigation was used to 'prevent soil exhaustion', but felt that a more likely explanation was that it 'enabled the plots to reseed themselves'. Harry W. Lawton and his colleagues (1976: 18–19, 38–9) challenged this idea in their detailed follow-up study, and presented a most provocative and insightful conclusion. They point out, more clearly than did Steward, that half of each plot abounded in yellow nut-grass and half in wild hyacinth. They also demonstrate that these two plants are nutritionally complementary, together constituting an ideal integrated food crop complex, and they present evidence from experimental studies done elsewhere, indicating that yellow nut-grass achieves its 'best yield' with a 'delayed harvest'. In other words, an annual harvest from both plots would result in yields less than a biennial harvest from one. Doubtless, fallowing and reseeding played a role as well. In any case, these people developed a most ingenious subsistence strategy that was predicated on encouragement.

While much of the ethnohistorical evidence for the encouragement of small herbaceous plants comes from the West and Southwest, data on the practice also come from other areas, specifically the Northeast. In this region plants valued for their berries were encouraged exclusively. The Penobscot of Maine, for one, were observed to occasionally burn fields where berries (species not identified) were grown (Speck 1940: 92). Writing about conditions in 1830 as extracted from historical documents, one researcher (Bromley 1945: 153) has reported that in eastern Massachusetts fires were used to preserve 'the blueberry barrens, which if left unburned would soon grow up to brush' (see also H. Russell 1980: 129). Without any apparent knowledge, or at least citation, of Penobscot practices, this second account not only provides its own ecological substantiation but also that of the former by noting that 'the present commercial blueberry barrens of eastern Maine have to be burned over every other year to insure maintenance of growth and yield' (Bromley 1945: 153, fn. 2).

Although the species of berry discussed in the first of these reports was not identified, it was, in all probability, blueberry. Of course, there exist no explorers' accounts of blueberry encouraging in this region, or elsewhere, at contact. There are, however, reports of strawberry encouraging, and there is at least one ethnological account as well. Wild strawberries (*Fragaria virginiana* Duchesne) were reported (Black 1978: 261) to be both common and collected widely by aboriginals in sub-Boreal Canada. Near the settlement of River Désert, also known as Maniwaki, 110 km north of Ottawa city, the plant 'is abundant in old pastures and fields' of the Eastern Ojibwa.

Cultivation

As was the case with post-contact era encouragement of small herbaceous plants, the preponderance of the ethnographic evidence of cultivation comes from California, the Great Basin, and the Southwest, but some comes from the Northwest Coast and the Plains. This distribution is undoubtedly a function of the limited and late intrusion of settlers whose ancestry can be traced to Europe, and the persistence of indigenous people who sustained themselves principally by hunting and gathering. At least sixteen groups from the Great Basin (Winter 1974: 79), nearly as many from California, some of the Western Pueblos, a couple of the Eastern Pueblos, a few other southwestern peoples, and various groups in other parts of the continent practiced cultivation. Although more than twenty different small herbaceous plants valued for either their tubers, bulbs, leaves, berries, or seeds were cultivated, only a few planting techniques were utilized. By virtue of being distinguished from encouragement by planting, there are only so many ways that cultivation can be carried out.

For the most part, evidence of cultivation, especially that from the arid quadrant of the continent, is in the form of simple statements noting something to the effect that 'wild plants were cultivated or planted'. The specifics of what was actually involved in cultivation or planting are rarely discussed. Accordingly, much, but certainly not all, of the ethnohistorical evidence of cultivation can be, and indeed is, listed in tabular form (Table 2.2). Of the few accounts that do mention specific planting techniques, none go into great detail but they do indicate that transplanting and broadcasting were the most common means of propagation. Planting by placing either seeds or cuttings of wild plants in holes seems to have been much less common.

Table 2.2. Indigenous groups for which there exists only limited ethnohistorical evidence for the cultivation of certain 'wild' plants

Group	Plant	Source
Algonquin	Wild strawberry	Black (1978: 261)
	Angelico	Asch (1994: 52, 56)
Iroquois	Jerusalem artichoke	Hedrick (1933: 28)
		Parker (1910: 105)
	Groundnut	Hedrick (1933: 29)
		Parker (1910: 105)
	Wild onion	Parker (1910: 105)
	Bowman's root	Asch (1994: 52, 58)
Tuscarora (NC)	Hoary puccon	Asch (1994: 52, 56)
Cherokee (NC)	Groundnut	Asch (1994: 52, 55)
	Red lobelia	Asch (1994: 52, 58)
	Vahl	Asch (1994: 52, 60)
	Indian poke	Asch (1994: 52, 60)
Seminole	Wild squash	Erwin and Lana (1956: 33, 36–7)
Creek	Groundnut	Havard (1895: 101)
Natchez	Chenopodium	Hudson (1976: 294)
Ojibwa (MN)	Spikenard	Asch (1994: 52, 56)
	Ginseng	Asch (1994: 52, 58)
	Indian cup plant	Asch (1994: 52, 60)
Dakota	Sweet flag	Asch (1994: 52, 54)
Fox (IA)	Glade mallow	Asch (1994: 52, 58)
Pawnee (OK)	Canaigre	Asch (1994: 52, 58)
Blackfoot (ALB, MT)	Tobacco	Lowie (1954: 26)
	Camas	Johnston (1969: 111)
	Breadroot	Johnston (1969: 111)
	Wild rice	Johnston (1969: 128–9)
Sarsi	Tobacco	Johnston (1969: 161–2)
		Lowie (1954: 26)
Crow	Tobacco	Lowie (1954: 26)
		Asch (1994: 52, 58)
Pueblos (general)	Tobacco	Castetter (1943: 327)
Laguna	Beeweed	E. Parsons (1920: 113)
Isleta	Husk tomato	Jones (1930: 36)
Acoma	Wild tomato	Winter (1974: 117–18)
	Beeweed	Winter (1974: 27, 123)
Zuni	Husk tomato	Winter (1974: 118)
		Winter and Litzinger (1976: 125)
		Stevenson (1915: 70)
	Wild tomato	Winter and Litzinger (1976: 125)
	Beeweed	Winter (1974: 123)
		Parsons (1920: 113)
	Culantro	Stevenson (1915: 66)
Hopi	Wild tomato	Winter (1974: 22, 117–18)
		Winter and Litzinger (1976: 125)
		Whiting (1939: 40)
	Wild dock	Whiting (1937: 16)
	Husk tomato	Winter and Litzinger (1976: 125)
	Wild tobacco	Whiting (1937: 16, 1939: 16, 17, 90)
	Wild potato	Whiting (1937: 16; 1939: 17, 90)
	Mint	Whiting (1937: 16)
Diné	Wild tomato	Winter (1974: 61, 64, 117–18)
	Wild sunflower	Winter (1974: 122)
	Tobacco	Winter (1974: 60, 61)

Table 2.2. *Continued*

Group	Plant	Source
	Wormwood	Winter (1974: 64)
	Fescue grass	Winter (1974: 64)
Havasupai	Devil's claw	Nabhan *et al.* (1981: 140–1)
Indéean	Devil's claw	Nabhan *et al.* (1981: 140–1)
O'odham	Devil's claw	Castetter and Bell (1942: 113)
		Nabhan *et al.* (1981: 140–1)
	Wild pepper	Winter (1974: 118)
	Panic grass	Nabhan and de Wet (1984)
Maricopa	Grass	Castetter and Bell (1951: 168)
Cocopa	Grass	Gifford (1933: 267)
Yuma	Grass	Forde (1931*a*: 113–14)
		Drucker (1941: 47)
	Amaranth (greens)	Winter (1974: 121)
Mohave	Grass	Kroeber (1925: 736)
	Herbs	Kroeber (1925: 736)
Cahuilla	Tobacco	Bean and Saubel (1972: 92)
	Chía	K. Anderson (1993: 156)
Pomo	Grass	K. Anderson (1993: 156)
Yurok	Tobacco	Heizer (1978: 650)
		Palmer (1929: 25, 195, 203)
Tolowa	Tobacco	Gould (1978: 130)
Maidu	Tobacco	Riddell (1978: 379)
Karok	Tobacco	Bright (1978: 181)
Hupa	Tobacco	Palmer (1929: 25, 195, 203)
Konkow	Tobacco	Riddell (1978: 379)
Eastern Miwok	Tobacco	Levy (1978: 403)
	Tarweed	K. Anderson (1993: 156)
Tipai-Ipai	Grass	Bean and Lawton (1973: xxv–xxvii)
		Shipek (1989: 162)
	Chenopodium	Bean and Lawton (1973: xxv–xxvii)
	Wild sunflower	Bean and Lawton (1973: xxv–xxvii)
Warihio	Amaranth	Gentry (1963: 87)
	Panic grass	Gentry (1942: 80; 1963: 88)
		Castetter and Bell (1951: 171)
Northern Mexican	Chia	Havard (1896: 44)
		Vaillant (1941: 133)
	Husk tomato	Yarnell (1965*b*: 667)

As mentioned earlier, the Hopi transplanted cattails. According to one first-hand account (Winter 1974: 125) those Hopi who lived along Oraibi Wash 'transplant it from Moencopi Wash' (see also Whiting 1937: 16, 1939: 17, 64). Exactly why this plant was cultivated in this manner is unclear, but it may have been because of its high production capabilities. Cattails can provide up to 6,165 kg per ha (Madsen and Lindsay 1977: 88). Another factor may have been ease of harvesting and transportation. Accounts of transplanting other species in different parts of the continent clearly indicate that, in general, the practice allowed for easier harvesting and shorter travel time between home and fields. It has been reported of the Chippewa of eastern Michigan (Gilmore 1933: 121) that even though Jerusalem artichokes had not been domesticated there was some 'transplanting . . . for the sake of convenience'. In California, 'tubers were replanted in areas near villages' (Lawton and Bean 1968: 23) of the Cahuilla, the Tipai-Ipai transplanted wild tuberous root plants 'to areas where they could be better tended (Bean and Lawton 1973: xxvi), and the Luiseño as well as the Tipai-Ipai 'transplanted wild onions and tobacco to more convenient locations' (Luomala 1978: 600).

Exactly what was involved in transplanting is far from clear. The literature provides next to nothing in terms of information on digging up wild plants, soil preparation, and the actual planting process. Of course, such details would vary greatly depending on the type of plant that was transplanted, its growing requirements, and the specific environments involved. For the most part, one can only conclude that small herbaceous plants were removed with care, transported in some type of container that facilitated retention of moist soil around the roots, and replanted with as much care as went into their removal.

In addition to transplanting, some small herbs appear to have been propagated by other vegetative means—specifically by the removal and subsequent planting of select parts that would resprout. Unfortunately, accounts of such a practice are exceptionally sparse and less than specific. A report of Diegueño and Luiseño practices (Bean and Lawton 1973: xxvi) stated only that these people 'knew exactly which of the native plants could be propagated by cuttings'. That such an observation has been made suggests that, at the very least, some indigenous people did indeed cultivate by means of vegetative propagation.

Evidence for planting cuttings is limited but, surprisingly, it is only slightly less abundant than that for planting seeds in holes that were then covered with soil. Perhaps not surprisingly, the evidence that does exist comes from the Southwest. In discussing Devil's claw (*Proboscidea parviflora* var. *parviflora*), ethnobotanist Gary Paul Nabhan and his colleagues (1981: 141) reported that the Akimel O'odham and the Tohono O'odham cultivated 'by planting in holes'. More specifically, these people 'planted in hills about six feet apart with four or five seeds to the hill, in holes about six inches deep, although the seeds were covered to a depth of only two inches' (Castetter and Bell 1942: 202; see also Bretting 1986; Bretting and Nabhan 1981).

Although evidence for seed planting is minimal, there is abundant evidence that historically, and presumably at contact as well, aboriginals sowed the seeds of small herbaceous plants by broadcasting or scattering. For the most part, statements about planting in this fashion are only that: seeds were 'sown', 'scattered', or 'broadcast'. Such observations have been made of the Shasta (S. Silver 1978*b*: 217), the Ningwi (Kelly and Fowler 1986: 371) and other Great Basin (C. Fowler 1986: 93; Steward 1941: 232; Thomas, Pendleton, and Cappanniri 1986: 207; Winter 1974: 19, 74), and southwestern people (Bohrer 1960: 199; Whiting 1937: 16, 1939: 16, 17, 90, 91). A few sources do point out that in some cases broadcasting was done in ways other than simply throwing seeds over the ground. Yumans in general scattered seeds of panic grass 'by blowing them from their mouths' (Dodge 1871: 419). This method was used specifically by the Cocopa and the Yavapai (Castetter and Bell 1942: 170, 175, 177). Finally, and perhaps most interestingly, it was found that the people of San Juan Pueblo baited bird snares, which were placed in their fields, with amaranth seeds (species not identified), some of which later sprouted and was subsequently harvested (Ford 1968: 192).

In contrast to this last case, which might be described as accidental planting, other southwestern groups are quite sophisticated in their planting techniques. Tomatillos or husk tomatoes or ground cherries (*Physalis philadelphia*) have long been cultivated by the Diné of northern Arizona and the Zuni (Bohrer 1960: 182; Winter 1974: 25, 65, 118). This plant is clearly not simply gathered, protected, or encouraged, as it is found well beyond other wild populations. Equally important, however, is the recent discovery that it is undergoing selection (Nabhan 1985*b*) and is well on its way through the domestication process.

In addition to sowing seeds, the activities of many indigenous peoples included a number of ways that contributed to the cultivation of certain small herbaceous plants. The Reese River Shoshone owned plots in which seeds were scattered (Thomas, Pendleton, and Cappanniri 1986: 267). The Cocopa did not own fields, *per se*, but 'individual plots were marked off with sticks' (Castetter and Bell 1942: 174). All of the sixteen groups that practiced cultivation in the Great Basin burned dead plants and brush prior to sowing tobacco, goosefoot, tansy mustard (*Sophia parviflora*), blazing star (*Mentzelia dispersa*), and rice grass (*Oryzopsis hymenoides*) (Winter 1974: 79; see also Kelly and Fowler 1986: 371; Winter and Hogan 1986: 128). The Blackfoot, Sarsi, and Crow of the Plains also burned areas prior to sowing tobacco (Lowie 1954: 26). Not only did western groups prune small cultivated herbs (C. Fowler 1986: 93) but at least one, the Shasta, also thinned the tobacco that they sowed (S. Silver 1978*b*: 217).

A major attribute of cultivation, if not the hallmark of the practice, was the actual planting process. This of course could not have been done unless extremely efficient means of harvesting were employed. In order to cultivate pigweed (*Amaranthus palmeri*) inhabitants of San Juan Pueblo harvested the seeds with great care, by breaking off the stalks with the seed-laden inflorescences remaining intact (Ford 1968: 191). The Cocopa both pulled up sown crowfoot grass (*Dactyloctenium aegypticum*) by the roots and broke off the top foot or so of the plant *before the heads began to shatter* (Castetter and Bell 1942: 175). According to their tribal chairwoman, the Chemehuevi of California 'harvest mature [*Amaranthus palmeri*] plants and dry them upside down with a bag banded around the inflorescence. When dry, they shake the seeds to the bottom of the bag, and sow them to increase the abundance of edible greens' (Nabhan 1979*a*: 131).

Far to the northwest, among the Kwakiutl of coastal British Columbia, come some of the more obscure and little-appreciated references to cultivation of wild plants. Franz Boas (1921: 186–91) reported 'gardens' in which natives harvested the roots of clover (probably *Trifolium wormskjioldii*) and cinquefoil, probably Pacific silverweed (*Potentilla anserina* ssp. *pacifica*). These were not gardens in the same sense as those to be discussed in Chapter 4, but rather specific locales where certain plants were handled in a manner more sophisticated than simple gathering and encouragement (Boas 1966: 24–8; Deur 1999: 142). Boas was quite clear in his descriptions, which portray ownership of gardens by women, a higher than usual density of plants within these plots than without, great disturbance by digging, and even specialized tools (see 1921: 146–50). These characteristics stand in stark contrast to his descriptions of at least seven other species which were subjected only to gathering or collecting (Boas 1921: 194–204). The singling-out of clover and cinquefoil can, therefore, only have been a function of their advanced status in the process of plant husbandry. Furthermore, recent studies have shown that these two species were transplanted (Deur 1999: 143; and forthcoming).

As was already discussed, of course, the most efficient seed harvesting carried out at contact may well have been that which involved rice in the western Great Lakes region. There, reaping was apparently so efficient that seeds had to be sown at the end of one season in order for a crop to be produced the following year. Although early evidence for both sowing and harvesting rice comes directly from only one locale, there are abundant ethnographic data indicating that the plant was cultivated throughout the area. Late in the 19th century, rice was found being cultivated in lakes on the Lac Courte Oreille reservation and in Barron and Forest counties—all in northern Wisconsin—and in lakes in far southwestern Ontario province (Jenks 1900: 1057–8). The practice extended across northern Minnesota as far as the Red River of the North, with some groups even weeding the crop by pulling out winter grasses that grew among the rice stalks. These claims were verified by at least one other observer (Coleman 1953), while yet another (Jenness 1931: 477, 1955: 42) found that the range of rice-sowing extended as far north and west as Lake Winnipeg.

Although the known limits of rice cultivation were south central Manitoba province on the west and northern Wisconsin on the east, the practice may have extended even farther to include the area north of Lake Erie. Numerous reports of rice having been sown exist for the Great Lakes region in general (see, for example, Densmore 1928: 314; Stickney 1896). At least two scholars (Aller 1954: 66; Linton 1924: 349) have even agreed that the cultivation of this plant may have been significant in terms of the beginnings of agriculture in the region, if not throughout North America.

In contrast, a few studies mention only that rice was harvested, suggesting, at least implicitly, that the plant was not sown or cultivated (Brown and Becker 1917; H. Wilson 1956; Winchell 1911). One (Quimby 1960: 143) even notes that the so-called 'sowing' was not always a deliberate activity but that some seeds were accidentally dispersed back into the water during harvesting. Such an assessment may be a function of observing recent rather than former activities. Indeed, only a few years ago two Ojibwa boys told an observer (Nabhan 1989: 116) that 'The way we harvest it, much of the seed gets knocked into the water anyway. We don't have to resow it'. Exactly why rice harvesting today does not require sowing is unknown. It may well be that there are fewer subsistence demands placed on the indigenous people of the region. Rather than needing to maximize the harvest, only a limited amount of grain is gathered simply to supplement purchased staples. In any case, it is quite clear that rice was cultivated in times past and that the practice was widespread among natives of the Great Lakes region when Europeans arrived.

Archaeological Evidence

Exactly what indigenous people in North America were doing in the way of manipulating small herbaceous plants and their habitats prior to European contact is not known. To be sure, however, a large number of archaeological projects conducted during the past century, and especially during the past twenty-five years, have turned up evidence that small herbs were not only gathered, but their growth was enhanced as well. Three types of evidence have been found by archaeologists and paleoethnobotanists, scholars concerned with pre-contact people and the plants they used. The most direct type of evidence is that of plant remains themselves; preserved seeds, stalks, and root parts (Asch and Green 1992: 10; Gremillion 1997*b*: 24–7). In some cases, such as with giant ragweed (*Ambrosia trifada*) and knotweed (*Polygonum erectum*), it is difficult to determine the degree to which plants had their growth manipulated, or whether they were simply harvested (Ford 1985*c*: 349; Wissler 1940: 21). In others, however, evidence of small herbaceous plants having been manipulated is not only apparent but irrefutable. For example, archaeologically excavated sumpweed or marsh elder (*Iva annua*) and sunflower (*Helianthus annuus* var. *macrocarpus*) seeds contain morphological variations sufficiently great to indicate that both plants were actually domesticated in North America (Crites 1993; Fritz 1997; Gremillion 1994, 1997*c*; Heiser 1985: 62; Schwartz 1967: 102; Watson 1988: 41; Yarnell 1978). Convincing evidence exists to suggest that goosefoot and a squash (*Cucurbita pepo* ssp. *ovifera*) were undergoing the domestication process as well (Cowan 1997; B. Smith 1984, 1985, 1989; Struever 1971). Finally, sheaves of goosefoot and amaranth seedheads constitute some of the earliest evidence that cultivation was practiced prehistorically (Fritz 1984; Gilmore 1931: 97).

The second type of evidence found by archaeologists to be indicative of the husbandry of small herbs involves the ecological settings in which botanical remains were found (Asch and Green 1992: 10; Camilli and Winter 1975: 153). For instance, the argument in favor of sumpweed having been cultivated, as well as having been domesticated, prior to contact is based in part on the discovery of ancient plant remains at sites well outside the range of the plant's natural distribution (Asch and Asch 1978: 301; Yarnell 1978: 291). The third type of evidence for the husbandry of herbs prehistorically is similar to the second type in that it involves disjuncts—plants found well beyond the geographical limits of their normal growth. In these cases, however, plants have been found *growing* rather than their remains having been unearthed. Indeed, this type of evidence was used in conjunction with excavated seeds by the first archaeologist–paleoethnobotanist, Melvin R. Gilmore (1930), to address the issue of wild plant manipulation on the continent.

The prehistoric use, and associated alteration, of small herbaceous plants, especially goosefoot, amaranth, sumpweed, sunflowers, and perhaps maygrass (*Phalaris caroliniana*), growing or grown near the confluence of the Tennessee, Ohio, Illinois, Missouri, and Mississippi rivers has long been known as the 'Eastern Agricultural Complex' (Jones 1936*b*), and more recently the 'Eastern Horticultural Complex' (M. Fowler 1992). All of the above-mentioned plants, as well as several others that played only minor roles in aboriginal subsistence schemes, had or have weed-like tendencies. They have been observed growing in disturbed, open habitats (Asch and Asch 1985: 159, 164, 1978: 308; Cowan 1978: 272; Heiser 1985: 62; B. Smith 1987: 29–31; Wissler 1940: 21; Woods 1986: 18; Yarnell 1965*a*: 80, 1978: 290), now thought to have been intentionally cleared (Delcourt *et al.* 1998). Although this complex became well developed after the arrival of the domesticated bottle gourd (*Lagenaria siceraria*) from Mexico (Ford 1981: 8–11), it has been widely accepted for some time, that horticulture was independently developed in the riverine lowlands of the Midwest (Chomko 1978; Chomko and Crawford 1978; M. Fowler 1971; B. Smith 1992; Struever 1964: 96–105; Struever and Vickery 1973; Watson 1985: 101; Yarnell 1993). One author (Heiser 1989) has concluded that a squash (*Cucurbita pepo*), was domesticated independently in the Mississippi Valley and Mexico. Ongoing research (Cowan 1997; Decker-Walters 1993; Perkl 1998) is not only substantiating his claims, but is expanding the area of domestication to include places as far east as Maine (Hart and Sidell 1997). For other cultigens, however, the domestication process has not been one of independence between Mexico and the Eastern Woodlands (Riley, Edging, and Russen 1990).

The archaeological evidence for the husbandry of small herbaceous plants is abundant and varied. Nevertheless, it is not trouble-free. One of two major problems with the data is that they cover three millennia. The second problem is that it is next to impossible to distinguish between protection, encouragement, and cultivation (Asch and Green 1992: 9). Knowing that at least two endemic plants, marsh elder and sunflower, were sufficiently changed morphologically and genetically to be considered domesticates, it is safe to assume that all forms of manipulation were carried out at one time or another prior to AD 1500 (B. Smith 1989: 1568). The studies conducted to date tend not to differentiate between practices. However, on the basis of statements, albeit brief ones, made by various researchers, it is clear that only limited evidence exists to substantiate protection with evidence of cultivation being most common (Crites 1987). Regardless of the specifics, however, the manipulation of small herbaceous plants and their habitats seems to have been widespread prehistorically (Fig. 2.4).

Protection

In a discussion of plant remains excavated from Salts Cave in the central part of what is today the state of Kentucky, it was noted that both goosefoot and maygrass were 'volunteer(s) in garden plots' (Yarnell 1976: 269; see also Yarnell 1969: 46). Similar observations were made of giant ragweed, remains

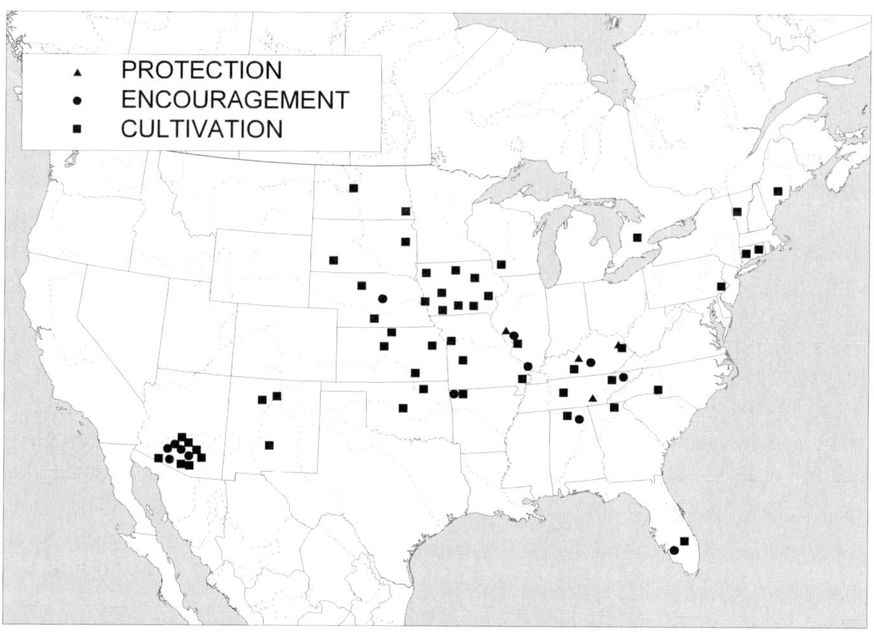

Fig. 2.4. Locations cited in the text and Table 2.3 for which there is archaeological and paleoethnobotanical evidence of small herbaceous plant husbandry.

of which were excavated in Cloudsplitter Cave in eastern Kentucky (Cowan 1985: 217), and wild sunflowers from the Koster site in the lower Illinois River valley (Ford 1985*c*: 348). Most recently, evidence retrieved from several cave sites from North Carolina to western Mississippi indicates that maypop (*Passifloria incarnata*) was 'maintained' along with other plants (Gremillion 1989, 1994, 1997*b*). For the Eastern Woodlands as a whole, it has been argued that a number of economically valuable plants were 'tolerated' in disturbed habitats near homes (Watson 1988).

Encouragement

Arguments for the encouragement of small herbaceous plants by aboriginals prior to the arrival of Europeans on the North American continent are, for the most part, little better than the obviously reserved claims for protection. Indeed, most researchers tend to view the proliferation of certain species growing in heavily disturbed areas near sites to be evidence that the species must have been encouraged there in times past. Interpretations of this nature constitute the bulk of the data on pre-contact plant encouragement. Such inferences have been drawn for sumpweed in the far southern tip of Illinois (Yarnell 1972: 338–9); sumpweed and sunflowers in the far eastern part of Tennessee (Chapman *et al.* 1982: 118); squash in southern Florida (Gilliland 1975: 35); goosefoot in central Kentucky (Watson 1974: 234), central Nebraska, and northwest Arkansas (Asch and Asch 1977: 7, 24, 25); beeweed in the Upper Sonoran Desert (Ford 1985*c*: 357); sunflowers throughout the entire Midwest (Yarnell 1972: 290); and a number of species in northeastern Alabama (B. Smith 1986: 51).

More definitive claims for prehistoric encouragement come from areas where certain botanical remains were actually found in abundance in archaeological contexts. Knotweed, it has been argued, was 'not a garden weed' but instead 'was deliberately encouraged' at sites in the lower Illinois River valley (Asch and Asch 1985: 185). Similar claims have been made for milkvetch (*Astragalus* sp.), little barley (*Hordeum Pursillum*), globemallow (*Sphaerlcea* sp.), plantain (*Plantago insularis*), tansy mustard (*Descurainia* sp.), bursage (*Ambrosia* sp.), maygrass, and goosefoot at four sites in southern Arizona (Gasser and Kwiatkowski 1991*b*: 209, 214–15, 219). Fire has been put forth as one agent most likely used in southwestern encouragement (Bohrer 1991: 233, 1992); irrigation is another (S. Fish 1985: 80).

Cultivation

References to the prehistoric cultivation of small herbs on the continent are much more numerous, and precise than those for either protection or encouragement. They also exist for a greater variety of plants, the growth of which was enhanced over a much wider area (Fig. 2.4). Much as is the case with the ethnographic reports of cultivation after European contact, most of the archaeological reports of the practice prior to contact contain only brief statements referring to certain wild plants being 'cultigens' or having been 'cultivated', 'planted', or 'sown'. One reference to 'transplanting' does exist, but this was made about the growing of sumpweed on dry upland areas where it is not found naturally, and not on the wet floodplains where it grows in the wild (Watson 1985: 109). In all likelihood, sumpweed seeds were sown in upland plots rather than the plants themselves having been uprooted, transported, and replanted. There also exists a reference to prehistoric soil manipulation and sowing. Sunflowers produce phytotoxins that accumulate in the soil (Rice and Wilson 1968). After two or three years of build-up, the phytotoxins begin to inhibit seedling growth and eventually result in the elimination of the plant community. To offset this problem, Ford (1974: 401) argued that prehistoric people would have either tilled the soil to facilitate the leaching of toxins or they would have scattered seeds in places where none fell naturally (see also Heiser 1985: 61). With the exception of these two inferential accounts, details about pre-European cultivation practices are lacking. For this reason alone evidence is presented in tabular form (Table 2.3) and not discussed in depth.

In addition to the archaeologically retrieved plant remains, a number of researchers have used biogeographical evidence, such as was done with documentary accounts of groundnut production, to support claims of pre-contact cultivation of small herbaceous plants (Johnston 1969: 128–9, 162). Because its natural range includes the southeastern coastal plain from North Carolina to central Texas, the Arkansas and middle Mississippi river valleys and western Missouri state, the presence of maygrass remains at sites in far northwest Georgia, eastern Tennessee, and Kentucky (Cowan 1978: 282), west central Illinois (Asch and Asch 1985: 188), and southwestern Wisconsin (Arzigian 1987: 239) has been attributed to cultivation. Similarly, the discovery of sumpweed remains at sites in eastern Kentucky, well to the east of the lower Mississippi Valley where it is found in the wild (Yarnell 1978: 291), and the finding of datura (*Datura meteloides*) seeds at sites in northern New Mexico, well above its Sonoran Desert natural range, has led scholars (Asch and Asch 1978: 301; Yarnell 1965*b*: 671–2) to the same conclusion.

Of course, there are claims of ancient cultivation of small herbs that are based entirely on present-day biogeographical considerations, irrespective of preserved archaeological remains. Among those plants which once were cultivated but have been found recently to be feral, none is more prominent than the sunflower. This plant which is nearly ubiquitous across the United States today appears to have originated on the Colorado Plateau and been introduced eastward (Heiser 1985: 57–8), first into the Great Plains and the Midwest (Yarnell 1978: 290), and then throughout the eastern half of the country (Asch and Asch 1985: 164). Along the way this plant seems to have

Table 2.3. Locales where remains of 'wild' plants recovered archaeologically have been considered evidence of prehistoric cultivation

Locale	Plant	Source
Southeastern Connecticut	Goosefoot	McBride and Dewar (1987: 308)
	Amaranth	McBride and Dewar (1987: 308)
Central Connecticut	Goosefoot	Snow (1980: 253, 334)
Vermont	Tobacco	Haviland and Power (1981: 160)
Southern Florida	Squash	Erwin and Lana (1956: 33, 36–7)
Northeastern Alabama	Goosefoot	B. Smith (1985: 56)
Eastern Tennessee	Sumpweed	Chapman *et al.* (1982: 118)
	Sunflower	Chapman *et al.* (1982: 118)
	Squash	Chapman *et al.* (1982: 118)
Northeastern Kentucky	Goosefoot	Jones (1936*b*)
	Maygrass	Jones (1936*b*)
	Sumpweed	Jones (1936*b*)
	Giant ragweed	Ford (1985*c*: 349)
Central Kentucky	Sumpweed	Yarnell (1969: 46, 1974: 117, 1976: 268)
		Watson (1974: 234)
	Sunflower	Yarnell (1974: 117, 1976: 268)
		Watson (1974: 234)
	Squash	Watson (1985: 101)
	Knotweed	Ford (1985*c*: 349)
	Maygrass	Ford (1985*c*: 349)
	Goosefoot	Ford (1985*c*: 349)
West central Illinois	Goosefoot	Struever (1964)
		Asch and Asch (1985: 172, 175–6)
		Asch (1994: 29)
		Buikstra *et al.* (1986: 530)
	Maygrass	Asch and Asch (1985: 189)
		Asch (1994: 29)
		Buikstra *et al.* (1986: 530)
	Giant ragweed	Asch and Asch (1985: 176)
	Sumpweed	Asch and Asch (1985: 164, 176)
		Asch (1994: 29)
		Buikstra *et al.* (1986: 530)
	Sunflower	Asch and Asch (1985: 164–5, 169, 176)
		Asch (1994: 29)
		Ford (1985*c*: 348)
	Squash	Asch and Asch (1985: 153–8)
		Asch (1994: 29)
	Knotweed	Asch and Asch (1985: 183–6)
		Asch (1994: 29)
		Buikstra *et al.* (1986: 530)
	Little barley	Asch and Asch (1985: 190–1, 194)
		Asch (1994: 29)
		Ford (1985*c*: 349)
	Tobacco	Asch and Asch (1985: 195)
		Asch (1994: 29)
Mississippi Valley	Jerusalem artichoke	Wissler (1922: 15)
Michigan	Rice	Ford and Brose (1975: 13)
Southwestern Wisconsin	Sumpweed	Arzigian (1987: 230, 239)
	Maygrass	Arzigian (1987: 232, 239)
	Sunflower	Arzigian (1987: 232, 236)

Table 2.3. *Continued*

Locale	Plant	Source
	Squash	Arzigian (1987:239)
	Tobacco	Arzigian (1987:232,239)
Iowa	Squash	Asch and Green (1992:11)
	Sumpweed	Asch and Green (1992:21)
	Sunflower	Asch and Green (1992:25)
	Goosefoot	Asch and Green (1992:55)
	Knotweed	Asch and Green (1992:56)
	Little barley	Asch and Green (1992:63)
	Reed canary grass	Asch and Green (1992:69)
	Tobacco	Asch and Green (1992:72)
Iowa–Nebraska	Sumpweed	Adair (1988:61)
	Cucurbits	Adair (1988:55)
	Sunflower	Adair (1988:7)
Western Missouri	Sumpweed	Watson (1985:108–9) Adair (1988:61)
	Cucurbits	Adair (1988:55)
	Sunflower	Adair (1988)
Northwest Arkansas	Goosefoot	Gilmore (1931:97) Fritz (1984)
	Amaranth	Gilmore (1931:97) Fritz (1984)
	Giant ragweed	Gilmore (1931:97)
Eastern Kansas	Cucurbits	Adair (1988:55)
	Sumpweed	Adair (1988:61)
	Sunflower	Adair (1988:71)
Northern Kansas	Cucurbits	Adair (1988:55)
	Sumpweed	Adair (1988:61)
	Sunflower	Adair (1988:71)
Southern Kansas	Cucurbits	Adair (1988:55)
	Sumpweed	Adair (1988:61)
	Sunflower	Adair (1988:71)
Northern Oklahoma	Sumpweed	Adair (1988:61)
New Mexico	Tobacco	Ford (1985*c*:359)
Southwest (general)	Panic grass	Minnis (1985:310)
	Little barley	Minnis (1985:310)
	Tepary bean	Minnis (1985:310)
	Devil's claw	Minnis (1985:310)
	Beeweed	Minnis (1985:310)
La Ciudad site (AZ)	Little barley	Bohrer (1991:230)
	Crucillo	Bohrer (1991:230)
Salt–Gila (AZ)	Amaranth	Bohrer (1991:231)
Frogtown (AZ)	Tobacco	Bohrer (1991:231)
Las Fosas (AZ)	Husk tomato	Bohrer (1991:231)
Columbia River Valley	Tobacco	Havard (1895:105)

evolved from a protected weed-like camp follower, to an encouraged, and then cultivated plant before becoming fully domesticated.

A number of other plants also were carried eastward, presumably as cultigens. The finding of disjunct stands of buffalo berry (*Shepherdia argentea*) in isolated locales in far eastern North and South Dakota, well beyond its range along the eastern flanks of the Rocky Mountains has been interpreted to be the result of deliberate, and ancient propagation (Gilmore 1930: 92). The same was claimed of a stand of wild tobacco, a Mexican plant, found in Ontario (Gilmore 1930: 93). At least two species found growing in an isolated locale near Sandusky, Ohio, far east of their natural ranges, have been interpreted as plants intentionally introduced by aboriginals at least before 1800 and probably prior to European contact (Moseley 1930: 170–1). They include prairie gentian (*Gentiana puberula*), the roots of which were used to make a tonic, and yellow nelumbo or water chiquapin (*Nelumbo lutea*), a lotus or water lily valued for its edible tubers and seeds. The discovery of this species in the lower Delaware and Connecticut river valleys has also led to the belief that it was introduced intentionally into the Atlantic coastal region (Havard 1895: 102).

Going in the opposite direction, from east to west, are four more plants, the presence of which in remote locales beyond the natural biogeographical range, has been attributed to cultivation. These include golden club (*Orontium aquaticum*), *Lithospermum caroliniense*, red lobelia (*Lobelia cardinalis*) and sweet flag (*Acorus calamus*). The first of these is an aquatic with deep bulbous roots and large pea-like seeds. It grows naturally in ponds along the Atlantic Coast from Massachusetts, but 'has also been found in isolated places further inland, near the site of Indian villages, where it is supposed to have been propagated by the natives' (Havard 1895: 102). The second plant has been found in only one locale in southwestern Colorado, well above and beyond its natural range in the lowlands of the East (Yarnell 1965*b*: 669). The last two have roots and flowers valued for medicinal purposes, and have natural ranges in the humid eastern half of the continent. Red lobelia has been found as well in the extreme southern part of Nebraska. Its presence there has been seen to be the result of introduction by natives in times past (Gilmore 1919: 129). Sweet flag is very common throughout the Atlantic coast region and 'is not adapted to disseminate itself or invade new territory' (Gilmore 1930: 90). Nevertheless, this plant has been found growing as disjuncts in several locales as far removed as northern North Dakota, far southern South Dakota, Nebraska, Kansas, and Oklahoma. '[E]veryone of these patches had its origin by intentional planting long ago by Indian medicinemen' (Gilmore 1930: 91; see also Black 1978: 258–9; and Ford 1985*c*: 362).

Overview

The evidence for the husbandry of small herbaceous plants by native North Americans at the time of European contact is, in some regards, sparse. To be sure, early explorers' accounts are few and they pertain mainly to those areas either on or readily accessible to the Atlantic coast. However, when viewed in concert with the archaeological, including the paleoethnobotanical, data that are abundant for the upper South, the Midwest, and the Great Plains; and the detailed ethnographic data from the Great Lakes area, the Southwest, the Great Basin, and California, it is clear that the native inhabitants in virtually every corner of the continent had a long and tenacious association with their immediate biophysical environment.

These people not only adapted their subsistence activities to the plant resources at hand, but they also adapted plants to fit their own needs. In addition to collecting or harvesting the produce of small herbaceous plants, they protected, encouraged, and cultivated several, and they even domesticated a few (B. Smith 1995). The exact number and the variety of plants manipulated may well never be known. Similarly, the precise plant parts valued, and the several uses to which they were put will undoubtedly remain beyond thorough comprehension. Suffice it to say that the evidence is overwhelming: native North Americans were masters when it came to flora and increasing its productive capabilities.

In the course of manipulating small herbs, these people also altered the locales, especially the soils, in which the plants grew or were grown. Burning, for example, not only affected the vegetation itself, but also the fertility of the plot involved. In some cases, the environmental impacts were temporary, while in others, they appear to have been permanent. Regardless of the degree to which the growing medium of the plants was altered, it seems never to have been negative. That is, environmental modifications that dealt with small herbaceous plants seem to have done nothing other than provide native North Americans with a greater, more dependable, and sustainable supply of food.

3

Husbandry of Large, and Woody Plants

The ability of native Americans to make plants produce yields greater than were capable under natural conditions was not limited strictly to small herbaceous plants, but extended beyond to include large herbs and woody plants, particularly giant succulents, trees, shrubs, and vines. Indigenous North Americans did not possess any domesticated trees prior to the arrival of the Europeans. Their husbandry of large, and woody plants for economic purposes, therefore, has not been well-recognized. Most of the research conducted on such activities during both pre-contact and recent times in the New World has involved various trees in Middle and South America (e.g., E. Anderson 1952: 89–106; Puleston 1968). Evidence presented here confirms that inhabitants of North America not only manipulated large, and woody plants prior to the arrival of the Europeans, but that the activity was widespread.

A number of terms, including 'forest-farming' (Douglas 1973), 'arboriculture' (Monkhouse 1965: 18), 'agrisilviculture' (Cox and Atkins 1979: 672), and 'silviculture' (W. Moore 1974: 199), have been used in reference to the cultivation of woody plants, particularly trees. None of these terms is particularly applicable here, however, as they all pertain specifically to practices involving domesticated species. Recently, some scholars (e.g., Denevan and Padoch 1988: 1) have begun to use the term 'agroforestry' in reference to sustainable management systems that involve either wild or domesticated trees, but in combination with the cultivation of other crops on the same unit of land, either simultaneously or sequentially. Agroforestry is, in effect, the use and management of woody species on fallow lands. Although this definition is meritorious for analyzing certain present-day practices, it is not applicable to pre-contact activities in North America for three reasons. First, it was developed and is used most frequently in reference to practices in the tropical rain forests, environments characterized by both an exceptionally high diversity of plants and rapid regeneration. Second, unlike the numerous groups of aboriginal people living in the tropics today, there are simply too few data on which to discuss this practice among North Americans prior to or at contact. Third, as will be discussed in Chapter 5, fallowing by North Americans *c.* AD 1500 may not have been as widespread and as important as previously thought. To be free of any constraints imposed by previously used definitions, no single or specific term is used here in reference to the manipulation of trees and their habitats.

This approach, of course, continues the theme established in the preceding chapter. Wild plants—be they small herbs or large, woody species—and the medium in which they are grown can be manipulated in a number of ways and to varying degrees. Plants can be protected, encouraged, or cultivated. Protection of large, and woody plants usually entails the simple practice of not removing certain species from dump heaps near houses (Havard 1895: 103; Hedrick 1933: 27; H. Maxwell 1910: 96–7) or leaving individual plants in otherwise cleared fields (R. White 1983: 21). The latter appears to have been the case, for example, with early 16th-century aboriginal farms in what is today far eastern Arkansas. There, the Gentleman of Elvas (1993: 114; see also 1907: 206), reported that 'In the open field were many walnut trees'.

Encouragement involves such practices as pruning, the periodic removal of small limbs principally to produce a small number of large fruits, rather than a large number of small fruits; pollarding, the cutting of branches to produce a dense crown; coppicing, the cutting of trunks near their bases in order to rapidly regenerate multiple trunks (K. Anderson 1991*a*: 21; Anonymous 1988); and thinning, the complete removal of some plants in order to ensure more robust growth of others (H. Russell 1980: 130). To illustrate, it has been recorded that the Cahuilla cut mesquites in order to improve both the growth of select branches and seed pods, and to provide easy access for harvesting the latter (Patencio 1943: 59). In one of the earliest reports of conditions in the New World, Verrazzano (1970: 136) noted, of wild grapevines on what is now the Delmarva Peninsula, that 'the bushes around them are removed so that the fruit can ripen better'. Finally, cut marks and growth patterns on some very old fir trees in southern Colorado indicate that prehistoric people at Mesa Verde were coppicing in order to increase the supply of roof timbers (Nichols and Smith 1965).

By definition, cultivation involves planting; and in the case of large, and woody plants, the creation of orchards (B. Turner 1992: 266). There currently exists very little direct evidence of native North Americans actually planting tree seeds; however, there is indirect evidence, and it seems most natural that certain trees would have been so propagated. For example, one historian of New York (Hedrick 1933: 23) claimed, although without providing details, that '[t]he only fruit tree cultivated by the Iroquois, if planting the seed can be called cultivating, was the black plum, *Prunus nigra*'. There is, in contrast, definitive evidence of transplanting. To illustrate, the Hopi are known to transplant rapid-growing and water-loving willows (*Salix* spp.) and cottonwoods (*Populus* spp.) along washes and near springs proximal to their homes. They did this so that they could conveniently obtain materials necessary for maintaining their houses, and, in the case of the latter species, for the manufacture of kachina dolls (Whiting 1937: 16, 1939: 17, 71–2).

Data on the husbandry of large, and woody plants at the time of contact are limited. In large part, this situation is a function of the practice not having been too important. There are, however, at least two other reasons why so little is known about it, and these involve the nature of the documentary data. First, observers and chroniclers in the 16th, 17th, and 18th centuries may not have fully realized what they were witnessing and therefore did not report it. Unlike the cultivated crops such as maize, beans, and squash, and cultivated small herbs such as chenopodium and tobacco, large, and woody plants live for several years. As such, the frequency of planting was markedly reduced and the chances that anyone would have actually observed the practice were minimal. Second, some of the documentary and ethnological information is confusing, and the origins of the cultivation of woody species are dubious, because it contains references to both New and Old World tree crops and a variety of manipulation practices. For example, Zenobius Membré (1903*b*: 175), one of the chroniclers of the 1681–2 expedition down the Mississippi River led by Robert Cavelier de la Salle, reported seeing 'plums, peaches, mulberry, apple, and pear trees'. He also reported that '[b]eans grow wild, . . . and runs up like ivy to the tops of the highest trees. The peach trees are quite like those of France, . . . the Indians have to prop up those they cultivate in their clearings. There are . . . many plum-trees and other fruit-trees, some known and other unknown in Europe' (Membré 1903*b*: 185). Similarly, Beverly (1947: 315) reported that 'Peaches and Nectarines I believe to be Spontaneous some-where or other on the Continent; for the *Indians* have, and ever had greater variety, and finer sorts of them than the *English*'.

On the basis of such accounts, it would be impossible to conclude that large, and woody plants were manipulated by native Americans prior to their exposure to introduced European practices. Some domesticated Old World trees were indeed introduced into Mexico and Florida by the Spaniards during the 1500s and diffused into the interior of North America, becoming established among several groups prior to the arrival of French and English explorers from the North and East in the 1600s (Brand 1939: 110; Gray 1947: 5, 7; Havard 1895: 98). This fact has given rise to the theory that many trees and their habitats were manipulated only during the past few hundred years (Brand 1939: 112). It is equally plausible, however, that Old World species were adopted quickly by certain groups because these people already had experience protecting, encouraging, and cultivating large, and woody plants. For the most part, if Old World species were mentioned, then the specific account is not accepted here as evidence that aboriginal farmers manipulated such plants prior to contact.

To be sure, the overwhelming proportion of produce from large, and woody plants consumed by indigenous people in North American came from the gathering of fruits and nuts from stands of naturally occurring species (Brand 1939: 112), with husbandry coming only very late (Ford 1985*b*: 5). The degree to which gathering has been perceived to have been dominant over husbandry is reflected in the writings of one early agricultural historian (Carrier 1923: 31), who wrote of the native American plum that aboriginals 'apparently made little or no effort to cultivate them', and echoed by a later scholar (Bennett 1955: 390), who commented that '[a]pparently no fruits were cultivated by the Indians' of New England. Conclusions such as these, although not substantiated, were probably based on a number of claims made by 16th- and 17th-century observers (e.g., Casañes 1927: 210; Castañeda 1907: 304; Champlain 1907: 89; Elvas 1907: 178; Marquette 1903: 15–16; Vega 1951: 59) that fruit- and nut-bearing trees not only grew prolifically without being cultivated but could have been even more productive if they were properly tended. Johannes Megapolensis (1909: 169), a missionary working among the Mohawk of New York in the 1640s, recorded of grapes, for example, that '[i]f people would cultivate the vines they might have as good wine here as they have in Germany or France'.

Although some scholars (e.g., Ford 1985*b*: 5) have discounted the existence of native husbandry of trees in North America, a few have done the opposite and argued for, often without providing details of, the existence of the practice. For example, it has been written that people in eastern Massachusetts 'took particular pains to protect chestnut groves from their annual fires' (Temple and Sheldon 1875: 46), that the 'Iroquois planted great groves of butternuts, chestnut, and varieties of hickory', and that in northern Georgia, 'the Cherokee nation in all its branches were nut growers' (Powell 1916: 132). Until now, there has been no systematic treatment of aboriginal manipulation of large, and woody plants and their habitats in North America.

Documentary Evidence

Although few, there are some surprising first-hand accounts from the 16th through 18th centuries indicating that large, and woody plants were manipulated in North America at the time of European contact. As was the case with the husbandry of small herbs, most of these accounts involve areas proximal to the Atlantic Coast. A few, however, do pertain to inland areas, but the practice seems to have been more common in the eastern half of the continent where large, and woody species are the dominant vegetation (Fig. 3.1).

Protection

In its most simple form, the husbandry of large, and woody plants consisted of nothing more than species of economic importance being left standing either in fields that were otherwise cleared and planted in crops such as maize, or along the margins of fields. The earliest such account was provided by the Gentleman of Elvas. Near the town of Apalache, in what is today the Florida

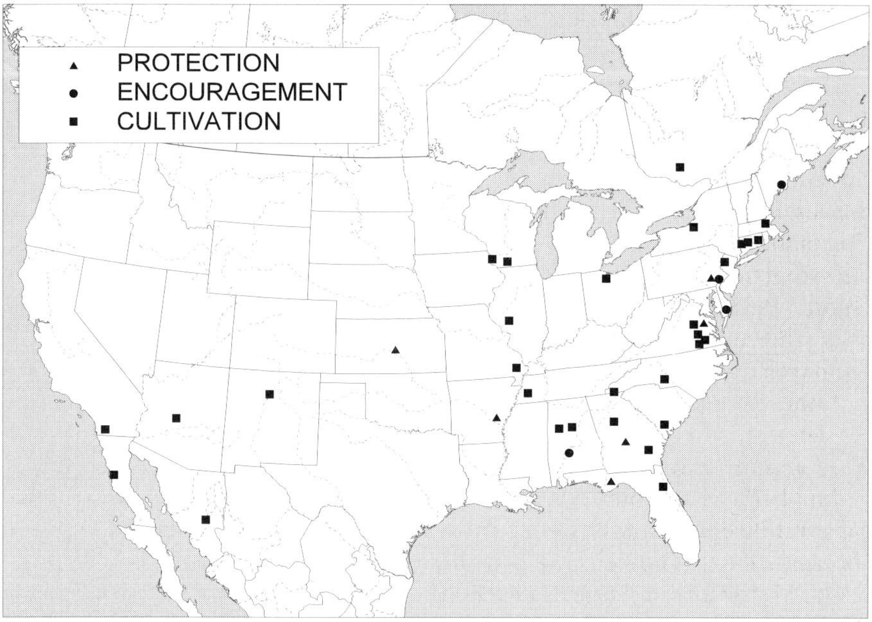

Fig. 3.1. Locations cited in the text for which there is documentary evidence, and biogeographical assessments of the husbandry of large, and woody plants.

panhandle, he reported that plums 'grow wild in the fields without being planted' (Elvas 1993: 72; see also 1907: 162). This description is, of course, quite similar to his previously discussed account of walnut protection in Arkansas. Contrary to one interpretation (L. Larson 1980: 214–15), it does not imply that trees were found in fallowed fields.

Throughout his travels, de Soto found great quantities of maize, beans, walnuts, and plums (e.g., Elvas 1993: 131, 1907: 226), the latter were reportedly 'better than any plums that are raised in Spain' (Elvas 1993: 170, 1907: 271). On the basis that trees were mentioned together with cultivated crops, that the former are known to have been left in otherwise cleared fields, and that they produced an abundance of fruit, at least two scholars have argued that the protection of trees was widespread in the Southeast. One (White 1983: 20), extrapolating from the 18th-century writings of Bernard Romans (1962: 308), claimed that wild nut trees were spared when forests were cleared for farming. The other, Carl O. Sauer (1971: 181), maintained this position in no small part on the premise that of the four species of plum native to the region, only *Prunus hortulana* is a large-fruited upright tree that does not form thickets. As its name implies, it is a garden or orchard form, quite distinct from the small-fruited, shrubby species that would not have attracted the attention, much less merited the praise, of the Spaniards (Sauer 1971: 290).

The comparison of New World to Spanish fruits by Elvas is curious in that at approximately the same time, but 1,000 km to the west, in what is today the state of Kansas, Pedro de Castañeda (1907: 334), one of the Coronado's chroniclers, reported that '[t]here were walnuts and . . . large quantities of prunes like those of Castile', and Juan Jaramillo (1940: 306) another of Coronado's chroniclers, noted 'plums of Castile'. Whether or not they were implying the protection of trees is debatable. However, in 1601 Juan de Oñate (1953*b*: 755), the colonizer and first governor of New Mexico, wrote that '[t]here were many beans, some gourds, and between the fields, some plum trees' along the Arkansas River in Kansas.

Although somewhat later and from areas more heavily affected by Europeans, two additional accounts suggest that natives protected certain trees within their fields. The first comes from the Delaware Valley where Kalm (1972: 41–2) found growing on 'old corn-fields' sumac (*Rhus glabra*), chestnut (*Fagus castanea*), black walnut (*Juglans nigra*), and mulberry (*Morus rubra*). The second comes from central Georgia where Bartram (1940: 57) 'observed, in the ancient cultivated fields, 1. diospyrus, 2. gleditsia. triacanthos, 3. prunus chickasaw, 4. callicarpa, 5. morus rubra, 6. Juglans exalata, 7. Juglans nigra, which inform us, that these trees were cultivated by the ancients'. He went on to note that '[t]hough these are natives of the forest, yet they thrive better, and are more fruitful, in cultivated plantations'.

It is not surprising that indigenous Americans would have protected both nut- and fruit-bearing trees in their fields, even though members of the species grew well and abundantly in the wild. Nut trees native to the eastern portion of the continent do not produce until they are fairly old, 20 to 40 years (USDA 1965). One study (Cowan 1985: 218–20) pointed out, therefore, that protection of older trees within fields would have preserved a good source of nuts, and the elimination of competitors through both clearing and weeding fields would increase the growth rate of young trees. Similarly, both plums and persimmons (*Diosyros virginiana*) are poor competitors with canopy-level trees, and do best in disturbed habitats (Havard 1895: 103; H. Maxwell 1910: 88, 96, 97). That growth advantages existed for younger trees may, of course, mean that encouragement and not simply protection was involved.

Encouragement

There exists only two definite early documentary references to the encouragement of large, and woody species. However, a third account intimates it and a fourth and most tantalizing account suggests that the practice might have been more common than would be prudently presumed and have included more than one species. The two undisputed descriptions and the first of the questionable ones involve grapes; the fourth, sassafras.

As has already been noted, Verrazzano (see also 1910: 187) referred to the encouragement of grapes by the removal of competing undergrowth. More than eighty years later, Champlain (1907: 61) made a nearly identical description on what he called the Isle de Bacchus, or what is today Richmond Island off the coast of Maine. He reported that 'the soil [is] cleared up, and many vineyards bearing beautiful grapes'. In what is today Alabama, the de Soto party enjoyed summer grapes 'as good as those of Spanish vines' (Sauer 1971: 289). They also found low vines of large and sweet grapes. On the basis that both muscadine and summer grapes are vigorous climbers, it has been suggested that this reference to low vines indicates that grapes were cut back or pruned (Sauer 1971: 289).

The final account for encouragement of large, and woody plants was made by Kalm (1972: 80). In speaking of sassafras (*Sassafras albidum* Nutt.), he noted that '[o]n old grounds, which are left uncultivated, it is one of the first that comes up, and as plentiful as young birches are on those *Swedish* fields, which are formed by burning the trees which grow on them'. Because it grows in stands or small groves, is conspicuous as second growth, has roots that spread laterally and are easily dug, and was important as a medicinal herb (Sauer 1971: 227), sassafras was more than likely encouraged by some native North Americans at contact.

Cultivation

Documentary evidence of cultivation, or the propagation of large, and woody plants, runs the gamut of credence from explorers' intuitive impressions that trees had been planted to direct, first-hand accounts of transplanting. Of the former group, which must be taken with great caution (Gartner 1997*b*: 337), five accounts exist; three deal with fruit trees in the middle of the continent, and two deal with grapes on either coast and separated in time by two centuries.

Along the banks of the Mississippi River in the late 1600s three priests reported trees that appeared to have been intentionally placed in certain locales by the aboriginals. Near the Wisconsin–Illinois boundary, Claude Allouez (1903: 72) saw trees 'planted designedly to form alleys more agreeable to the sight than orchards'. Membré (1903*a*: 161), who has already been discussed, saw 'little groves, as if planted by hand'. Providing somewhat more detail, Anastasius Douay (1903: 202) reported seeing 'some of the finest trees in the world, set as regularly as though they had been planted by man. Among them were many mulberry and other fruit trees'.

Writing of southern California in 1769, Father Juan Crespí reported seeing wild grapevines 'which look as if they had been planted' (Bolton 1927: 4). Although this comment is rather vague and late, it is similar to an earlier and more definitive description made on the other side of the continent. Near Chesapeake Bay, yet another priest, Juan de Carrera (1953: 130, 138–9) found in 1572 'a very beautiful vineyard, as well laid out and ordered as the vineyards of Spain. It was located on sandy soil, and the vines were laden with fair white grapes, large and ripe. Also within the vineyard there was a great number of plum, cherry, and persimmon trees'. The reference to other trees suggests that they were protected, the simplest form of tree manipulation. The reference to white grapes suggests they were scuppernongs, a variety of southern muscadines (*Vitis rotundifolia*). Sauer (1971: 225, 289) maintained that a few such plants might be found growing in the wild, but not a stand of them. Hence, cultivation of grapes was most probable.

The credibility of these early descriptions as accounts of the husbandry of large, and woody plants is improved markedly by reports of trees in gardens. Referring to inhabitants somewhere along the Georgia or Carolina coast in the early 1500s, Pedro Mártir de Anglería (1944: 515) noted that 'They cultivate orchards, and they have many types of fragrant plants, and they like to cultivate gardens; in the orchards they cultivate trees: in particular one that is called *corito*, that produces a delicious fruit, as large as a small peach. The other is called *guacomine*; it gives a fruit larger than the peach'. According to one writer (L. Larson 1980: 215), the *corito* was probably persimmon and the *guacomine* the plum. Not too far away, in northeastern Florida, in 1562 Jean Ribaut (1927: 74) said in his discussion of fields and gardens that 'There grovith many walnuttrees hazeltrees and smale cerytrees'.

In the early 1600s, John Smith (1907: 90) noted that in Virginia, '[b]y the dwellings of the Savages are some great Mulbery trees'. Although this passage does not confirm cultivation of trees (see, for example, H. Maxwell 1910: 96), other passages made by Smith strongly suggest that he was indeed referring to an advanced form of tree husbandry. For example, he went on to note that Virginia natives had 'plenty of fruits as well planted as naturall' (J. Smith 1907: 95). Strachey (1953: 119) echoed Smith's assessment by saying '[b]y their dwellings are some great Mulberry trees', but added, 'and these in some parte of the Country are found growinge naturally'. These statements strongly suggest cultivation.

In New England, he reported seeing 'groves, gardens, and corne fields' (J. Smith 1907: 192). That two of the three items involved planting suggests that the third one did as well. He provides better evidence, however, when he noted that elsewhere in the region he observed 'many corne fields and delightful groves', 'many faire high groves of mulberie trees gardens', and lands 'planted with corne; groves, mulberries, salvage gardens' (J. Smith 1907: 204). Finally, but more than a century later, Kalm (1972: 42) observed '*Morus rubra*, the mulberry tree, on fields, hills, and near the houses' of Indians in the Delaware Valley. The mulberrys growing on fields and hills may only have been encouraged, but those growing near houses probably did so because they were planted.

The best possible evidence for the cultivation of large, and woody plants would, of course, be an eyewitness account of someone actually observing the practice. About as close as we come to having this type of data are four somewhat imprecise passages from opposite sides of the continent. Two of the earliest such accounts come from missionaries who reported on conditions in the Southwest during the first half of the 17th century. At several locales in the Southwest, Alonso de Benavides (1965: 36) claimed that among the crops he saw 'planted' were some Old World fruits such as apricots and some decidedly New World species, specifically 'prickly pears' (*Opuntia* spp.) and 'pitahaya' (*Stenocereus* sp.). Similarly, his counterpart reported that the Névome (Pima Bajo) of Sonora, Mexico:

plantaban junto a sus casas un género de viñas, de una planta que los españoles llaman lechuguilla, porque en su forma es semejante a la lechuga; pero mucho más fuerte sus hojas y que para criarse y sazonarse, ha menester uno o dos años; y cuando llegado a sazón la cortan y asada la raíz con algo de sus hojas, les sirve de sustento, que es sabroso y dulce; y hacen de ellas molidas un modo de cajetas, como de conserva. Y cuando se corta una de esas raíces deja ya brotados otros renuevos y cogollos; y así plantado una vez este género de viña, les dura por muchos años (Pérez de Ribas 1944, vol. 2: 150).

plant next to their houses a kind of vineyard of a plant that the Spaniards call lechuguilla, because its form resembles that of a lettuce. Its leaves are, however, much stronger. It takes one or two years for it to grow and mature. When it matures, they cut it. Baked with some of its leaves, the root serves them as food that is tasty and sweet. They grind it and make a sort of jelly like a preserve. When one of these roots is cut, its leaves already started other renewals and shoots. Thus, once this type of vineyard is planted, it lasts them for many years (see also Pérez de Ribas 1968: 6).

This plant is clearly a species of *Agave* (Dobyns 1988: 50).

Writing a century later the famous Franciscan missionary Junipero Serra made an observation near the Pacific coast of the northern Baja peninsula that echoes Pérez de Ribas. Serra (1967: 68) wrote that he 'climbed a hill from which we could observe on the next one, a plantation of mescales, so dense, ranged row on row, that space could hardly be found for any other vegetation at all'. According to Nabhan (1985*a*: 47) this description 'leaves no doubt that these succulents were intentionally managed'. Indeed, the only way they could have been arranged in rows is by planting, hence they were cultivated.

On the other side of the continent, John Lawson (1937: 95) wrote that in North Carolina during the early 1700s: 'I knew two Trees of this Wood among the Indians, which were planted from the Acorn, and grew in the Freshes, and never saw anything more beautiful of that kind'.

Although direct, first-hand observations of seed planting are nearly nonexistent, there are at least three specific and definitive accounts of transplanting and a great deal of ecological evidence for cultivation of large, and woody plants at the time of contact. During the late 1600s, John Josselyn

(1972: 60–1) noted of wild cherries in one New England locale that 'they grow in clusters like Grapes, . . .' and that 'they are also good for Fluxes [juice]. Transplanted and manured, they grow exceedingly fair'. Beverly (1947: 136) made a somewhat similar observation in Virginia during the early 18th century, writing that the indigenous people 'had planted the Cuttings of their Vines at *Michaelmas*, and had Grapes from those very Cuttings, the Spring following'. Later in that century James Adair (1968: 48–9) wrote 'There is a species of tea, that grows spontaneously, and in great plenty, along the sea-coast of the two Carolinas, Georgia, and East and West Florida, which we call *Yopon*, or *Cusseena*: The Indians transplant, and are extremely fond of it'. His observations were confirmed by Bartram (1940: 291) who wrote of Jore, a town near the Georgia–North Carolina border, in the 1770s, that 'here I observed a little grove of the Cassine yapon, which was the only place where I had seen it grow in the Cherokee country; the Indians call it the beloved tree, and are very careful to keep it pruned and cultivated'.

Bartram's observation does not explicitly mention transplanting, but this stand of youpon (*Ilex vomitoria*) being the only one in the area and well beyond its natural coastal plain habitat is suspect. Adding to this suspicion are references made by another late 18th-century traveler, Benjamin Hawkins (1848: 30, 45), who noted at two northern Alabama Creek towns, Took-au-bat-che and Oc-fus-kee, respectively, that 'they have but a few plum trees, and several clumps of cassine yupon', and that 'they have some peach trees in the town, and the cassine yupon, in clumps'. Given that Bartram's and Hawkins' accounts, of a species that was known to have been transplanted, were made well outside of the tree's normal range, at least four scholars (Fairbanks 1979: 129–30; Ford 1985*c*: 382; Hudson 1979: 8; Merrill 1979: 40, 42–3, 57–8) have concluded that youpon was cultivated widely across the Southeast.

The transplanting of large, and woody plants into the interior parts of the Southeast involved not only species from the coast, but at least one from the West as well. In a footnote to his accounts of conditions in central Georgia, Bartram (1940: 57) commented of the 'chicasaw plum' that 'I never saw it wild in the forests, but always in old deserted Indian plantations. I suppose it to have been brought from the S.W. beyond the Mississippi, by the Chicasaws'. This interpretation is supported by ecological evidence. *Prunus angustifolia* has a natural range that extends from the eastern slopes of the southern Rocky Mountains eastward only as far as the Mississippi River (Havard 1895: 103).

Much as was the case with the archaeological evidence for the husbandry of small herbaceous plants, biogeographical evidence exists to support the premise that cultivation of large, and woody plants was widespread *c.* 1492. In at least six cases, economically important species have been found growing in proximity to aboriginal sites, and well outside their normal range. Butternut (*Juglans cinerea*) has been found growing at River Désert, Ontario. The natural range of this tree is in the Great Lakes area to the south (Black 1978: 259–60). Wild crab apples are reported to have been observed by some early explorers along the New England coast. Although this plant is a New England native, it is not naturally abundant near the seashore (H. Russell 1980: 85).

There is suspicion that the Iroquois may have planted the Kentucky coffee tree (*Gymnocladus canadensis*) and the American chestnut (*Castanea dentata*). The former is common along the shores of Cayuga and Seneca lakes, but has been found, as has the latter (G. Day 1953: 339), growing in groves near the sites of native villages (Hedrick 1933: 27). Similarly, there is some suspicion that aboriginals in the Midwest and the South cultivated pecan (*Carya illinoensis*). This tree grows naturally as isolated plants in stands of other species in the valleys of the Texas and Louisiana coastal plain. That it is found in groves in Missouri and Illinois (Sauer 1971: 291) and Alabama and Georgia (Havard 1895: 104) suggests that it was carried north and east very early on. Concrete evidence confirming this notion is lacking however (Flack 1970: 48). Similarly, the discoveries of a cactus (*Opuntia rafinequii*) valued for its fruit, New Jersey tea or redroot (*Ceanthus herbacea*), the leaves of which were used to make a tea, and Indian currant (*Symphoricarpus orbiculatus*), with its medicinal berries, near Sandusky, Ohio, and well east of their natural range, have been interpreted as evidence of planting by early native Americans (Moseley 1930). Finally, a species of grape (*Vitis arizonica*) found growing near Prescott, Arizona, has been 'accepted as conclusive evidence of its culture by the Pueblo Indians' (Havard 1895: 104). Of course, present-day scholars would argue that the Anasazi did not live in the Prescott area.

In sum, the documentary and biogeographical data suggest very strongly that the husbandry of large, and woody plants was common at contact. More recent and more ancient data, that involving ethnology and archaeology, respectively, lend additional support for this conclusion.

Ethnographic Evidence

Assessments of native American husbandry of large, and woody plants during the 19th and 20th centuries are similar to the first-hand explorers' accounts in that they are about as numerous, and frequently contain confusing mention of Old World species. They are, however, different in at least three other respects. They include groups in what may be considered the more remote parts of the continent, the West, including the relatively little-affected mountain and desert peoples, particularly those in Northwest Mexico (Fig. 3.2). They contain a fair number of accounts of encouragement. Finally, and most importantly, they provide much insight into indigenous practices.

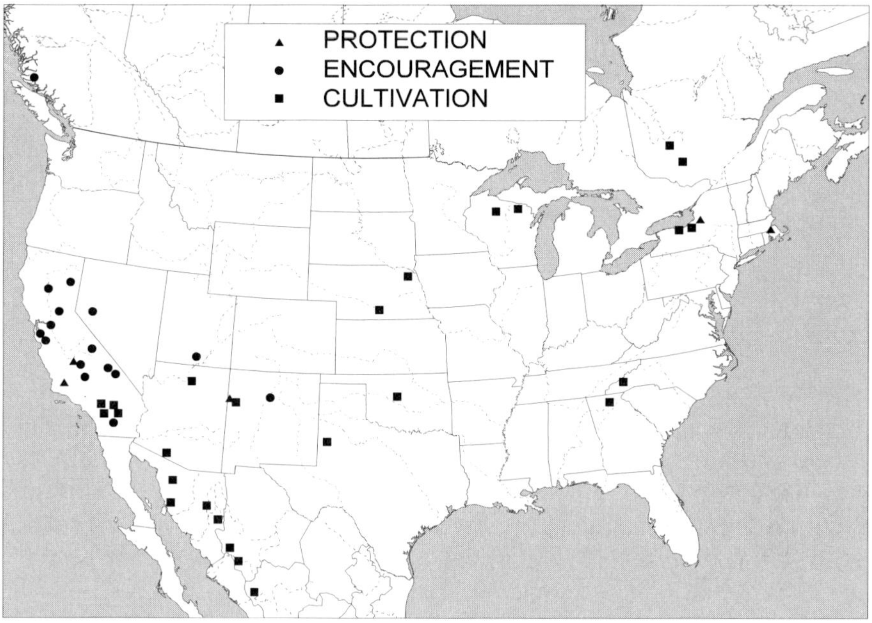

Fig. 3.2. Locations cited in the text for which there is ethnographic evidence of the husbandry of large, and woody plants.

Protection

Only five ethnographic references exist to suggest that native North Americans protected large, and woody plants. The first is the previously mentioned description of remnant groups in Massachusetts going to some length to protect chestnut groves from annual fires that presumably were intentionally set (Temple and Sheldon 1875: 46). The second comes from neighboring upstate New York. During the 1800s, the Iroquois are reported to have tended plums that escaped from cultivation. This activity is said to have taken place on the outskirts of villages (Hedrick 1933: 27), areas notorious for having been seriously disturbed by humans. The third account is from New Mexico where a fraternity of the Zuni were said to 'maintain a bed' of cane cactus (*Opuntia arborescens*) 5 km from the main pueblo. Exactly what was involved in maintenance is not stated, but the members are reported to have limited access to the plants, decorated them with eagle feathers, and sprinkled them with meal. They also used the branches for whipping themselves during ceremonies (Stevenson 1915: 95). The final two reports come from California, where certain groups are known to have practiced 'judicious harvesting' in order not to deplete the stock of willow (*Salix* sp.) shoots used in basket-making (Ortiz 1993: 199–200), while others bestowed spiritual properties to oak (*Quercus* sp.) groves, thereby ensuring that acorns were not over-exploited (McCarthy 1993: 225–7).

Encouragement

Every example of historic encouragement of large, and woody plants comes from the far western part of the continent—the Southwest, the Great Basin, California, and British Columbia. One informant reported that the Maidu of northern California commonly removed competing vegetation from around elderberry (*Sambucus* sp.) bushes and removed dead branches in order to promote better growth of the fruit (Bean and Lawton 1973: xxvii). The precise method of clearing away of other species was not reported, but it was probably by pulling or possibly burning. The use of fire for encouraging the growth of large, and woody plants was certainly not uncommon in this part of the continent. Indeed, the Southern Sierra Miwok of Yosemite, the San Juan Ningwi, the North Fork Mono, and the Maidu are known to have burned patches of redbud (*Cercis occidentalis*), willow, lemonade berry (*Rhus trilobata*), sour berry (species not identified), and maple (*Acer* sp.), respectively, in order to promote hardy and straight growth of stems used for basketry fibers (K. Anderson 1991*a*: 18, 1991*b*: 149–50, 1993: 159, 163; C. Fowler 1986: 94). The North Fork Mono also burned white oaks (*Quercus douglasii*) to get knot-free wood for making cradleboards (K. Anderson 1993: 159–60). Their Western brethren burned stands of black oak (*Quercus* sp.) to obtain straight wood for making stirring sticks and spoons, while the Chukchansi Yokats fired buttonwood (species not identified) to get straight shoots to be used as arrow shafts (K. Anderson 1993: 159). The Cahuilla also used fire, but here they burned trees to ensure better fruit. According to one informant (Patencio 1943: 69): 'The bugs that hatched in the top of the palm trees, they made the tree sick, and no fruit came. After the trees were set afire and burned, the bugs were killed and the trees gave good fruit'.

The Southern Sierra Miwok coppiced and pruned redbud (K. Anderson 1991*b*: 150–2) while the Yokuts-Mono and Paiute only pruned the plant. Other Mono pruned sourberry (K. Anderson 1993: 160, 163), and the Pomo, pruned willow (Ortiz 1993: 202). Each of these activities was for the sole purpose of obtaining straight shoots for basket-making. As was discussed earlier, the Cahuilla are also known to have pruned trees in natural mesquite groves, and individuals even marked the trees they had altered, thereby claiming private ownership (Patencio 1943: 58). Pruning has also been reported among the Death Valley Shoshone (C. Fowler 1986: 94–5). These people routinely broke the tips off of piñon (*Pinus monophylla*) branches for the purpose of producing more cones, and hence nuts, in subsequent years.

Perhaps the most complex form of encouragement reported in the ethnographic literature dealing with large, and woody plants involves coppicing.

Northern Paiute women, particularly those of the Walker River branch in western Nevada, regularly cut first-year sprouts of various willows (*Salix* spp.). This annual cutting from the bases of trunks near the roots promoted vigorous and straight growth. The harvested parts were then used in the manufacture of baskets (C. Fowler 1986: 94, 95). Coppicing was also practiced by the Southern Miwok in order to ensure straightness. These people manipulated mock orange (*Philadelphus lewisii* ssp. *californicus*), and wild rose (*Rosa californica*) for making arrows, elderberry (*Sambucus caerulea*) for making flutes, and redbud, willow, dogwood (*Cornus* spp.), maple (*Acer macrophyllum*), deerbrush (*Ceanothus integerrimus*), and buckbrush (*Ceanothus cuneatus*) for making baskets (K. Anderson 1991: 22).

If coppicing is the most complex form of encouragement of large, and woody plants, then some of the activities reported and assessed by Ford (1968: 194–6) during his work at San Juan Pueblo, have to be some of the most intriguing. Ford discovered first that non-flowering, straight branches of the shrub Apache plume (genus and species not identified) were cut off and used in the making of snares, arrows, and brooms. Although he was less than definite on the point, he intimates that cutting select branches was beneficial to the growth of the plant, and therefore may have been tantamount to pollarding. Next, he found that the roots of a yucca (species not identified) were frequently excavated and used in the manufacturing of soap. Although there is nothing new or exceptional in this observation, Ford did note that old and unhealthy plants were preferred. He argued, then, that eliminating the weakest plants in any one stand of yucca encouraged the other plants to grow and multiply more rapidly: tacitly, it was thinning. Finally, he discovered that certain cacti (species not identified) were encouraged to grow in disturbed areas, and they were fertilized, however inadvertently. Specifically, he found that seeds that were eaten passed through people's digestive systems intact, and were deposited with other fecal matter in areas near the edge of the village. These seeds then sprouted and grew luxuriantly in the most fertile of habitats. Although this particular activity, as well as the two others reported by Ford, may not seem to be 'encouraging' in a literal or deliberate sense, it is something more than protection, and something less than cultivation.

Finally, much like his accounts of clover and cinquefoil encouraging, Boas (1921: 209–10, 216–18, 221–2) reported huckleberry (*Gaylussacia* sp.), viburnum (*Viburnum* sp.), and gooseberry (*Ribes* sp.) bushes growing in 'gardens', or 'patches'. Indeed, the viburnum patches were so large, regularly shaped, conterminous, and devoid of other species (Boas 1966: 24–8) that they can probably be best described as fields or orchards (Deur, forthcoming). That they were privately owned (Boas 1921: 1345–8) and distinctive from those locales in which berries from seven other species of woody perennial were simply collected (Boas 1921: 204–9, 211–15, 218–21) adds credence to this interpretation. The descriptions of Boas are not sufficiently strong to indicate cultivation, but they do imply that something other than gathering or even protecting was taking place.

Cultivation

Numerous ethnographic accounts confirm that large, and woody plants were indeed deliberately planted and tended by native North Americans. That many of these accounts include, almost exclusively, New World plants, and involved what might be considered unusual cultivation techniques, is taken as evidence that the practice was native to the continent and employed at the time of European encounter. In some cases, these accounts lack details, but instead, contain simple statements indicating that certain large, and woody plants were artificially propagated. In addition to the previously mentioned passage about the Cherokee (Powell 1916: 132), who also planted smooth alder (*Alnus Serrulata*), the bark of which was used as a medicinal (Asch 1994: 52, 54), eight such reports of six other peoples are presented here.

In discussing the Tipai-Ipai, one source (Shipek 1989: 163, 164) noted that oaks (*Quercus* spp.), wild plums (*Prunus* sp.), manzanita (*Arctostaphylos* sp.), ceanothus (*Ceanothus* sp.), 'wild' grapes and sage (*Artemisia* sp. and *Salvia* sp.) 'were planted' for food, while willow (*Salix* sp.), sumac (*Rhus* sp.), and elderberry (*Sambucus* sp.) were planted to stabilize riverine areas. In writing about the Névome of Sonora, Mexico, one source (Dobyns 1988: 52) reports that agave is 'still grown', while another (Pennington 1980: 177) claims that an unspecified agave is 'cultivated in at least 50 percent' of their gardens. Of the Rarámuri, and Odami (Tepehuan) who live in the mountains of southern Chihuahua and northern Durango, Mexico, respectively, it has been written that '[t]he large flat-leafed cactus (subgenus *Platyopuntia*) is planted near the houses and dwelling caves' (Bennett and Zingg 1935: 44, 38), and '*Opuntia* cactus are occasionally cultivated in garden plots' (Pennington 1983*b*: 308). Comanche women in West Texas are said to have had as their duty 'the planting and cultivating of maize and a few fruits and berries' (Fitzhugh 1926: 96). Across the continent in Quebec, Canada, an Ojibwa informant at Barrière reported that she was responsible for, by virtue of having 'planted', chokecherries (*Prunus virginiana*) that are found in the vicinity (Black 1978: 260). Similarly, it was reported that prairie sagewort (*Artemisia frigada*) was 'planted' by aboriginal people in Forest County, Wisconsin and used as a fumigant to revive comatose patients (Asch 1994: 52, 54).

Although details about planting are lacking in some sources, others indicate that it mostly involved seeds, sometimes involved cutting and transplanting, and always resulted in groves or thickets rather than formal orchards with trees in rows. Another Ojibwa woman, this time one who lived at River Désert, Quebec, said that she had intentionally planted from seeds several years ago the Canada plum trees found near her abandoned house

(Black 1978: 259). As has already been mentioned, Iroquois farmers planted tree seeds (Hedrick 1933: 23) in groves (Powell 1916: 32). This tradition was found across the continent, on the Plains, in the Southwest, California, and northwestern Mexico. Oral histories, supported by biogeographical evidence of certain trees' existence well beyond their natural range, hold that some Menominee of Wisconsin planted hop-tree (*Ptelea trifoliata*) seeds brought in from Kansas (Asch 1994: 52, 58), that stands of crab apple (*Malus ioensis*) in northeastern Nebraska were established by seeds carried home by the Ponka (Gilmore 1930: 91–2), that Pawnee planted small groves of black walnuts in southern Nebraska (Gilmore 1930: 91), and that thickets of wild plums in Oklahoma resulted from seeds brought there by these same people (Gilmore 1930: 92). Whether or not the thickets of wild plums found along streams throughout the Plains and utilized by native people (Wedel 1941: 2, 5, 1986: 21) were cultivated, is uncertain. One thing that is clear, however, is that the Zuni also claim to have planted groves of wild plums (Fig. 3.3) in sandy areas (David Rhode, personal communication 1988).

For southern California there exist no fewer than three separate accounts of the Cahuilla having planted, by seed, groves of palm trees. In two cases (Chase 1919: 16; Bean and Saubel 1972: 149), these involve stands planted in higher elevations, while in the third case (Patencio 1943: 94, 101) it was

Fig. 3.3. A grove of 'wild' plum trees planted by the Zuni of far western New Mexico. (Photograph by author.)

written of several settlements in low-lying areas with abundant water that 'people carried the seed to their homes, and palm trees grew from this seed in many places'. Also from California there exist reports of both mesquite (*Prosopis* sp.) and Torrey Pines (*Pinus torreyana*) having been planted from seeds (Shipek 1989: 163).

The Odami are known to have grown avocado trees (*Persea americana*) in some locales. These were typically planted as individuals near habitations, or 'in clusters apart from habitations, particularly along the edges of fields' (Pennington 1969: 95, 94). Along with the previously mentioned cactus, the Odami's immediate neighbors to the north, the Rarámuri, have also been reported as having cultivated a number of trees by various means. In addition to their having 'occasionally grown . . . native fruits' such as cherry (*Prunus serotina*), Campbell W. Pennington (1983*a*: 280, 1963: 71–2, 74) reported that they planted, by seed, avocados, a wild plum (*Spondias mombin*), and zapote (*Achras zapota*), the seeds of which were scraped before being planted. Another wild plum (*Prunus americana*), and another wild cherry (*Prunus capuli*), were propagated by either seed or slips cut from wild trees, and mulberry trees (*Morus microphylla*) were propagated exclusively by slips. All of these 'cultivated native fruit trees are usually located in the fields' (Pennington 1963: 72).

Planting by slips is also known to have been done in order to propagate living fence rows, stands of closely spaced trees between fields and stream channels in the Sonoran Desert. These features aid in diminishing lateral erosion and hence field destruction, and, by slowing floodwaters they increase sedimentation and therefore help maintain soil fertility of fields (Nabhan and Sheridan 1977). Such fence rows involving willow, cottonwood, mesquite, and other trees continue to be planted by Tohono O'odham (Nabhan *et al.* 1982: 128). The Tipai-Ipai also planted fences of prickly pear (*Opuntia* sp.), from vegetative cuttings, around their houses (Shipek 1989: 164).

The only ethnographic evidence found for native Americans truly transplanting (i.e., uprooting and moving entire plants), also comes from the Southwest and southern California. As mentioned earlier, it has been written of the Hopi that: '[c]ottonwoods grow along washes and near springs, often being transplanted . . .' (Whiting 1939: 71), and '[s]everal species of willow grow on the reservation. Young shrubs are transplanted into convenient washes' (Whiting 1939: 72; see also Whiting 1937: 16). Further south, the world's premiere authority on agaves, Howard Scott Gentry (1972: 101) said that he saw *Agave mupheyi* 'cultivated at Sells, Ariz., and a Tohono O'odam reported the same had been brought in from the neighboring mountains'. He also noted that between Quitovac and Caborca, Sonora, Mexico, agave was '[c]ultivated in a yard whose owner stated that his plants had been brought from the nearby hills' (Gentry 1972: 101). Finally, it has been reported of the Seri who live along the coast of the Gulf of California that near Punta Sargento *Agave subsimplex* 'and two kinds of prickly pear (*Opuntia phaeacantha* and *O. violacea*) were said to have been planted . . . many years ago. . . . Isolated populations of these species occurring there in the 1980s may have been derived from these transplants' (Felger and Moser 1985: 225, 271). To be sure, certain agaves are easy to transplant. According to one source (Crosswhite 1981: 59), propagation involves stripping a plant of its 200 or so bulbils and putting them aside in the shady environment of a sack or basket for a week or two until the cut surfaces have healed over. They are then planted in loose soil and watered occasionally until they become established. Finally, Tipai-Ipai elders once recalled that both mesquite and jojoba (*Simmondsia* sp.) were 'transplanted' in times past, but no one now retains the necessary knowledge (Shipek 1989: 165).

Overall, the ethnographic data add considerably to our understanding of native husbandry of large, and woody plants; they shed light on the various ways in which such plants were protected, encouraged, and cultivated, and they help fill in blank spots on the map. The final picture, however, becomes more complete when prehistoric evidence is considered.

Archaeological Evidence

There actually exists very little evidence of large, and woody plants having had their growth manipulated by North Americans in prehistoric times. This condition may, in part, be a function of the relative unimportance of the practice. It may also be partly attributable to the elusive nature of the data, and the misperceptions of scholars. The precise nature of archaeological evidence of the husbandry of large, and woody plants is not well understood and hence is not treated in texts that deal with archaeological methods. Also, it may well be that prehistorians have not considered the practice to have been significant and therefore simply dismissed it. The little evidence that does exist for such prehistoric plant husbandry in North America suggests that the practice may have been more common than previously thought, particularly in the Southwest and the Southeast (Fig. 3.4) and that prehistorians have not found much evidence simply because they have not looked for it. Once evidence is discovered, it seems to become quite prevalent (see e.g., Nabhan 1995).

Protection

The first of only two known works that treat prehistoric protection of trees involves carbonized shells of butternut (*Juglans cinerea*) found at sites in the Yazoo Delta of Mississippi (Cutler and Blake 1985: 117). Because these sites are near the margins of the trees' natural range and because butternuts are uncommon in the area, it has been suggested that 'viable nuts, brought from afar, could have fallen into the village middens, sprouted, and grown to trees

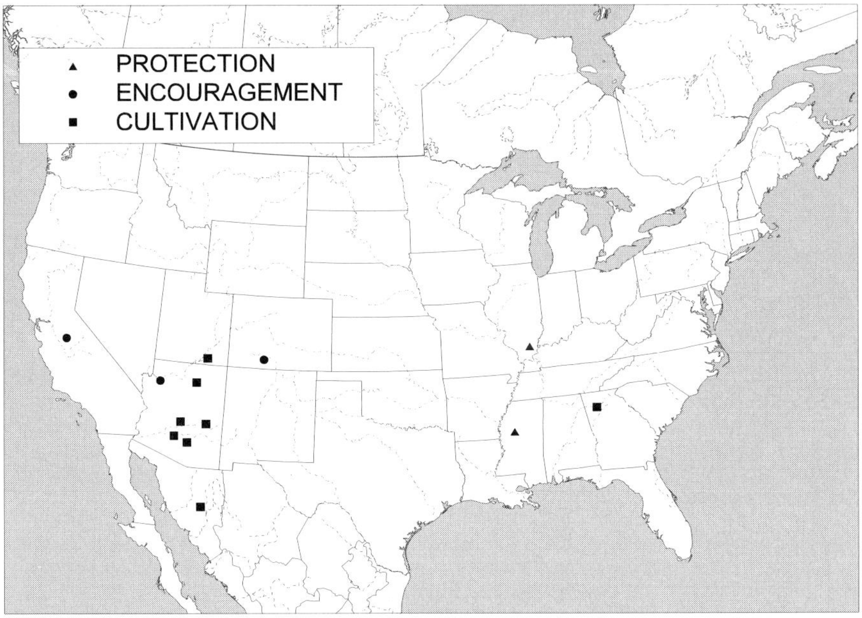

Fig. 3.4. Locations cited in the text for which there is archaeological evidence of the husbandry of large, and woody plants.

which bore fruit' (Cutler and Blake 1985: 118). This is tacitly an argument in favor of protection.

The second case of prehistoric protection is similar to that of the first in that it involves macrofloral remains. In this case, however more than one species is involved. Flotation analysis of sites in the Black Bottom area of southern Illinois produced remnants of persimmon, cherry, and plum. Identified as 'pioneer trees in the secondary succession of old fields' that begin producing only after several years, these species are thought to have been grown at the edges of fields that contained subsistence crops (Blakeman 1974: 145, 147–8). There, they would have benefited from the disturbed habitat, and would have been maintained in order to decrease travel time to fruit sources.

Encouragement

Three examples of pre-contact encouragement have been reported, but there are probably others which await identification. One is that of the previously mentioned coppicing on Mesa Verde. A truly remarkable discovery, made in the end of the 1960s, revealed the manipulation of Douglas firs (*Pseudotsuga menziesii*) during the 13th century AD (Nichols and Smith 1965). At least two dead trees in what now is a national park were found to have stubs of limbs that show stone-ax cut marks, and signs of having their trunks deliberately and permanently bent over. The argument put forth claims that Mesa Verdeans first removed all the branches from the saplings before bending the trunks parallel to the ground and tying them in that position. As new branches sprouted, all but one was removed from each tree. During the course of ten to fifteen years, new branches were systematically removed, thereby allowing what would otherwise be a sizable lateral branch on each tree to assume apical or terminal dominance. At the end of this period the remaining branch was removed and another branch allowed to grow. Presumably the process went on until the trees died or people abandoned the area. Although only two trees were found to contain signs of such coppicing, it is probable that the practice was widespread in the area as there are a seemingly infinite number of fir stands in the region (Nichols and Smith 1965: 63–4). To be sure, the practice makes ecological sense. Branches, especially those given apical dominance, grow much faster than trunks. This is particularly true if trees have to first sprout from seeds and establish root systems. By coppicing in the manner just described, indigenous people at Mesa Verde were able to cut a branch suitable for use as a roof timber in half the time it would take them to harvest a comparable trunk. The practice would also not result in soil erosion as the entire biomass would not be removed during harvesting. By any standard, the coppicing of Douglas firs has to be one of the more unusual and sustainable agricultural practices of aboriginal North Americans.

The second example also involves coppicing and comes from the P. A. Hearst Museum of Anthropology at the University of California, Berkeley. Finished products made by various California peoples and curated on shelves and in displays, including clapper sticks made from elderberry, arrows made from snowberry, and baskets made from branches of maple are 'too straight, too uniform, and too lacking in lateral branch scars to be from undisturbed plants in the wild'. Coppicing is attributed to having been the cause (K. Anderson 1993: 160, 162).

The third example of encouraging woody plants involves a split-twig animal figurine made approximately 3,500 years ago and found in a cave in Grand Canyon. It was made by wrapping a single, two-meter long stalk of squawbush (*Rhus trilobata*). Now, squawbush does not normally produce long and straight stalks. Indeed, under natural conditions the bush is low-growing, with irregular branches. When burned, however, stands of squawbush produce long, straight sprouts the following spring (Bohrer 1983). The stick figurine could only have been made from plants that were burnt. Accordingly, encouragement seems plausible. Given this case, one can only wonder how many other museums have artifacts, made by other peoples, which show similar signs of plant encouragement.

Cultivation

The evidence of prehistoric cultivation of large, and woody plants in North America ranges from archaeological inferences based on certain cultural artifacts to actual relict 'orchards'; there exists no definitive archaeological data, especially in the eastern half of the continent (Asch 1994: 62). The former type of data are associated with the cultivation of youpon in the southeastern United States. The documentary evidence in concert with the biogeographical data clearly indicate that the species was transplanted. The distribution and antiquity of sites that contain certain dippers or cups made of shell and known to have been used exclusively for the drinking of a tea brewed from youpon leaves indicates that perhaps as early as 1000 BC youpon was not only cultivated in inland locales well beyond its normal range, but that it may have constituted an important element in cultures of the Eastern Woodlands (Milanich 1979).

Understanding of pre-European cultivation of large, and woody plants in North America has improved markedly in recent years. For nearly a century it has been argued, albeit without hard evidence, that certain species of *Opuntia* and *Agave* were 'cultivated' since 'time immemorial' in Mexico (Havard 1895: 116, 1896: 34; Wissler 1922: 15). Few people, if any, questioned the accuracy of these statements because they made intuitive sense given the large number of documented cases in which indigenous people were confirmed as having planted the species (e.g., Bennett and Zingg 1935: 44). In the 1960s one archaeologist, Jesse D. Jennings (1966), suggested that the practice was not only ancient but that it extended north of Mexico into the Southwest. Specifically, along two tributaries of Glen Canyon in southern Utah, he found small pure stands of prickly pear (*Opuntia* sp.) in areas otherwise devoid of the species (Jennings 1966: 21–3). Given the artifacts found during his investigations, including the discovery of a small irrigation system, and at least two other cultigens—amaranth and beeweed—growing nearby, he concluded that the prickly pear comprised 'a relict patch of this domesticate' (Jennings 1966: 48).

Jennings' findings are actually little more than speculation, which may well account for their apparent lack of acceptance. Few prehistorians cite this specific interpretation, but that may change given new findings elsewhere in the Southwest. Vorsila L. Bohrer (1991: 228–30), the dean of southwestern archaeobotanists, suspects that cholla (*Opuntia* subgenus *Cylindropuntia*) may have been cultivated. The tender buds of this plant have long been used as a vegetable and some plants have been found growing near abandoned ruins in the Jemez Mountains of New Mexico, far higher in altitude than their natural range. Also in New Mexico, other studies have concluded that *Agave parryi*, found in eastern (Minnis and Plog 1976: 305–7) and northern (Dosh 1988) parts of the state, well outside of its natural range, and also restricted to archaeological sites, may be relicts of cultivated plants. That finding, however, was not new, but a follow-up on the ideas of Gentry (1972: 443) who noted that clones of *Agave murpheyi* were often found in association with prehistoric archaeological sites in the southern part of the state, and never observed in extensive or dense populations. Given that *Agave murpheyi* is not a natural invader of disturbed habitats, but instead is easily propagated by planting bulbils that form on the fruiting stalks (Crosswhite 1981: 59), it is accepted that this species was transplanted in prehistoric times (Gentry 1972: 443; Miksicek 1984: table II.2.4).

This conclusion has been reinforced most convincingly lately with evidence, supplied by Suzanne K. and Paul R. Fish, and John H. Madsen, from the Santa Cruz Valley. There, on the bajada west of the Tortolita Mountains near Marana, north of Tucson, wild stands of agave are absent today. The natural range of agave is currently much higher in elevation. Curiously, however, agave remains have been found in several very large, 30 m diameter, prehistoric roasting pits (P. Fish, S. Fish, and Madsen 1984: 95), and tabular basalt knives, reminiscent of those used by historic southwestern natives for cutting agave (Castetter, Bell, and Grove 1938), are common on the surface (P. Fish, S. Fish, and Madsen 1984: 88; S. Fish, P. Fish, and Madsen 1992*b*: 83–4). Added to this evidence are the remains of numerous 'rock piles', initially identified as 'agricultural' features (Czaplicki and Rankin 1984: 6, 10, 24–8, 60–2, 68), and later confirmed as having been used in the cultivation of agave prehistorically. At least 112 loci (S. Fish, P. Fish, and Madsen 1990: 204) covering 5.1 sq km and containing 42,000 piles (P. Fish, S. Fish, and Madsen 1984: 86; S. Fish, P. Fish, and Madsen 1992*b*: 85) have been identified. More than forty fields are less than 2 ha in areal extent, but a few cover more than 50 ha (S. Fish, P. Fish, and Madsen 1990: 204–5). Constructed between AD 1150 and 1300, many of the rock piles have been reduced to rock scatters over the past 800 years (S. Fish, P. Fish, and Madsen 1992*b*: 86, 75). Originally they were comprised of fist-sized cobbles stacked 75 cm high and 1.5 m in diameter (S. Fish *et al.* 1985: 110).

Agaves were planted in the rock piles. The rocks apparently protected the plants from predation by burrowing rodents (S. Fish, P. Fish, and Madsen 1992*b*: 80). More importantly, however, they helped conserve scarce moisture. Many are topographically situated so as to intercept runoff (S. Fish, P. Fish, and Madsen 1989: 252–4, 1990: 203). Furthermore, the uneven, porous rock surfaces allow infiltration of rainfall in contrast to surrounding hard-packed and impermeable ground surface. Rocks then acted as mulch, slowing evaporation of soil moisture by blocking capillary action and preserving higher moisture levels beneath (S. Fish *et al.* 1985: 109).

The response by plants to the microhabitats of prehistoric rock piles is demonstrated by the continued preferential growth of certain large species (Fig. 3.5). That plants grow better on rock piles than on unmodified surfaces has long been quantified in one study in which root biomass in soil directly beneath rock piles was compared to adjacent control plots. Root weight in

Fig. 3.5. A saguaro (*Carnegiea gigantea*) cactus growing in the preferential habitat created by a prehistoric rock pile near Marana, Arizona. (Photograph by author.)

the rock pile soils averaged 2.7 times the weight of roots in the control samples (S. Fish *et al.* 1985: 109).

It has been estimated that as many as 102,000 agaves were grown on the Marana rock piles at any one time, and that 10 percent or 10,200 plants were harvested annually (S. Fish *et al.* 1992*b*: 86), providing 40.8 metric tons of food each year (S. Fish *et al.* 1985: 112, 110). This production is impressive, but it probably was not unusual, or unique to the Santa Cruz Valley. Other studies have found prehistoric rock piles (e.g., Downum 1993: 16, 35, 36, 4), often in association with tabular knives (Fig. 3.6; e.g., Rankin 1989*b*: 969) and relict stands of *Agave murpheyi* (Fig. 3.7), well beyond its natural range today, as far north as the Tonto, Verde, Agua Fria, and New River basins in Central Arizona (Antieau 1977: 8; Bohrer 1991: 229; Dove 1970: 21; Homburg 1997; Homburg and Ciolek-Torello 1997; Homburg, Ciolek-Torrello, and Shelley

Fig. 3.6. A prehistoric tabular basalt knife being used by archaeologist Adrianne G. Rankin to cut a leaf of *Agave murpheyi* growing on an ancient rock pile site in the New River Valley north of Phoenix, Arizona. (Photograph by author.)

Fig. 3.7. A surviving stand of *Agave murpheyi* growing on a prehistoric rock pile site, and being inspected by archaeologists David E. Doyel, left, and Adrianne G. Rankin, right, in the New River Valley, far outside its natural range. Note the bulbils which can be transplanted thereby producing clones. (Photograph by author.)

n.d.; Rankin 1989*b*: 965–79), and numerous locales in between (Dart 1983: 404–10; Debowski *et al.* 1976: 90, 108, 109, fig. 23; Doyel 1993: 35–6; Hodgson, Nabhan, and Ecker 1989). Indeed, the *Agave murpheyi* found at the New River sites are thought to be descendants of plants cultivated in prehistoric times (Vorsila Bohrer, personal communication 1988).

Recent discoveries of relict stands in unusual archaeological and biogeographical contexts has spurred many archaeologists to look for other types of evidence for the husbandry of large, and woody plants. Disproportionately high frequencies of agave fragments at the Frogtown site have been used to suggest cultivation there (Miksicek 1984: 588). Exceptionally high counts of cholla pollen have been envisaged as possible evidence of cultivation at the La Ciudad site as well as other sites in the Salt–Gila valleys (S. Fish 1984: 120, 130–4). Stones or pits more robust than usual, have, similarly, been proffered as indicators of the cultivation of Mexican crucillo (*Condalia warnockii* var. *Kearneyana*) at one Hohokam site (Bohrer 1991: 230). Surely, future work will carry this line of research to new conclusions (Bohrer 1992).

Overview

Although native North Americans did not have domesticated trees either prior to or at the time of the arrival of Europeans, they certainly husbanded large, and woody plants, sometimes modifying their habitats markedly. Evidence of virtually every type of activity—protection, encouragement, and cultivation—exists from ethnohistorical, ethnological, and archaeological records. In concert with biogeographical data it is clear that the practices were common from coast to coast. Also, and not inconsequentially, a number of most unusual practices were developed.

To be sure, there is nothing spectacular or noteworthy in protecting certain species. That practice is about as elementary as any activity involving plants. However, when it came to encouraging the growth of some large, and woody plants, indigenous North Americans developed some most clever strategies that not only demonstrate their resourcefulness and ingenuity, but also a keen appreciation and understanding of the biophysical environment and plant ecology. Without reiterating, suffice it to say that the coppicing of Douglas firs (*Pseudotsuga menziesii*) stands as evidence that Mesa Verdeans at least, and undoubtedly countless other peoples as well, knew exactly what they were doing. As for cultivation, the last step before full-blown agriculture and the development of plants dependent on people for their survival, the propagation of agaves through cloning with bulbils suggests as well that southwestern peoples, and presumably others, were actively and deliberately experimenting with large, and woody plants.

The degree to which native North Americans were successful with their husbandry of such plants is perhaps most evident in the long continuation of the practice among these people, and the transfer of certain species and practices to, and the adoption of some by people of European ancestry. For example, evidence indicates that youpon (*Ilex vomitoria*) was cultivated prehistorically and at contact. Today, it is transplanted as an ornamental throughout much of the southeastern and south-central parts of the continent by people commonly, although sometimes inappropriately, called 'Anglos'. Similarly, but certainly more importantly in terms of subsistence, is the indigenous contribution of plum cultivation to 'Anglo' settlers. As the native population decreased in the eastern half of the continent during the 1700s and 1800s, settlers of European extraction not only occupied lands formerly held by aboriginals, but they typically maintained the plum stands they found there. Also, when these later settlers moved on, they often carried with them native trees which were planted in new home orchards (Sauer 1971: 290). Whether or not these people recognized the contribution of their forerunners is unclear. One fact is, however, certain; the legacy of native North Americans' husbandry of large, and woody plants is much richer than previously thought.

4

Gardens

The relationship between people and plants is certainly an ancient one, extending back in time to the earliest days of human existence. In addition to the obvious consumption of plants, humans have affected plant growth, genetic structure, and, most importantly for present purposes, the habitats in which plants grow. Two factors are crucial in understanding people–plant relationships, time and space or distance. The longer people interacted with a particular species, the more change it and its habitat underwent. Similarly, the closer people lived to or with a plant, the greater the change it and its habitat experienced. In other words, species and habitats near places of human habitation were manipulated more than those located further away (Ford 1985*b*: 6; Hurt 1987: 11).

As Chapters 2 and 3 illustrate, native inhabitants of North America have greatly affected literally scores of species and their habitats. Some plants and habitats were only slightly modified, suggesting relatively recent and ephemeral associations with people. Others, however, were altered markedly, demonstrating long and tenacious associations. For the most part, plants and habitats in the latter category were spatially close to, if not an integral part of, human habitation sites. This proximity led to the development of what was probably the earliest and, arguably, remains the most important type of agricultural feature on the landscape—the garden.

The importance of gardens as early agricultural settings was first noted by Darwin (1958: 55–7). It was, however, the late and great botanist Edgar Anderson (1952: 136–51), who, after years of studying people–plant relationships, developed the model to explain their role. Anderson formulated the dump-heap hypothesis which states that middens near permanent or nearly permanent residential sites were ideal habitats for domestication to occur. The disposal of domestic waste water made these locales perennially moist; the disposal of inorganic trash improved their soil texture; and the disposal of organic matter (garbage) resulted in increased soil fertility. In effect, dump-heaps have as their present-day equivalents, composts. Add to this, the diversity of plant remains discarded, and the constant observation they must have received due to their proximity to dwellings, and middens take on very special properties. Many discarded plant remains not only sprouted but experienced exceptionally luxuriant growth. The diversity of plants led to hybridization. Finally, the attention dump-heaps received facilitated the selection of advantageous traits.

Anderson's ideas precipitated so much research by archaeologists that the term 'horticulture' took on an entirely new meaning. In its original context, and as it is still used in plant science, and geography, horticulture is the practice of cultivating fruits, vegetables, and ornamental plants (see Janick *et al.* 1974; B. Turner 1992: 267). Sometimes it is seen as small-scale agriculture (Holder 1991: viii; Scarry 1993: 7). It is, however, an end in itself. In its current archaeological context, horticulture is the incipient cultivation of indigenous plants, which are typically seed-bearing (see Keegan 1987*d*, M. Fowler 1992: 2; Cowan and Watson 1992). It is a phase in the transition from gathering to agriculture (see Chapter 2).

The expansion of the term 'horticulture' has been paralleled by what can only be considered an explosion in the number of adjectives applied to gardens. The terms 'urban' (D. Page 1986), 'house-lot', 'residential' (Killion 1990), 'home' (González-Jacóme 1985), 'household' (Niñez 1984), 'kitchen' (Ginsberg 1987), 'dooryard' (Kimber 1973), and 'dooryard orchard' (Herlihy n.d.) have all been used, with each investigator making a case for why hers or his term is most appropriate. In reality, each of these adjectives is acceptable as they all imply and connote essentially the same idea. Gardens, regardless of how various authors qualify them, have certain common characteristics. Condensed from a list compiled by Clarissa T. Kimber (n.d.), a noted biogeographer and authority on gardens, these are:

(1) small size, usually only a few hundred square meters;
(2) great plant diversity, polycropping, involving numerous species of trees, shrubs, and herbs;
(3) variety of produce, including, for example, foods, fibers, medicines, intoxicants, poisons, dyes, construction materials, and ornamentals, extracted from virtually all plant parts—roots, stems, leaves, and flowers;
(4) constant management, however casual or inadvertent, including weeding, thinning, pruning, hand-irrigating, and manuring (often with human excrement);
(5) experimentation with newly introduced plants;
(6) high productivity;
(7) high labor inputs annually, but work is spread throughout the year, and some is done casually (e.g., irrigating by disposing of dish water), so that little labor is expended at any one time;
(8) located near, often around houses.

Of course, gardens vary considerably for a number of environmental, economic, and social reasons (McCarry 1990), and for those in the humid tropics, where most of the research on gardens has been carried out, the above list of characteristics can be expanded. Given the comparatively short growing season and the limited state of knowledge about gardening in North America, this list will suffice for present purposes.

Garden Plots

To be sure, not much is currently known about North American gardens *c.* AD 1500. Furthermore, what is known is often conveyed in such a way that it is ambiguous at best, and confusing and misleading, at worst. Two recent cases serve as illustrations. Writing about subsistence practices in New England and the northern Plains, respectively, it was stated only that 'families planted gardens' (Nabokov and Snow 1992: 122), and they had 'generous garden plots' (Iverson 1992: 90). What, exactly, is one supposed to interpret from such statements? Clearly, readers' understandings will be conditioned by preconceived notions about what constitutes a garden.

Distinctions need to be explicitly made between gardens and fields. A number of scholars (Holley 1978: 37–8; Ford 1979: 237; Woods 1986: 17; Maxwell and Anschuetz 1992: 39–44) have made such distinctions but without detailing the criteria involved. It would appear, however, that in general fields are agricultural lands with characteristics different from those outlined above. They are usually large, monocropped, for one product, managed infrequently, rarely subjected to experimentation, require much labor at certain times but relatively little over the course of a year, and are located at some distance from places of residence (B. Turner 1992: 265–7). There are of course, agricultural lands that possess some characteristics of gardens and some of fields. In at least two cases, certain types of prehistoric fields—ridged fields of the upper Midwest (see Chapters 6 and 12) and pebble-mulched fields of New Mexico (see Chapter 7)—have been envisaged as made up of numerous and small garden beds and pebble-mulched gardens, respectively (D. Lightfoot 1990: 3–4; Peske 1966: 190).

The difference between a garden and a field is clearly subjective, and it can be culturally dependent (Gleason 1994: 2). For present purposes, only those cultivated lands that conform to a great majority of the characteristics enumerated above are considered here. Other types of plots, such as the garden bed/ridged fields and pebble-mulched garden/fields, will be discussed later. Little systematic research has been conducted on native gardens of North America. It remains a fruitful avenue for future study. Although the currently available information is limited, that which does exist is quite insightful.

Documentary Evidence

European explorer–reporters who traversed parts of the continent from contact through the 18th century made several observations of native gardens.

Some descriptions are better, that is they provide more information, than others. Some conform explicitly to the general characteristics outlined previously, while some are vague and only mention that gardens were seen. For example, writing of natives in Virginia, Percy (1969: 140) noted only that in 1608 'there was a Garden of Tobacco, and other fruits and herbs'. Similarly, Beverly (1947: 123) in 1705 referred only to 'natural Plants . . . transplanted into a Garden'. This latter observation, of course, refers to cultivation of non-domesticated plants as well as gardening.

So many traditional domestic activities actually take place outdoors (e.g., cooking and washing) that many scholars argue it is often difficult to distinguish where houses end and where gardens begin (e.g., Wilhelm 1975: 82). Similarly, it has been noted (Kimber 1973: 8) that garden plants sometimes are so much like surrounding natural vegetation that it is hard to determine the outer limits of a garden. It has even been noted that gardens constitute a climax vegetation in their own right (Alcorn 1984: 325). All of these characteristics are particularly true where trees are involved. Indeed, some of the accounts discussed in Chapter 3 are equally relevant here. Mártir de Angleria's (1944: 515) account of the Georgia–Carolina coast and Smith's (1907: 90; see also Strachey 1953: 119) comments about Virginia, refer to gardens while simultaneously describing orchards, and Smith's (1907: 192, 204) discussion of Plymouth, Pérez de Ribas' (1944: 150) description of Sonora, Benavides' (1965: 36) comments about New Mexico, and Kalm's (1972: 42) appraisal of the Delaware Valley, all mention orchards being located near houses. Kalm (1972: 343) even refers to both gardens and orchards together, but the late date in conjunction with the inclusion of apple trees and watermelons—both Old World crops—renders this statement suspect.

The most convincing evidence for gardens is that which refers explicitly to small plots located: (i) near houses, (ii) away from large fields, or (iii) both. In every such case vegetables are mentioned or implied for gardens, staple grains are mentioned or implied for fields. Nine observations exist which refer only to gardens near houses. These come mainly from the Southwest, but also from Canada down the east coast to the Southeast and along the upper Missouri River (Fig. 4.1). The earliest citing was by Coronado's chronicler, Castañeda. Upon observing Acoma pueblo, a town located atop an imposing sandstone butte, and referred to in the vernacular as the 'sky city', he wrote: 'On the top they had room to sow and store a large amount of corn, and cisterns to collect snow and water' (Castañeda 1907: 312; Winship 1904: 39). Castañeda's observations apparently were echoed by the 16th-century historian Baltasar de Obregón (1928: 16) who in 1584 relayed that Acoma 'has, on its site and settlement, fields of maize, beans, calabashes, tobacco, and cotton. There also wells and reservoirs'. Given the promontory setting of Acoma, the use of terms such as 'on', the continued use of cisterns in the town, and the limited space between houses, there can be little doubt that these are references to gardens. In a 1599 deposition before Juan de

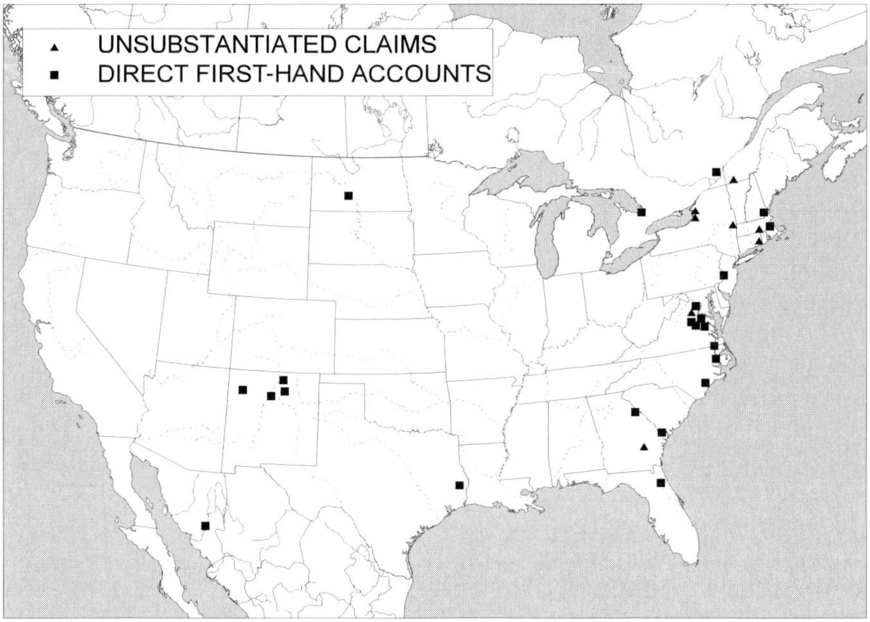

Fig. 4.1. Locations cited in the text for which there is documentary evidence of gardens.

Oñate, one native, Jusepe Gutiérrez, testified about conditions at an unnamed pueblo east of the Rio Grande. He said that '[t]he houses were built . . . close together, along narrow streets, like alleys. However, in some places between the houses there were fields of corn, calabashes, and beans' (Velarde 1966: 324). A witness in a later testimony echoed the same description, saying that 'all the space between the houses was utilized for farming' (Valverde 1953*b*: 858). Strachey (1953: 79), who borrowed from and embellished Smith (1910*c*: 367; see Rountree 1989: 4), wrote of aboriginals in Virginia in 1612 that '[a]bout their howses, they haue commonly square plotts of cleered grownd, which serve them for gardeins, some 100. some 200. foote square, wherein they sowe their Tobacco, pumpeons, and a fruit like vnto a musk-million'. Champlain (1907: 68) found 'many cabins and gardens' together along the St. Lawrence River in 1613, a fact supported later when, in 1638, the Jesuit Father Paul le Jeune (1898: 79) reported that it is not unusual for a Huron to make 'a small garden near his cabin'. Adair (1968: 406) noted that in the Southeast in 1775 'Every dwelling-houfe has a fmall field pretty clofe to it', and Jean Baptiste Trudeau (1952: 379) noted in 1796 that 'the Cheyenne built these same permanent huts, around which they cultivated little fields of maize and tobacco'.

As for references to both fields and gardens, Champlain (1907: 244), Smith (1907: 193; 1910*c*: 363), and Adair (1968: 406) made such observations.

Perhaps surprisingly, there are as many accounts of gardens, houses, and fields together as there are gardens by themselves, or with either houses or fields. Furthermore, of these five accounts, some are quite early, dating to before 1600. Writing of northeastern Florida in 1563, Jean Ribault (1927: 73), commander of the first French expedition in the region, said: 'About there howses they laboure and till there ground, sowing fildes with a grayn called Mahis, wherof the[y] make there meale, and in there gardens the[y] plant beans, gourdes, cowekcumbers, citrons, peasen, and many other simples and rootes unknon unto us'. Commenting further on labor inputs he (Ribault 1927: 84) noted 'gardens which the[y] dresse verry dilligently'. In his 1585 watercolor painting of the village of Secota on the Pamlico River, North Carolina (Fig. 4.2), Governor John White portrayed both gardens and fields as well as houses. European engraver Theodore de Bry reproduced White's work, embellishing it somewhat (Fig. 4.3). Although de Bry's portrayal is fanciful, in no small way a function of his never having set foot on North America, it is nevertheless accurate. Thomas Heriot (1812) who traveled with White wrote the captions for de Bry's volume, noting that 'The houses are farther apart and have gardens (marked E), in which they grow tobacco. . . . They also have . . . fields [marked F] where they sow their corn' (Lorant 1946: 264; see also Gilmore 1919: 128). Near Plymouth harbor in 1603, Martin Pring (1905: 58–9) reported that he 'saw certaine Cottages together, abandoned by the Sauages, and not farre off we beheld their Gardens and one among the rest of an Acre of ground, and in the same was sowne Tobacco, Pompions, Cowcumbers and such like; and some of the people had Maiz or *Indian* Wheate among them. In the fields . . .'

Later in the 17th century, 1687 to be exact, Henri Joutel (1906: 175) a chronicler for LaSalle reported of Indians in east Texas that 'their Cottages . . . are in the Midst of their Fields and Gardens'. Finally, Bartram (1940: 169), in discussing native conditions in North Carolina in 1773, said 'They plant . . . a small garden plot at each habitation, consisting of a little Corn, Beans, Tobacco, Citruls, &c. Their plantation, which supplies them with the chief of their vegetable provisions, such as Zea, Convolvulus batata, Cucurbita citrulus, Cuc. lagenaria, Cuc. pepo, Cuc. melopepo, Cuc. verrucosa, Dolichos varieties, &c. lies on the rich prolific lands . . . two miles distant' (see also Waselkov 1997: fig. 9.5, 186).

Two things stemming from these accounts are worthy of discussion. One is apparent and needs only qualification, the other is subtle and warrants much discussion. The immediately apparent point is the spatial distribution (see Fig. 4.1). As seems to always be the case, early explorers' accounts come principally from the Eastern Seaboard. But, how representative are a few dots on a map? According to at least two sources, they are most representative. One

Fig. 4.2. Watercolor painting of the village of Secota, on the Pamlico River, North Carolina, by Governor John White in 1585. (Provided courtesy of the Library of Congress, Neg. No. LC-USZ62-583.)

author, Herndon (1967: 284) has concluded that Strachey's description applied to all natives living in eastern Virginia and northeastern North Carolina. He also determined that Adair's comments pertained to places throughout the Southeast, as far west as the lower Mississippi Valley (Herndon 1967: 290). A second writer (H. Russell 1961: 60) found, on the basis of both Pring's and Champlain's observations that gardens were used in parts of Vermont, New Hampshire, and southern Maine.

In addition to direct accounts of gardens by early explorers and these two regional assessments, there exist several other works that refer to native gardens at contact. Although they are not well-documented, these works

Fig. 4.3. Woodcarving of the village of Secota made in 1590 by Theodore de Bry from White's painting (see Fig. 4.2). (Provided courtesy of the Library of Congress, Neg. No. LC-USZ62-544.)

conform to the regional picture portrayed above. Gardens are claimed to have been common in New England (H. Russell 1980: 145, 146), especially among the Western Abenaki of southern Quebec, New Hampshire, and Vermont (G. Day 1978*a*: 153), the Mahican of the Hudson Valley (Brasser 1978: 199), the Iroquois of southern Ontario and upstate New York (Fogelson and Kutsche 1961: 114; Stites 1904: 23, 30), the Potomac who resided along a river of the same name (Clayton 1973: 185), and throughout Georgia (C. Jones 1873: 42). In sum, sufficient documentary evidence exists to indicate that at the time of the great cultural encounter, gardens were common all the way from southeastern Canada to the mouth of the Mississippi River. References to gardens in more remote locales suggest that gardens were used wherever people lived in permanent settlements.

The broad range of environments in which gardens were located, and the large number of different types of plants reported growing in them, either alone, or in various combinations, hints at the probability that gardens served more than one purpose. Indeed, three types of gardens can be identified. First, are those gardens in which one type of crop was raised exclusively. The second type of garden is that in which a variety of vegetables were grown together and these were different than the staple grain—maize—which was grown separately and exclusively in distant fields. Finally, are those gardens in which were grown the same crops as were raised in fields.

Evidence for gardens in which special crops were raised alone is limited, and less than perfect. The only account of such a garden is from the interpretative text accompanying de Bry's engraving of White's watercolor painting (Gilmore 1919: 128; Lorant 1946: 264). However, this account involves the raising of tobacco, a crop that was smoked and required much attention. Given its wide distribution and high value, it is highly likely that tobacco was grown in specialized gardens to a greater extent than documentary evidence suggests.

Gardens in which a number of vegetables were raised but from which the staple grain was excluded seem to have been the most common among native North Americans at contact. The accounts of Ribault (1927: 73), and Pring (1905: 58–9) most clearly make this distinction. The accounts of Percy (1969: 140), and Strachey (1953: 79) do not mention maize, but the omission of this crop implies that staples were not grown in the gardens they saw. In all likelihood, this segregation of vegetables in gardens and staples in fields reflects a spatial zonation based on a gradient of agricultural intensification.

Stated simply, by factoring in: (i) yields, (ii) product values, (iii) production costs or labor inputs, (iv) transportation rates or travel time, and (v) distance from home or settlement, one can construct a series of production curves for each crop, and, hence, a land use map (Chisholm 1968). In general, crops that have high yields per unit area, high values, and require much effort are grown close to farmers' homes because daily travel to and from fields, and the transportation of the crop back to the settlement after harvest, takes time, a valuable resource, that could be spent on other aspects of production. Crops that have low yields, value, and labor requirements can be, are, and have been, grown in more distant and remote places.

Explaining why some crops were grown exclusively in gardens while others were grown in fields, makes it difficult to explain those cases in which crop redundancy was reported. Bartram (1940: 169) revealed that many of the crops grown by natives in large fields were also grown in gardens. Velarde (1966: 324) did not mention fields, but his inclusion of maize indicates that staples usually raised in fields were sometimes grown in gardens. Why was this so? No one really knows but two possibilities exist, and indeed each might be applicable under certain circumstances.

It has been suggested that garden plots were planted earlier in the year than fields, and, hence, were harvested earlier. Garden yields, therefore, could have been used to sustain people until the larger crop was harvested (Herndon 1967: 286, 290; C. Jones 1873: 42). This is certainly a compelling argument. Popular myth holds that aboriginal farmers harvested crops in order to have food through the winter. Although this is true, it is not the entire story. The leanest time of the calendar year for agricultural societies is not the winter, but, paradoxically, the growing season. Winter follows the harvest almost immediately, and planting follows winter. Stored food decreases at a rather regular rate. It is actually abundant during the winter, less so during the spring, and diminishing in the summer. A small crop one year can, therefore, result in supplies running out well before the next year's harvest. The period in which a crop is in the field maturing can be the time of greatest food shortages. Modifying a line from the ancient mariner's rhyme: 'food, food, everywhere, and nary a bite to eat!'

A second possible explanation for why a similar complex of crops is grown in gardens and fields involves the prevention of crop losses in consecutive years. Gardens as small as those described by European explorers clearly could not have produced a sufficient amount of food to feed the huge numbers of people they also reported. Large fields were essential, and although they can produce more food than small gardens, they are also more susceptible to crop failures. Hazards to fields include such things as frosts, pests, droughts, and even destruction by enemies. Gardens, on the other hand can be protected relatively easily, and individual plants can even be nurtured. For example, plants can be covered with straw, leaves, and blankets in order to prevent frost damage, and they can be hand-watered in times of drought. Birds can be frightened away, and insects and worms can be picked-off. Finally, soils can be fertilized with household wastes. As a result of the great care that gardens tend to receive, a small, but dependable crop can be produced each and every year. If the seeds from these plants are not eaten, they can be used to plant fields the following year. In other words, in spite of what may happen to a field crop, gardens can provide a dependable seed crop year after year (S. Fish, P. Fish, and Downum 1984: 69–70). Other lines of evidence support this conclusion, as well as others concerning native North American gardening.

Ethnographic Evidence

Just as is the case with documentary accounts, much of the ethnographic evidence for gardens is extremely vague, consisting principally of brief statements. For example, in discussing Hopi cultivation of sunflowers, Nabhan (1979*c*: 26) said '[w]hen the plant is cultivated, it is more often in dooryard

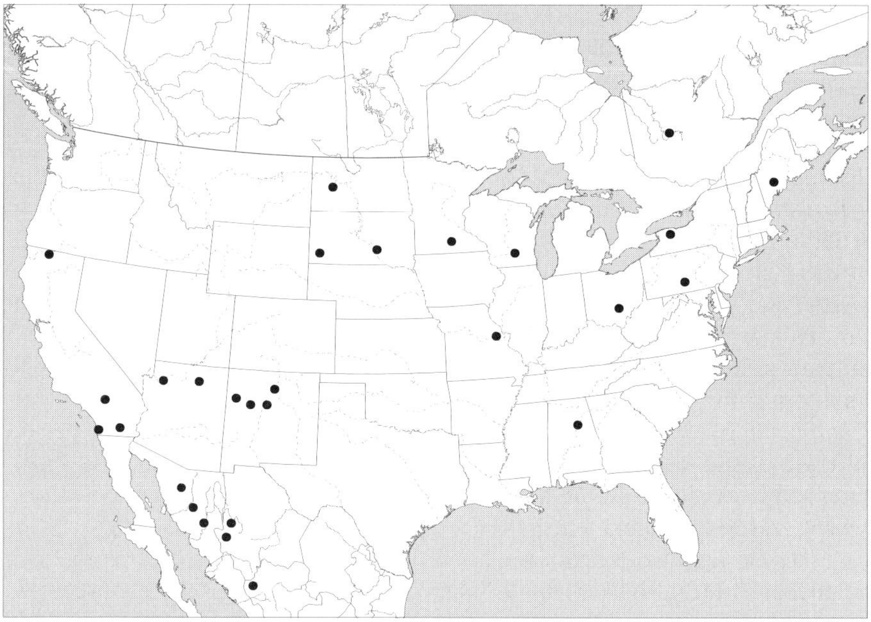

Fig. 4.4. Locations cited in the text for which there is ethnographic evidence of gardens.

gardens rather than in the typical agricultural fields away from the villages'. Ethnographic evidence of native American gardening also sometimes overlaps that for protection, encouragement, and cultivation of non-domesticated plants. One such example comes from California where a Tipai-Ipai woman reported that 'her people always cleared a small spot near their dwelling to plant some of the greens and seeds and roots that they liked, just the things that grow wild' (Bean and Lawton 1973: xxvi). Another involves the Zuni who 'cultivated' at least three 'wild' plants—culantro (*Coriandrum sativa* L.), ground cherry (*Physalis fendleri* A.), and cane cactus (*Opuntia arborescens*)—in 'gardens' (Stevenson 1915: 66, 70, 95).

Passing references to gardens being maintained by aboriginals during the 19th and 20th centuries come from numerous and spatially diverse places (Fig. 4.4). Tobacco gardens are mentioned for the Shasta of northern California (S. Silver 1978*b*: 222). Gardens with undescribed crops have been seen at Palm Springs, California (Patencio 1943: 56). The Warihio of northwestern Mexico have gardens with numerous crops (Gentry 1963: 86–8). And, the Supai of northern Arizona (Nabhan 1979*c*: 24), like their Hopi neighbors and the Iroquois of New York (Waugh 1916: 20–1), are said to raise sunflowers in gardens. Further to the east the Sioux of the Dakotas (Culbertson 1952: 98; Nurge 1970: 82), the Chippewa of Wisconsin (Stowe 1940: 10, 11), the

Shawnee of the Mississippi Valley (Voegelin 1940: 519), the Creek of the Southeast (Hudson 1976: 292), unnamed native inhabitants of Ohio (R. Jones 1983: 13), and the Ojibwa of northern Ontario (Rogers and Black 1976: 11, 12) raised a multitude of crops in gardens. These accounts are problematic, however, in that some contain references to Old World crops, and it is unclear if some reporters were using the term to describe true gardens or only as a convenient label for small fields. None of the eastern groups mentioned relied on agriculture to any great extent during historic times, gaining much of their subsistence from collecting and hunting. These gardens, therefore, might simply have been small-scale attempts to ensure food supplies at critical times of the year, as discussed earlier. In any case, these reports suggest that numerous people did engage in agricultural pursuits that involved something other than extensive fields of staple crops.

At least five historic accounts confirm this notion. These reports are ambiguous but in addition to mentioning what are clearly gardens, they also refer explicitly to separate maize fields. As is often the case with relatively recent relations, most deal with native people who lived in the West. References to both native gardens and fields involve the Tohono O'odham of northern Sonora (Nabhan *et al.* 1982: 136), San Juan (Ford 1968: 189), and Jemez (Bandelier and Hewitt 1937: 97) pueblos, New Mexico, and an unnamed group in Pennsylvania (S. Fletcher 1950: 38). Referring to the continent as a whole, one writer (Stout 1914: 52) even generalized long ago that several crops 'were sometimes grown along with the corn, but were often grown in gardens'.

Gardens and fields cited together can be better accepted as evidence of aboriginal gardening when accounts of such also put the gardens close to places of habitation. From across the continent come three such reports. The Cahuilla of southern California (Lawton and Bean 1968: 28), the Arikara of the upper Missouri (Will and Hyde 1964: 101), and the Penobscot of Maine (Speck 1940: 91) each are said to have had fields at some distance, as well as gardens near their houses.

Without doubt the most thorough ethnographic reports of aboriginal gardening in North America come from people living in and around the Sierra Madre Occidental of northwestern Mexico, specifically the Rarámuri, the Odami, and the Névome. Each of these groups has done everything imaginable in terms of gardening. For simplicity the following discussion is organized according to first garden plants and their uses (following the theme of Chapters 2 and 3), second gardening techniques, and third garden landforms.

Described as a 'jumble of vegetation' (Pennington 1980: 165–6), native gardens of northwestern Mexico are intercropped, having a variety of plants growing together simultaneously without any areal plan or organization (see also Graham 1994: 72–3, fig. 40). Névome gardens are less orderly and not as carefully weeded as the gardens of their neighbors who are of mixed aboriginal and Spanish ancestry (Pennington 1980: 166). This implies that natives place value on plants that others might perceive as useless. Névome gardens also differ from their *mestizo* counterparts by a fewer number of edible plants and a greater number of ornamentals (Pennington 1980: 165). This suggests that aboriginal peoples in the area had a keen sense of the aesthetic, and placed a high priority on visually attractive plants (see also Farrington and Urry 1985: 146). Ornamentals found in Névome gardens include *Canna* sp., the climbing vine *Bougainvillea* sp., the red-flowered tree *Delonix regia*, *Dahlia coccinea*, *Zinnia elegans* (Pennington 1980: 165), and marigolds (*Tagetes erecta*) (Laferriére and Asdall 1992: 164). Not grown for their inherent beauty, but rather for the beauty that comes from their processing, are plants valued for their dyes. These include *Indigofera suffruticosa* and possibly even *Amaranthus cruentus*, which is also valued for its seeds and can be consumed as a green.

In addition to the ornamentals, some plants are actually grown for other non-dietary purposes. One tree commonly found in Névome gardens, *Piscidia mollis*, has bark that is used in the manufacture of a poison for the purpose of fish stupefaction. Two other plants, *Antigonon leptopus* and *Nicotiana tabacum* are grown in order that their dried leaves can be smoked.

A number of the plants found in Névome gardens are grown specifically to be eaten. These vary from crops that constitute an essential part of the total diet to those that are probably used only for their taste. Of this latter type of plant, the most conspicuous are the chilies. *Capsicum annuum* tastes a variety of ways depending on when it is harvested. While still immature, these chilies are sometimes picked as the mild *chili verde*. Left to ripen, they become considerably hotter and are referred to as *chili colorado*. *Capsicum annuum* var. *minimum* has a small, pea-sized fruit that is extremely hot, especially when ripened and dried. These *chiltepins* are tasty but are so hot they can only be consumed in very small amounts. All the chilies are used as condiments, principally in bean and meat dishes.

Several Névome and a few Rarámuri garden plants are used as greens. In addition to *Amaranthus cruentus*, *A. palmeri* and *Chenopodium album*, and *Lepidium viginicum* have leaves that are prepared by boiling and then consumed like spinach. *Amaranthus cruentus* is often found as an escapee, growing well in trash deposits. *Amaranthus palmeri* and *C. album*, on the other hand, grow in the wild and often invade gardens. These plants are tolerated because they grown rapidly, and profusely during the early part of the summer rainy season. This is the time of the year when the staple crops, especially maize, are in the field and stores from the previous year's harvest are running low. In effect, these *quelites* harvested from gardens are used as an emergency food source during critical periods in years following poor harvests.

An unidentified maguey (*Agave* sp.) is cultivated in more than half of all Névome gardens. The leaves of these plants are cut and then roasted in a pit before being consumed.

The greatest variety of plants found in the gardens of the Névome involve those whose fruits and tubers are eaten. These include the trees *Pithecolobuim dulce* and *Achras zapota*, and the shrubs *Psiduim guajava*, *Lycopersicon esculentum*, *Cucurbita maxima*, *C. mixta*, and *C. moschata. Ipomea batatas* is cultivated by the Névome while their Rarámuri neighbors cultivate *Solanum tuberosum* in their gardens. They also grow *Opuntia* sp. for its fruit.

Although not normally considered as garden crops, those plants valued for their seeds or grain are not uncommon in northern Mexican gardens. In addition to *Amaranthus cruentus* which is grown by the Névome, *Amaranthus retroflexus*, *A. palmeri*, and *Helianthus annus* are raised in gardens by the Rarámuri. Typically, the seeds of the amaranths are popped while those of the sunflower are dried, ground, and used in the making of small cakes.

The cultivation of such a wide variety of plants for so many reasons is not simple and straightforward. A great deal of knowledge and care is needed. The most common form of attention afforded plants in the gardens of northwestern Mexico's indigenous people is irrigation. For this purpose domestic waste water is often used as is water hand-carried from nearby streams and springs (Pennington 1963: 669, 1969: 87). The Névome hand-irrigate most of their ornamentals that are often grown in old or cracked ceramic vessels. They also water *Lycopersicon esculentum* and *Nicotiana tabacum*. There is no record of the Rarámuri irrigating this latter plant but Odami women have been known to hand-irrigate tobacco in their gardens. Although not documented, it might well be that contemporary and prehistoric peoples in the region hand-irrigated *quelites*. Cheno-ams are ephemerals that do not sprout in dry years but sprout quickly and grow profusely at the onset of the summer rains during wet years. On the basis that hand irrigation was used on some crops and, assuming that ancient inhabitants depended on *quelites* in the years following poor staple harvests, it seems likely that these people would have irrigated their *quelites* in order to insure any food at all.

As has been noted, *Amaranthus creuentus* and all the cucurbits have been found growing in trash deposits, and the connection between human refuse and gardens is a very old one (E. Anderson 1952: 136–50). Also, numerous human activities, including the elimination of body wastes, typically occur in gardens (Kimber 1973: 8). Dogs and turkeys are not, and in the past have not been, infrequent inhabitants of house yards and gardens (Alcorn 1984: 130–1). The Odami even manure gardens with bat guano collected from nearby caves (Pennington 1969: 87). In sum, it appears that gardens were well-fertilized, by both human, and other animals even if explicit references to such an activity are few.

While irrigation and fertilization are most important activities in northern Mexico, where water is scarce and the soils are low in organic matter, one cannot overlook the importance of assisting in the propagation and maintenance of the various plants. There is no evidence of weeding being practiced by the Névome or the Odami. That the Rarámuri weed their gardens (Pennington 1963: 67), however, suggests that the practice is, and probably was, widespread.

Weeding, fertilizing, and especially irrigating are labor-intensive activities. People who expend energy on such tasks have much invested and, accordingly, their investment is often visible in the form of permanent garden features. Unlike gardens in humid tropical areas, those of arid northwestern Mexico typically have some type of enclosure. Rarámuri construct fences as a matter of personal preference (Pennington 1963: 65). The Odami enclose many of their gardens with fences and low walls in order to keep out unwanted animals (Pennington 1969: 88–9). There is no evidence of the Névome or their neighbors using fences, however, it is likely that these people employed the practice in historic and contact times. The simplest type probably involves the use of split stakes or the stalks of *Fouquieria splendens* driven into the ground vertically and adjacent to each other. Another related technique is to pile brush, especially that of thorny genera such as *Acacia*, *Mimosa*, and *Prosopis* in a roughly circular area. Such features are usually only about 1 m high and can be used to pen small animals such as turkeys.

A somewhat more elaborate technique of enclosing gardens in northwestern Mexico is to build a low stone wall two to four courses high (Pennington 1969: 88; Laferriére and Van Asdall 1992: 164; Graham 1994: 34, fig. 19). Enclosures this low do not keep animals out of the cultivated space but they do act as deterrents to pedestrian traffic and they probably keep very young children from damaging the plants. Low walls also assist in the containment of mulch materials that otherwise could blow or be washed away, and irrigation water that might otherwise be lost. Going hand in hand with enclosure is small size. Gardens of the Odami average only 13 m square (Pennington 1969: 87). Those of the Névome measure approximately 4 by 5 m (Pennington 1980: 165). Although small, size might well be deceiving. There is little evidence to support this supposition, but these gardens might be rotated in a manner not unlike that of swidden plots in a typical slash-and-burn agricultural system. For one thing, being both small and well-demarcated, these gardens are certainly distinctive from their tropical counterparts which are often large and almost indistinguishable from the house yard proper and even the surrounding natural vegetation. Furthermore, the Rarámuri (Pennington 1963: 65), the Odami (Pennington 1969: 72), and the Névome (Dunnigan 1983: 222) are all known to move their corrals (used for penning large Old World animals) periodically and plant new gardens in the well-fertilized abandoned corral site. The possibility of this practice having evolved out of relocating refuse middens or dump-heaps (e.g., E. Anderson 1952: 136–50) in prehistoric times cannot be ruled out.

Although not an 'agricultural landform', in the true sense of the term (Golomb and Eder 1964), there is one other important characteristic of native gardens in northwestern Mexico that merits discussing—their locations relative to the houses themselves. Personal observations have revealed that gardens in the region tend to be located on the south sides of houses. Such locations might be intended to protect young plants from the northerly winds in the early spring. However, the location may also have something to do with the availability of summer sunshine. By locating gardens on the north or shady side of houses, horticulturists would be reducing the number of hours the gardens would be in the sun each day.

In addition to being located on the south sides of houses, gardens of indigenous people in the *sierras* tend to be located in places slightly lower in elevation than, or downslope from, their owners' houses. Such locations are, undoubtedly, a function of drainage and runoff. Rainfall in the region is, of course, low. During the late summer, however, heavy thunderstorms are common and can result in large amounts of runoff in brief periods of time. Locating gardens below houses helps prevent flooding of the dwellings and it facilitates the washing away of household refuse and watering of the crops.

When speaking of native American gardening, especially in terms of small plots, enclosures, and irrigation during historic times, there is perhaps no more easily recognized and novel case than that of the famed 'waffle gardens' of Zuni pueblo. Doubtless, a great deal of the recognition afforded these gardens is the direct result of a series of striking photographs taken in the late 19th and early 20th centuries (Fig. 4.5). These photographs have been reproduced and published in a number of sources (e.g., Cushing 1920: plates I, XVI, XVII; Ferguson and Hart 1985: 46, 84; Ladd 1979: 493, 497), and in agreement with the old adage, each is 'worth a thousand words'. Indeed, these photographs have had to speak for themselves as there exists actually very little in the way of literature that covers them in any detail. Granted there has been some discussion, but in relation to their being known, these gardens remain poorly understood (Perramond 1994: 31–2, 37–9).

Waffle gardens are so-called because in a very real sense they resemble waffles (Baxter 1882: 81). The ridges that give these gardens their unique appearance are approximately 10 cm high and wide. About this there is no disagreement. However, there does exist some controversy about the materials used, how the beds were constructed, and their size. Some researchers claim simply that the ridges were made of clay (Ladd 1979: 498) or mud (Ferguson and Hart 1985: 38). Others (Ford 1985*d*: 61) say they were constructed by hoeing together sand and manure. Most authors imply, and most of the historic photographs show that the base of the garden beds was the same elevation as the natural ground level. At least one scholar (Bohrer 1960: 182) has noted, however, that each garden bed was 'built-up' approximately 10 cm with the ridges being yet another 10 cm higher. This same writer reported that each square of the waffle gardens she saw varied between 50 and 60 cm on a side, while another said they ranged from 37 to 60 cm (S. Clark

Fig. 4.5. A 'waffle garden' at Zuni Pueblo, New Mexico, *c.* 1910. (Photograph by Jesse L. Nusbaum. Provided courtesy, and published by permission, of the Museum of New Mexico, Neg. No. 8742.)

1928: 242). Another (Ford 1985*d*: 60), however, has claimed that they were smaller, ranging between 30 and 50 cm square. While yet another (Ferguson 1985: 60), took the middle ground and claimed they averaged 'about 45 cm (18 in) square'.

As for spatial configuration each garden bed was typically six or more squares long and two squares wide, with a space approximately 50 cm wide between beds. This space provided easy and convenient access not only to each square, but to every plant (G. Stewart 1940*b*: 337, 338). Every researcher who has studied waffle gardens agrees that the principal reason for the ridging, and the pathways between beds is to facilitate manual irrigation. That is, water historically has been carried in clay jars on bearer's heads from the river nearby and poured on the gardens with gourd dippers (Holmes 1912: 35; Ladd 1979: 497, fig. 12). Each square required 'one or two scoops of water' (Ferguson n.d.), applied every morning during the growing season (Stevenson 1904: 353). Among those watered were the ancient southwestern crops beans and maize (Ford 1985*d*: 60), pumpkins, squash, cotton (Ferguson and Hart 1985: 38), and tomato (Bohrer 1960: 182), the historically introduced New World crops, onions (Stevenson 1904: 353; Bohrer 1960: 182) and melons (Ladd 1979: 497), and a number of non-domesticated plants that were either protected, encouraged, or cultivated, including coriander, ground cherry, amaranth, and some undescribed herbs (Bohrer 1960: 182; Ladd 1979: 497; Stevenson 1904: 353).

In addition to requiring irrigation, the gardens also provided protection for plants. One source (Brandt 1995: 297) reported that the low earth ridges surrounding each square not only retained irrigation water but also minimized evaporation. Another (Ferguson and Hart 1985: 38) noted that they regulated temperatures and reduced damage from the wind. Exactly how they performed both functions has not been specified. What is known, however, is how the gardens themselves were protected. Nearly all of those who have written about the Zuni waffle gardens, and nearly all of the early photographs, indicate that the gardens were enclosed with either low (less than 2 m high) adobe walls or fences made of thin tightly spaced upright juniper posts (see Fig. 4.5). Only one source (Ford 1985*d*: 60) posits an explicit reason for enclosure—to keep out children, dogs, and livestock. The latter, of course, involve Old World species, but children and dogs could certainly have been a problem in gardens prior to *c.* AD 1500.

The Zuni waffle gardens were at one time quite numerous and conspicuous. They were located immediately west of the pueblo along the floodplain on both sides of the river (Fig. 4.6). Today they are nearly non-existent and almost invisible. Their demise has been well-documented (Bohrer 1960: 181–2; Ferguson 1985: 123; Ferguson and Hart 1985: 38) and attributed to the lack of interest on the part of younger generations (Ladd 1979: 498). Waffle gardens used at Zuni today number only two. One is very small, used only periodically, and is located behind a house south of the river. The other (Fig. 4.7) is somewhat larger, more permanent, and is located north of the river, but is obscured somewhat by a fence and two houses. This particular garden is in many respects like its older counterparts. There are, however, at least three differences. First, the metal buckets that some authors (Bohrer 1960: 182; Ladd 1979: 497) said replaced gourd dippers and clay pots, have been replaced by plastic ones. Second, the controversy over whether or not garden beds were level with the natural surface or raised seems to have reached a median. Current gardens appear to be terraced: that is, on the slightly sloping surfaces the upslope part of the garden is at ground level while the downslope side is built-up. Third, and perhaps most importantly, the size of the waffle squares has increased. The average now seems to be approximately 1 m on each side. This increase in size may be related to the overall decrease in number, in that waffle gardens may well be reaching that point at which they are simply more trouble than they are worth. Stated another way, it no longer makes sense to invest large amounts of labor in preparing, maintaining,

Fig. 4.6. A view north of the Zuni Pueblo from the south side of the river *c.* 1919. Note the abundance of 'waffle gardens' and their enclosures. (Photograph by Jesse L. Nusbaum. Provided courtesy, and published by permission, of The Denver Public Library, Western History Department, Neg. No. 312.)

Fig. 4.7. One of only two 'waffle gardens' remaining at Zuni Pueblo in 1988. Note the relatively large size of the individual grids (compared to Fig. 4.5), the plastic bucket used in manual irrigation, and the dark soil indicating that each grid had been watered only moments earlier. Note also the trees of an unidentified species that could indicate the husbandry (cultivation) of woody plants. Appeared originally in Doolittle (1992*a*: 396, fig. 5). (Reprinted by the permission of the Association of American Geographers.)

tending, and irrigating a garden when one can simply buy, at a relatively low price, what they need at a store.

In addition to being important in their own right, the waffle gardens of Zuni are significant in regard to the gardens of two other puebloan peoples, the Hopi to the west in Arizona, and the Acoma to the east. Similarities between Zuni and Hopi gardens have long been known (e.g. Hodge 1959: 97), as has the importance of gardening to Hopi culture (Dozier 1954: 358). In no small way, similarities were based on morphology and function. When recently shown a photograph of Zuni waffle gardens (G. Stewart 1940*b*: 333), one Hopi gardener who will remain unidentified (personal communication 1993) responded that they were identical to those made by his grandfather. Hopi gardens were also irrigated, but not solely with water transported in jars on bearers' heads. In some cases, water was carried in vessels attached to the ends of long poles (Holmes 1912: 35). Furthermore, manual irrigation did not involve water taken directly from a river. Instead water was carried from either reservoirs that were spring-fed (Forde 1931*b*: 365–6, 391–3), or the springs themselves (Titiev 1944: 16; Colton 1960: 92).

Although little is actually known about the Zuni gardens, and even less is known about Hopi gardens, next to nothing is known about gardens at Acoma pueblo. What is known, however, is most significant. A photograph taken from atop the church in the 1880s clearly shows that waffle gardens identical to those at Zuni were used historically at Acoma (Garcia-Mason 1979: 460, fig. 12). These gardens were adjacent to residences and on top of mesa. This photograph, therefore, not only indicates that gardens were important at yet another locale, but it also substantiates the previously discussed 16th-century claims of Castañeda (Winship 1904: 39), and Obregón (1928: 16).

Archaeological Evidence

There currently exists as much speculation as substance when it comes to claims made about pre-European gardens in North America. In no small way this condition is a function of the material remains of the people involved and the nature of archaeological research. Nomadic people who subsisted largely by collecting did not have gardens of any type. Those who stayed in one place long enough to create middens or dump-heaps had what might be considered prototype gardens. The archaeological discovery of such features, often containing abundant, varied, and obviously manipulated plant remains, has resulted in scholars working on the development of agriculture in North America to either speculate about gardens explicitly, or at least suggest them, however implicitly. For example, it has been openly and clearly claimed that prehistoric women in Vermont stayed home and 'tended the gardens' while men traveled to hunt and fish (Haviland and Power 1981: 160; for an example

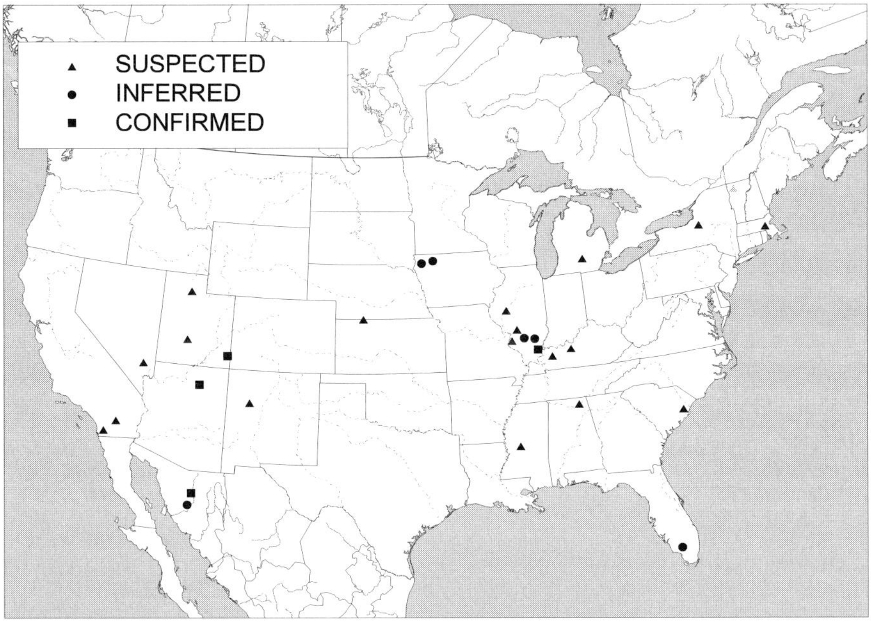

Fig. 4.8. Locations cited in the text and Table 4.1 for which there is archaeological evidence of gardens.

of the important role that women performed in gardening elsewhere (see Keys 1996). In a more subtle case, that of an archaeological report of excavations in northwest Mississippi, gardens are not mentioned but the cover illustration is an artist's rendition of the site showing what are unmistakably three gardens within the confines of the palisade (Williams and Brain 1983).

At least two brief statements about prehistoric gardens, one of which dates back to the 19th century, were made without reference to either time or location, covering all of the continent and its human history (E. Sturtevant 1885: 444; Wissler 1967: 573–4). A few others (Ford 1979; 1985*b*: 10; 1985*c*: 353) focus on a large region, the Midwest, and cover less than a millennium—350 BC to AD 250. For the most part, however, passing references and statements inferring to ancient gardens are time and place specific.

Such statements span nearly 5,000 years and come from three of the four corners of the continent and numerous places in between, especially the Great Basin and the Mississippi–Ohio river valleys (Fig. 4.8). For instance, 'house gardens' are claimed to have been used in south Florida shortly after 3000 BC (Keegan 1987*b*: 336), and a type of 'kitchen garden' is posited as having been used shortly before European contact by the Cahuilla and Tipai-Ipai

Table 4.1. Locales where archaeologists have inferred that gardens existed in pre-European times

Locales	Source
Vermont	Haviland and Power (1981: 160)
Northeast	W. Fowler (1954: 48)
New York	Ritchie (1969: 276)
Michigan	Hinsdale (1928: 31, 36)
South Florida	Keegan (1987*b*: 336)
Northeast Alabama	B. Smith (1984: 165, 1985: 55–6, 69, 1986: 51)
Central Mississippi	Williams and Brain (1983: cover)
Kentucky	Wissler (1940: 21)
	Yarnell (1974: 117)
	Watson (1974: 234)
	Cowan (1985: 238)
Illinois Valley	Yarnell (1976: 266, 269, 270)
Mississippi Valley	M. Fowler (1971: 125)
Cahokia (IL)	Woods (1986: 19)
Nebraska–Kansas	Wedel (1986: 116)
New Mexico	Wills (1988: 475)
Northern Utah	Aikens and Madsen (1986: 160)
Central Utah	J. Jennings (1986: 116)
Southern Nevada	D. Fowler and Madsen (1986: 175, 177)
Southern California (2)	Lawton and Bean (1986: 18)

of southern California (Lawton and Bean 1968: 18). For the sake of brevity, references to brief passages that infer the existence of prehistoric gardens are listed in tabular form (Table 4.1).

In addition to the abundant speculation about prehistoric gardens, there is some physical evidence, albeit sketchy, that tends to confirm what many archaeologists have claimed. This evidence takes four forms. Listed in order from the most tentative to the most convincing, these are: sites with unusual locations and spacing of houses; inhabited refuse areas with signs of cultivation; locales with cultivation soil signatures; and sites with relict garden features (see e.g., Miller and Gleason 1994).

There is an old adage that says: 'absence of evidence is not necessarily evidence of absence.' This saying might well apply to prehistoric gardens in North America. A recent reassessment of sites in the Valley of Sonora, northwestern Mexico has revealed an interesting pattern of habitation structures and vacant spaces on sites, and a provocative, albeit tenuous, suggestion about gardens based largely on ethnographic parallels (Doolittle 1992*a*: 82, 87, 89–90). Of a total of thirty-eight relatively little-damaged and largely intact sites with fewer than nine houses, most were located along the northern sides of mesa tops overlooking arroyos, and away from the edges that overlook the north–south flowing river. Given that the mesa tops are not flat, but slope slightly toward the river, and toward the arroyos on the south sides of mesas, thirty-two or 84 percent of these small sites have most of their houses on the northern and upslope sides. Clearly, there was a tendency not to build houses on the downslope portions of mesa tops. It was in these areas that gardens are posited as having been located.

A cursory survey of present-day houses and gardens in the same valley supports this suggestion. It revealed that gardens tend to be located on the south sides and downslope from houses. In such locales, gardens are better protected from either late or early freezes, exposed to more sunlight, and receive more water via runoff than if they were located on the northern and upslope sides of houses.

Some village sites on the Big and Little Sioux rivers in northwestern Iowa can also be interpreted as containing gardens. Oval and rectangular in shape, and covering approximately 0.8 ha, these sites consist principally of refuse accumulated to a depth (or height) of more than 3 m. They each have remains of ten to twenty houses, and excavations revealed 'evidence that corn, beans, and other crops were cultivated' (Wedel 1961: 98–9). Whether or not the agricultural people who lived in these villages actually maintained gardens, in the truest sense of the term, can be debated. However, midden-village sites of this size undoubtedly had areas between and around houses where economically important plants grew profusely. Given the ecology of middens and the agricultural knowledge these people had, combined with what is suspected about featureless spaces on Sonoran sites, it is likely they maintained gardens.

Not too far away, in southern Illinois, several other multiple component sites in two locales have characteristics that have been interpreted as evidence of gardens. Sites along the Big Muddy River (Denny and Woods 1981: 69; Woods and Denny 1980: 171) and in the Saline Valley (Woods 1982: 1405) contain soils that are exceptionally high in both organic matter and trash. Their distinctiveness from surrounding soils, high fertility, and association with crop remains indicates that they were not only culturally enriched, but that enrichment was a function of cultivation. Their association with places inhabited for a prolonged period means that these soils are the vestiges of prehistoric gardens.

Areas within archaeological sites that contain no structures, possess abundant agricultural refuse, and artificially enriched soils may be indicative of ancient gardening, but they are not convincing. Physical remains of garden features in proximity to dwellings are, in contrast, much more compelling. Sites with such evidence have been found in the Ohio–Mississippi Valley, and, especially, the Southwest (see Fig. 4.8). The former area is known for archaeological earthworks, the latter are characterized by features made of rocks.

Analyses of aerial photographs of the Kincaid site in far southern Illinois revealed traces of what were undoubtedly gardens. The site, which was occupied between AD 900 and 1500, was surrounded by a palisade. Within this structure, a small area measuring 50 by 60 m and consisting of parallel rows that were probably ridges, was detected (Holley 1978: 45). Details concerning such things as the height of, and distances between ridges have not been reported. However, these features seem to be quite similar to the so-called 'garden beds' mentioned earlier and discussed in Chapters 6 and 12. What makes these particular features different from others is their location within the walled confines of a habitation site. Clearly, these were gardens in the truest sense of the term.

The remaining physical evidence of gardens comes from the Southeast and the Southwest (see Fig. 4.8). Ongoing archaeological excavations of the Creek townsite of Fusihatchee indicates that houses were widely spaced and used only briefly. Being semi-subterranean, abandoned houses were filled with trash and converted into gardens (Waselkov 1997: 185, 187). Three distinct types of garden features have been reported from the Valley of Sonora. The first of these are simply low rectangular rock enclosures, two to four courses high and measuring approximately 4 by 5 m. Thirteen such features were found on seven sites (Doolittle 1992*a*: 86, 88–9, figs 4–8). Originally, it was thought that these enclosures were filled with earth and used as house platforms. However, the lack of fill is too great to have resulted from deflation or erosion, and the walls are in too good a shape; these enclosures were probably never filled. The interpretation that these features are garden borders is borne out by ethnographic parallels from the Odami, who not only garden as discussed previously but also have identical enclosures (Pennington 1969: 88, 89, fig. 13).

In addition to rectangular, multiple-course rock enclosures at least ten small single-course rock rings were found on six Sonoran sites. These rings measure between 1 and 2 m in diameter and involve rocks set into the ground endwise and adjacent to each other (Doolittle 1992*a*: 83–4, figs 4–5). These features were initially interpreted as having been the foundations of bulb-shaped thatch and plaster granaries like those found in caves further to the east in the Sierra Madres (Lister 1958: 31, plate 5; Sayles 1936: 19, plate VIIIa) or hearths. They were reconsidered, however, when further study revealed that there is no other evidence for such granaries in Sonora, that no one has assessed the foundations of those in the mountains of Chihuahua, that no evidence of charcoal was found in those that were excavated archaeologically, and none of the rocks show signs of having been burned. The latest suggestion is that these rings were used for growing ornamentals (Doolittle 1992*a*: 83). Such features have not been reported in the ethnographic literature of northern Mexico. Somewhat similar rings of rocks used for the protection of ornamentals have, however, been seen near houses further south in central Mexican villages. That the prehistoric rock rings in Sonora were used for growing ornamentals is supported by their locations; all are proximal to house remains.

The last type of physical evidence of prehistoric gardens in Sonora is low rock walls. Measuring only a course or two high, but 10 to 15 m in length, these walls are located parallel to the edges of seven mesa tops perpendicular to the gentle (1 percent) slopes (Doolittle 1992*a*: 85, 87, figs 4–7). Either by intentional fill or by natural depositional processes, these walls have collected sediment, resulting in wide, flat, fertile, surfaces suitable for gardening. Perhaps not surprisingly, these walls are located only a short distance downslope of houses, where household refuse would be carried by runoff, and on the sunny south sides of houses. These walls are too low and short to have been used for defensive purposes. They could have been used for leveling site surfaces for house construction. However, a terrace built for a house could also serve the double function of shoring up a garden plot (see also Gleason 1994: 6).

Rock features similar to, but more elaborate than, those just discussed have also been found at a few sites in the Hovenweep region of southeastern Utah. Referred to there as 'mesa top terraces' (Winter 1976: 119–22) these features are formed by multiples of up to six walls, 5 to 8 m apart, that are connected with a few perpendicular walls. The overall configuration then is one of rock grids that abut room block walls. Pollen samples collected from these features revealed that maize, beans, cheno-ams, milkweed, and an unidentified solanaceous plant were cultivated in these plots (Winter 1978: 94). Given their proximity to houses, small areal extent, and diversity of crops grown, these features are irrefutable vestiges of gardens, although experiments to grow certain crops in them met with little success (Litzinger 1976).

Also at Hovenweep are features referred to as 'springside irrigated gardens' (Winter 1976: 103–10). These are composed of rock walls, morphologically not unlike those discussed above except they are located just below the habitation sites on the sides, rather than the tops of the mesas. In these locales, gardens were not only fertilized by refuse washed down from the houses and lands above, and watered by means of runoff, but they were also watered via springs. Experiments indicated that earth was probably added manually as sedimentation processes were quite slow (Noisat 1975, 1976). To retain sediment below the springs, series of low rock walls or alignments were built parallel to the slopes, with some connecting walls running perpendicular. Given the steepness of these slopes, walls were typically less than 2 m apart (Winter 1976: 106–9; figs 23–26). They were found in association with at least four habitation sites and were not very extensive—involving between a few and several score square meters.

Between Hovenweep and the Valley of Sonora are several sites with features that have been identified as prehistoric gardens. Small plots supported by low retaining walls were found at Chaco Canyon (Windes 1978: 69–72, figs 30–32, 95–96, figs 41, 42; see also Toll *et al.* 1985: 80, 92). At the Citadel and Ridge ruins near Sunset Crater in northern Arizona, are series of rock walls

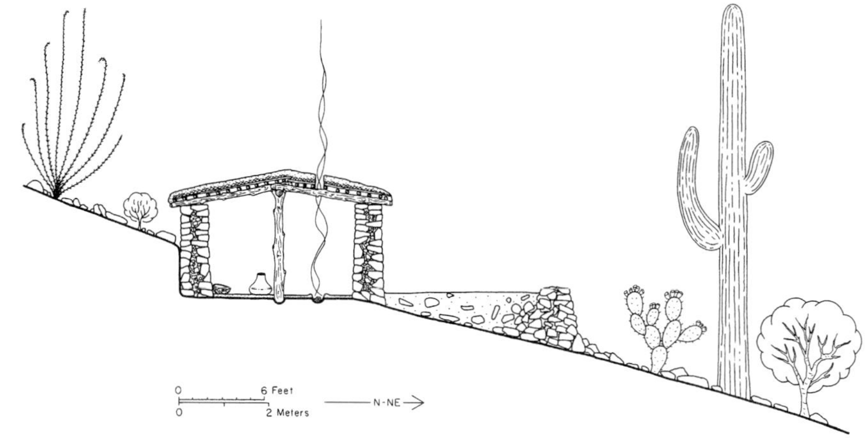

Fig. 4.9. Hypothetical reconstruction of a house and terraced garden on a *cerro de trincheras* north of Tucson, Arizona. Appeared originally in Downum (1993: 65, fig. 4.11). (Provided courtesy of Christian E. Downum, and reprinted by permission of the University of Arizona Press.)

on slopes below habitation sites (Stewart and Donnelly 1943*a*: 42–4; Colton 1960: 97). Similarly, in northern Sonora and in southern Arizona, particularly in the Santa Cruz Valley near Tucson, are some fascinating sites called '*cerros de trincheras*', or terraced hills (Downum 1993: 65; Downum, Fish, and Fish 1994: 282–3). The areas behind, or upslope of some of the walls, both at Sunset Crater and on the *cerros de trincheras*, contain an abundance of trash from houses located above (Fig. 4.9). There is little doubt that the garden interpretation is correct. However, these rock walls that support the planting surfaces are sufficiently substantial to merit separate discussion. They will, therefore, be treated in more detail when terraces are discussed (Chapter 8).

Clearly, gardens are difficult items with which to deal. Separating the husbandry of non-domesticated plants, be they small, herbaceous, or large, and woody species, from true gardening is perhaps more artificial than real. Also, given that some gardens are irrigated by means of canals, some are terraced, and some are both irrigated and terraced, makes distinguishing between agricultural landforms a difficult exercise. The evidence presented here, however, regardless of typological overlap, demonstrates one point very well. Indigenous Americans have, and have long had, gardens. Wherever they lived in permanent or semi-permanent settlements, these people gardened. The ethnographic and the archaeological evidence demonstrate the great time depth involved. The documentary evidence confirms that gardens were cultivated at the time of contact.

Seed Beds

It is widely accepted that gardens serve a variety of purposes, and the evidence provided here supports that notion. However, of all the functions that gardens are known to serve, the one which has received the least amount of scholarly attention, and, hence, is little understood, involves the growing of crops to be planted later in fields. The garden production of seed as insurance against total crop loss has already been discussed. Attention now turns to crops that are germinated in gardens and later transplanted in fields where they mature and are harvested.

Transplanting from gardens to fields typically involves what are known as 'seed beds'. To date, there exists but one scholarly article that systematically addresses seed beds (Vargas Rivero 1983); more work is definitely needed. For now, the following discussion of seed beds, in general, is extracted from this seminal piece.

Seed beds are small areas, rarely more than 1.5 m wide, that are well-marked or bordered, and often elevated. They are small so that each plant is within easy reach; hence they are always less than two arms-length wide. Because of their small size, seeds are planted very close together. Indeed, the principal idea behind seed beds is to germinate as many plants as possible in the least amount of space. Transplanting is critical because at some stage of growth the plants will begin to compete with each other for soil nutrients and water, and eventually, if not transplanted, kill each other. It is also important because it guarantees that all of the plants grown in fields will mature. Seed beds are usually conspicuous and well-defined for protective purposes. Finally, many seed beds are raised above the normal ground surface. This raising contributes to both protection and accessibility or the ease with which plants can be reached, and it is often a function of bringing in the best soil available, and enriching it further with fertilizers.

Evidence for the use of seed beds by native North Americans at contact is sparse. Nevertheless, there are some rather accurate and descriptive first-hand accounts, some 19th- and 20th-century observations that serve as parallels, and some physical remains that have been revealed archaeologically.

Documentary Evidence

Reports of seed beds by early explorers come exclusively from the Northeast (Fig. 4.10). They number ten, deal with at least four different crop plants, and involve some most unusual practices. Perhaps not surprisingly, given both the extent to which wild plants were manipulated and the popularity of smoking, tobacco was reported twice as having been raised in seed beds and then transplanted.

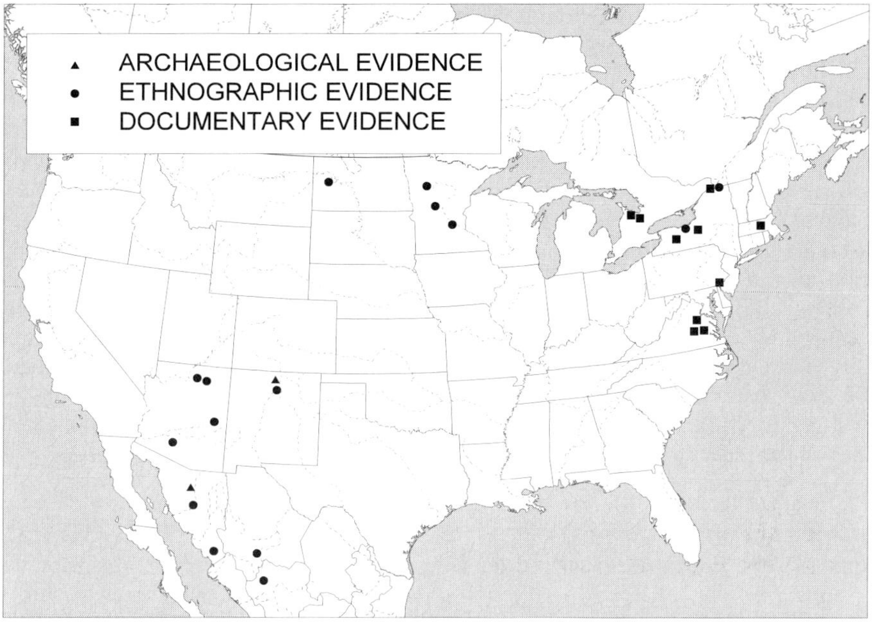

Fig. 4.10. Locations cited in the text for which there is evidence of seed beds in aboriginal gardens.

Heriots' (1812: 597) previously discussed (Chapter 2) 16th-century account of tobacco being sown apart has been interpreted by some scholars (Schmidt and Ross 1925: 47) as evidence of seed beds. Although it is not stated as such, it is possible this interpretation was also based on other accounts. For example, describing conditions in Virginia in the early 1600s, George H. Harris (1812: 242) reported that '. . . tobacco ſeeds are firſt ſown in beds, where having remained a month, the plants are tranſplanted into little hillocks, like thoſe in our hop-gardens . . .'. Peter Lindestrom (1925: 181) reporting of 17th-century native practices in the Delaware Valley wrote of tobacco that '1. Its seed is sown in a bed, like a cabbage seed. 2. When the plants are so large that they should be transplanted, then the tobacco plantation is prepared with hills . . . one plant is set in each hill' (see also Lantis 1929: 61).

Writing a few decades later and of New England Josselyn (1833: 261–2) provided more details about seed beds noting of tobacco that:

> [I]t is sowen in April upon a bed of rich mould sifted, they make a bed about three yards long, or more according to the ground they intend to plant, and a yard and a half over; this they tread down hard, then they sow their seed upon it as thick as may be, and sift fine earth upon it, then tread it down again as hard as possible they can, when it hath gotten four or six leaves, they remove it into the public ground. . . .

The layering of seeds and earth, the use of soils with high organic content, and the great care and elaborate attention bestowed on seed beds is no more apparent than in those used in the germination of cucurbits. In describing the cultivation of pumpkins in the early 1600s, Sagard-Théodat said:

The Huron women in season go into the neighboring forests to gather a quantity of rotten wood powder around old stumps; then having prepared a large bark box they make a layer in it of this powder on which they sow the pumpkin seed; afterward, they cover it with another layer of the same dust and again sow seeds, up to two, three, and four times, as much as they wish, in such a way nevertheless that there still remains four or five good fingers of empty space in the box, in order to leave room for the shoots of the seeds. Afterward, they cover the box with a large bark and put it on two poles suspended in the smoke of the fire, which heats gradually the powder and then the seeds so much that they sprout in a very few days; being well grown and ready for planting they take them by bunches with their powder, separate them, and then plant them in the places prepared . . . (Kinietz 1940: 19; see also Quimby 1960: 114).

A similar, but less detailed, account of aboriginal Canadian seed beds was provided by another Jesuit priest, Joseph François Lafitau (1977: 55). Writing in the 1680s, and familiar with the writings of Sagard-Théodat, he noted:

They prepare special fields for their pumpkins and melons, but before sowing them in these fields, they plant the seeds in a preparation of black, light soil between two sheets of bark and place them above their hearths where they germinate (see also Heidenreich 1978: 380; Trigger 1969: 28; Waugh 1916: 16).

A yet later; even less detailed, although similar, report was made by still another member of the Society of Jesus. Pierre François Xavier de Charlevoix (1923: 112) conveyed the following about Huron seed beds in the 1700s. 'Turnsoles [sunflowers], water melons, and pompions [gourds], are first raised in a hot-bed and afterwards transplanted.'

It is unclear if the heat alluded to in this last account was artificially or spontaneously generated. It is evident, however, that heat was known by native Americans to be an important factor in the germination process, and appropriate means were taken to ensure the production of a good seed crop.

Perhaps the most unusual of the seed bed relations in regard to heating, came from Virginia. There, in 1705, Beverly (1947: 145) observed of potatoes that 'People bury 'em under Ground, near the Fire-Hearth, all the winter, until the Time comes, that their Seedings are to be set'. Although one could argue that this passage describes storage rather than a seed bed, *per se*, it is clearly storage in a specific locale, with special properties or conditions, for a certain reason—replanting not consumption. This account, therefore, described what can be considered a seed bed.

Stretching the definition of a seed bed further are two accounts that describe the soaking of corn kernels prior to planting. One, that of Sagard-Théodat (1964: 8), says only that in the 1620s the Huron sowed maize '. . .

which they have first picked out, sorted, and soaked in water for a few days . . .'. Soaking softens the pericarp and thus expedites germination (Janick *et al.* 1974: 425). The second account, Kalm's (1935: 106) description of Iroquois cultivation of maize in the 1750s is much more elaborate, informative, and entertaining, and, it provides another reason for soaking (see Parker 1910: 26; Starna, Hamell, and Butts 1984: 204).

> In order to prevent crows, maize theives, and squirrels from scratching and picking up the newly planted maize, the following procedure is used: The root of *Veratrum Linn. hort. Cliff 468*, which is *Helleborus albus J.B.*, grows in quantities in swamps and small streams. This root is cooked in water and the water allowed to cool. The maize which is to be sowed is put into the water in the evening and allowed to stand and soak until morning. The kernels which have been soaked are then planted. The squirrels, maize theives, crows, and other animals destructive to maize, who pick up the planted kernels, become so dizzy from one or two kernels that they tumble over and frighten the others, who no longer venture to return (Kalm 1935: 106).

Regardless of the reasons for soaking, these relations indicate that aboriginal Americans at contact did prepare seeds for planting. The nature of the containers used for soaking was not mentioned, but in all likelihood they were ceramic pots. Such vessels may seem to be a far cry from true seed beds, but they share some characteristics with bark boxes, and, therefore, are so considered (Heidenreich 1971: 175).

In sum, documentary evidence reveals that indigenous people of northeastern North America used seed beds *c.* AD 1500. The features considered seed beds were not always located within the confines of gardens as usually envisioned; the bark baskets of the Huron, for example, were hung over fires within longhouses (Heidenreich 1978: 380; Quimby 1960: 114; Trigger 1969: 28). They were, however, integral aspects of gardening, and other forms of evidence not only lend support for the notion, but indicate that seed beds were used throughout the continent.

Ethnographic Evidence

Much as is the case with documentary evidence, relatively recent ethnographic accounts of seed beds are few, and refer to a variety of different specific practices. They do, however, come from diverse parts of the continent (see Fig. 4.10) and they provide more detail on what one would have little trouble as identifying as true seed beds. The ensuing discussion progresses from the simplest to the most insightful accounts.

The Gila branch of the Akimel O'odham (Castetter and Bell 1942: 155), the Indé (Buskirk 1986: 59, 65–6), and the inhabitants of San Juan Pueblo, New Mexico (Ford 1968: 189) are known to have soaked corn kernels in water prior to planting. The Hopi also soaked cotton seeds in jars of water for several days prior to planting (Beaglehole 1937: 41). San Juan informants relayed that spruce seeds were soaked along with the corn and this helped repel worms. They also reported that soaking made seeds stronger and had the latent function of accelerating germination (Ford 1968: 189).

The Iroquois of upstate New York also soaked their maize seeds in recent times. However, the practice is different from both that of their southwestern contemporaries and their ancestral counterparts. These people are reported to have soaked kernels in a mixture of water and various herbs which had been previously boiled. As is the case with puebloan practices, water facilitated germination while the other ingredients provided protection from worms and bothersome birds (Waugh 1916: 18). This soaking, however, took place after the water and herbs cooled, and it lasted only one hour. Upon their removal, the seeds were placed in a basket while still moist and left until they began to sprout (Waugh 1916: 19).

In 1850, a traveler on the upper Missouri River wrote that aboriginals in what is now North Dakota had an interesting way of accelerating the germination of maize kernels. These people put the seeds in bundles of hay, which presumably was moist, and suspended them in skin bags over fires in their hearths. They later planted only those seeds which had begun to sprout (Culbertson 1952: 98).

Not too far away, in northern Minnesota, the Hidatsa sprouted squash seeds in much this same manner. According to one observer (G. Wilson 1917: 68), who provided more details, these people first laid out a piece of tanned buffalo hide, 0.75 m long and 0.45 m wide, fur side up. They then wet grass leaves and spread them out on the fur. Next, they wet the squash seeds and laid them on the grass. The seeds were then covered with sage and buck brush leaves, and the skin folded over into a bundle measuring 0.45 m by 0.4 m. This bundle was then hung on a drying pole over, but a little to one side, of a fire. After two days the bundle was taken down and opened. At the end of another day the seeds had sprouted and were 2.5 cm high. They were then planted.

On the Red Lake Reservation in northern Minnesota, indigenous people mixed their maize kernels with an unidentified moss in a box of unspecified dimensions. The box was then soaked in water, presumably a lake or stream, 'for a time', and then set in a warm place until the seeds sprouted (Biggar 1918: 126). These were then planted. Yet other aboriginals of the old Northwest are reported to have filled willow baskets with maize seeds and then poured water through them. The baskets were subsequently placed in a warm spot to start germination (Biggar 1918: 126).

Reports of permanently situated seed beds are few and limited to northwestern Mexico and the American Southwest. There is one report from the northern border of New York which relates that native farmers planted seeds which had already begun to sprout because they 'had been previously covered in the earth, and permitted to remain 'til germinating' (Delafield 1943: 140), however, it is not too precise. The best ethnographic reports of seed beds are those from the Rarámuri, the Odami, the Névome, and the Warihio. The Rarámuri are described as having planted chili (*Capsicum annuum*), tomato (*Lycopersicon esculentum*), and tobacco (*Nicotiana tabacum*) seeds in 'shallow baskets'. Those containing chili seeds were kept in the sun and watered until the sprouts were 5 cm high. Baskets containing tobacco seeds were kept in the shade until the sprouts were 7.5 cm high. At the appropriate times, germinated seeds of each species were transplanted to 'garden plot[s]' (Pennington 1963: 67–8). The Névome are reported to have planted tobacco in old *ollas*, ceramic pots, and transplanted when the sprouts reached a height of 20 cm. They are also said to have planted tobacco seeds in 'a small shaded bed', but details were not provided (Pennington 1980: 174).

Clearly, the most interesting seed beds are those of the Warihio. Near their houses these people build platforms measuring approximately 1.5 m by 2 m, raised 1.5 m above the ground. These platforms are made of four Y-shaped poles standing upright across the tops of which branches are laid. A course or two of rocks rings the top of the platform, and is filled with earth (Fig. 4.11). Called '*tapancos*', these are used for germinating chilies (Gentry 1963: 112, plate 33b).

The Odami also use seed beds, but theirs are built directly on the ground. Photographic evidence (Fig. 4.12) suggests they are approximately 1 m wide and 6 to 8 m long, and bounded by rocks stacked two or three courses, or approximately 30 cm, high (Pennington 1969: 92, fig. 14). The area between the rocks is filled within a few centimeters of the top with earth. Seed beds such as this one are located in full sunlight. The principal crop seeded is tobacco, but tomatoes and chili are also grown. Prior to planting, the earthen fill is loosened and thoroughly soaked. Seeds are then sown and covered lightly with dirt. When the tobacco is 15 to 20 days old, and when the chilies are approximately 10 cm high, they are transplanted to larger plots in gardens (Pennington 1969: 91). The best evidence of seed beds in the Southwest comes from northern Arizona, where at least in one locale, Lemeva Spring (Fig. 4.13), Hopi gardeners manually irrigate sprouts to be transplanted (Forde 1931*b*: 366, Plate XLV, fig. 2).

Hard evidence for the use of truly permanent seed beds by aboriginal North Americans in historic and recent times is geographically limited. However, the similarity between some widespread practices, such as the soaking of seeds prior to planting, indicates that several crops were germinated and then transplanted in gardens across the continent. The similarities in germination techniques between the early explorer accounts and the ethnographic record helps fill spatial as well as temporal gaps in the picture. Following this reasoning, one should have few problems in predicting

Fig. 4.11. A *tapanco*, or elevated seed bed used by the Warihio of northwest Mexico. After Gentry (1963: plate 33b). (Drawn by, and published by permission of, Diane Doolittle.)

Fig. 4.12. A seed bed used by the Odami of northern Mexico for germinating tobacco plants. Appeared originally in Pennington (1969: 92, fig. 14). (Provided courtesy of Campbell W. Pennington, and reprinted by permission of the University of Utah Press.)

Fig. 4.13. Seed beds at Lemeva Springs, Second Mesa, on the Hopi Reservation of northern Arizona. (Photograph by author.)

where physical evidence of prehistoric seed beds could be found—the Southwest.

Archaeological Evidence

Physical evidence of seed beds is extremely scarce, and tentative. The paucity of such evidence may well be attributed to their not having been so identified by prehistorians. Indeed, given that so few researchers have even considered the existence of gardens, much less their importance, there is no reason to think they would have considered the possibility that seed beds were used and can be identified. It is highly likely that field workers have long looked at remnants of seed beds, shrugged their shoulders, and then reported them as 'features of unknown function'.

There are only two locales where physical remains of possible prehistoric seed beds have been observed (see Fig. 4.10). The first is in the Valley of Sonora. There, two parallel rows of rocks approximately 6 m long, 1 m apart, with rocks set partially within the ground and adjacent to each other, were found slightly downslope from the remains of ancient houses. The alignments were too close to the edge of the mesa on which the habitation site was located to be remnants of walls from two houses set close together. The ends

Fig. 4.14. A possible prehistoric seed bed in northern New Mexico. Located on a mesa top site overlooking the Chama River, this feature is strikingly similar to seed beds used by the Odami of northwestern Mexico (cf. Fig. 4.12). Note that the person in this photograph, archaeologist Timothy Maxwell of the Museum of New Mexico, can easily reach the center of the feature, demonstrating that every seedling could have been carefully tended. (Photograph by author.)

of these walls are not connected, but erosion along the top of the mesa could certainly have destroyed any end walls. Given their location on a habitation site, downslope from houses, and their morphological similarity to seed beds used today by the Odami, who live not too far away, these alignments could quite feasibly be remnants of an ancient seed bed (Doolittle 1992*a*: 85, 86, figs 4–6).

The second site for which there appears to be physical remains of a prehistoric seed bed is in the Chama Valley, New Mexico. On one of several sites that have garnered much attention because of other agricultural features (discussed in Chapter 7) there was found a feature that looks strikingly like Odami seed beds (Fig. 4.14). It measures approximately 5 m long and 1 m wide. Unlike the suspected ancient seed bed in Sonora, this one is closed on both ends and, therefore, is largely intact. All that is lacking is a second and possibly a third course of rocks which would raise the height of the feature. Of course, the interpretation of this and the Sonoran feature as a seed bed is speculative. More research is clearly needed.

Overview

Overall, several conclusions can be drawn about gardening in North America *c.* AD 1492. Three, however, stand out as being especially significant. First, gardens were widespread and important. Gardening was an integral subsistence activity, wherever people resided permanently. Second, gardening varied considerably. In all probability, no two gardens were alike, and the reasons for their existence were multiple. Third, seed beds were common garden features. Many important crops were started in beds and later transplanted.

Studies of gardening conducted in other places and focusing on different times demonstrate that the practice is labor-intensive. Unlike in distant fields, yields in gardens are very high, but so too are labor inputs. The spatial dimension of gardens in respect to fields cannot be ignored. Studies also reveal that in spite of its importance, gardening is not well-understood and it is not fully appreciated. More research needs to be done on gardening, especially in North America. This chapter not only synthesizes data currently available on native North American gardening, it discusses everything that is known, and in detail. A more fruitful line of future research could hardly be found.

Part III

Rainfed Systems

5

Plain Fields in Warm Environments

Husbandry of non-domesticated plants—be they small and herbaceous or large, and woody—and maintaining gardens are both forms of agriculture, but clearly they are not the types of activities that most people think of immediately in terms of food production. Much more apparent is the cultivation of staple crops, domesticated grains that comprise the greatest part of a peoples' diet.

The growing of cereals, especially on reasonably large, or at least discrete, tracts of land, is contingent on certain environmental conditions. Of specific importance are ample but not excessive moisture, a sufficiently long frost-free or growing season, naturally fertile soil, and gentle terrain. Foremost among these is moisture, particularly that which occurs in the form of rain. Although the amount of moisture on a field can be regulated artificially by applying more water (see Part IV) or removing some (see Part V), the most basic form of agriculture is that which is dependent solely on precipitation falling directly on fields and which requires no drainage. This form of agriculture, known as 'rainfed', is certainly straightforward. It is, however, not without ambiguities.

Several types of cultivation systems can be classified as rainfed agriculture. Slash-and-burn shifting cultivation, sometimes referred to as swidden agriculture, as practiced in the humid tropics (Manshard 1974: 52–67), is one type that falls into this category. Others include such activities as mixed farming in the Mediterranean (Oram 1979: 214–16), and mechanized wheat production on the Great Plains (Turner and Brush 1987: 9). The term 'rainfed agriculture' is made even more confusing by some scholars' use of the term 'dry farming'. Most commonly, dry farming is defined as a form of cultivation practiced in arid or semi-arid areas, and involving certain soil moisture retention techniques, such as mulching (W. Moore 1974: 70–1). Under this definition, the growing of agave in rock piles (Chapter 3) is a form of dry farming. For those working in moisture-deficient environments, dry farming is cultivation without any attempts to provide additional water, such as by means of irrigation canals; it is agriculture dependent solely on precipitation that falls directly on cultivated plots and stored in the soil (Lawton and Wilke 1979: 3–4). In other words, it is the same thing as rainfed agriculture. If there is a distinction between dry farming and rainfed agriculture, it is that the former is the latter, but the latter is not necessarily the former (see Chapter 7). Although no pun is intended, the term 'rainfed agriculture' does have certain umbrella-like qualities.

In addition to problems with terminology, rainfed agriculture is problematic in terms of landscape modification and the archaeological record. In some cases, rainfed agriculture involves extensive reworking of the land, and, as a result, leaves substantial physical remains. Abandoned terraces which were not irrigated but used primarily to create nearly level planting surfaces in mountainous environments stand as relicts of such past agricultural activities (Spencer and Hale 1961). In other cases, however, such as where terrain is not a problem, rainfed agriculture may result in no appreciable traces being left on the present-day landscape. To illustrate, one recent study revealed that only 10 percent of the rainfed fields have features on the earth's surface that would be identifiable after abandonment (S. Russell 1978: 40). If no other evidence of prehistoric agriculture is encountered, such as subsurface differences detected by means of remote sensing or seeds of domesticated cultigens uncovered during excavations, and assuming that ancient peoples did cultivate, researchers working in these areas usually infer that some form of rainfed agriculture was practiced (Lawton and Wilke 1979: 4).

Given problems with terminology, environmental alterations, and the archaeological record, those native North American agricultural activities that relied on direct precipitation but left substantial vestiges on the landscape are not discussed here. They are covered later, in those chapters dealing with specific agricultural landforms. Only those forms of cultivation which were dependent entirely on direct rainfall and which left little or no physical remains of fields on the present-day surface are discussed in this chapter. Borrowing from, and combining both geographical (Monkhouse 1965: 238; W. Moore 1974: 167) and English language definitions, 'plain' fields are generally flat, level, or gently rolling or undulating, and lack embellishment. Not without consequence, these practices and fields were limited to locales with a longer than adequate growing season.

Documentary Evidence

Not surprisingly, most explorers' accounts of indigenous rainfed agriculture come from the humid eastern half of the continent (Moodie and Kaye 1969: 525, fig. 2; Spinden 1915: 273, plate III, 1917: 188), but a few do come from the arid Southwest. Documentary evidence of this form of farming is abundant and of varying detail. There exist textual references that mention fields and cultivated areas, and those that describe planting procedures, cultivation techniques, and harvesting activities. There are even old maps which portray fields, either under cultivation or abandoned.

Textual references to fields come in six forms. There are, to begin with, what might best be considered allusions that do not mention fields, *per se*, but suggest that they existed. For example, writing of indigenous farmers near the confluence of the Illinois and Mississippi rivers in 1704, J. F. Buisson de St. Cosme (1861: 77) said only that they 'cultivate the earth'. Similarly, Alonso de Leon (1968: 114) observed in 1689 that natives in east Texas 'plant much corn, beans, and squash' (see also Weber 1992: 153). Accounts like these actually say very little. However, they do say more than statements such as '[t]hey have great store of Maiz', a comment made by Robert Juet (1909: 18), chronicler of Henry Hudson's 1609 trip up the valley named after him. References to crops alone do not necessarily mean that the possessors were the growers. Furthermore, even if such inferences are drawn, accounts of this nature say nothing of specific agricultural practices. Allusions to fields may allow one to determine specific places where cultivation took place. However, such accounts are so numerous that at best all they can be used for prudently is to delimit areas in which agriculture was widespread and common (e.g., Doolittle 1992*c*: fig. 1, 388; Weatherwax 1954: fig. 18, 52; Wissler 1922: figs 1 and 3, 2, 20).

Much more informative and far less numerous than allusions are explicit references to fields and cultivated areas. These are found in five forms: (1) simple, but clear, mention of 'fields' only, (2) accounts of numerous fields in proximity to each other, (3) relations of expansive cultivated areas, (4) reports of large fields, and (5) reports of small fields. Some simple observations of fields such as the Gentleman of Elvas' (1907: 162) mention of 'trees that grow in fields' (Chapter 3) in Florida, and Champlain's (1907: 244) notice of both the 'gardens and the fields' (Chapter 4) along the St. Lawrence have already been discussed. There are, however, at least three dozen more such references (Table 5.1, Fig. 5.1).

To illustrate, Hudson reported in his 1609 log of riverine travels in New York that he saw crops 'growing in the fields' (De Laet 1909: 49). Further in the interior and nearly a century later Antoine de la Mothe Cadillac (1947: 64) noted that families on Beaver Island in Lake Michigan 'have fields of Indian corn'. Not all simple accounts of fields are so brief. A few contain information that confirms a dependence on direct rainfall, even though one might suspect otherwise. For example, in 1582 Antonio de Espejo (1966: 220) reported that the Piro at one pueblo in the upper Rio Grande Valley, New Mexico 'have fields planted with corn, beans, calabashes, and tobacco (*piciente*) in abundance. These crops are seasonal, dependent on rainfall'.

In addition to mentioning fields under cultivation at the time, it was not unusual for explorers to mention fields that at some time in the past had been cultivated, but were abandoned by the time they arrived on the scene. Typically, references to formerly cultivated fields were made well after the native farmers first encountered Europeans; the implication being that Old World pathogens took their toll on the indigenous population, thereby reducing the number of farmers. A good example of this comes from the Southeast. There,

Table 5.1. Locales for which there exist simple accounts of plain fields indicative of rainfed agriculture

Locale	Observer	Date	C/A[a]	Source
Cale (FL)	Elvas	1539	C	Elvas (1907: 156, 1993: 65)
Apalache (FL)	Elvas	1539	C	Elvas (1907: 162, 1993: 72)
	Ranjel	1539	C	Ranjel (1904: 82, 1993: 268)
Cuacoco (FL)	Ranjel	1539	C	Ranjel (1904: 64, 1993: 259)
Caliguen (FL)	Ranjel	1539	C	Ranjel (1904: 71, 1993: 263)
Florida	Ribaut	1563	C	Ribaut (1927: 73)
Pensacola (FL)	Sigüenza y Gongora	1686	C	Sigüenza y Gongora (1963: 24)
Georgia Coast	Bartram	1773	A	Bartram (1940: 65)
Little River (GA)	Bartram	1773	A	Bartram (1940: 57)
Georgia/Carolinas	Adair	1775	C	Adair (1968: 408–9)
Charlotia (NC)	Bartram	1773	A	Bartram (1940: 98)
Pacaha (AR/MS)	Biedma	1544	C	Biedma (1905: 26–7, 1993: 239)
Casqui (AR/MS)	Ranjel	1539	C	Ranjel (1904: 141, 1993: 302)
	Elvas	1541	C	Elvas (1907: 206, 1993: 114)
Quizquiz (AR/MS)	Biedma	1544	C	Biedma (1905: 25, 1993: 238)
Red River (LA)	Custis	1806	C	Flores (1984: 154)
Red River (TX/OK)	Fernández	1788	C	Fernández (1967: 322)
	Merino	1804	C	John and Wheat (1991: 173)
Cuarac Pueblo (NM)	Perea	1633	C	Perea (1937: 131)
Pecos Pueblo (NM)	Vargas	1692	C	Vargas Zapata (1940: 122)
	Tamarón y Romeral	1760	C	Tamarón y Romeral (1954: 51)
	Dominguez	1776	C	Dominguez (1956: 213)
A Piro pueblo (NM)	Espejo	1582	C	Espejo (1966: 220)
A Tigua pueblo (NM)	Espejo	1582	C	Espejo (1966: 221) Obregón (1928: 323)
San Gabriel Pueblo	Brondata	1601	C	Valverde (1953*a*: 626)
	Velasco	1601	C	Velasco (1953: 610)
Rio Grande pueblos	Espinosa	1601	C	Valverde (1953*a*: 634)
	Dominguez	1776	C	Dominguez (1956: 175, 213, 217)
Acoma Pueblo (NM)	Espejo	1584	C	Obregón (1928: 325)
	Dominguez	1776	C	Dominguez (1956: 194)
Acomita (NM)	Pérez de Luxán	1582	C	Pérez de Luxán (1966: 201)
Zuni Pueblo (NM)	Dominguez	1776	C	Dominguez (1956: 201)
New England coast	Champlain	1605	C	Champlain (1907: 71)
	Dermer	1619	A	Dermer (1905: 251)
	Levett	1624	A	Levett (1905: 263)
	Higginson	1630	A	Higginson (1929: 308)
	Johnson	1633	C	Johnson (1910: 85)
Massachusetts coast	Smith	1614	C	J. Smith (1910*b*: 192)
Eastham (MA)	Bradford and Winslow	1620	A	Bradford and Winslow (1844: 153–4)
Cape Cod	Bradford and Winslow	1621	C	Bradford and Winslow (1844: 171)
Plymouth (MA)	Pring	1603	C	Pring (1905: 59)
	Bradford and Winslow	1621	A	Bradford and Winslow (1844: 165)

Table 5.1. *Continued*

Locale	Observer	Date	C/A[a]	Source
New Hampshire coast	Smith	1614	C	J. Smith (1910*b*: 193)
Maine coast	Smith	1614	C	J. Smith (1910*b*: 203)
Hudson Valley (NY)	Hudson	1609	C	De Laet (1909: 49)
Kendaia (NY)	Burrowes	1779	C	Cook (1887: 46)
	Barton	1779	C	Cook (1887: 9)
	Beatty	1779	C	Cook (1887: 29)
Chemung (NY)	Livermore	1779	C	Cook (1887: 184)
	Shute	1779	C	Cook (1887: 271)
Seneca Lake (NY)	Sullivan	1779	C	J. Sullivan (1887: 299)
	Barton	1779	C	Cook (1887: 9)
	Hubley	1779	C	Cook (1887:158)
	McKendry	1779	C	Cook (1887: 204)
	Roberts	1779	C	Cook (1887: 245)
Newton (NY)	Beatty	1779	C	Cook (1887: 28)
Conadasaga (NY)	Fellows	1779	C	Cook (1887: 89)
Kanandaqua (NY)	Jenkins	1779	C	Cook (1887: 174)
Chinesee (NY)	Livermore	1779	C	Cook (1887: 188)
	McKendry	1779	C	Cook (1887: 205)
Tioga (NY)	Jenkins	1779	C	Cook (1887: 176)
Allegheny River (NY)	Norris	1779	C	Cook (1887: 237)
	Dearborn	1779	C	Cook (1887: 78)
Cheoquock (NY)	Jenkins	1779	C	Cook (1887: 173)
St. Lawrence Valley	Champlain	1613	C	Champlain (1907: 244)
Lake St. Clair (MI)	Sabrevois	1714–17	C	Kinietz (1940: 322)
Detroit River (MI)	Cadillac	1721	C	Cadillac (1947: 12)
Eastern Michigan	Perrot	1642	C	Perrot (1911: 304)
	Perrot	1642	C	Perrot (1911: 322)
Western Michigan	Allouez	1667	C	Allouez (1899: 27)
Beaver Island (MI)	Cadillac	1721	C	Cadillac (1947: 12)
Illinois	Allouez	1677	C	Allouez (1903: 78)
Muskingum River (OH)	Doddridge	1782	C	Doddridge (1925: 203)
Missouri River (ND)	La Vérendrye	1738	C	Gaultier de Varennes (1980: 61, 64)
	Tabeau	1795	C	Tabeau (1939: 69)
	Biddle	1810	C	Biddle (1978: 531)

[a] C, fields under cultivation at the time account was made; A, fields had been cultivated but were abandoned at the time account was made.

in 1773, Bartram (1940: 57, 65, 98) mentioned seeing in three different places 'ancient cultivated fields', 'an ancient field', and 'old fields'. The later term was so widely used by explorers that it has come into the contemporary literature as standard parlance (e.g., R. Brown 1948: Butzer 1990*a*: 49; H. Russell 1980: 141).

Agriculture in many locales appears to have been sufficiently important to native North Americans that they cultivated several plots. This fact was not lost on many early European explorers who were often so impressed with the number of fields that they reported the same (Table 5.2), especially in the Southwest, the Southeast, and New England (Fig. 5.1). For example, Rodrigo

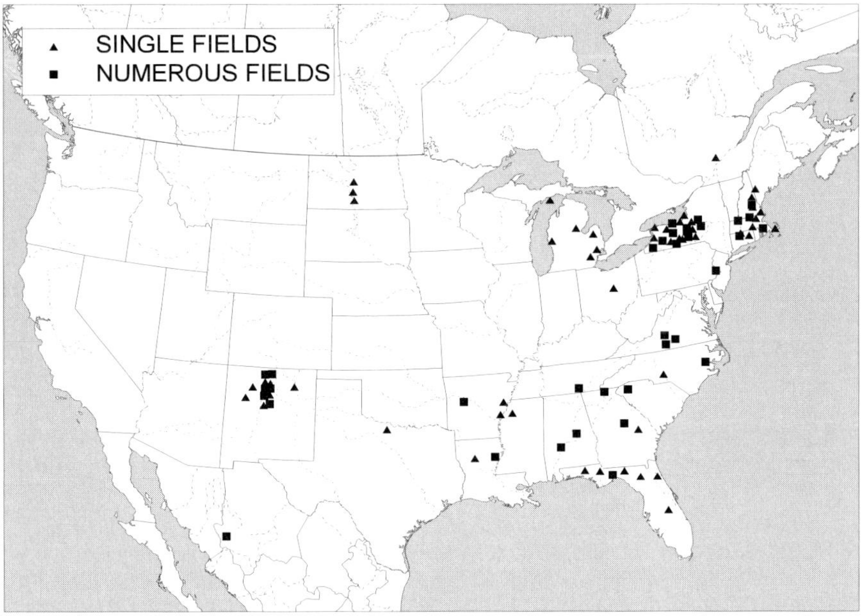

Fig. 5.1. Locations cited in the text and Tables 5.1 and 5.2 for which there is documentary evidence of either single fields or numerous fields that were in all probability dependent solely on direct rainfall.

Table 5.2. Locales for which there exist accounts of numerous plain rainfed fields in proximity to each other

Locale	Observer	Date	C/A[a]	Source
Apalache (FL)	Cabeza de Vaca	1528	C	Cabeza de Vaca (1946: 29)
	de Soto	1540	C	Vega (1951: 292, 1993: 280)
Ocute (GA)	Elvas	1543	C	Elvas (1907: 270, 271, 1993: 168, 170)
Jalameco (SC)	Ranjel	1539	C	Ranjel (1904: 102)
Cape Fear River (NC)	Lawson	1714	C	J. Lawson (1937: 121)
New River (VA)	Fallows	1671	A	H. Maxwell (1910: 83)
	Clayton	1671	A	H. Maxwell (1910: 83)
Roanoke River (VA)	Byrd	1724	A	H. Maxwell (1910: 82)
SC/NC/GA/TN	Schneider	1783	C	Fogelson and Kutsche (1961: 96)
Chiaha (TN)	Elvas	1540	C	Elvas (1907: 179, 1993: 88)
Talisi (AL)	Elvas	1540	C	Elvas (1907: 186, 1993: 95)
SW Alabama	Romans	1775	A	Romans (1962: 309)
Acuera (LA)	Ranjel	1539	C	Ranjel (1904: 161)
Tanico (AR)	Elvas	1540	C	Elvas (1907: 217, 1993: 124)
Kansas	Oñate	1601	C	Oñate (1908: 261)
A Tigua pueblo (NM)	Bustamante	1581	C	Bustamante (1966: 129)
	Oñate	1598	C	Oñate (1953*a*: 319)
San Felipe Pueblo (NM)	Espejo	1581	C	Espejo (1966: 231)
				Obregón (1928: 291)

Table 5.2. *Continued*

Locale	Observer	Date	C/A[a]	Source
San Miguel Pueblo (NM)	Gallegos	1581	C	Gallegos (1966: 82)
A Quires pueblo (NM)	Espejo	1582	C	Espejo (1966: 223)
Sierra Madre (MEX)	Falcón Mariano	1777	C	Falcón Mariano (1979: 109–10)
New England coast	Champlain	1605	A	Champlain (1907: 71)
	Smith	1614	C	J. Smith (1910*b*: 204)
Cape Cod	Bradford and Winslow	1620	C	Bradford and Winslow (1844: 161–2)
Taunton Valley (MA)	Bradford and Winslow	1621	A	Bradford and Winslow (1844: 206)
Rhode Island	Williams	1643	C	R. Williams (1973: 128)
Allegheny River (NY)	Brodhead	1779	C	Cook (1887: 308)
Teaoga River (NY)	Sullivan	1779	C	J. Sullivan (1887: 303)
	Beatty	1779	C	Cook (1887: 34)
	Campfield	1779	C	Cook (1887: 56)
	Dearborn	1779	CA	Cook (1887: 70, 74)
	Norris	1779	C	Cook (1887: 229)
Chemung (NY)	Blake	1779	C	Cook (1887: 39)
	Elmer	1779	C	Cook (1887: 85)
	Fellows	1779	C	Cook (1887: 89)
	Jenkins	1779	C	Cook (1887: 170)
	Webb	1779	C	Cook (1887: 287)
Seneca Lake	Hubley	1779	C	Cook (1887: 158)
	Campfield	1779	C	Cook (1887: 59)
	Fellows	1779	CA	Cook (1887: 90)
	Dearborn	1779	C	Cook (1887: 77)
	Norris	1779	C	Cook (1887: 233, 237)
Kendaia (NY)	Dearborn	1779	C	Cook (1887: 73)
	Grant	1779	C	Cook (1887: 111)
	Hubley	1779	C	Cook (1887: 159)
	Jenkins	1779	C	Cook (1887: 174)
	Roberts	1779	C	Cook (1887: 245)
Newton (NY)	Webb	1779	C	Cook (1887: 287)
Kanandaqua (NY)	Fellows	1779	A	Cook (1887: 90)
	Hubley	1779	A	Cook (1887: 160)
Chinesee (NY)	Hubley	1779	C	Cook (1887: 163)
	Dearborn	1779	C	Cook (1887: 76)
	Norris	1779	C	Cook (1887: 235)
	Shute	1779	C	Cook (1887: 272)
	Beatty	1779	C	Cook (1887: 32)
Shanandaque (NY)	Campfield	1779	C	Cook (1887: 59)
	Dearborn	1779	C	Cook (1887: 74)
	Fellows	1779	A	Cook (1887: 90)
	Hubley	1779	CA	Cook (1887: 161)
	Norris	1779	C	Cook (1887: 234)
Kannawaloholla (NY)	Dearborn	1779	C	Cook (1887: 73)
	Fellows	1779	C	Cook (1887: 89)
	Norris	1779	C	Cook (1887: 232)

[a] C, area under cultivation at the time account was made; A, area had been cultivated but was abandoned at time account was made.

Ranjel (1904: 102), one of de Soto's chroniclers, reported in 1539 that he saw 'many very fine fields' of maize at Jalameco in what is today western South Carolina. Similarly, Pedro de Bustamante (1966: 129), who accompanied the Chamuscado–Rodriguez expedition, testified that indigenous people at one Rio Grande pueblo in 1581 'had many fields of corn, beans, and calabashes', and Smith (1910*b*: 204) reported 'many corn fields' in Massachusetts during 1614.

The references to crops in fields suggests that all of the numerous fields were cultivated simultaneously. However, there exist four cases in which it is most evident that although fields were numerous, they were not all continuously under cultivation. Champlain (1907: 71) provides the earliest example of this when he said that at one place along the New England coast in 1605, indigenous people had 'several fields entirely uncultivated, the land being allowed to remain fallow'. A short time later and a little further south—1636 at Narragansett Bay—the Governor of Connecticut, John Winthrop Jr. (1863: 515) wrote that natives 'have every one 2 feilds, which after the first 2 yeares they lett one feild rest each yeare, & that kepes their ground contin(u)ally in hart'. Not too far away, in 1671 the Reverend John Clayton mentioned 'land that has been cultivated by the Indians and left fallow, which is generally overrun with what they call "broom grass" ' (H. Maxwell 1910: 83). Finally, and more than a century later in the Sierra Madre of northwestern Mexico the Franciscan missionary Joseph Agustín Falcón Mariano (1979: 110) said:

> When the soil is good a cleared field is used for two or three years. After three years have passed these fields no longer produce, and must be fallowed until wild growth returns to cover them. The Indians can then return to clear them again.

Also problematic is the spatial relationship between fields reported in multiples: were they contiguous or separated by uncultivated land? To illustrate, Williams (1973: 128) noted that aboriginals in southeastern Massachusetts or Rhode Island in 1643 'sometimes having fields a mile or two, or many miles asunder, when the worke of one field is over, they remove house to the other'. In contrast, Martin Schneider, a Moravian missionary reported in 1783–4 of the Cherokee that 'every Family has its own Field, yet they begin fellowshiply on one End, & continue so one after the other till they have finished it all' (Fogelson and Kutsche 1961: 96). Both of these reports suggest individual ownership, the former clearly indicates that fields were disjunct, the latter implies contiguity of fields, and communal labor (see also Cadillac 1947: 12; Elvas 1993: 170; 1907: 271).

In some cases, reports of multiple fields in one locale are of such a nature to suggest that fields were not only contiguous, but so numerous that sizable areas were cultivated. Two such accounts serve as examples, one from very early on in Georgia reports that fields were currently in use at the time, the second, from Massachusetts, is later, reports abandonment, but indicates early conditions similar to those in the Southeast. Writing of Georgia in 1543, the Gentleman of Elvas (1907: 270, 1993: 168) said '[t]he land of Ocute is more strong and fertile than the rest, the forest more open; and it has very good fields along the margins of the rivers'. William Bradford, the second governor of the Plymouth colony, and Edmund Winslow, who traveled with him, wrote of the Taunton River in 1621 that '[t]he ground is very good on both sides, it being for the most part cleared. Thousands of men have lived there . . . not long since; and pity it was and is to see so many great goodly fields, and so well seated, without men to dress and manure the same' (Bradford and Winslow 1844: 206).

For numerous locales, fields, *per se*, were not only mentioned by early explorers, but exceptionally large tracts of land were reported to have been under cultivation at one time. Reports of expansive areas being devoted to rainfed agriculture vary in nature from rather general qualitative phrases to precise quantitative measurements (Table 5.3, Fig. 5.2). An example of a qualitative description of expansive cultivated tracts come from the Northeast. Champlain (1907: 66) wrote of New England in 1605 that '[a]ll along the shore there is a great deal of land cleared up and planted with Indian corn'. Two examples of quantitative descriptions involve important river valleys, one in the North, the other in the south. In 1524 Verrazzano

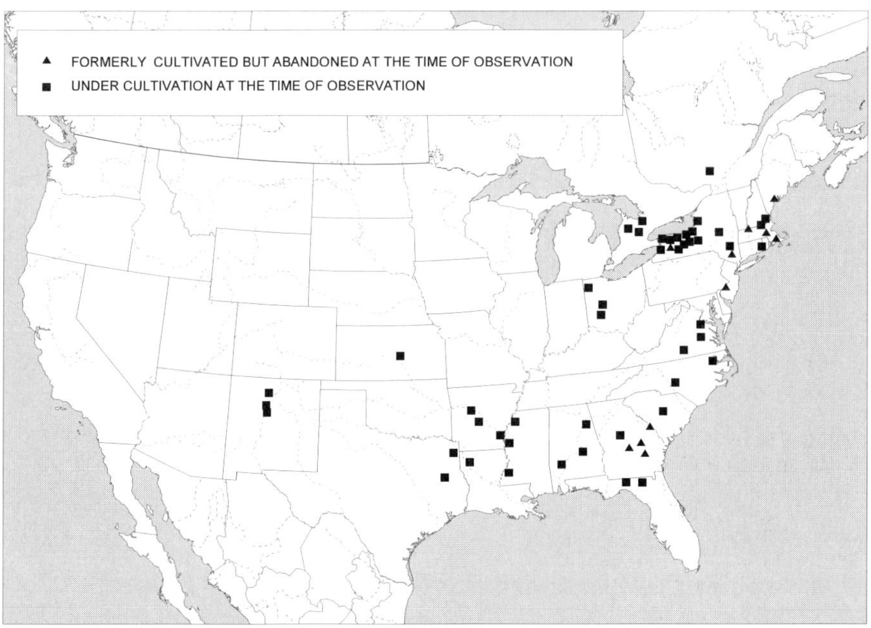

Fig. 5.2. Locations cited in the text and Table 5.3 for which there is documentary evidence of extensive areas devoted to rainfed agriculture.

Table 5.3. Locales for which there exist accounts of extensive areas devoted to rainfed agriculture

Locale	Observer	Date	C/A[a]	L/N[b]	Source
Osachile (FL)	de Soto	1543	C	N	Vega (1951: 169, 1993: 184)
Apalache (FL)	de Soto	1543	C	N	Vega (1951: 182, 1993: 194)
Oakmulge (GA)	Bartram	1773	A	N	Bartram (1940: 68, 307)
Cuscowilla (GA)	Bartram	1773	C	L	Bartram (1940: 169–70)
Alachua (GA)	Bartram	1773	A	L	Bartram (1940: 173)
Ocone River (GA)	Bartram	1773	A	L	Bartram (1940: 306)
Savanna River (GA/SC)	Bartram	1773	A	L	Bartram (1940: 258)
Canos (SC)	Vandera	1566	C	L	Bandera (1990: 298, 301)
Charlotia (NC)	Bartram	1773	C	N	Bartram (1940: 96)
Secota (NC)	White	1585	C	L	Lorant (1946: 191)
Pamanches (VA)	Archer	1610	C	N	Archer (1969: 93)
Staunton River (VA)	Lederer	1670	C	L	Lederer (1958: 27)
Kecoughtan (VA)	Smith?	1612	C	N	Strachey (1953: 67)
Creek Country (GA/AL)	Hawkins	1798	A	N	Hawkins (1848: 28)
Coca (AL)	Elvas	1540	C	L	Elvas (1907: 184, 1993: 93)
Tombigbee (AL)	Elvas	1541	C	L	Elvas (1907: 194, 195, 1993: 105)
Casqui (AR/MS)	Elvas	1541	C	N	Elvas (1907: 206, 1993: 114)
Natchez (MS)	Membré	1681	C	L	Membré (1903*b*: 177)
Cappa (AR)	Joutel	1687	C	L	Joutel (1906: 175)
Tanico (AR)	Elvas	1541	C	L	Elvas (1907: 218, 1993: 125)
Nilco (AR)	Elvas	1542	C	N	Elvas (1907: 226, 248, 1993: 131, 149)
Red River (LA)	Glass	1808	C	N	Glass (1985: 48)
Red River (TX/OK)	Parilla	1759	C	L	Newcomb and Field (1967)
	Glass	1808	C	N	Glass (1985: 50)
East Texas	Joutel	1687	C	L	Joutel (1906: 141, 142–3)
Quivira (KS)	Oñate	1601	C	L	Oñate (1953*b*: 755)
A Tigua pueblo (NM)	Alvarado and Padilla	1540	C	L	Alvarado and Padilla (1940: 183)
San Felipe Pueblo (NM)	Escalante and Barrando	1581	C	N	Escalante and Barrando (1966: 141, 142)
	Pérez de Luxán	1582	C	L	Pérez de Luxán (1966: 172)
New Jersey	Kalm	1749	A	L	Kalm (1972: 254)
Connecticut	Gookin	1674	C	LN	Gookin (1806: 180, 184–6)
Boston	Champlain	1605	C	L	Champlain (1907: 66, 67)
New England coast	Champlain	1606	C	L	Champlain (1907: 95)
Cape Cod	Bradford and Winslow	1620	A	L	Bradford and Winslow (1844: 130–1)
Kingston (MA)	Bradford and Winslow	1620	A	L	Bradford and Winslow (1844: 167, 168)
Southern Maine	Levett	1624	A	L	Levett (1905: 263)
Hochelaga/Montreal	Cartier	1535	C	L	Cartier (1906*b*: 63)
Hudson Valley	Verrazzano	1524	C	N	Verrazzano (1910: 192–3)
	De Vries	1638	A	N	De Vries (1909: 205)
Albany (NY)	Kalm	1749	C	L	Kalm (1972: 343)
Niagara (NY)	Denonville	1687	C	L	Denonville (1850: 147)
An Onondaga village (NY)	Greenhalgh	1677	C	N	Stites (1904: 25)
Chemung (NY)	Burrowes	1779	C	N	Cook (1887: 44)
	Hubley	1779	C	L	Cook (1887: 154)
	Norris	1779	C	N	Cook (1887: 231)
	Shute	1779	C	N	Cook (1887: 273)

Table 5.3. *Continued*

Locale	Observer	Date	C/A[a]	L/N[b]	Source
Middletown (NY)	Burrowes	1779	C	N	Cook (1887: 44)
Yoghroonwago (NY)	Brodhead	1779	C	N	Cook (1887: 308)
Kanandaqua (NY)	Sullivan	1779	C	L	J. Sullivan (1887: 299–300)
	Livermore	1779	C	L	Cook (1887: 187)
	McKendry	1779	C	L	Cook (1887: 205)
Newton (NY)	Beatty	1779	C	L	Cook (1887: 27)
	Hubley	1779	C	L	Cook (1887: 156)
Chinesee (NY)	Sullivan	1779	C	L	J. Sullivan (1887: 301)
	Hubley	1779	C	L	Cook (1887: 163)
	Jenkins	1779	C	N	Cook (1887: 175)
	Livermore	1779	C	N	Cook (1887: 188)
Cayuga (NY)	Sullivan	1779	C	N	J. Sullivan (1887: 303)
Shanandaque (NY)	McKendry	1779	A	L	Cook (1887: 205)
Kannawaloholla (NY)	Hubley	1779	C	L	Cook (1887: 157)
Teaogo (NY)	McKendry	1779	C	N	Cook (1887: 205)
Lake Simcoe (ON)	Champlain	1615	C	L	Champlain (1907: 284)
Huronia	Sagard-Théodat	1624	C	L	Sagard-Théodat (1939: 104)
Lake Huron	Perrott	1642	C	L	Perrott (1911: 110)
Xenia (OH)	Withers	1780	C	N	Withers (1908: 306)
Springfield (OH)	Withers	1780	C	N	Withers (1908: 308)
Auglaize–Maumee Valley (OH)	Wayne	1794	C	N	Howe (1898: 545); Manypenny (1972: 84)

[a] C, area under cultivation at the time account was made; A, area had been cultivated but was abandoned at the time account was made.

[b] L, account is a qualitative assessment for which no measurements were provided; N, account is a quantitative assessment which contains linear or area measurements.

(1910: 192–3) noted that indigenous people living along the Hudson River not only practiced agriculture, but that their 'fields are from xxv to xxx leagues wide'. Just a few years later, the Gentleman of Elvas (1907: 206; see also 1993: 114), in describing the native agricultural landscape of the Arkansas side of the Mississippi River, said '[t]he Governor marched two days through the country of the Casqui . . . , the greater part of the way lying through fields'.

As was the case with individual fields and multiples of fields, some accounts of extensive agricultural landscapes referred not to crops in the field at the time the observation was made, but instead to areas farmed in times past and long since abandoned. Such accounts come mainly from later dates. To illustrate, during his travels through New Jersey in 1749 Kalm (1972: 245) 'found large tracts covered with young trees, only twenty, thirty, or forty years old: these tracts, I am told, the *Indians* formerly had their little plantations in'. This account is clearly qualitative, making no mention of the size of the formerly cultivated tract. Undoubtedly, the best example of a quantitative description of expansive abandoned agricultural land is provided by Bartram, who wrote in 1775 of the area near what is today Macon, Georgia:

Fig. 5.3. A 1591 woodcarving of the town of Pomeiock, in what is now Hyde County, North Carolina. Made by Theodore de Bry, it depicts multiples of large, nearly contiguous fields that cover an extensive area around the settlement. This illustration is a composite, based on three separate watercolor paintings made by Governor John White in 1585 (see Lorant 1946: 243, 200, 190, 191). (Provided courtesy of the Library of Congress, Neg. No. LC-USZ62-54015.)

On the east banks of the Oakmulge, this trading road runs nearly two miles through ancient Indian fields, which are called the Oakmulge fields: they are the rich low lands of the river. . . . Their old fields and planting land extend up and down the river, fifteen or twenty miles from this site (Bartram 1940: 68).

To be sure, some very sizable tracts of land are reported by early explorers to have been cultivated by native North Americans.

Not only did aboriginals have fields, often in multiples, and not infrequently such that vast expanses were cultivated, but in many cases the individual fields themselves were quite large (Fig. 5.3). In some cases it is difficult to determine if European explorers were describing large individual plots or multiple contiguous plots as one field. For example, in his early history of de Soto's expedition, Garsilaso de la Vega (1951: 169; see also 1993: 184) said that about Osachile 'they found great fields of corn, beans, and a type of squash'. Does 'great' refer to large, numerous, or some combination of both? In an account reminiscent of Schneider's previously discussed description of communal planting of several contiguous plots, Le Petit (1900: 137) said of the Natchez in 1730: 'Each year the people assemble to plant one vast field with Indian corn, beans, pumpkins, and melons, and then again they collect in the same way to gather the harvest'. Is this really a reference to 'one', single field?

Although it is difficult in some cases to differentiate between numerous fields, expansive cultivated tracts, and truly large individual fields, there do exist several accounts that indicate some native North American farmers at contact did indeed have very large rainfed fields (Table 5.4, Fig. 5.4). A number of accounts refer specifically to 'large fields' (e.g., J. Adair 1968: 406; Joutel 1906: 138), most commonly with 'maize' or 'corn' growing in them (e.g., Kalm 1972: 344; Smith 1910*b*: 205). In at least one case, however, it was reported that the Piro 'have large cotton fields' (Gallegos 1966: 83). Further verification for the utilization of large fields by aboriginal farmers comes in three forms, simultaneous reference to value and magnitude of fields, accounts of single large fields, and some actual field measurements.

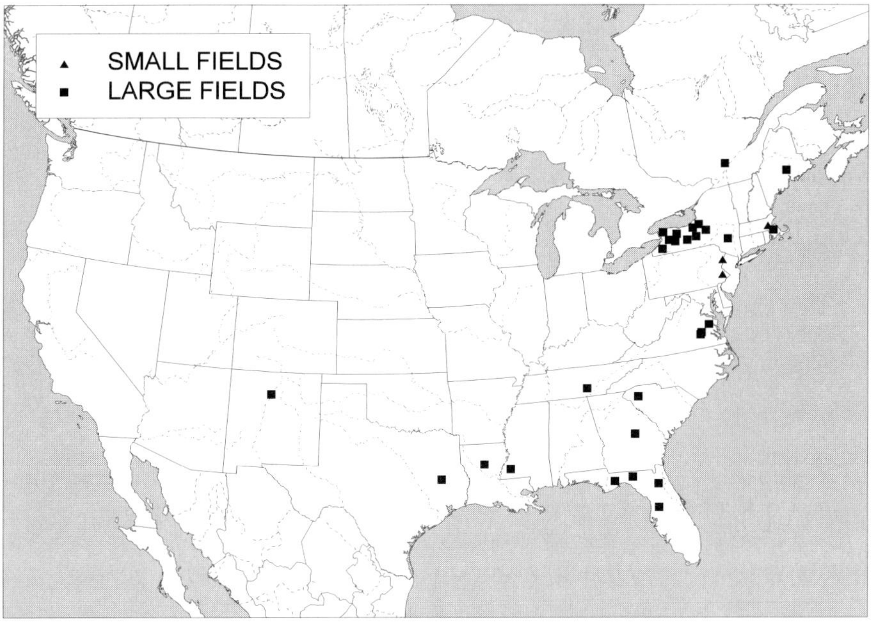

Fig. 5.4. Locations cited in the text and Table 5.4 for which there is documentary evidence of the sizes of agricultural fields early explorers actually saw.

Table 5.4. Locales for which there exist explorer's accounts of large plain rainfed fields

Locale	Observer	Date	C/A[a]	L/N[b]	Source
West Florida	Cabeza de Vaca	1528	C	L	Cabeza de Vaca (1946: 22, 25)
Osachile (FL)	de Soto	1543	C	L	Vega (1951: 169, 1993: 184)
Apalache (FL)	de Soto	1543	C	L	Vega (1951: 182, 1993: 194)
Georgia	Adair	1775	C	L	Adair (1968: 406)
W South Carolina	Adair	1775	A	N	Adair (1968: 225)
Jamestown (VA)	Archer	1607	C	N	Archer (1969: 93)
Virginia	Smith	1608	C	N	J. Smith (1910*c*: 363)
Florida	Romans	1775	A	L	Romans (1962: 330)
Chiaha (TN)	Elvas	1540	C	L	Elvas (1907: 180, 1993: 89)
Natchez (MS)	Le Petit	1730	C	L	Le Petit (1900: 137)
Red River (LA)	Freeman	1806	C	L	Flores (1984: 175)
Angelina River (TX)	Joutel	1687	C	L	Joutel (1906: 138)
A Piro pueblo (NM)	Gallegos	1581	C	L	Gallegos (1966: 83)
New England coast	Smith	1614	C	L	J. Smith (1910*b*: 205)
St. John's River (ME)	Giles	1689	C	L	Giles (1869: 14)
Hochelaga–Montreal	Cartier	1535	C	L	Cartier (1906*b*: 59)
Albany (NY)	Kalm	1749	C	L	Kalm (1972: 344)
Teaoga (NY)	Sullivan	1779	C	L	Cook (1887: 297)
	Dearborn	1779	C	N	Cook (1887: 70)
Koneghsaws (NY)	Sullivan	1779	C	L	Cook (1887: 300)
Chemung (NY)	Barton	1779	C	L	Cook (1887: 6, 7)
	Burrowes	1779	C	N	Cook (1887: 44)
	Dearborn	1779	C	N	Cook (1887: 71)
	Fellows	1779	C	N	Cook (1887: 87)
	Nukerck	1779	C	L	Cook (1887: 216)
Seneca Lake (NY)	Fellows	1779	CA	L	Cook (1887: 90)
Kanandaqua (NY)	Dearborn	1779	C	L	Cook (1887: 74)
	Fellows	1779	A	L	Cook (1887: 90)
	Hubley	1779	C	L	Cook (1887: 160)
	Norris	1779	C	L	Cook (1887: 234)
Middletown (NY)	Burrowes	1779	C	LN	Cook (1887: 44, 45)
Anyayea (NY)	Burrowes	1779	C	L	Cook (1887: 47)
	Dearborn	1779	C	L	Cook (1887: 74)
	Fellows	1779	A	L	Cook (1887: 90)
	Hubley	1779	CA	L	Cook (1887: 161)
Kannawaloholla (NY)	Dearborn	1779	C	L	Cook (1887: 73)
	Fellows	1779	C	L	Cook (1887: 89)
Swahyawanah (NY)	Dearborn	1779	C	L	Cook (1887: 77)
	Norris	1779	C	L	Cook (1887: 236)
Allegheny River (NY)	Livermore	1779	C	L	Cook (1887: 189)

[a] C, area under cultivation at the time account was made; A, area had been cultivated but was abandoned at the time account was made.
[b] L, account is a qualitative assessment for which no measurements were provided; N, account is a quantitative assessment which contains linear or area measurements.

In his 1535 description of Mount Royal, or Montreal, Cartier (1906*b*: 59) said 'we began to finde goodly and large fields, full of such corne . . .. In the midst of those fields is the citie of Hochelaga, placed neere, and as it were joyned to a great mountaine that is tilled round about'. Clearly, this reference leaves no doubt that Cartier saw numerous large fields. He even used the word 'fields' twice, and reports that they were widespread and of high quality. Additional evidence of large fields is found where only one such field was reported. In 1689, a captive of Maine natives, reported seeing 'a large cornfield' (Giles 1869: 14). Similarly, but commenting on earlier native agricultural practices, an 18th-century traveler through Florida claims that he 'saw a ſpacious old field' (Romans 1962: 330). Both of these accounts refer to large fields in the singular.

Finally, there exist two accounts that actually report field sizes, both come from early 17th-century Virginia. In a statement that was echoed if not plagiarized outright by Strachey (1953: 79) in reference to gardens (see Chapter 4), Smith (1910*c*: 363) said '[t]heir houses are in the midst of their fields . . .. Some 20 acres, some 40. some 100. some 200. some more, some lesse'. Archer (1969: 93) confirms the average field size by noting of aboriginal fields that '[t]he platt of grownd is bare without wood some 100. acres, where are set beans, wheate [maize], peaze, Tobacco, Gourdes, pompions'.

In contrast to reports of large fields, numerous fields in proximity to each other, and vast expanses of cultivated areas, stand a mere three explorers' accounts of 'small' fields. Not without consequence, these reports come from areas of the Northeast which were densely settled by indigenous people at the time of contact, underwent rapid population decline, and were then colonized quickly and intensively by settlers from Europe. These accounts, therefore, may well indicate only a small fraction of a much greater earlier picture. In discussing their 1620 travels beyond some lands 'where the Indians had formerly planted their corn', Bradford and Winslow (1844: 130–1), said they 'went on further and found new stubble, of which they had gotten corn this year . . . [p]assing thus a field or two, which were not great, we came to another, which had also been new gotten' (Bradford and Winslow 1844: 132–3). Two other accounts of diminutive fields cultivated by natives come from the late 1700s in the Delaware Valley. Near what is today Philadelphia Kalm (1972: 77) said they 'have little corn fields'. Further upriver near where Swedes had already colonized he said that '[t]he *Indians* had their little plantations of maize in many places' (Kalm 1972: 257, see also 245).

A largely untapped resource concerning native North American agriculture, as well as land use, and environmental change, are maps drawn by early European explorers. Until recently, those made by Champlain in the early 1600s, one of which is of Plymouth Harbor and clearly portrays numerous fields with crops, have been the most widely used. Unfortunately, however, they have been used less in an analytical vein (e.g., Mrozowski 1994: 51, 54, 55, figs 7 and 8), and more as illustrations on the pages of textbooks (e.g.,

R. Brown 1948: 14). More cartographic research remains to be done and one scholar is showing how it should be done.

In a recent paper, Gregory A. Waselkov (1997) used a number of early 18th-century maps of the Southeast to better understand native agriculture. Previously unpublished maps show Chickasaw fields scattered between villages (Waselkov 1997: 181, fig. 9.1), fields along rivers (Waselkov 1997: 182, fig. 9.2, 183, fig. 9.3), and even abandoned floodplain fields (Waselkov 1997: 184, fig. 9.4, 189, fig. 9.8) in Alabama. Hopefully, his work will prove to be seminal. For now, its immediate significance lies in substantiating the textual claims made by others.

Descriptions of agricultural fields by explorers, be they textual or cartographic, clearly provide much insight into aboriginal farming *c.* AD 1500. They do not, of course, provide a complete picture, especially in regard to specific rainfed practices. Fortunately, early European travelers also recorded information on field preparation, planting, cultivation, and harvesting.

At least thirty accounts of native North Americans preparing their fields prior to planting exist for North America (Table 5.5). The locales for which there are such descriptions include the Gulf coast, the upper Rio Grande Valley, the Tidewater area of Virginia, New England, and southern Canada (Fig. 5.5). Descriptions of preparation include mention of burning weeds which have overgrown the field since the previous harvest, tilling the field to turn-under the weeds and break-up, thus aeriating the soil, or both. To illustrate, in his description of fields along the Angelina River in east Texas, Joutel (1906: 138) saw 'Weeds which had been some Time before burnt by the Natives'. Accounts of tillage only are the most widespread, coming from every corner of the continent. Despite their locational diversity, however, they are remarkably similar. For example, in his history of Espejo's expedition past Rio Grande pueblos in 1584, Obregón (1928: 322) noted that '[t]hey till the soil with hoes like the Mexicans'. More than 3,000 km away and fifty years later Williams (1973: 170, 171) said that natives in southern New England 'break up their fields . . . [with] Howes of shells and Wood'. One of the earliest accounts of tillage provides similar information, but in much greater detail and comes from the far Southeast. In a narrative accompanying some paintings he made during the 1560s, Jacques Le Moyne (1968: 44), who traveled with René Laudonniére, noted: 'The Indians cultivate the earth diligently; and the men know how to make a kind of hoe from fishes' bones, which they fit to wooden handles, and with these they prepare the land well enough, as the soil is light' (Fig. 5.6).

Preparation by burning and tilling were not mutually exclusive practices, although tillage by itself was most common. In many cases, native farmers both burned and tilled, and several early Europeans made note of it. For instance, in his 1699 report of farming practices among the Pascagoula of coastal Mississippi, the French chronicler Pénicaut made the following statement:

Table 5.5. Locales for which there exist accounts of land preparation prior to planting

Locale	B/T[a]	Observer	Date	Source
St. Augustine (FL)	T	Le Moyne	1564	Le Moyne (1968: 44)
	BT	Laudonniére	1564	Laudonniére (1975: 15)
	T	De Bry	1590	Bennett (1968: 45)
Apalache (FL)	BT	Diáz Vara Calderón	1675	Diáz Vara Calderón (1936: 13)
Florida	T	Romans	1775	Romans (1962: 119)
Pascagoula (MS)	BT	Pénicaut	1699	Swanton (1946: 310)
Natchez (MS)	BT	Le Page du Pratz	1758	Swanton (1946: 310)
Angelina River (TX)	B	Joutel	1687	Joutel (1906: 138)
East Texas	T	Joutel	1687	Joutel (1906: 142–3)
	T	Espinosa	1722	Espinosa (1927: 156)
A Tigua pueblo (NM)	T	Espejo	1584	Obregón (1928: 322)
	T	Ortega	1601	Valverde (1953*a*: 660)
Rio Grande Pueblo	T	Gallegos	1581	Gallegos (1966: 102)
Virginia	T	Barlow	1584	Barlow (1906: 234)
	T	Heriot	1587	Heriot (1812: 596)
	T	Smith	1608	J. Smith (1910*c*: 357)
	T	Smith	1612	Strachey (1953: 118)
	T	Spelman	1613	Spelman (1910: cxi)
Delaware Valley	BT	Zeisberger	1776	Mahr (1955: 211, 212, 233)
New England coast	T	Champlain	1605	Champlain (1907: 62)
Nauset Harbor (MA)	BT	Champlain	1605	Champlain (1907: 71)
Rhode Island	T	Williams	1643	R. Williams (1973: 170, 171)
	T	Winthrop	1660s	Mood (1937: 127)
Hudson Valley	BT	Van der Donck	1656	Van der Donck (1841: 58)
St. Lawrence River (QB)	T	Cartier	1535	Cartier (1906*b*: 68)
	BT	Champlain?	1609	Lescarbot (1977: 274)
	T	Champlain	1613	Champlain (1907: 244)
Ontario	BT	Lafitau	1717	Lafitau (1977: 54)
Huronia	T	Sagard-Théodat	1623	Sagard-Théodat (1964: 104)
	BT	Charlevoix	1761	Charlevoix (1923: 111)
Detroit River (MI)	T	Cadillac	1721	Cadillac (1947: 12)
Illinois	T	De Gannes	1721	De Gannes (1934: 340) Liette (1947: 122)

[a] B, burning; T, tillage.

> The savages have flat, bent sticks, which they use to hoe the ground, for they do not know how to work it as is done in France. They scratch the soil with these crooked sticks and uproot with them the canes and weeds which they leave on the earth in the sun during fifteen days or a month. Then they set fire to them (Swanton 1946: 310).

Similarly, Lescarbot (1977: 274) said of the Armouchiquois (southern New England groups) and others in 1606–7 that 'All thefe people doe till their land with a woodden picke-axe, weed out the weeds and burne them'.

Tillage complete, aboriginal farmers at contact then planted their fields.

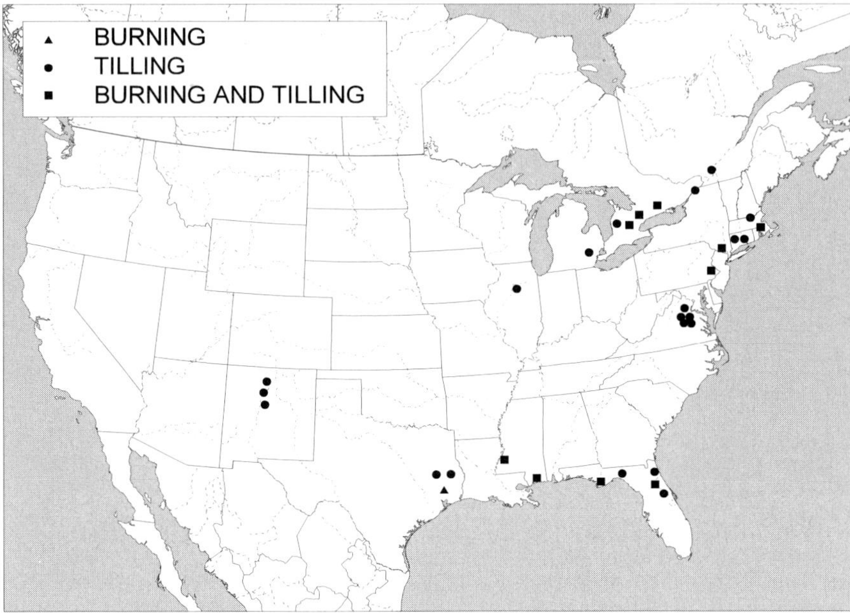

Fig. 5.5. Locations cited in the text and Table 5.5 for which there is documentary evidence describing how indigenous farmers prepared fields for planting.

Although the process of planting may seem straightforward, it was quite complex. Numerous differences, some obvious, others subtle, can be identified from across the continent. Also, several nuances did exist. Planting was normally done with a digging stick, or dibble (Fig. 5.6). Le Moyne (1968: 44) reported that at St. Augustine native farmers went 'first with a stick, and make holes'. Pénicaut (Swanton 1946: 310) noted of the Pascaugoula that 'they have a stick as large as the arm, pointed at one end, with which they make holes in the earth 3 feet apart'. In at least one case, however, it is documented that native farmers sometimes had to exert much energy in planting and get down on their hands and knees. Within three leagues of what is today Quebec City one Jesuit missionary, Sebastian Rasles (1847: 27) noted in 1723 that in order to plant Indians had 'to make with their fingers, or with a small stick, different holes in the earth'.

The most common and widely cultivated crop was, of course, corn or maize (*Zea mays*). The cultigen has a bewildering array of varieties, each with different growth characteristics and requirements. There exist varieties suited for nearly every environmental condition imaginable (see, e.g., Mangelsdorf 1974; Weatherwax 1954; Wellhausen *et al.* 1951). Perhaps because the early European explorers in North America did not understand anything about maize when they first encountered it, its obvious economic importance to native peoples made them take special notice. At least a few of these astute

Fig. 5.6. Timucua tillage and planting as depicted in a 1591 woodcarving made by Theodore de Bry. This illustration, which also shows the communal nature of agricultural activities, was based on an earlier painting by Jacques Le Moyne (1968: 44–5), which has long been lost (see Lorant 1946: 77). This illustration has a checkered history. It is somewhat inaccurate as it shows European-style plow marks, most likely an addition by the woodcarver (Axtell 1992: 210) who had a reputation for taking certain liberties and embellishing his work almost fancifully (see Figs 4.3 and 5.3). It was also used inappropriately by Joseph François La Fitau in 1724 to illustrate Huron agriculture (after W. Fowler 1960; W. Sturtevant 1968; Trigger 1976: 35). (Provided courtesy of the Library of Congress, Neg. No. LC-USZ62-31869.)

observers noted different varieties, which were planted at different times. For example, in writing of the Asinai of eastern Texas in 1691 the Franciscan missionary Francisco Casañas de Jesús Mariá (1927: 211) said: 'Among the seed which the Indians plant at the proper season, is corn of two kinds, which they plant in abundance. One kind matures in a month and a half and the other in three months'. Providing more detail, and including a third variety is a report from Virginia in 1584 by M. Arthur Barlow (1906: 234). He noted that the natives' 'corne, which is very white, faire and well tasted, and groweth three times in five months: in May they sow, in July they reape, in June they sow, in August they reape; in July they sow, in September they reape'.

Some accounts suggest that the observers did not actually see maize being planted at different times but they clearly indicate fields in different stages of growth. Near Nauset Harbor in Massachusetts, for example, Champlain (1907: 71) remarked that in 1605 he 'entered a field planted with Indian corn . . . [which] was in flower, and five and a half feet high. There was some less advanced, which they plant later'. Along the banks of the Mississippi River, near the confluence of the Arkansas River, the Jesuit missionary James Marquette (1903: 51) commented in 1673 that native farmers had 'corn in abundance, which they sow at all seasons; we saw some ripe; more just sprouting, and more just in the ear, so that they sow three crops a year'.

The raising of more than one crop per year was clearly a common and widespread practice, but it is perplexing. The previous accounts all indicate that something other than true multiple cropping—cultivating two or more separate crops in the same field each year, one being planted after the other is harvested (Wilken 1987: 253)—was being done. However, other accounts such as William Hilton's (Swanton 1922: 63), which says only that Creek farmers along the Georgia coast in 1664 'have two or three crops of Corn a year', leave the issue open for debate. Although it lacks detail, this could possibly be a reference to true multiple cropping. Furthermore, the previously cited report of Barlow (1906: 234), for all its detail, has room for speculation. The field from which the first crop was harvested in July, could have been planted in the third crop later that same month, while the second crop was planted in a separate field.

There exists only four reports that explicitly describe true multiple cropping; one of northeastern Florida in 1564, one for Virginia in 1587, and one for eastern Texas in 1722, and one for Kansas in 1601. In discussing the first area Laudonniére (1975: 15) said:

> The Indians sow their corn twice a year, in March and June, replanting the same soil. This corn, from the time it is sown until the time it is harvested, is in the ground about three months, and the land is allowed to rest during the other six months of the year.

Writing of Virginia, Heriot (1812: 597) noted that the natives:

> raifed out of one and the felf-fame ground two harvefts or of-comes: for they fow or fet, and may at any time when they think good, from the midft of March until the end of June: fo that they alfo fet when they have eaten of their firft crop. In fome places of the country notwithftanding they have two harvefts, as we have heard, out of one and the fame ground.

As for double cropping in Texas, Isidro Felis de Espinosa (1927: 156), a Franciscan missionary, said in 1722:

> They plant two crops each year, the first at the end of April. . . . This crop is gathered at the end of May . . .. Upon the same ground, after clearing it anew, they plant small but covered with grain [*sic*]. This crop is gathered at the end of July.

The growth periods seem exceptionally short, but the reference to replanting the same field is certainly clear. Finally, in his description of one place along the Arkansas River in what is today Kansas, Oñate (1953*b*: 755) said he was:

> surrounded everywhere by Indian cornfields . . .. The land was so fertile that even though the corn had just been harvested, there was a second crop about six inches high, without other cultivation or preparation of the soil than pulling up the grass and

making some holes in which the corn was planted . . .. These were not irrigated fields, but depended on seasonal rains.

Raising more than one crop per year, be it by means of true multiple cropping or by sequentially planting different fields at different times, is a function, in large part, of the length of the growing or frost-free season. It should come as no surprise, therefore, that all of the explorers' accounts which mention two or three crops all come from southern and coastal localities. There exists not so much as a hint of anything other than a single crop from upstate New York and places further north (Table 5.6).

That latitude, and hence the length of the growing season, had an affect on cropping practices seems to be substantiated by the manner in which planting occurred. In fact, the revealing evidence involves the number of seeds planted in each hole. The various early relations indicate that as few as two (Laudonniére 1975: 15) and as many as ten (Champlain 1907: 327) corn seeds or kernels were planted together, the average being approximately five. The South had remarkable variability within a limited region. However, as one moves northward three discrete areas stand out in regard to the number of seeds planted per hole. Accounts from the Tidewater area of Virginia reveal that three to six seeds were planted, with four to five being both most common and the average. From the Delaware Valley through New England the range is three to seven, with five to six being most common, and five being average. Finally, from Huronia down the St. Lawrence Valley between eight and ten seeds were planted in each hole with nine being typical and average (Table 5.6, Fig. 5.7). The exact reason for this inverse relationship between seed number and length of growing season is unclear. Indeed, this is the first time the phenomenon has been identified. As a tentative working hypothesis, it is suggested that germination was less certain in colder climes and that an increase in the number of seeds planted increased the chances of every hole sprouting at least one plant.

In addition to multiple cropping, documentary relations indicate that native North Americans at the time of first European contact also practiced intercropping—the raising of two or more species or types of crops interspersed in one field simultaneously (Innis 1980: 2). At least half of the early accounts of planting mention intercropping, most typically involving maize, beans, and squash, and they involve virtually every corner of the continent. All of the accounts of intercropping are quite similar, but two, one by an anonymous chronicler of activities in Virginia in 1610, and another by John Winthrop, Jr., the governor of the Connecticut colony in 1662, stand out as exceptionally illustrative:

They use to put their wheat into the ground, fiue cornes in one spit of earth, and two beans with them: which wheat cornes multiplying into diuers stalks, grow up twelve, or fourteene foote high: yeelding some foure, fiue or six ears, on euery stalke; and in euery eare, some fiue hundred, some six hundred, some

Table 5.6. Locales for which there exist accounts of planting of crops dependent on direct rainfall

Locale	Communal	Intercrop	Seeds/hole	Crops/year[a]	Observer	Date	Source
St. Augustine (FL)	yes	yes	–	1	Le Moyne	1564	Le Moyne (1968: 44)
	–	–	2	2	Laudonniére	1564	Laudonniére (1975: 15)
Apalache (FL)	yes	–	–	1	Díaz Vara Calderón	1675	Díaz Vara Calderón (1936: 13)
Florida	–	–	4–5	–	Romans	1775	Romans (1962: 119)
Georgia coast	–	–	–	2 or 3	Hilton	1664	Swanton (1922: 63)
Cucowilla (GA)	yes	–	–	–	Bartram	1773	Bartram (1940: 170)
Southeast	–	yes	6–8	2S[a]	Catesby	1731	Swanton (1946: 308)
Pascagoula (MS)	–	yes	7–8	–	Pénicaut	1699	Swanton (1946: 310)
Natchez (MS)	–	–	–	3–4	Membré	1681	Membré (1903*b*: 185)
	yes	–	–	–	Le Petit	1730	Le Petit (1900: 137)
MS/AR rivers	–	–	–	3S	Marquette	1673	Marquette (1903: 51)
MS/AR/LA	–	–	–	3	D'Ablon	1674	D'Ablon (1899*b*: 99)
Red River (LA)	–	–	–	3	Harpé	1719	Harpé (1958: 254)
Arkansas River (KS)	–	–	–	2	Oñate	1601	Oñate (1953*b*: 755)
East Texas	yes	–	–	2S	Casañas	1691	Casañas (1927: 211)
	yes	–	–	2S	Espinosa	1722	Espinosa (1927: 156)
Sierra Madre (MX)	–	yes	4–5	1	Falcón Mariano	1777	Falcón Mariano (1979: 108–10)
Virginia	–	–	–	3S	Barlow	1584	Barlow (1906: 234)
	–	yes	4	2	Heriot	1587	Heriot (1812: 596, 597)
	–	yes	4	–	Smith	1608	J. Smith (1910*c*: 357, 359)
	–	yes	5	–	Anonymous	1610	Anonymous (1844: 12)
	–	yes	3–5	–	Smith?	1612	Strachey (1953: 118)
	–	yes	4–5	–	Spelman	1613	Spelman (1910: cxi)
	–	–	4–5	2S	Harris	1600s	G. Harris (1812: 242)
	–	yes	3–5	–	Beverly	1705	Beverly (1947: 144)
	–	yes	5–6	–	Adair	1775	Adair (1968: 409)

Staunton River (VA)	–	–	–	3	Lederer	1670	Lederer (1958: 27)
Delaware Valley	–	–	6–7	–	Lindestrom	1655	Lindestrom (1925: 179)
	–	yes	4–5	–	Zeisberger	1776	Mahr (1955: 214)
New England coast	–	yes	3–4	1	Champlain	1605	Champlain (1907: 62)
Nauset Harbor (MA)	–	–	–	2S	Champlain	1605	Champlain (1907: 71)
Plymouth (MA)	–	–	5–6	–	Rasieres	1628	Cronon (1983: 43)
Rhode Island	yes	–	–	–	Williams	1643	R. Williams (1973: 170)
Connecticut	–	yes	4–5	–	Winthrop	1660s	Mood (1937: 127, 128)
Hudson Valley	–	yes	–	–	De Vries	1655	De Vries (1909: 219)
	–	yes	5–6	–	Van der Donck	1656	Van der Donck (1841: 158)
St. Lawrence River	–	yes	4	–	Champlain?	1609	Lescarbot (1977: 274)
	–	–	10	–	Champlain	1616	Champlain (1907: 327)
	–	–	8–9	–	Rasles	1723	Rasles (1847: 27)
Ontario	–	yes	9	–	Lafitau	1717	Lafitau (1977: 54, 55)
Huronia	–	–	9–10	–	Sagard-Théodat	1623	Sagard-Théodat (1964: 103)
	yes	yes	–	–	Charlevoix	1761	Charlevoix (1923: 112)
Illinois	yes	–	–	–	De Gannes	1721	De Gannes (1934: 339–40) Liette (1947: 122)

[a] S, crops were planted in separate fields at different times within the same growing season rather than the second crop being planted in the same field after the first crop had been harvested.

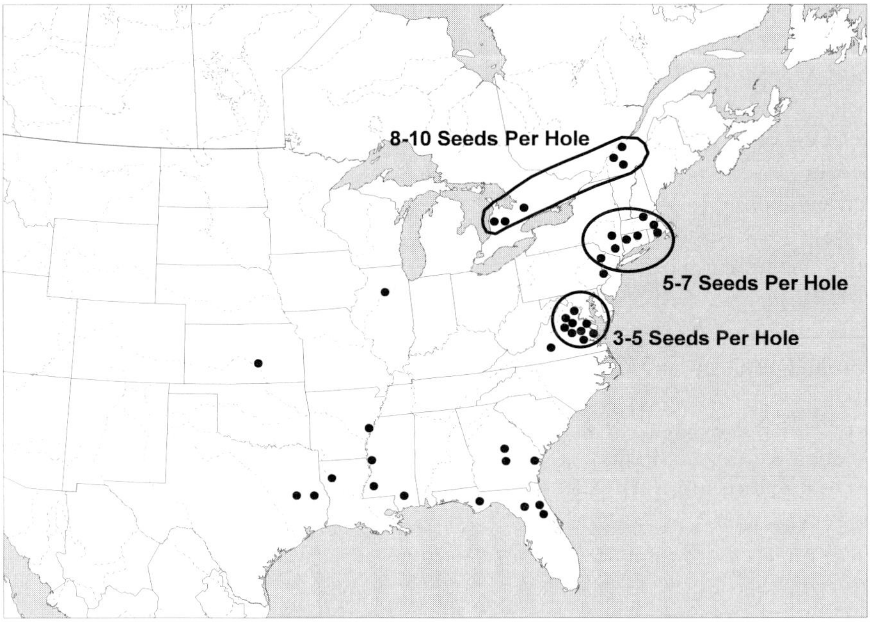

Fig. 5.7. Locations cited in the text and Table 5.6 for which there is documentary evidence describing how aboriginal Americans actually planted crops.

seauen hundred cornes: the two beenes, runne upon the stalkes of the wheat, as our garden pease upon stickes, which multiplie to a wonderous increase (Anonymous 1844: 12).

The Indians . . . will plant a kind of Beans with the Corne (they are like those here called French Beans or turky Beans) and in the Vacant places . . . , they will plant Squashes and pumpions, loading the Ground with as much as it will beare; The Stalks of the Corn serveing in stead of poles for the Bears to Climb up, which otherwise must have poles to hand upon (Mood 1937: 128).

Grown together, maize, beans, and squash are advantageous in that they promote each other's growth (Harwood 1979: 86). By growing earliest, maize stalks facilitate the later climbing of beans—as indicated clearly in the above accounts. In turn, the beans are beneficial to the maintenance of soil fertility. Maize is very hard on soil nitrogen, it depletes it quickly in the course of promoting rapid growth (Janick *et al.* 1974: 314). Beans, as is the case with all legumes, support colonies of essential nitrogen-fixing bacteria on their roots (Thorne 1979: 93–4). These bacterial colonies replace the nitrogen removed by maize. The squashes complete the complex with their large and numerous leaves that have allelopathic properties. The leaves very effectively cover the ground thereby reducing the soil erosion by dissipating the impact of raindrops. Shade created by the leaves reduces soil moisture evaporation. It also inhibits weed growth due to minimized photosynthesis at ground level

(Gliessman 1984: 166). Weeds are further reduced by toxic chemicals produced by squash leaves that are washed into the soil by rainfall (Chacón and Gliessman 1982: 1–11). In sum, these three cultigens—maize, beans, and squash—have qualities that make them mutually symbiotic when intercropped.

The process of planting intercrops is not, as one should expect, an easy task. Indeed, it may well be more than three times as difficult to plant maize, beans, and corn together than it would be to plant these crops individually. Given different germination periods, and rates of growth, timing is all important. It is certainly more than can be done by one person. To compensate for labor difficulties native North Americans helped each other plant. At least nine early Europeans were sufficiently impressed by the activity to document it (see Table 5.6), but from only the Gulf Coastal Plain, and the far North (see Fig. 5.7). Some accounts, such as that by Father Le Petit (1900: 137) which was used previously in the context of large fields, were simple statements of people assembling to plant. Some were more detailed. For example, at St. Augustine some people made holes with digging sticks, as mentioned earlier, 'in which the others place the beans, or grains of maize' (Le Moyne 1968: 44). According to Gabriel Díaz Vara Calderon (1936: 13), the Bishop of Cuba who visited Florida in 1675, it was a division of labor based on gender, 'the man goes along opening the trench [*sic*], the woman follows sowing' (see Fig. 5.6). Regardless of who did what, planting involved more than one person and a report from eastern Texas in 1691 provides the best insight into the communal nature of the activity:

> At planting time, they come together and plant whatever each one has to plant, according to the size of the family—beginning first at the home of the grand *xinesí*. . . . Next, they plant the corn and other crops for the *caddí*. Then they work for the other officials and the old men. In this way they continue working from the highest to the humblest until each has planted what he needs for the year (Casañas 1927: 217).

Once the crops were planted they could not be ignored or left untended, lest birds and other predators would eat the crop. In order to provide for crop protection native North American farmers built various structures in or near their fields. These structures were always made of wood and other perishable materials so they left precious little evidence recognizable to prehistorians today. In one case, for example, archaeological investigations uncovered postmolds arranged in a small rectangular pattern in what was thought to be a prehistoric field. These features were identified as remains of a watcher's stand (William I. Woods, personal communication 1996). Writing about the Piro of one Rio Grande pueblo in 1582, Espejo (1966: 220) said 'in each planted field the worker has a shelter, supported by four pillars, where food is carried to him at noon and he spends the siesta; for usually the workers stay in their fields from morning until night'. In his history of Espejo's journey Obregón (1928: 325) reported that at Acoma pueblo in 1584 'each Indian has a shack in his field'. Exactly what constituted a 'shack' is open for debate, but it is likely, on the basis of Espejo's own account, that it was a reference to an arbor, a structure with a roof but no walls.

Perhaps the best example of a field structure used by farmers for protecting their crops was provided in the previously mentioned painting of Secota (Fig. 4.2) by White and woodcarvings (Figs 4.3 and 5.3) by de Bry. In the accompanying text Heriot noted: 'In the cornfields they set up a little hut on a scaffold, where a watchman is stationed (F). He makes a continual noise to keep off birds and beasts which would otherwise soon devour all the corn' (Lorant 1946: 264). With greater brevity Williams (1973: 163) said only that Rhode Island natives in the 17th century 'put up little watch-houses in the middle of their fields, in which they, or their biggest children lodge, and earely in the morning prevent the Birds, &c.', and Adair (1968: 408) said of indigenous farmers in Virginia during the 18th century that 'each owner raifes an high fsaffold, to over-look this favourite part of their vegetable poffeffions. . . . This ufually is the duty of the old women'.

In addition to building shacks, arbors, scaffolds, or some combination of each and then manning them, at least one group devised a system of controlling pests that did not involve constant human attention. In the late 17th century, John Banister reported something most ingenious developed by the Powhatan of what is today North Carolina. He noted that:

> [t]hey keep the crows from their corn, by tying strings of bark from pole around the field, which is all the fence the outward Indians make where there are no hoggs: they also hang Gourds on the top of poles, in which build small birds, of the swallow or Marsh kind, that . . . will not let them rest near the place of their aboard (Rountree 1989: 49).

Although fields needed constant tending in order to prevent predation, there were at least two other tasks that needed to be done in order to ensure the production of a good harvest, weeding and mounding. The previously discussed squash leaves which produce ground shade and toxins that inhibit weed growth, do so only once the squash has reached a rather mature stage. Between planting and that time, however, weeds can be a real problem and, therefore, have to be removed. There exist a few reports (Table 5.7), mainly from the Southwest (Fig. 5.8). These tend to be very simple. For example, in 1601 one Marcelo de Espinosa testified at the hearings recorded by Francisco de Valverde (1953*a*: 641) that farmers at San Gabriel Pueblo 'plant, weed, and cultivate the land'. Weeding seems to have been done with a hoe not unlike, if indeed not the very same hoe as, that used during field preparation. To wit, Espejo's claim that farmers at a Tigua pueblo 'cultivate the land in the rainy season, using hoes like the Mexicans' (Obregón 1928: 323) is nearly identical to his earlier discussed description of tillage prior to planting.

Hoes were not only used in tillage and weeding, but they were also used to build mounds around individual corn plants. Maize, being a tall plant with a

Table 5.7. Locales for which there is documentary evidence of weeding, mounding, or both activities

Locale	Observation[a]	Observer	Date	Source
Southeast	B	Catesby	1731	Swanton (1946: 308)
East Texas	M	Espinosa	1722	Espinosa (1927: 156)
Rio Grande pueblos (NM)	W	Gallegos	1581	Gallegos (1966: 102)
A Tigua pueblo (NM)	W	Espejo	1584	Obregón (1928: 323)
San Gabriel Pueblo (NM)	W	Espinosa	1601	Valverde (1953*a*: 641)
Sierra Madre (MX)	W	Falcón Mariano	1777	Falcón Mariano (1979: 109)
Georgia/Carolinas	M	Adair	1775	Adair (1968: 409)
Virginia	M	Heriot	1587	Heriot (1812: 596)
	B	Smith	1608	J. Smith (1910*c*: 358)
	B	Smith?	1612	Strachey (1953: 114, 118–19)
	B	Harris	1600s	Harris (1812: 242)
	B	Beverly	1705	Beverly (1947: 144)
Delaware Valley	M	Lindestrom	1655	Lindestrom (1925: 179)
	B	Zeisberger	1776	Mahr (1955: 211, 216)
New England coast	B	Champlain	1605	Champlain (1907: 62)
Plymouth (MA)	M	Rasieres	1628	Cronon (1983: 43)
	W	Anon.	1700s	Anon. (1804: 101)
New England	B	Wood	1634	W. Wood (1967: 79, 106)
	B	Winthrop	1660s	Mood (1937: 127)
Rhode Island	B	Williams	1643	R. Williams (1973: 170, 171)
Hudson Valley	B	Van der Donck	1656	Van der Donck (1841: 158)
St. John's River (ME)	B	Giles	1689	Giles (1869: 21)
St Lawrence River (QB)	M	Champlain?	1609	Lescarbot (1977: 274)
	M	Champlain	1616	Champlain (1907: 95)
Iroquois (ON)	B	Lafitau	1717	Lafitau (1977: 54, 55)
Huronia	W	Sagard-Théodat	1623	Sagard-Théodat (1964: 104)
Illinois	M	De Gannes	1721	De Gannes (1934: 340) Liette (1947:122)

[a] W, weeding; M, mounding; B, both weeding and mounding.

remarkably shallow and not so extensive root system needs to be propped up, otherwise it will blow over quite easily. Being a true domesticate—dependent on humans for its propagation and survival (P. Smith 1976: 10)—maize has evolved such that it has aerial and support roots that grow out of the sides of the stalk a few centimeters above the normal ground surface (D. Parsons 1981: 13). Once the plant is sufficiently tall that these roots start to form, earth must be piled up around them, creating, in effect, a buttress (Fig. 5.9). This process is known as mounding or hilling (Weatherwax 1954: 68–72; Wilken 1987: 136–44). There exist a handful of accounts of mounding from all over the eastern half of the continent (Amsden 1929: 16, fig. 3.3; Pargellis 1959:

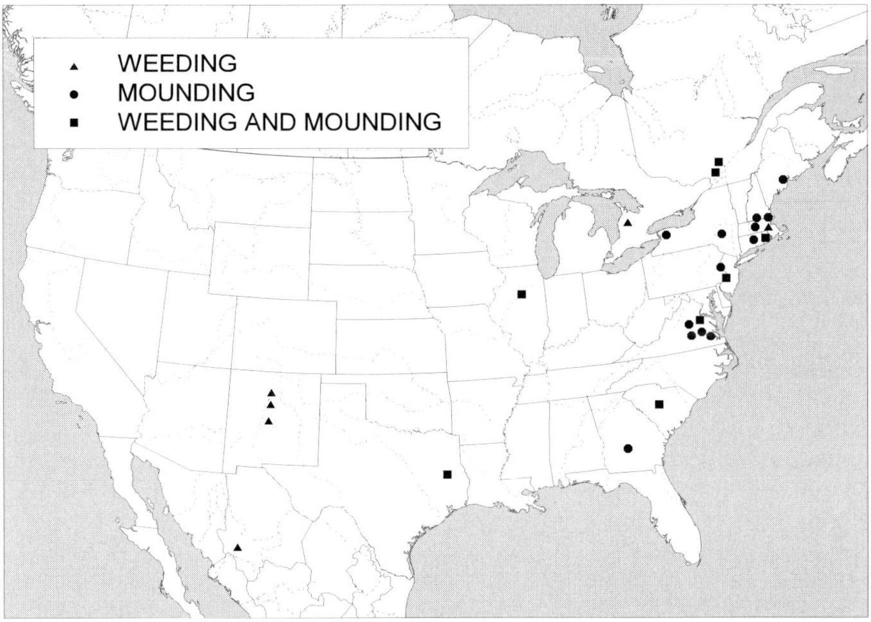

Fig. 5.8. Locations cited in the text and Table 5.7 for which there is documentary evidence describing how natives weeded, mounded, or both.

Fig. 5.9. Recently mounded or hilled maize plants in northern Mexico. (Photograph by author.)

239). Some are quite simple and refer only to either a single 'hill' (e.g., Espinosa 1927: 156) or multiples of 'hills' (e.g., Lindestrom 1925: 179). Others are a little more complex, saying things such as 'they hill up their corn' (De Gannes 1934: 340).

Some explorers claimed that crops were planted in hills; that is the hills were made prior to planting. Isaac de Rasiers, a Dutch traveler in New England during the late 1620s, for instance, wrote that the natives 'make heaps like molehills, each about two and a half feet from the other, which they sow or plant in April with maize, in each five or six grains' (Cronon 1983: 43). It is possible that chroniclers who claimed crops were planted after the mounds were formed inferred process from pattern. They may have seen crops growing in mounds rather than actually having witnessed the planting process. It is also possible that in some cases the native farmers did not level their fields during tillage each year, but instead left the mounds more or less intact and replanted them. There is at least one very good reference to this having been done in Huronia in 1632:

> [E]very year they sow their corn thus in the same holes and spots, which they freshen with their little wooden spade, shaped like an ear with a handle at the end. The rest of the land is not tilled, but only cleansed of noxious weeds, so that it seems as if it were all paths, so careful are they to keep it quite clean (Sagard-Théodat 1939: 104).

Regardless of the extent to which indigenous farmers have planted in mounds, the origin of mounding and the genesis of mounds seem to be related to weeding (Linton 1924: 347). There are numerous cases from North America where aboriginal people were observed by Europeans to be mounding their maize after it sprouted. And in these cases, it was almost always associated with weeding. Two passages serve as excellent examples. In Virginia during the early 1600s it was noted that after the seeds are 'covered with earth: they keep it weeding from time to time, and as the ſtalk grows high they keep the mould about it like the hillocks in a hop-garden' (G. Harris 1812: 242). Later during that century Winthrop reported that in New England:

> [a]fter the Corne is growne up, the length of a hand, it wilbe time to weed about it, which is done with a broad how, which cuts up the Weeds, and looseneth the Earth, and this Labour is so often performed as the Weeds do grow up in any Quantity.
>
> When the stalke beginneth to grow high, they draw (at the second weeding) a little Earth about it, and afterwards, as it groweth higher, and puteth forth the Eare, they draw so much Earth about these Stalkes, they maketh a little hill, like hopp hills, using as the same manner, as they do hopp grounds with broad Howes (Mood 1937: 127).

The references to hop hills in both accounts is interesting in two respects. First, it shows the parallels explorers were drawing from activities with which they were familiar in Europe. Second, it suggests that later writers might well have read and, hence were influenced by earlier writers. The best example of this comes from Virginia and involves Smith and Strachey (Rountree 1989: 4). In 1608, Smith (1910*c*: 358) said 'Their women and children do continually keepe it with weeding, and when it is growne middle high, they hill it about like a hop-yard'. Writing his history just a few years later, Strachey (1953: 118–19) said 'their women and children doe contynually keepe the grownd with weeding, and when the Corne is growne middle highe, they hill yt about like a Hoppeyard'.

There is some evidence (e.g., J. Adair 1968: 409) that corn hills were arranged in rows, and suggestions (Waselkov 1977: 514; Woods 1987: 284) that they were connected together thereby forming ridges. There is also at least one document that hints at the possibility that the Névome of Sonora planted in ridges, *à la* Fig. 5.6 (Pennington 1980: 18). These, however, are far overshadowed by numerous accounts of mounding into individual hills.

In sum, there exist numerous, informative documentary accounts of indigenous rainfed agriculture on plain fields in North America *c.* AD 1492. Together, as a corpus, they provide a wealth of information that shows that this form of agriculture was neither simple nor straightforward. Diversity, complexity, and subtlety are most apparent. Rainfed agriculture involving plain fields was widespread and common, especially in the eastern, humid, and wooded half of the continent. Some explorers did report seeing it in the arid Southwest, but there it was less important than other forms of agriculture (see Part IV). In the Eastern Woodlands, by contrast, it was the dominant and most important form of agriculture.

Ethnographic Evidence

Understanding of native rainfed agriculture at contact is enhanced only slightly by ethnographic studies undertaken in the 19th and 20th centuries. In large part this is the result of massive population declines, and the increasing spread of settlers of European ancestry who took over lands once worked by aboriginal farmers. By AD 1800 there were few natives left in the Eastern Woodlands and those who survived were not practicing agriculture on the same scale or level of intensity as their ancestors. As a result of this, combined with the large number of highly detailed early explorers' accounts of indigenous cultivation practices, recent writers have been able to portray rather accurately what was going on in terms of agriculture in the East largely by means of documentary sources. Indeed, there exists a very large number of historical ethnographies written of aboriginal North Americans which at least mention rainfed agriculture and use as evidence one or more early descriptions (Table 5.8). These range from detailed assessments (e.g., Swanton 1946) and brief overviews of specific peoples or regions (e.g., Snow 1978*a*,*b*), to general, continental-wide syntheses (e.g., C. Thomas 1903) and college-level textbooks (e.g., R. Brown 1948), to local histories (e.g., Temple

Table 5.8. Historical ethnographies of native North Americans which mention or discuss rainfed agriculture and are based on early European explorers' accounts

Locale	People	Source
North America	numerous groups	Palmer (1929)
		Driver and Massey (1957)
		Sauer (1971, 1980)
Eastern Woodlands	numerous groups	Carr (1893)
		C. Thomas (1894, 1903)
		R. Brown (1948)
		Wissler (1922)
Southeast	numerous groups	Swanton (1946)
		Hudson (1976)
		T. Silver (1990)
NW Florida	Seminole	Reynolds (1957)
	Apalachee	Hann (1988)
Carolinas	Catawba	Merrell (1982)
N Georgia/E Tennessee	Cherokee	Thomas (1961)
Alabama	Creek	Swanton (1922)
Alabama/Mississippi/Tennessee	Chickasaw	Malone (1922)
		Gibson (1971)
S Alabama–Mississippi	Choctaw	R. White (1983)
Louisiana	Natchez	Williamson (1940)
East Texas	Caddo	Swanton (1942)
		Newcomb (1961)
		Perttula (1992)
	Hasinai	Griffith (1954)
Oklahoma	Wichita	Newcomb (1961, 1976)
		Newcomb and Field (1967)
Rio Grande Valley (NM)	Puebloan	Bandelier (1937)
Tidewater (NC/VA)	Algonquian	Willoughby (1907)
		Blackburn (1929)
		McCary (1957)
		Feest (1978)
Chesapeake Bay	Powhatan	E. Turner (1985)
		Rountree (1989)
		Sale (1991)
N Virginia	Potomac	Clayton (1973)
Virginia–New England coast	Algonkians	Brasser (1971)
		Jennings (1975)
New England	numerous groups	Willoughby (1935)
		G. Day (1953)
		Wilbur (1978)
		Vaughan (1965)
		Snow (1980)
		H. Russell (1980)
		Salisbury (1982)
		Cronon (1983)
		Patterson and Sassman (1988)
Southern New England	numerous groups	Hallett (1954)
		Salwen (1978)
Connecticut	numerous groups	Trumbull (1818)
		De Forest (1964)
Connecticut–Rhode Island	numerous groups	Cammisa (1984)
Narragansett Bay	Narragansetts	Simmons (1978)
Cape Cod	Nausets	Brierley (1991)

Table 5.8. *Continued*

Locale	People	Source
E Massachusetts	Wampanoag	Weinstein (1985)
	numerous groups	Willoughby (1927)
W Massachusetts	Pocumtucks	Temple and Sheldon (1875)
N New England coast	numerous groups	Willoughby (1927)
Maine	numerous groups	Eckstrom (1919)
S Maine	Pennacooks	Allen (1849)
Androscoggin Valley (ME)	Abenaki	Starbird (1928)
Coastal Maine	Abenaki	Snow (1978*a*,*b*)
Canada	numerous groups	Jenness (1955)
E Canada	Micmac	Wallis and Wallis (1955)
Vermont	Mahican	G. Day (1965)
NE New York	Mohawk	Carse (1949)
New York/Ontario	Iroquois	Stites (1904)
		Hunt (1940)
Huronia	Huron	Trigger (1969)
		Heidenreich (1971, 1978)
Michigan	Potawatomi	Deale (1958)
Great Lakes area	numerous groups	Parkins (1918)
		Kinietz (1940)
Ohio Valley	Shawnee	Callender (1978*a*)
Illinois	Illinois	Callender (1978*b*)
Missouri River (SD)	Arikaras	Denig (1930, 1961)
Missouri River (ND)	Mandan	Will and Spinden (1906)
		Bowers (1950)
Upper Missouri (SD–ND)	Mandan, Hidatsa, Arikara	Ewers (1968)
Plains	numerous groups	Holder (1991)

and Sheldon 1875), popular magazine articles (e.g., Brierly 1991), and even newspaper pieces (e.g., Blackburn 1929). Furthermore, there are at least thirty rather detailed histories of aboriginal agriculture which are based nearly exclusively on early reports, and for which rainfed agriculture was dominant (Table 5.9). As is the case with the general historical ethnographies that contain references to rainfed agriculture, these mainly involve the eastern half of the continent (Fig. 5.10).

Ethnographic studies that contain information about rainfed agriculture on plain fields are few. Those that do exist involve mainly people on the Plains (Oliver 1962), and in the Southwest, particularly northwestern Mexico (Fig. 5.11). For the Plains these studies can be classified into one of three groups. Works in the first group are in some respects like historical ethnographies from the East, syntheses of earlier observations. In these cases, however, the sources are 20th-century composites of 19th-century accounts. One of these ethnographic overviews involves an entire state (Hinsdale 1928). Others deal with specific groups, particularly the Wichita (Newcomb and Field 1967), the Pawnee (R. White 1983), and the Mandan (Will and Spinden 1906). The most thorough, comprehensive, and overall most important of these studies is that by George F. Will and George E. Hyde (1964). It deals exclusively with agri-

Table 5.9. Histories of native North American agriculture which include discussion of rainfed practices and are based on early explorers' accounts

Locale	People	Source
North America	numerous groups	E. Sturtevant (1885)
		Carr (1895)
		Holmes (1912)
		Carrier (1923)
		Schmidt and Ross (1925)
		Fitzhugh (1926)
		C. Gregory (1929)
		Lantis (1929)
Southeast	numerous groups	Gray (1933)
		Core (1967)
		Herndon (1967)
		Murphy and Hudson (1968)
		L. Larson (1980)
Florida	Timucua and Apalache	Spellman (1948)
	Timucua	Covington (1978)
Georgia/Tennessee	Cherokee	Fogelson and Kutsche (1961)
Louisiana	numerous groups	Williamson (1940)
Lower Mississippi River	numerous groups	Waselkov (1977)
Southern New England	numerous groups	Butler (1948)
New England	numerous groups	Willoughby (1906)
		Bennett (1955)
		H. Russell (1961, 1969, 1980)
		Connole (1976)
		P. Thomas (1976)
		Kupperman (1979)
Maine	numerous groups	C. Day (1954)
Vermont	numerous groups	G. Day (1965)
NE New York	Mohawk	Onion (1964)
New York/Ontario	Iroquois	Parker (1910)
		Waugh (1916)
		Hedrick (1933)
		Sykes (1980)
New Jersey/Pennsylvania	Delaware	Mahr (1955)
Pennsylvania	numerous groups	S. Fletcher (1950)
Ohio Valley	Shawnee	Voegelin (1940)
Great Plains	numerous groups	Carlson (1992)

culture along the upper Missouri River as practiced by natives and observed by travelers throughout the 1800s. The travelers whose accounts of indigenous agriculture were gleaned by Will and Hyde include George Catlin (1841), Alexander Henry (Henry and Thompson 1897), Pierre-Jean De Smet, Henry A. Boller, and Washington Mathews (1877), Maximilian (1906), F. V. Hayden, Meriwether Lewis, William Clark, and John B. Dunbar (1880), the latter made famous by the recent film *Dances with Wolves*.

The second group of studies containing ethnographic evidence of agriculture are true ethnographies. Of five such studies, three involve groups, the Pawnee (Weltfish 1965), the Caddo (Weltfish 1937), and the Sauk (Black

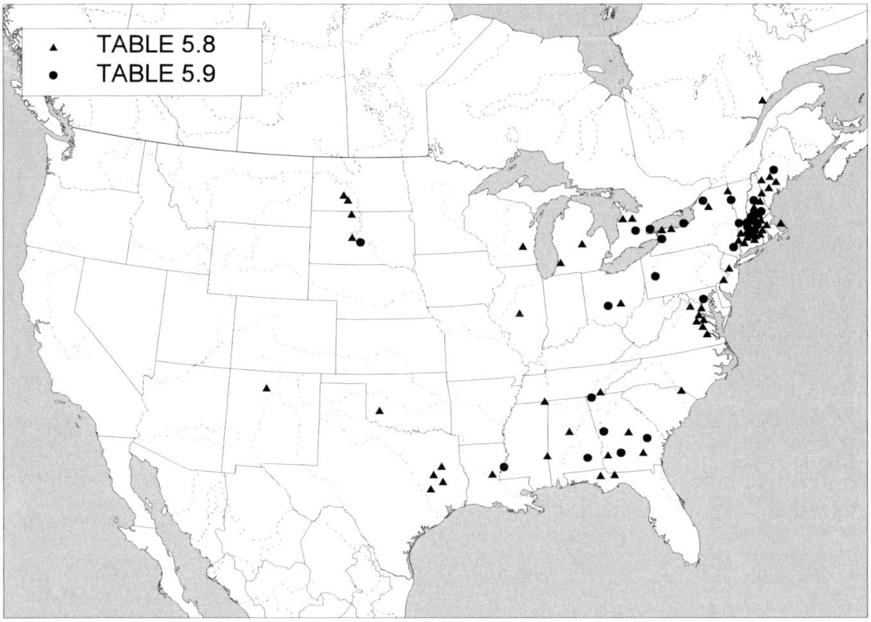

Fig. 5.10. Locations cited in the text and Tables 5.8 and 5.9 where recent scholars have relied on documentary evidence nearly exclusively in portraying native rainfed agriculture.

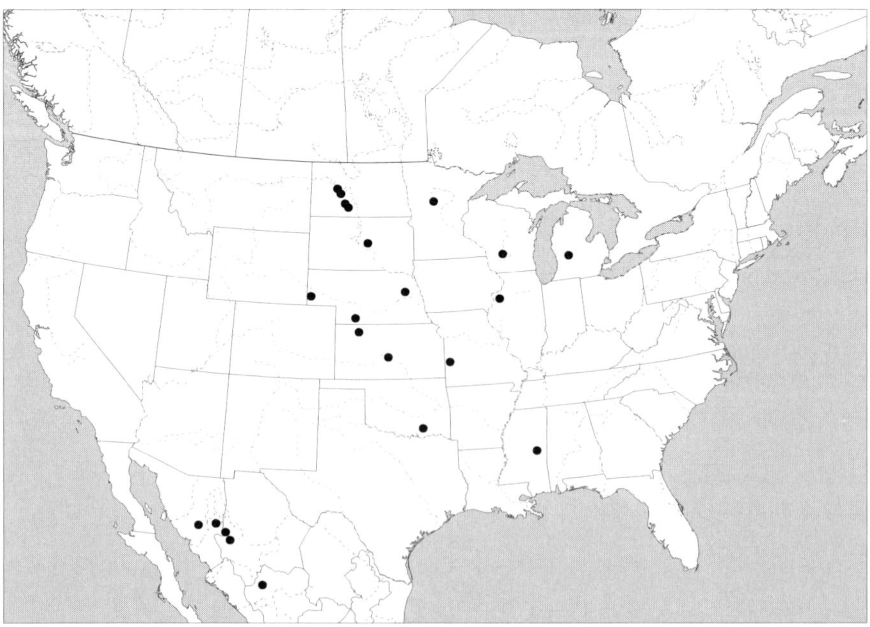

Fig. 5.11. Locations cited in the text for which there are detailed ethnographic studies that include discussion of native rainfed agriculture.

Hawk 1964), the fourth involves a state (Winchell 1911), and the last is a detailed study of the agricultural practices of one group, the Hidatsa (Wilson 1917).

The third group of studies which include references, albeit often not cited specifically, of native cultivation are histories. One is rather general and continent-wide (Biggar 1918). Another is a history of a particular people—the Cheyenne (Grinnell 1923)—while two others deal with aboriginal Plains dwellers in general (Lowie 1954; Wedel 1941). The quantity and quality of data that are contained in each of these three groups of sources vary considerably. The data sometimes appear contradictory, but when viewed in concert can be used to reconstruct a rather accurate picture that then helps understand what was going on earlier.

Without doubt, the greatest point of apparent conflict involves fields, their size, number, and spatial relationship. The Arikara, Mandan, and Hidatsa, who are often discussed together are described as having cultivated both 'little patches' and 'a large field (1,200 acres)' (Will and Hyde 1964: 98, 63). The exact dimensions of a patch are not clear. Arikara fields were claimed to vary from 0.2 to 0.4 ha (Will and Hyde 1964: 102). According to one source Mandan fields were each approximately 0.4 ha (Lowie 1954: 20), however, another says they were either 1.2, 1.6, or 2 ha (Maximilian 1906: 226; Will and Hyde 1964: 99). The Omaha, Pawnee, and Cheyenne also had small fields, but of various sizes. The term 'patches' was used to describe Cheyenne fields near the Black Hills (Grinnell 1923: 253). The Omaha are said to have had fields between 0.2 and 1.2 ha (Will and Hyde 1964: 81). One source says the Pawnee had fields that ranged between 0.2 and 0.6 ha (Weltfish 1965: 103), while another says they varied between 0.2 and 1.2 ha (Lowie 1954: 20). Another source says simply that they had 'fields' (Dorsey and Murie 1940: 75), while yet another claims the Pawnee to have cultivated several small plots (R. White 1983: 159). For the Plains as a whole one authority (Wedel 1941: 9) says that individual fields of indigenous farmers varied from 0.1 to 1.2 or 1.6 ha.

On the whole, this last claim seems quite conservative. It says a great deal yet not much, simultaneously. Much more insightful is an observation made at the turn of this century by Gilbert Livingston Wilson (1917: 24) about fields, their sizes, and variations. He noted that the largest field cultivated by the Hidatsa measured 165 by 80 m, or approximately 1.3 ha. However, according to his informants, this field began small and was enlarged each year until it reached the size he observed. Presumably, this field may have even increased in size during following years. Nevertheless, changes in its size over time lends credence to variations noted on the order of 0.1 to 2 ha. It may well be that specific areal assessments reflect conditions only at one moment in time. Regardless of how small is small, however, one thing should be evident in these descriptions; fields are almost always mentioned in the plural.

Indeed, the Hidatsa are said to have had 'numerous corn fields' (Will and Hyde 1964: 101).

That fields were small but numerous raises the issue of the total extent to which the landscape was domesticated. Evidence indicates that environmental transformation for agriculture was quite extensive in the Upper Plains, and that it involved the bottomlands or floodplains of rivers. The Omaha (Will and Hyde 1964: 81), the Pawnee (Lowie 1954: 20; Weltfish 1965: 103), the Hidatsa (G. Wilson 1917: 9, 15), specifically, and Plains people, in general (Wedel 1941: 9), are said to have cultivated valley bottoms. The Arikara in one place, cultivated an island 'three miles long, and covered with fields' (Will and Hyde 1964: 101). Within the area occupied by the Mandan there was 'a valley . . . covered with maize plantations' (Will and Hyde 1964: 102). In Michigan 'many acres bordering the river were covered with luxuriant maize' cultivated by aboriginals in the 1830s (Hinsdale 1928: 34). But how many is many? In his autobiography, Black Hawk (1964: 88) says the Sauk had 320 ha of cornfields extending along a river for 3.2 km in 1833. Similarly, the Ojibwa of Minnesota are reported to have cultivated 340 ha of bottomland around AD 1900 (Winchell 1911: 693, 664). And, as late as 1794, extensive cultivated tracts were reported from the Maumee and Augllaize valleys of Ohio and Indiana, respectively (Manypenny 1972: 84).

Elsewhere on the Plains, descriptions of fields are less detailed, but informative nonetheless. The Caddo had fields described in the context of city streets; 'half a block', by 'a block and a half' (Weltfish 1937: 39). According to Catlin (1989: 410, 396) cornfields of the Sioux were 'handsome', and those of the Creek, who lived south of the Arkansas River, were 'immense'. The Arapaho are said to have been agriculturalists before they became horsemen (Elkin 1940: 207; Eggan 1955: 36). If nothing else, these reports indicate that sizable tracts or extensive areas of land were cultivated on the Plains in late historic times. They also mirror and therefore substantiate earlier explorer accounts (Newcomb and Field 1967).

The annual preparation of native fields on the Plains involved tillage with hoes (Fig. 5.12A) typically made of a short wooden handle to which was attached a blade made from the scapula of either a bison (J. Mathews 1961: 449; Maximilian 1906: 276; W. Strong 1933: 277; Wedel 1941: 9), or an elk (Will and Hyde 1964: 82), or from stone (Grinnell 1923: 252). In addition to hoeing, the Pawnee also pulled up weeds by hand and cut small brush with axes (Weltfish 1965: 102), and the Hidatsa dug up some plants by the roots with digging sticks (G. Wilson 1917: 10–15). Both groups burned the cut or uprooted vegetation before planting crops. Rakes made of willow twigs and deer antlers (Fig. 5.12B) were also used by native people of the upper Missouri (Will and Hyde 1964: frontispiece).

Planting, unlike as described in many documentary accounts from the East, appears not to have always involved intercropping. The Hidatsa

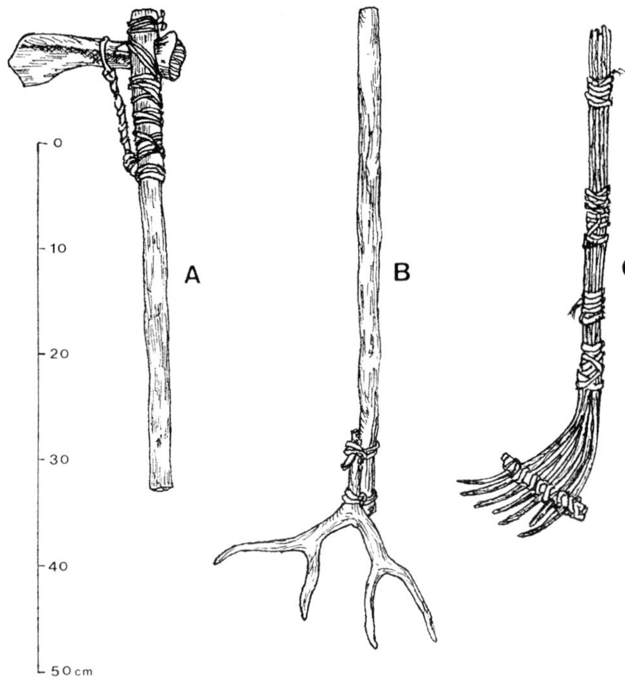

Fig. 5.12. Hidatsa agricultural implements. (A) Scapula-blade hoe; (B) antler rake; (C) twig rake (after Wilson 1917: 12, fig. 3; 14, figs 4 and 5). (Drawn by, and published by permission of, Diane Doolittle.)

(G. Wilson 1917: 82), the Pawnee (Weltfish 1965: 103), and the Caddo (Weltfish 1937: 30, 38) did raise beans and maize together. Even these people, however, raised various cucurbits in separate fields (G. Wilson 1917: 82; Weltfish 1937: 31, 38, 1965: 103). Other groups for which there are reports of crops grown in separate fields include the Omaha, the Mandan (Will and Hyde 1964: 82, 99), and various peoples of Nebraska (Gilmore 1915: 182). There exists no evidence that multiple cropping was practiced on the Plains historically, indeed the short period of time between the last frost of each year and the first, mediates against the practice (Will 1924).

The actual planting of crops was done with digging sticks in previously constructed mounds or hills (e.g., J. Mathews 1961:449; Weltfish 1937: 30, 38), and always in the same spot year after year. Some sources say that maize was planted each year in the same mounds (G. Wilson 1917: 22, 69; Winchell 1911: 490). Others say that the spaces between the mounds were not tilled and that only the previous year's mounds, themselves, were prepared prior to planting (Biggar 1918: 127–8; Will and Hyde 1964: 79). In either case, it is clear that each mound was reused and/or remade annually. Although there exists no documentary evidence for this practice on the Plains, accounts from elsewhere plus these suggest that replanting mounds was not uncommon in the center of the continent in pre-contact times.

Weeding was a common practice on the Plains, sometimes being carried out twice a year (Wedel 1941: 10; Will and Hyde 1964: 82, 91). It was typically done well after the maize had sprouted (Biggar 1918: 129; J. Mathews 1961: 451; Weltfish 1937: 32, 39); the Sauk even waited until it was at least knee high (Black Hawks 1964: 194); and the Caddo until it was waist high before they weeded a second time (Weltfish 1937: 32, 39). The Hidatsa (G. Wilson 1917: 26), the Omaha (Will and Hyde 1964: 82), the Mandan (Maximilian 1906: 276), the Sauk, and the Caddo (Weltfish 1937: 33, 39) used hoes in such a way that weeding either resulted in the original formation of mounds, increased the size of, or was instrumental in the maintenance of eroding mounds (Fig. 5.13). Whatever the exact scenario, weeding, mounding, planting, and tillage were interrelated activities. Weeding with hoes contributed to the formation of mounds which were further enhanced during tillage. They were then planted and subsequently reworked again during later weedings.

Fig. 5.13. Hidatsa women weeding and mounding maize and squash with a bone hoe (after F. N. Wilson from Wissler 1922: 22, fig. 5, and G. Wilson 1917: 14a). (Redrawn by author.)

Fig. 5.14. Plains women scaring away crows from scaffolds built in maize fields. Note also that earth has been mounded into hills at the base of each plant (Schoolcraft 1851*b*). (Provided courtesy, and published by permission, of the Ohio Historical Society, Neg. No. OHS 2888a.)

Whereas weeds posed a botanical threat to crops on the Plains historically, birds posed a zoological threat, as presumably they did in pre-European times as well. To scare birds away, native peoples of the Plains in the 19th and early 20th centuries used exactly the same techniques, for which there is documentary evidence, as their earlier counterparts further east. They built field houses that functioned to shelter individuals, usually, but not always, girls, who stayed near the crops. Evidence of such comes from only one group, the Mandan (Will and Hyde 1964: 94), but more than likely this was a pan-Plains phenomenon.

Plains dwellers also built scaffolds in their fields and on which people were stationed (Fig. 5.14). These features were very much like those reported earlier from the East (see Figs 4.2, 4.3, and 5.3). On them young Pawnee (Dunbar 1880: 276), Mandan (Will and Hyde 1964: 94), and Hidatsa (G. Wilson 1917: 25–31) females would either beat drums or two sticks together in order to scare away birds.

Ethnographic evidence of rainfed agriculture on plain fields in the Southwest is limited. The paucity of such information may well be a function of the relative abundance and hence importance of other forms of agriculture practiced by some native peoples today. To be sure there are numerous studies of pueblos in New Mexico and Arizona, but their agricultural emphasis is on those activities that involve manipulation of soil (Chapters 7 and 8) and water (Chapters 9 and 10). There are a few studies of Indéean agriculture which at least mention the importance of rainfed fields (e.g., Baldwin 1965: 59; Buskirk 1949: 59; Goodwin 1942: 30–1, 163; Opler 1941: 374), but since these people were late prehistoric or early historic arrivals, the best that can be said is that certain groups of their predecessors living in higher elevations where rainfall was dependable and adequate probably cleared small fields in the forest, planted them with maize and other crops, and weeded them with hoes (Pool 1985, 1994: 93).

Clearly, the best ethnographic evidence of rainfed agriculture on plain fields in the Greater Southwest comes from northwestern Mexico. There, in the Sierra Madres, the Rarámuri, the Odami, and the Névome, have long traditions of being dependent on rainfed agriculture. Recent ethnographies have, accordingly, not failed to discuss this important aspect of their subsistence activities. Much as is the case with Plains peoples, the cultivation practices of these distinct groups are quite similar, doubtless a function of like biophysical environments.

Fields of the Rarámuri and the Odami have both been described as 'patches' (Bennett and Zingg 1935: 26; Pennington 1969: 59), implying a relatively small areal size. Preparation of these fields prior to planting among all three groups involves plowing with a wooden ard pulled by a team of draft animals (Graham 1994: 10, 12, fig. 4; Laferriére and Van Asdall 1992: 162; Pennington 1963: 52, 1969: 62–6, 1980: 144). Obviously, a European-introduced practice, plowing probably replaced aboriginal tillage done with hoes. In no small way this assessment is conditioned by the facts that hoes were commonly used elsewhere, and that these peoples still use aboriginal tools for planting and weeding. Digging sticks, not unlike those described previously, are used by all three groups (Bennett and Zingg 1935: 29; Pennington 1963: 54; 1969: 66, 1980: 148, 149). Typically, planting involves two people, one punching a hole in the ground, another dropping in seeds and covering the hole. Intercropping maize, beans, and squash is common among the Rarámuri (Bennett and Zingg 1935: 26; Pennington 1963: 55) and the Névome (Laferriére and Van Asdall 1992: 166). The Odami plant some beans and corn together, but far more typical is the cultivation of crops in separate fields (Pennington 1969: 67, 59). Only those Névome who live in the lower elevations, where conditions are warmer and the growing season is longer, manage to plant and harvest two crops per year per field (Pennington 1980: 148).

Once the crops have been planted and began to sprout, weeds become a problem in Mexican fields, as they do elsewhere. The Rarámuri, Odami, and Névome all weed with hoes, and as a result, create hills or mounds around the individual crops (Bennett and Zingg 1935: 29–30; Laferriére and Van Asdall

1992: 168; Pennington 1963: 55, 1969: 69). Today, hoe blades are made of steel for the most part, but the use of some one-piece wooden hoes (Pennington 1969: 69) is certainly a throwback to pre-European practices and times. Interestingly, some Névome still use *coas* for weeding and hilling (Pennington 1980: 155). *Coas* are much like hoes, but the blades are not attached perpendicular to the handle; they are attached in-line and in some respects are similar to shovel blades. The *coa* is an aboriginal tool, but one indigenous to Mesoamerica, not North America (Donkin 1970). Its use in Sonora demonstrates that the Névome live close to that all-important indigenous cultural boundary.

As for structures associated with rainfed fields, a number of fascinating activities are carried out. The Rarámuri and the Névome both build field houses, or small shelters near the cultivated areas (Pennington 1963: 57, 1980: 153). The Névome and the Odami both construct scarecrows (Laferriére and Van Asdall 1992: 168, 169; Pennington 1980: 153, 1969: 71). These are traditional human-like figures which may actually be of Old World, rather than New World origin. However, the Odami and the Rarámuri both employ two strategies in their attempts to ward off predatory birds which are unquestionably of indigenous origin. One strategy involves hanging a dead bird from a pole in each field, the idea being that live birds would learn from the fatal mistake made by their forerunners and avoid the field. The second strategy involves planting a number of poles in and on the margins of each field, and running string made of *ixtle* fiber between each pole. The intent of this strategy is to disrupt the flight of birds, thereby creating an environment which they would find unattractive. The degrees to which these strategies are effective are not known, but the practices and the features are well-documented (Bennett and Zingg 1935: 27; Pennington 1963: 57, 1969: 710). The Rarámuri also send children into the fields to scare away birds (Graham 1994: 13).

Outside of the Plains and northwestern Mexico there is but one historic assessment of native agriculture, and it involves the Choctaw (Campbell 1959). In the 1820s, physician and naturalist Gideon Lincecum compiled ethnographic notes about these remaining people in eastern Mississippi (see Fig. 5.11). In some respects, he did not say much, but his comments do shed some insight and demonstrate continuity. He noted for example 'little corn and bean fields', suggesting small parcels that were intercropped. He noted, too that planting was done with a digging stick and involved communal labor (Campbell 1959: 16).

The ethnographic evidence, although coming from areas different from the documentary evidence, tends to substantiate the early explorers' observations. The descriptions of fields, their sizes and associations with one another; and assessments of field preparation, planting, and cultivation activities are all supportive. Combined, the ethnographic evidence and the documentary evidence illustrate that rainfed agriculture was practiced far and wide on fields which did not involve the construction of any features anticipated to be permanent. Across much of North America rainfall was adequate and temperatures were sufficiently moderate to allow for a rather basic type of agriculture at the time of European contact.

Archaeological Evidence

By definition, one should not expect to find physical remains of plain fields (Waselkov 1997: 179). However, some archaeological evidence of fields does not involve features, *per se*, and some involves features that were not intended as permanent land improvements. Of the former type, one can include such evidence as correlations between locations of habitation sites and arable lands (e.g., P. Thomas 1985), and locations of agricultural implements, particularly lithic hoes and flakes from their resharpening. As for the latter type, the best examples are remnants of corn hills that are products of the previously discussed weeding and mounding activities. Other forms of physical evidence, such as analysis of fossil pollen (e.g., S. Fish 1994); opal phytoliths (e.g., Dunn 1983; Starna and Kane 1983), and soil structure and chemistry (e.g., Eidt 1984) remain promising, but to date have proven to be more hopeful than helpful. Being quite buoyant, pollen from crop plants can blow into non-field areas (Raynor, Odgen, and Hayes 1972) and non-crop pollen can blow onto fields. Problems exist with distinguishing phytoliths of certain crops from those of other non-cultivated plants (Doolittle and Frederick 1991). And, a sufficient number of independent studies have not been conducted to corroborate certain soil analyses (Knapp 1985: 371, 372).

Identifying plain fields archaeologically is, to be sure, tantamount to looking for the proverbial needle in a haystack. Indeed, it is so problematical that the task may be counterproductive. Out of necessity, or perhaps frustration; most prehistorians have contented themselves with inferring that agricultural people occupying sites near fertile, well-watered, frost-free lands must have farmed those lands. A number of archaeological studies from the Southeast to the Upper Missouri, refer to certain tracts of land having been cultivated by means of direct precipitation solely on the basis of proximity to habitation sites (Table 5.10, Fig. 5.15). Some of these studies involve several sites and broad areas, some are rather location specific, and yet others link individual sites, and even structures, to particular parcels. Clearly, the scale of such studies varies considerably.

At the broadest end of the spectrum are the regional assessments such as those by the late, but distinguished Plains archaeologist Waldo R. Wedel (1941, 1961, 1986). Writing as early as the 1940s he said of prehistoric villages throughout the region that 'it can be assumed that sufficient arable ground

Table 5.10. Locales for which rainfed agriculture has been inferred on the basis of correlations between archaeological sites and arable land

Locale	Source
Northeast	Ford (1974: 400)
New England	Snow (1980: 25, 47, 56)
Vermont	Haviland and Power (1981: 8–9, 159)
Hudson Valley	Salwen (1975: 57)
Mohawk Valley	Bond (1985)
North Carolina	J. Coe (1979: 159)
Central Georgia	Martin, Quimby, and Collier (1947: 376)
Eastern Woodlands	J. Griffin (1967: 189, 190)
Southern Indiana	Muller (1986: 254, 258)
Southern Illinois	Holley (1978)
	Hargrave *et al.* (1983: 335)
	Woods (1987; 1993)
Mississippi Valley	Lynott *et al.* (1986: 61, 62)
Mississippi–Missouri Valleys	Chapman and Chapman (1983: 71)
Kansas	M. Adair (1988: 97–8)
Upper Missouri Valley	Meyer (1977: 5)
Great Plains	Wedel (1941: 11–13)
	Wedel (1961: 29, 80, 81, 95, 160, 238)
	Wedel (1986: 33–5, 114, 116–20)
Four Corners	Prudden (1903: 234, 247)
	Winter (1975: 111)
	Osborne (1976: 43, 44)
	Matson (1991: 41, 59, 84, 86, 90, 214–16)
Western New Mexico	Wills (1988: 468)
Southern New Mexico	Minnis (1985: 75–6)
Northern Arizona	Henderson (1979: 26, 42–4)

was available in the bottom lands close at hand' (Wedel 1941: 13). His opinion did not change over the next half-century; if anything it was supported by more detailed environmental assessment (Wedel 1986). In a similar vein, but speaking of Eastern Woodlands cultures, another scholar wrote:

Maize, beans, and squash do well on the sandy and light loam soils found in the bottom lands of the great rivers of the East. There is a high correlation between the presence of these soils and the location of Mississippian sites . . .. These soils could be tilled with digging sticks and with hoes of flint, shell, and bore (J. Griffin 1967: 189).

At what might be considered the mid range of studies associating sites with agricultural lands, two archaeological assessments will serve as examples, one from the Midwest and one from the Northeast. Investigations by geographer and archaeologist William I. Woods around Rend Lake Reservoir in southern Illinois revealed that the 'site distribution exhibits a high correlation with areas of Sharon soils . . .. A subsistence base relying heavily on agriculture is strongly indicated' (Woods and Denny 1980: 183). Given that rainfall in the area is neither too little nor too great, a reliance on direct precipitation is also

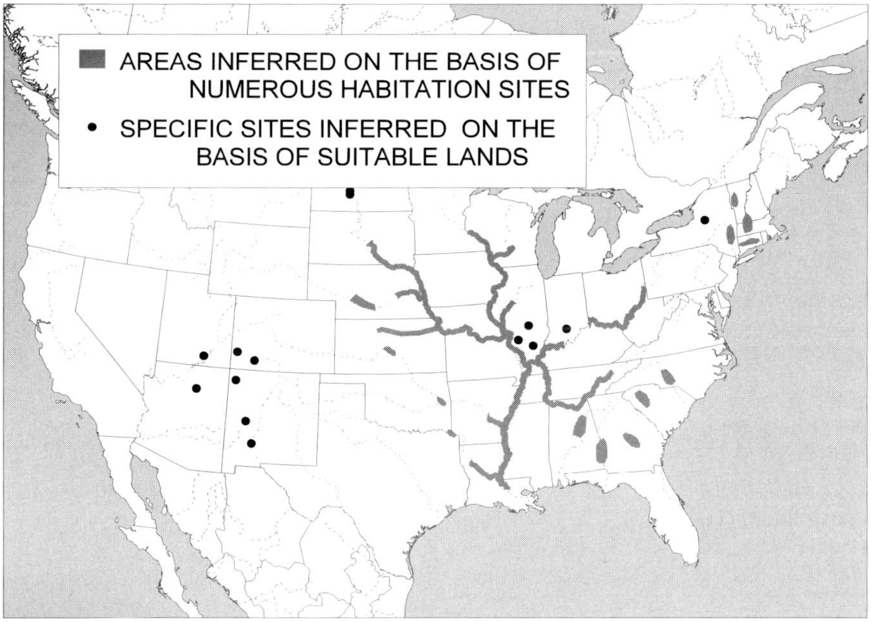

Fig. 5.15. Locations and areas cited in the text and Table 5.10 where rainfed agriculture has been inferred on the basis of correlations between archaeological sites and suitable lands.

presumed. Similarly, a study of the relationship between seven prehistoric settlements in the Mohawk Valley of upstate New York, and soils within a 1 km radius of each site, concluded that the '[s]oils most often settled were the most productive soils in the area' (Bond 1985: 37). Here, productivity was based on estimates of the number of bushels of maize per acre (transformed to liters per hectare) made by the United States Department of Agriculture, Soil Conservation Service (Bond 1985: 29).

Similar studies, coming to nearly identical conclusions, were also undertaken in the Southwest (e.g., Osborne 1976: 43, 44; Van West 1994: 131–86). However, in this part of the continent most correlations between environmental conditions and sites that were dependent on direct rainfed agriculture involve very small tracts and individual structures. Doubtless, the scale of study is a function of spatially limited areas suitable for this type of cultivation; most are in mountainous locales (e.g., Matson 1991: 59). Also, there is a proliferation of remote and isolated sites comprised exclusively of remnants of single-room structures. These features are known throughout the region as 'field houses' (Ward 1978*a*) even though in many cases there is no corroborative evidence of fields, only inference. For example, in a study of archaeological sites in the Coconino National Forest, northern Arizona, T. Kathleen

Henderson (1979: 26) reported that: 'Single structures occurred regularly on mesa tops at the highest elevations and with direct facings. This is undoubtedly a function of their association with mesa top field areas'. The single structures referred to here are later in the report called 'field houses', mapped as such, and inferences about cultivation drawn from their presence (Henderson 1979: 42a–46b).

The association of environments and sites is a convenient, yet tenuous way of attempting to understand landscapes of cultivation. The previously discussed cases make sense intuitively but remain unconfirmed. Supportive evidence that certain locales were indeed cultivated prehistorically in North America can be found in the presence of agricultural implements. The most common form of cultivation tool reported archaeologically are blades of stone hoes (e.g., Pauketat and Woods 1986: 116–22, fig. 8). Some of these are pointed or triangular-shaped (Fig. 5.16A, B), and should probably be considered warrens. Others have straight edges (Fig. 5.16C). In some cases, hoes were made of shell (Fig. 5.16D, E), thereby confirming the accounts of some early Europeans. In other cases, especially on the Plains, they were made of bison or elk scapula, as reported by some ethnographers (Driver and Massey 1957: 224, fig. 3.6). In addition to having different shapes and being made of different materials, hoes of different sizes have been found archaeologically. Larger, and more robust models are thought to have been used for initial ground-breaking; medium sized hoes were used for building corn mounds; and small ones for weeding (W. Fowler 1954: 44). Also recovered archaeologically are lithic spade blades, most of which have straight edges (Fig. 5.16F) but some have rounded and others have pointed edges. It is thought that since women used these tools, they also quarried and manufactured them (W. Fowler 1976: 9). Finally, some unusual lithic artifacts identified as corn planters have been recovered. Some of these appear to have been shaped so as to facilitate being held by hand (Fig. 5.16H, I). Assuming that such agricultural implements were used exclusively and hence abandoned in fields, their locations demarcate parcels farmed in times past. In those cases where cultivation tools were discovered archaeologically in habitation sites, it can be inferred with some confidence that arable lands nearby were probably ancient fields. Perhaps not surprisingly, the geographical distribution of such implements as revealed through a synthesis of archaeological studies (Fig. 5.17) is quite similar to that for the correlation of lands and sites (see Fig. 5.15).

Hoe blades are reported extensively from areas in which rainfed agriculture was feasible. They are particularly prevalent from the Northeast, the Midwest, and the higher elevations of the Southwest.

Every type of implement is common throughout New England (Wilbur 1978: 29–31). They were recovered during projects undertaken as early as the 1930s (Willoughby 1935: 171) and involving, particularly, valleys including those of eastern Massachusetts and Rhode Island (W. Fowler 1954: 46), the

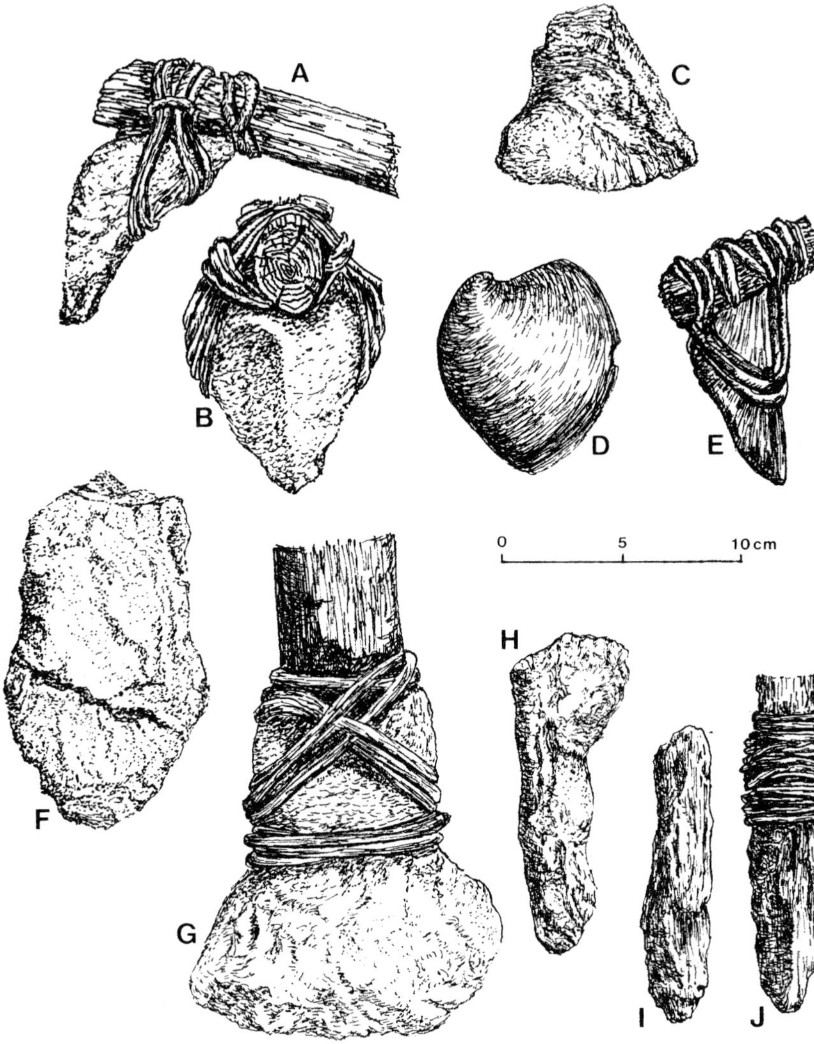

Fig. 5.16. Various agricultural implements recovered archaeologically from places that were dependent entirely on direct precipitation for producing a crop. (A) side, and (B) front views of a triangular stone hoe blade hafted to a wooden handle (after W. Fowler 1960: 40, fig. 5); (C) front view of a straight-edged, unhafted stone hoe blade (after W. Fowler 1954: 45, fig. 15, 1963: 26, fig. 13); (D) front view of an unhafted shell hoe blade, and (E) side view of a shell hoe blade hafted onto a wooden handle (after H. Russell 1969: 12, fig. 6); (F) unhafted, pointed stone shovel blade (after W. Fowler 1963: 26, fig. 13, 1974: 5, fig. 3, 1976: 10, fig. 7); (G) a straight-edged stone shovel blade hafted onto a wooden handle (after W. Fowler 1954: 47, fig. 16; H. Russell 1969: 16, fig. 10); (H) hand-held stone corn planter, (I) unhafted stone corn planter, and (J) stone corn planter hafted onto a digging stick (after W. Fowler 1954: 48, fig. 17, 1960: 42, fig. 7; H. Russell 1969: 17, fig. 11). (Drawn by, and published by permission of, Diane Doolittle.)

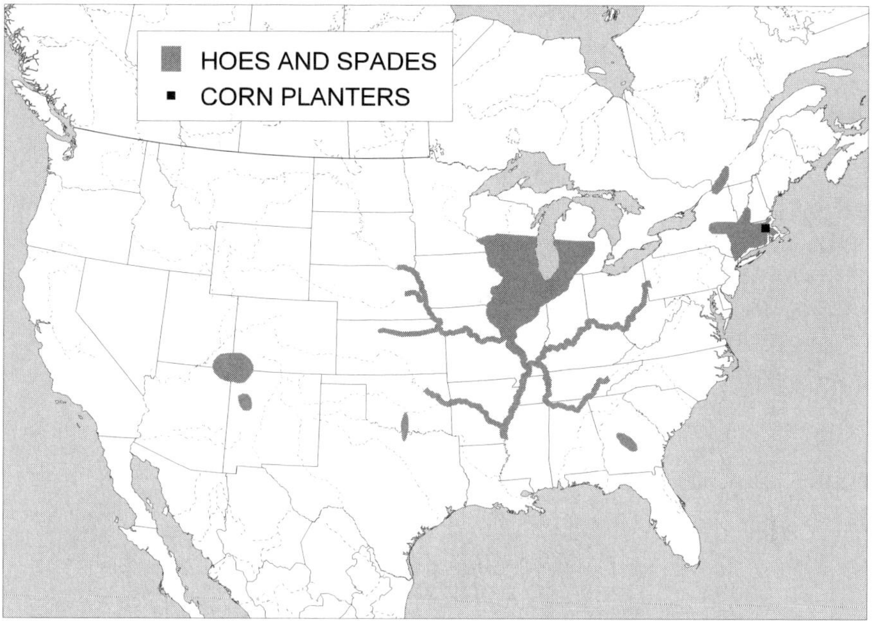

Fig. 5.17. Locations and areas cited in the text where rainfed agriculture can be inferred on the basis of the archaeological discovery of ancient farming tools.

Connecticut (Daniel 1962: 16, 18), Hudson (Salwen 1975: 57), Mohawk, and St. Lawrence (W. Fowler 1960: 41) rivers. These implements have been compared with those reported by early Europeans (H. Russell 1969), and with those recovered in the Midwest (W. Fowler 1974).

As for the central part of the continent (Fig. 5.17), hoes and spades were found in the Mississippian culture area, extending down the Wisconsin River valley to the Mississippi and then as far as the confluence of the Arkansas, and in the Missouri River valley from South Dakota to near the core of the culture area—Cahokia (Chapman and Chapman 1983: 79; Pauketat and Woods 1986: 116–22, fig. 8). Hoes were also found in other Mississippian areas including the entire lengths of the Ohio, Tennessee, and Arkansas valleys (Winters 1981: 28–9), and along the bottom lands of smaller valleys in Oklahoma, Kansas, and Nebraska (Wedel 1961: 95, 109, 139, 141–2, 173, 187, 212). They were also found along small streams in the Ozarks (Jurney 1981: 25, 28; J. Brown 1984: 42, 45). Some very early studies also report stone hoes from eastern Tennessee (Thruston 1890: 220–1), and central Georgia (C. Jones 1873: 302). Finally, earlier peoples, those of the Hopewell culture, used hoes in the same areas as their later Mississippian counterparts, but they also did so as far north as central Wisconsin and Michigan (Quimby 1960: 100). There is little doubt, especially given the environmental conditions, that the type of agriculture associated with the locations of the implements just discussed was dependent entirely on direct rainfall (W. Fowler 1954).

There exist remarkably few examples of prehistoric lithic hoes found in those areas of the Southwest where rainfed agriculture was feasible. A few have been found in northeastern Arizona (Kidder and Guernsey 1919: 125–6), Mesa Verde in Colorado (Rohn 1971: 246–8), in northern (Neely 1993*a*), and northwestern (Morris 1939: 138–9) New Mexico. Were the paucity of hoes not problem enough, the few that have been found are of two distinctive types, thereby fueling multiple interpretations. Some of the Four Corner hoes are rather small, and ground and polished. Others are somewhat larger, crude, and chipped. It has been suggested that the former might be ceremonial, and that the later had multiple uses (see Rohn 1971: 246–8). Regardless, everyone agrees that they were agriculturally related, however indirectly, and that the principal form of agriculture in the region whence they came involves direct precipitation.

The presence of stone agricultural implements, especially those found *in situ*, is certainly more compelling evidence of ancient rainfed agriculture than the use of site–environment correlations alone. It is not perfect however, as tools can be carried away from fields, and they can be used for non-agricultural purposes (e.g., semi-subterranean house excavation). The best archaeological evidence for rainfed fields that did not involve what are normally thought of as permanent features on the landscape, are vestiges or relicts of what are typically considered temporary modifications—specifically, mounds created during weeding which also support maize stalks. Known widely as 'corn hills', thousands of such features have been reported from New England through Wisconsin (Fig. 5.18), most during the late 19th and early 20th centuries. Corn hills were reported mainly from Michigan and Wisconsin, with no fewer than 145 locations cited in the *Wisconsin Archaeologist* alone. Of the former, it has been noted that such features 'are found by the sides of nearly all our lakes' (Peet 1887: 395). Similarly, of the latter, it has been stated that '[t]here was not a township in southern Michigan without one or more trails with camping grounds and cornfields' (Hinsdale 1928: 33). Although details are lacking, one treatment of corn hill sites says they are found in Illinois, Ohio, New York, and other neighboring states (Stout 1914: 51), presumably Indiana and Pennsylvania. Despite their distinctiveness and distribution, a mere forty-two localities have begun to be subjected to any type of publication (Table 5.11).

Descriptions of these corn hills match rather closely documentary and ethnographic assessments. There exist qualitative references to 'patches' (C. Brown 1909: 100) and 'extensive' (Hinsdale 1928: 31) areas of hills. Likewise, there are quantitative estimates of field sizes that range from 0.8 ha (Laidlaw 1917: 91) to 160 ha (Hibbard 1904: 146). One study (Heidenreich 1978: 381) found an average of 6,180 corn hills per hectare. Some accounts describe corn hills as being 'irregular' (Lapham 1855: 19), 'haphazard' (Hinsdale 1928: 32), or 'scattered' (Hallowell 1921: 233) over field surfaces. Others are reported as

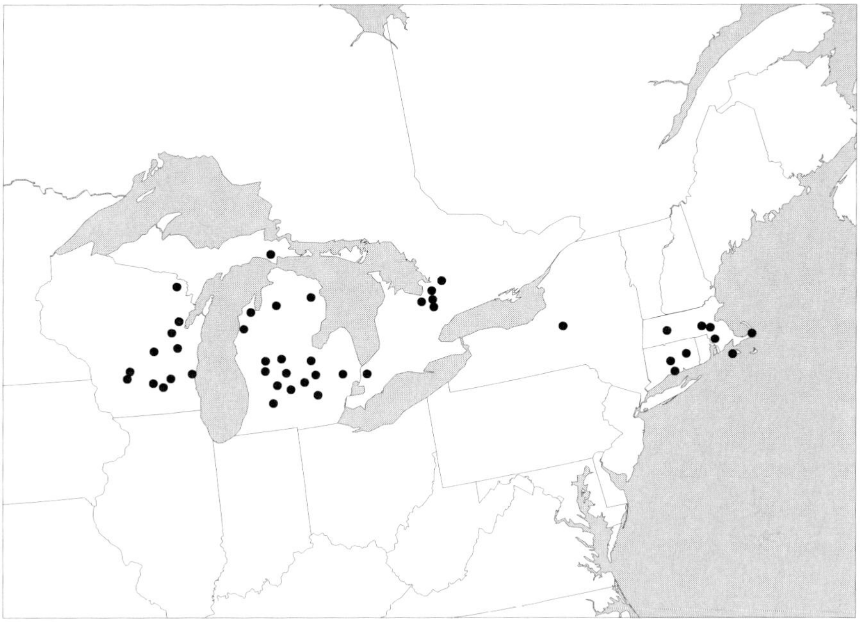

Fig. 5.18. Spatial distribution of corn hill sites cited in the text and Table 5.11.

Table 5.11. Ancient corn hill localities

Locale	Comment	Source
Milwaukee, WI	Irregular, 4 ha	Lapham (1855: 19, pl.VIII); C. Brown (1909: 97)
Waukesha, WI	6 ha, Carroll College	Lapham (1855: 27, pl.XVIII); Gallagher (1992: 96, 99, 100); Doolittle (1992*c*: 395)
Menasha, WI	Regular	Lapham (1855: 61)
Honey Creek, WI	Numerous, irregular	Lapham (1855: 70)
Sauk Prairie, WI	400 acres, regular	Hibbard (1904: 146); West (1907: 194)
	Several hundred acres	Stout (1914: 51)
Monona, WI	Few	Hibbard (1904: 147)
Marquette, WI	Numerous patches	C. Brown (1909: 99, pl.2.I, 103)
Calumet Co., WI	Patches	C. Brown (1909: 100)
Green Bay, WI	Irregular patches	C. Brown (1909: 101, pl.2.III, 104)
Lake Kashkonong, WI	40–60 acres, not arranged in definite rows	Stout (1914: 51)
Menominee R, WI–MI		Hinsdale (1928: 34)
Harbor Spring, MI		Hinsdale (1928: 33)
Alcona Co., MI	Many acres	Hinsdale (1928: 34)
	$18\frac{1}{2}$ acres	Hinsdale (1928: 35)
	25 acres, conical, irregular 2 ft across, 4 ft apart, 6–12 in high	Hinsdale (1928: 36)
St. Ignace, MI	5–6 acres	Hinsdale (1928: 31)
Charlotte, MI	Extensive, arranged in rows	Hinsdale (1928: 31)

Table 5.11. *Continued*

Locale	Comment	Source
Montcalm Co., MI	3 ft apart, 1.5 ft wide, 6 in high, acre	Hinsdale (1928: 31)
Romeo, MI	Few acres	Hinsdale (1928: 31)
Barry Co., MI	Several acres, not in rows, haphazard	Hinsdale (1928: 32)
Williamston, MI	10–15 acres	Hinsdale (1928: 32)
Kent Co., MI	Many acres	Hinsdale (1928: 32)
Grand Rapids, MI	Many acres	Hinsdale (1928: 32)
Manistee Co., MI		Hinsdale (1928: 32)
Kalkasca Co., MI		Hinsdale (1928: 32)
Pewamo, MI	Large field	Hinsdale (1928: 32)
Kalamazoo, MI	3 or 4 miles square, very regularly laid out	Hinsdale (1928: 32)
Livingston Co., MI	Many fields	Hinsdale (1928: 32)
Saginaw Valley, MI	13 fields, one of about 100 acres	Hinsdale (1928: 33)
Washtenaw Co., MI		Hinsdale (1928: 33)
Corunna, ON		Lee (1958: 10–11)
Simcoe Co., ON	Tiny Twp., 3 sites	A. Hunter (1899: 13, 37, 41)
	Tay Twp., 10 sites	A. Hunter (1900: 58, 65, 67, 69, 73, 75–8)
	Medonte Twp. 4 sites	A. Hunter (1901: 76, 85, 87)
Victoria Co., ON	Fenelon Twp, 2 acres	Laidlaw (1917: 91)
Creemore, ON	1 ft high, but eroded, 40 in wide, 55 in apart	Heidenreich (1974: 385)
	2000/acre, fertilized and not cultivated between	Heidenreich (1974: 392)
Oneida, NY	Four times larger than modern hills	Schoolcraft (1851*a*: 57)
New Londen, CT		H. Russell (1961: 60)
Mohegan, CT	8–10 acres, irregularly scattered	Hallowell (1921: 233)
Pomfret, CT	Numerous, small mounds	Delabarre and Wilder (1920: 204)
Northampton, MA	Definite rows	Delabarre and Wilder (1920: 214)
	3 ft apart	Delabarre and Wilder (1920: 221)
	2 ft wide, 8–12 in high	Delabarre and Wilder (1920: 222)
		Howes (1946: 68–9)
Holbrook, MA		H. Russell (1961: 60)
Dover, MA		H. Russell (1961: 60)
Martha's Vineyard, MA		H. Russell (1961: 60)
Taunton, MA	Covered with rows	Delabarre and Wilder (1920: 210)
	30 acres, 80,000 hills	Delabarre and Wilder (1920: 213)
	Perfectly regular rows	Delabarre and Wilder (1920: 213)
	3.5–5 ft apart	Garman *et al.* (1997: 70)
		Delabarre and Wilder (1920: 214)
Cape Cod	40 hills found during excavations	Mrozowski (1994)

'regular' (Hibbard 1904: 146) or laid out 'in almost perfectly regular parallel rows, nearly evenly spaced' (Delabarre and Wilder 1920: 213), a description that suggests to some authors (Garman *et al.* 1997: 70) that the hills might not be native. The spacing between corn hills varied from one place to another, although with no discernable regional pattern. In some locales, hills were approximately 1 m apart (Hinsdale 1928: 31) whereas elsewhere they were 1.5 m apart (Delabarre and Wilder 1920: 214). Corn hills varied in size, again with no apparent areal differentiation, from as much as 1 m (Heidenreich 1974: 385) to as little as 50 cm (Hinsdale 1928: 31) in diameter. None were more than 30 cm high. One report says they were 'conical' (Hinsdale 1928: 36), but the one noting that 'erosion and cattle tramping over them had worn them down' (Heidenreich 1974: 384) seems more accurate.

That various erosional processes have had an impact on relict corn hill sites is obvious. Several reports of their existence contain statements to the effect that the features are greatly reduced from what they must have been, and are now only barely visible. In one case (Heidenreich 1974), they were difficult to see unless the sun was low on the horizon thereby casting faint shadows (Fig. 5.19). In another case, they were made evident as subtle differences in soil temperature resulted in snow melting faster on the mounds than in the interstitial areas (Fig. 5.20). Were erosion of the hills themselves not problem enough, it is also evident that the areas covered by them are not as extensive as they once were. Natural processes, such as being covered by aeolian deposits (cf., Mrozowski 1994) has undoubtly destroyed or at least hidden some fields. Human activties, such as plowing associated with the expansion of relatively modern agricultural practices seems to have had the greatest impact. For example, a 19th-century settler in Huronia relayed that on his farm ancient 'cornhills in abundance were to be seen before the ground had been cultivated long enough to obliterate them' (Hunter 1899: 75). The construction of buildings, roads, and other forms of development associated with urbanization have also taken their toll on corn hills. In at least one case (Mrozowski 1994) natural processes buried a corn hill site.

The limited spatial distribution of corn hills (Fig. 5.18), especially in respect to the broad distribution of mounding (see Fig. 5.8), is certainly curious, even suspicious. Why have corn hills been found archaeologically only in New England and around the Great Lakes and not elsewhere? Is there something special or unusual about the physical environments of the region that is responsible for either their use or their preservation? Could they perhaps be natural and not agriculturally related? These questions can best be answered in reverse order.

Although there exists a number of archaeological reports of corn hills, there have been only three systematic and detailed investigations of these features to date. Geographer Conrad E. Heidenreich (1974) proved conclusively that corn hills near Creemore, Ontario were not natural. They were not the result of tree falls, animal burrowing, ants, or soil expansion due to reported freezing and thawing, but indeed were the result of agriculture in pre-European settlement times. Heidenreich analyzed soils on the hills themselves, in the interstices between hills, and in 'natural' areas away from the hills. He conducted textural, or particle-size analysis, calculated loss-on-ignition values to evaluate organic content, and he measured soil

Fig. 5.19. Relict corn hills near Creemore, Ontario. (Provided courtesy, and published by permission of, Conrad E. Heidenreich.)

Fig. 5.20. Ancient corn hills on the campus of Carroll College, Waukesha, Wisconsin, 1902. Visibility of these features is enhanced by differential snow melt. Urban development (in the background) undoubtedly obliterated many more such hills. (Photograph by George R. Fox.) (Published by permission of the State Historical Society of Wisconsin, Neg. No. WHi: (X3) 444498.)

compaction. Heidenreich's thorough assessment revealed that sandy loams, rather than the more common and extensive loamy sands were the preferred soils for cultivation. The hills were a result of mounding; they were shaped out of topsoil and wood ash created where the fields were initially cleared in part by burning. Soils in between hills were more compacted than those of either the hills themselves, or the uncultivated areas, indicating that hills were used year after year, and that the areas between them were used as paths within the fields.

The other two studies of corn hills involve the recently discovered Sandy Point site on Cape Cod. There, archaeologists uncovered sandy overburden from two 6 m by 6 m areas and revealed no fewer than forty corn hills. The hills measured 70–90 cm in diameter and were spaced 1 m apart. No heights were reported, but the hills were 'eroded' (Mrozowski 1994: 49). Some of these corn hills and surrounding soils were cross-sectioned and subjected to micromorphological examination. This aspect of the research concluded that unmounded soils around and under the hills were light colored, excessively drained sands and loams. The corn hills themselves were dark in color, of fine texture, and with a relatively high amount of organic matter (Curric 1994: 63, 66, 68). One of the more intriguing findings was the discovery of hoe marks in the micromorphologically analyzed thin sections that were not visible to the naked eye in the field (Currie 1994: 68, 69, 71, figs 6 and 8). Clearly, the hills were made by hoeing, and they involved additions. Radiocarbon dating placed the ages of these hills at AD 1100–50 (Mrozowski 1994: 59).

That the corn hills systematically studied were located on sandy soil is suggestive that there may be some environmental factor responsible for their use, specifically control of soil moisture. Such a conclusion, however, is unwarranted. 'The mounds near Creemore lie on an *elevated* spur of land' (Heidenreich 1974: 379; emphasis added). As noted earlier, corn hills were at one time found quite commonly near the shores of lakes in Wisconsin (Peet 1887: 395), and those at Sandy Point were within 50 m of the Atlantic Ocean (Mrozowski 1994: 48, fig. 1). Corn hills have been reported from both well-drained uplands, and comparatively moist low-lying areas. The far northern location of these relict features might suggest that there is some environmental factor other than soil or moisture that is responsible for their use, specifically temperature control. Such a conclusion, while plausible and possible, is not totally correct. Although relict corn hills were found near the northern limit of cultivation (see also Moodie and Kaye 1969), it must be remembered that mounding was reported by both early European explorers and ethnographers in more southern locales as well. Indeed, mounding was reported everywhere agriculture was practiced by aboriginal North Americans. It was simply a consequence of general tillage.

The question of archaeological evidence of corn hills being restricted to the Great Lakes and New England areas cannot be answered by investigating the function of these features. The answer lies principally in their survival and the wisdom of some very perceptive early scholars. A large debt of gratitude is owed to Increase A. Lapham, Andrew F. Hunter, Charles E. Brown, and Wilbert B. Hinsdale for identifying features that other individuals missed, for recognizing both their significance, and the danger of their disappearance, and for reporting them in publications. Had these individuals lived in a later time and worked elsewhere, Fig. 5.18 would undoubtedly look different than it does.

All these individuals mentioned that corn hills were rapidly disappearing in their day. Can we assume then that they are all gone now? No. Even though mechanized agriculture and urbanization has spread considerably during the past century, there probably remain a number of locales where ancient corn hills continue to exist. Indeed, some were found in eastern Massachusetts after the area as a whole had been settled for 300 years and, presumably, transformed (Delabarre and Wilder 1920). Although these have disappeared recently (T. Riley 1994: 104), the discovery of buried corn hills at Sandy Point demonstrates that ancient fields are extant. Some were found in a reasonably developed area as recently as two decades ago (Heidenreich 1974). Others must exist. Scholars who assume ancient corn hills are now all obliterated are putting blinders on themselves unnecessarily, and doing us all a great disservice in the process.

In summary, the documentary, ethnographic, and archaeological evidence, when viewed separately, and especially in concert, indicates that rainfed agriculture involving no permanent landscape modifications was widely practiced in North America at the time of the first European contact. It was not practiced everywhere, but it was common in specific environments across the continent. In the Southwest, it was restricted to the higher elevations. On what is considered the Plains, it was actually limited to the valley bottoms. Throughout the Eastern Woodlands rainfed agriculture was practiced in all types of local environments, but in the South and the Midwest it was more typical of bottomlands than uplands, although some of the latter types of locales were cultivated. It was found in both bottom lands and uplands across the Great Lakes region and through New England. If there is one common denominator to native rainfed agriculture it is that it was particularly prevalent, indeed the dominant form of agriculture, in the wooded areas.

The Notion of Swidden

Cultivation of woodlands, especially on the scale reported by some early explorers, requires before anything else that the land be cleared of trees. Although total clearance, or the removal of every tree, is not necessary, and the process can take place incrementally (see Doolittle 1984*a*), some, if not most or nearly all, trees must eventually be removed. The removal of trees is a two part process. Trees must first be felled. This can be, and indeed was, done with axes. As was the case with lithic hoes, stone axes are common archaeological artifacts across North America (e.g., J. Brown 1984: 36–7; Wilbur 1978: 28–9). The second part of the clearing process involves removal of the felled trees. Abundant evidence exists to show this was done with fire (Driver and Massey 1957: 226, map 38; Martin 1973).

That fields were cleared of trees by the use of axes and fire has resulted in many writers describing rainfed agriculture on plain fields as 'slash-and-burn' (Brasser 1971: 64; Fausz 1985: 227; Feest 1978: 258; Fenton 1978: 297; Fogelson and Kutsche 1961: 94; L. Larson 1980: 206, 209; Morissonneau 1978: 390; Peske 1966: 192; Pyne 1982: 47; Watson 1988: 41–2; Winters 1981: 30). To be sure, fields were slashed and burned, but to what extent does slashing and burning imply what is normally thought of as slash-and-burn agriculture. To many people slash-and-burn agriculture is synonymous with 'swidden' agriculture which involves 'shifting cultivation', or the periodic rotation of fields (Day 1978*b*: 795; R. Ford in S. Williams 1971: 16; 1974: 403–4; Hargrave *et al.* 1983: 335; Heidenreich 1978: 380–1; Mulholland 1988: 147; Patterson and Sassaman 1988: 115; Sale 1991: 301; Scarry 1993: 7; Starna, Hamell, and Butts 1984: 197; Stiger 1979: 138; Weatherford 1988: 83–4; R. White 1983: 23–4; Wolkomir 1995: 102; Yarnell 1974: 117). Indeed, some authors even draw explicit parallels between aboriginal North American rainfed agriculture with swidden cultivation as practiced in tropical rainforests today (R. Brown 1948: 13; Hann 1988: 127; Salisbury 1982: 33; M. Williams 1989: 31). Key components in swidden agriculture are the sequential abandonment of fields, regrowth of natural vegetation, reclearing, and recultivation of the same fields. Abandonment results in the shifting of fields and hence results in rotation. The regrowth of natural vegetation during the intervening period is, therefore, fallowing. Numerous scholars who have written about native rainfed agriculture in North America mention neither 'slash-and-burn', nor 'swidden', but do refer to field rotation and fallowing (Blackburn 1929; Cammisa 1984: 68; Carrier 1923: 97; G. Day 1953: 341; Hallett 1954: 60, 62–3; Salwen 1978: 57; Snow 1980: 47; Waselkov 1977: 517). The degree to which such claims imply the practice of slash-and-burn shifting cultivation, or swidden is debatable. However, there exists such a large number of works that mention this form of agriculture in the context of native North American agriculture that the notion needs to be discussed if not questioned.

Swidden may be one of the oldest and most widespread forms of agriculture on earth (D. Harris 1977). It is also one of the most studied. Detailed investigations of swidden have been conducted in Southeast Asia (Spencer 1966), Oceania (Conklin 1957), Africa (de Schlippe 1956), South America (D. Harris 1971), and Central America (Horst 1989). There exists a bewildering variety of swidden systems, each with its own peculiarities resulting from environmental, and cultural vagaries, and name (Grigg 1974: 57–74; Manshard 1974: 52–63; Ruthenberg 1976: 14–18). Despite numerous differences, however, there are certain common characteristics shared in this general type of cultivation strategy. Swidden is typically identified with warm, humid forests, environments known for poor soils in addition to abundant rainfall and luxuriant vegetation. Soils in such environments tend to be infertile because high rainfall leaches nutrients to depths beyond the root zone. To compensate, vegetation recycles itself rather quickly; the bulk of the natural biomass is above-ground woody vegetation.

In order to cultivate such environments, farmers have to clear the forest and fertilize the soil. Both of these activities are done by slashing-and-burning. Slashing is usually done a year or two before planting. Trees on the chosen plot are girdled; using an ax farmers chop a ring of bark from around each tree at a level approximately 1 m above the ground. Once the trees are dead, they are felled and burned, but burning does not result in total clearance. Stumps remain rooted in the ground and fallen trunks and large branches lie charred on the surface. Burning does, however, create a nutrient-rich layer of ash. Crops are planted in this ash, between the stumps and logs (Fig. 5.21).

Swidden fields are typically cultivated only a few years, frequently just one, and then abandoned. There are two reasons for such rapid abandonment.

Fig. 5.21. A typical swidden field in Belize, Central America. Note the stumps, and fallen trunks and large branches remaining after burning. Note, too, the ash layer on the surface and small size of the cleared area. (Photograph by author.)

First, both cropping and leaching remove nutrients from the ash rapidly, thereby resulting in yields the second and subsequent years which are significantly lower than those of the first year of cultivation (Nye and Greenland 1960). Second, and every bit as important as fertility loss, but much less-known and appreciated, weed encroachment is an increasingly greater problem through time as the natural vegetation tries to regenerate itself. A point is reached after only a few years when it is easier for the farmers to clear a new plot from a mature forest than fight the weeds and the succession sequence (Stromgaard 1989).

By virtue of being abandoned, swidden plots lie fallow as natural vegetation takes over. Short-lived, rapidly reproducing grasses, 'weeds', are the pioneer species. These are followed by shrubs and bushes which, in turn, are followed by trees that are long-lived and reproduce slowly. The longer a particular plot remains undisturbed, and succession allowed to run its course, the closer the vegetation will come to approaching a so-called 'climax' state. In tropical environments this can take up to twenty years. In more temperate or cooler climates it can take more than 150 years (Sarmela 1987: 242).

Assuming farmers continue to clear and abandon plots, a time will come when a formerly cultivated plot is cleared again and recultivated. The frequency of cultivation, or the number of years a specific plot is cultivated in comparison to the number of years it lies fallow (Joosten 1962) is, all other factors held constant, largely a function of the number of people being sustained from a given region (Boserup 1965). Generally, when the population is small, and, hence, an abundance of land exists, the frequency of cultivation is low. The area under cultivation is quite small and the area in fallow is great. These conditions allow much time for regrowth of natural vegetation and, accordingly, a 'forest fallow' form of swidden. With a larger population, the amount of land under cultivation to that in fallow at any given period is proportionally greater, and the amount of time available for regrowth on abandoned plots is less than in forest fallow systems. Appropriately, swidden systems of this nature are referred to as having a 'bush fallow'. When the population of a region is sufficiently large that the frequency of cultivation is great, regeneration of natural vegetation is extremely limited and the system of cultivation is described as having a 'grass fallow'. In sum, forest fallow systems are characterized by a few small plots with many stumps; clearing is partial. Grass fallow systems of agriculture have numerous large fields or extensive tracts of cultivated land with no remaining stumps; clearing is total. According to Ester Boserup (1965, 1981), a prominent authority on the correlates between population, technology, and frequency of cultivation, digging sticks and axes are associated with long fallow swidden systems. Hoes are used only when fallows are relatively short (Boserup 1965: 24)

True to its name, slash-and-burn shifting cultivation involves slashing, burning, and field rotation. Fields can, of course, be slashed and rotated without being burned, or burned and rotated without being slashed. It is also possible that fields can be slashed without being burned and rotated, burned without being slashed and rotated, or rotated without being slashed or burned. True swidden, however, requires all three activities. The extent to which native North American rainfed agriculture at the time of European contact was a swidden system requires that evidence of slashing, burning, and field rotation be scrutinized. Burning, however, is a complex topic to deal with in and of itself; it was used for a multitude of reasons. As was pointed out earlier (see Chapter 2), it was used in the husbandry of non-domesticated plants. It was also used to clear undergrowth of forests in order to: enchance the growth of forage and therefore improve hunting, drive game, drive away reptiles and insects, facilitate travel, and improve visibility; and for offensive and defensive purposes in times of war (G. Day 1953: 334–9; H. Maxwell 1910: 86–94; E. Russell 1981, 1983; Thompson and Smith 1971). In order to address the swidden issue, therefore, evidence for burning has to be related directly to slashing and planting. Similarly, evidence for the felling of trees by means of slashing as part of swidden cultivation has to be restricted to those situations which involved burning and planting as well. Forests can be cleared for a number of non-swidden activities including the preparation of buildings sites, and the harvesting of building materials and fuel wood (G. Day 1953: 329–30).

There exists a mere twelve documentary references to slashing, burning, and planting. All but one of these descriptions come from the eastern half of the continent (Fig. 5.22). Perhaps the most commonly cited, and plagiarized (Strachey 1953: 118; see Rountree 1989: 4), early account to be used as evidence of native swidden is that provided by Smith of practices in Virginia in 1607. He reported:

> The greatest labour they take, is in planting their corne, for the Country naturally is overgrowne with wood. To prepare the ground they bruise the barke of the trees neare the roote, then doe they scortch the roots with fire that they grow no more.
>
> The next yeare with a crooked peece of wood the beat up the weeds by the rootes, and in that they plant their Corne (J. Smith 1910*c*: 357).

In the same area a few years later, in 1613, Spelman provides a similar account, but one that contains additional information. He noted:

> They take most commonly a place about their howses to sett their corn, *wh*ich if ther be much wood, in that place the[y] cutt doune the great trees sum half a yard aboue the ground, and ye smaller they burne at the roote pullinge a good part of barke from them to make them die, and in this place they digg many holes (Spelman 1910: cxi).

Champlain provided some informative accounts of clearing. Two from Canada place emphasis on burning, one even downplays slashing on the basis of having inadequate tools. Writing of his brief observations of a group of Al-

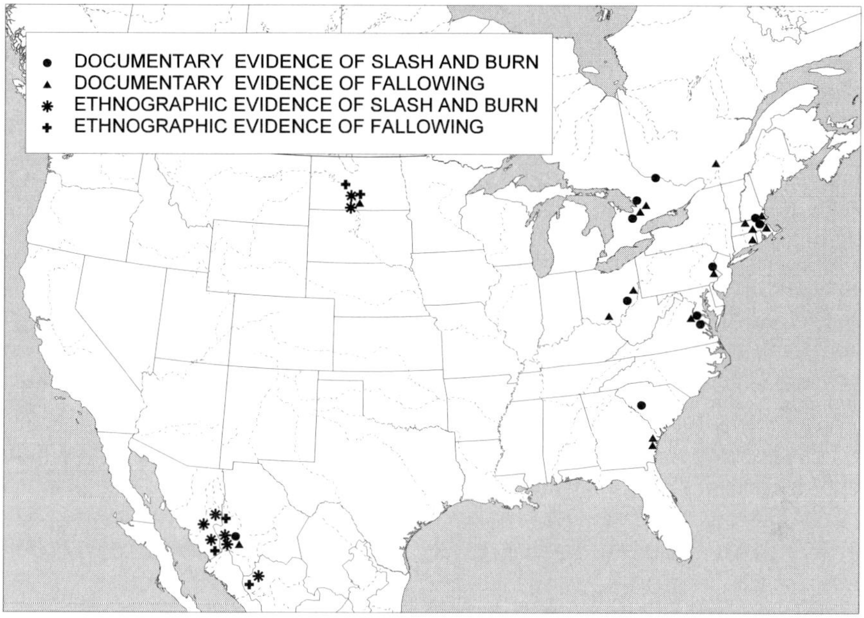

Fig. 5.22. Locations cited in the text for which there is either documentary or ethnographic evidence of forests being slashed, burned, and planted, and for which there is documentary evidence of fallowing or short-term use of fields.

gonquins who practiced only casual agriculture (Conrad Heidenreich, personal communication 1996) along the Ottawa River in 1613 Champlain said:

> When they wish to make a piece of land arable, they burn down the trees, which is very easily done, as they are all pines, and filled with rosin. The trees having been burned, they dig up the ground a little, and plant their maize (Champlain 1907: 244).

Of activities in Huronia in 1616 he relayed:

> They clear it [the land] up with great difficulty, since they do not have the implements adapted to this purpose. A party strip the trees of all their branches, which they burn at their base in order to kill them. They clear carefully the land between the trees, and then plant their corn (Champlain 1907: 327).

Champlain's comment about not having tools to fell trees is curious. There is certainly archaeological evidence of stone axes and, as will be seen, Champlain referred to chopping down trees elsewhere. Furthermore, there is a slightly later account from Huronia which reiterates some of what Champlain said, but includes reference to slashing. In 1623 Sagard-Théodat commented:

> Clearing is very troublesome for them, since they have no proper tools. They cut down the trees at the height of two or three feet from the ground, then they strip off all the

branches, which they burn at the stump of the same trees in order to kill them, and in course of time they remove the roots. Then the women clean up the ground between the trees thoroughly, and at distances a pace apart dig round holes or pits. In each of these they sow (Sagard-Théodat 1964: 8, 1939: 103, 1865: 133).

The earliest, and in many respects the most comprehensive account of slashing, burning, and planting by natives of North America comes from around Gloucester Harbor. Chronicling activities there in 1606 Champlain wrote:

Some of the land was already cleared up, and they were constantly making clearings. Their mode of doing it is as follows: after cutting down the trees at the distance of three feet from the ground, they burn the branches of the trunk, and then plant their corn between these stumps, in course of time teering up also the roots (Champlain 1907: 92).

The six relations just presented were all quite early and and date within seventeen years of each other. The following six accounts are, in contrast, somewhat late—150 years later—and well within the time when European settlers had a significant impact on the landscape. Regardless, they indicate that slashing-and-burning was a viable agricultural activity. Writing of the Southeast in 1775, Adair (1968: 405) said that native farmers:

deadened the trees by cutting through the bark, and burned them, when they either fell by decay, or became thoroughly dry . . .. By the aforefaid difficult method of deadening the trees, and clearing the woods, the contented natives got convenient fields in procefs of time . . .. Now, in the firft clearing of their plantations, they only bark the large timber, cut down the fapplings and underwood, and burn them in heaps; as the fuckers fhoot up, they chop them off clofe by the ftump, of which they make fires to deaden the roots, till in time they decay (see also Williams 1930: 435).

Of the Delaware Valley in 1749, Kalm repeated an earlier observation of Champlain's stating:

When the *Indians* intended to fell a thick strong tree, they could not make use of their hatchets, but, for want of proper instruments [saws?], employed fire. They set fire to a great quantity of wood at the roots of the tree, and made it fall by that means (Kalm 1972: 220).

He went on to say that:

if the ground where they intended to make a maize-field was covered with trees, they cut off the bark all round the trees with their hatchets, especially at the time when they lose their sap. By that means the tree became dry, and could not take any more nourishment, and the leaves could no longer obstruct the rays of the sun from passing. The smaller trees were then pulled out by main force, and the ground was a little turned up with crooked or sharp branches (Kalm 1972: 220–1).

The first portion of Kalm's statement does not mention slashing, but only burning. It shares this similarity with comments made by an anonymous writer in Massachusetts sometime in the late 1700s. That writer noted that natives:

> make a fire round the foot of every tree, as many trees as ftanding on the ground they intended to clear, until the bark of the tree is burnt through, for trees are killed very eafy in this manner. They planted while trees are ftanding after they are killed; and as foon as trees are fell, they burnt it of fuch length that they might roll the logs together, and burnt them up to afhes, thus they do till they get it quite clear (Anonymous 1804: 101).

The second part of Kalm's statement does not mention burning, but only slashing. It has this in common with observations made by David Zeisberger in the Ohio Valley in 1776:

> Their hatchets . . . were not used for splitting wood but only to kill trees, as no more is necessary than to chop through the bark, if this is done at the right time. . . . Trees were killed wherever they wished to have space for planting (Zeisberger 1910: 28–9).

Although both the anonymous report from Massachusetts and Zeisberger's report from Ohio are incomplete in terms of slashing and burning, together, and in the context of Kalm's account, they do portray a picture of forest clearing as part of swidden cultivation.

The final documentary account of slashing-and-burning in reference to agricultural cleaning comes from far to the southwest, but form the same late time period. Writing of the Rarámuri in 1777 a Franciscan priest said:

> The Indians have fields called *rosas*, on the sides of this canyon, even though these are steep and rugged. In February and March these slopes are cleared of brush and brambles, which are left until May or early June, when the rabble is burned. They take care to insure that all the waste burns and that the soil is evenly coated with ashes. These fields are then planted (Falcón Mariano 1979: 109–10).

Ethnographic evidence of slashing-and-burning in order to clear forested areas for agriculture is limited, in no small way because this land extensive practice grew increasingly difficult to employ as the non-native population grew in number and expanded across the continent. Between 1800 and today, the only descriptions of the practice come from reasonably remote, sparsely inhabited areas. Perhaps not surprisingly the only place where slashing and burning is currently practiced is in the mountains of northwestern Mexico, where indigenous groups remain dominant.

Girdling trees, letting them die and then dry for a year or so, before burning followed by planting is not uncommon among those Névome who live in the village of Nabogame close to the Continental Divide (Laferriére and Van Asdall 1992: 162). A similar set of activities was found among those Névome who lived at Onavas, Sonora in the 1960s (Pennington 1980: 144–5). In the 1930s the Warihio of southern Sonora and western Chihuahua were described as practicing '*milpa*' agriculture which required clearing of the forest (Gentry 1963: 74). Milpa agriculture was also discussed of the Rarámuri at this same general time period, and details provided mention both slashing and burning (Bennett and Zingg 1935: 50; Graham 1994: 10, 12) as did their southern neighbors, the Odami (Pennington 1969: 59, 61, 62).

The only place other than the mountainous southwestern portion of the continent from which there is ethnographic evidence of slashing-and-burning is the upper Missouri Valley, home of the Hidatsa. There, large trees were girdled and felled and brush was heaped before burning. At least two rather detailed descriptions of these activities were provided early in the 20th century (Will and Hyde 1964: 77; Wilson 1917: 13).

Archaeological evidence of slashing-and-burning is fugitive at best. Although a number of researchers have made claims about pre-European swidden in North America, none have anything that can be even remotely construed as solid archaeological evidence; other types of information are typically relied upon. A statement made about agriculture at the Blain Village site in Ohio will serve in the way of example:

> [T]he higher terraces *could* have been farmed by slash and burn techniques. That one of the many forms of slash and burn farming was practiced by the villagers is supported by two pieces of *circumstantial* evidence: (1) While some animals [remains of which were found during excavations] preferring a mature Deciduous-Forest habitat zone were hunted, the majority of species taken prefer Forest Edge, Second Growth, or Prairie habitat zones. Such zones develop in areas that have been burned over to clear land for farming. (2) Historic records indicate that Prairie conditions prevailed in the immediate site locality when the area was opened for White settlement in the late eighteenth century (Prufer and Shane 1970: 251). (Emphasis added.)

Although there is some archival and ethnohistorical evidence of slashing-and-burning for the purpose of clearing forests for agriculture, such do not necessarily indicate swidden. The other key element in true swidden agriculture is the periodic shifting of fields. Abandonment, fallowing, and reuse, or field rotation is essential for swidden. Evidence of such activities are of varying quantity and quality.

As is evident in Tables 5.1–5.4, there exist at least twenty-two separate accounts of fields having been abandoned at the time they were observed by European explorers or settlers. With few exceptions these can be dichotomized into early 17th-century accounts from New England and late 18th-century accounts from the Southeast. The former area was heavily populated between AD 1500 and 1600. It then suffered major demographic reductions as a result of plagues, and statements to this fact accompany nearly every chroniclers' account of abandoned fields in the region. The latter area was also heavily populated, but perhaps not as much so as the coastal areas of the Northeast. Nevertheless, many explorers of European ancestry entered this area only in the late 1700s and their reports of abandoned fields are also accompanied by statements of depopulation, quite in contrast to the early accounts of the de Soto expedition. Accounts of abandonment, therefore, must not be accepted as evidence of slash-and-burn agriculture, at least without exercising great caution (Patterson and Sassaman 1988: 115).

To be conservative, evidence of field rotation must be restricted to explicit descriptions of fallowing, short-term or temporary use of fields, and field reuse. As was discussed earlier, there are only four documentary accounts that clearly mention 'fallowing' (Champlain 1907: 71; Falcón Mariano 1979: 110; H. Maxwell 1910: 83; Winthrop 1863: 515). There are another thirteen early European accounts of fields being cultivated for a relatively brief period before being allowed to revert back to natural vegetation. Only two of these accounts mention precise lengths of cultivation periods. Ten state that decreases in soil fertility were the cause for abandoning fields and moving cultivation elsewhere. One says weed encroachment was the chief factor and three state that shortages of locally available fuelwood were also a, if not the, factor responsible for changing cultivation locations. These accounts come mainly from the Southeast, New England, and French Canada, but also from the Ohio, Missouri, and Delaware valleys (Fig. 5.22).

The earliest accounts of short-term rainfed agriculture come from the Georgia coast in 1570. Speaking of natives and their agriculture in this area, one Jesuit missionary, Antonio Sedeño (1946: 424) said 'the land is so miserable, they move with their huts, from time to time, to seek other lands that can bear fruit' (see also L. Larson 1980: 208). Another priest, Juan Rogel (1946: 476) reported that 'the land will not support it [permanent agriculture], because it is weak and miserable and tires very quickly. And thus . . . they spread out and they move so often' (see also L. Larson 1980: 207).

Writing of the natives of New England in 1634, William Wood (1967: 13) said they 'plant corne eight or ten yeares in one place'. In his dictionary of native languages in Rhode Island written in 1643, Williams (1973: 167) lists words for 'New ground', and 'Fields worne out'. An anonymous writer in the same general area but later, in the 18th century, added that '[w]hen they find that their fields will fail, they are to prepare another piece of land' (Anonymous 1804: 101).

These observations are paralleled by similar ones made by two Jesuits who served in Huronia in 1639, by a priest who served near Montreal in 1716, and by Lafitau who observed the Iroquois in 1712–17. François du Peron (1898: 153) wrote that the land 'produces only ten or twelve years at most; and when ten years have expired, they are obliged to remove their village to another place'. Jean de Bréheuf (1898: 7) noted that aboriginals 'change their abode sometimes—certainly when there is no longer sufficient wood for their fires, or when the land, long tilled, produces scanty crops'. Similarly, but later, L. A. de Bourbon (1900: 25) reported:

> that these savages could no longer remain in their village, because the soil was exhausted and the woods too far away; and that it was absolutely necessary for them

to settle elsewhere. (As the savages cultivate nothing but Indian corn, which greatly impoverishes the soil, their lands cannot last them long.)

Confirming these reports, Lafitau (1977: 69–70) noted of native Canadians that: '[a]s the Indians never manure their ground and do not let it lie fallow, it is soon exhausted (and worn out). Then they are forced to move their villages elsewhere and make new fields in new lands'.

Also from the late 18th century come reports of short-term cultivation and allied problems elsewhere. Relating conditions in the Ohio Valley, George H. Loskiel (1794: 66) noted '[w]hen the strength of the soil is exhausted, they remove their plantations'. Zeisberger (1910: 44) reported that indigenous farmers cultivated bottomland 'because it yields abundant crops for many years. When, however, their fields begin to grow grass they leave them and break new land, for they regard it as too troublesome to root out the grass'.

Two 18th-century accounts, one mid and one late, are interesting in that they not only mention short-term cultivation and the causes for abandonment, but they also say something about new lands being cleared for cultivation. Among the 'Rricara' (Arikara) of the Missouri Valley, Tabeau (1939: 69) reported 'that they cultivate only new lands, being forced to change their habitation often for want of wood which they exhaust in five or six years'. In the Delaware Valley Kalm (1972: 299) noted that nativess 'cleared new ground, as soon as the old had lost its fertility'. He also reported that:

the Indians make use of; that is, to sow uncultivated grounds, as long as they will produce a crop without manuring, but to turn them into pastures as soon as they can bear no more, and to take in hard new spots of ground, covered since time immemorial with woods, which have been spared by the fire or the hatchet ever since the creation (Kalm 1972: 300).

These thirteen accounts, individually and together, are telling. They indicate that decreasing soil fertility and weed encroachment were problems. However, they also indicate that fields were cultivated for more than a few years and that non-agricultural factors also precipitated movement. Finally, the reference to clearing of what might be considered virgin forests and the lack of references to returning to formerly cultivated and later overgrown fields casts a shadow of doubt over the notion of swidden cultivation as being a common practice at the time of contact. Ethnographic evidence of short-term cultivation, fallowing, and field rotation is mixed on this point.

As in the case of clearing woodlands by slashing and burning for rainfed agriculture, evidence for short-term use of fields comes form the Upper Missouri Valley and the mountains of northwestern Mexico (see Fig. 5.22). The Mandan of North Dakota had in the early 19th century relocated their fields after the soil became exhausted and yields declined (Henry and Thompson 1897: 323–4). Later in the century, and on into the early 1900s, the Mandan as well as the Arikara and Hidatsa around Fort Berthold continued to do the same thing, but they also reused old fields after allowing them to lie fallow a few years (Mathews 1877: 11, see also Will and Hyde 1964: 98). One writer, (G. Wilson 1915: 43) once reported that the Hidatsa had no regular system of rotation (see also Will and Hyde 1964: 83), but later noted that they fallowed fields for two years (G. Wilson 1917: 114).

In northwestern Mexico, the Odami are described as preferring newly cleared land for cultivation, abandoning fields after a few years and allowing them to revert back to a wooded vegetative state (Pennington 1969: 59, 62). The Névome of Nabogame have a variety of rotation systems—some fields are cultivated annually with no fallow, some are cultivated one year and fallowed the next, and some are cultivated one year and left fallow for several (Laferriére and Van Asdall 1992: 162). Whether or not fallowed fields become reforested is not known. However, this system of cultivation is suspected to be quite recent, as the amount of cleared land has been increasing during the past few years (Laferriére and Van Asdall 1992: 172). The Warihio also abandon plots when soil fertility declines, but the fallow sequence, if one exists, has not been reported (Gentry 1963: 74). As with the Névome, this system of cultivation may not be all that old (Gentry 1963: 84–5).

Archaeological evidence of short-term field use, fallowing, and rotation is almost non-existent. There exists only one study that has produced material remains suggestive of such. In the previously mentioned study of field houses in one part of northern Arizona's Coconino National Forest, archaeologists found that each structure was used for only a brief period and few were used contemporaneously. On the basis of environmental conditions, especially edaphic qualities, and chronological sequencing, it was concluded that field rotation occurred between AD 900 and 1300 in response to soil exhaustion (Henderson 1979: 43–51). Given the forested environment, it is plausible that some form of slash-and-burn shifting cultivation was practiced. However, the reliance on field houses as evidence of swidden has been questioned (Wilcox 1978), and it has been demonstrated that forest litter can be burned, and fields planted and rotated without removal of the trees themselves (A. Sullivan 1982).

The paucity of concrete evidence, reservations about such, and alternative scenarios for shifting cultivation have not deterred prehistorians from speculating on field rotation. Some unsubstantiated qualitative claims for shifting cultivation come from the Mississippi (Butzer 1990*a*: 34) and Ohio valleys (Muller 1986: 214, 1987: 260). Elsewhere, claims tend to be quantitative, suggesting either years in cultivation or years in fallow. These come from other places in the eastern half of the continent and seem to be based on documentary assessments, although few are mentioned as such. For the central Mississippi Valley short fallows of only one or two years (Waselkov 1977: 517), and four to six years (Hargrave *et al.* 1983: 335) have been suggested. During the 1970s farmers in Huronia estimated that they could not get a crop after four to six years without the use of fertilizers. This, combined with documentary evidence has led one prominent scholar (Heidenreich 1971: 181, 188, 1978: 380, 381), to conclude that prehistoric natives in the area cultivated for a similar period and then waited thirty-five to fifty years for the forest to recover before they cleared and planted again. The fifty-year recovery rate has also been applied to the Mohawk Valley, but here as a maximum. It has been suggested that fields may have been reused after the forest reached a less-than-climax stage of regrowth (Bond 1985: 23). A period of reuse between ten and twenty years has also been suggested for the Hudson Valley (Salwen 1975: 57).

The pattern proffered, with little in the way of concrete evidence, for New England as well as the Southeast parallels that from Huronia and New York. It has been suggested that prehistorically, people in these regions cultivated plots for eight to ten years, and then left them fallow and allowed the forest to regrow for between twenty-five and fifty years (P. Thomas 1976: 13; Cammisa 1984: 68). Nearly identical figures were used in reference to what the Choctaw might have done prior to contact (R. White 1983: 24).

The claims just cited for prehistoric swidden agriculture are just that—claims. Essentially, they only reify the long-standing notion that Eastern Woodland natives must have practiced some type of slash-and-burn shifting cultivation. Indeed, a handful of scholars have questioned this notion and have been rather ardent in their suggestions to the contrary. Interestingly enough, opposition points of view come from the same regions as claims of support. Arguments against swidden, based on the lack of firm evidence and the gossamer nature of the claims in support of the notion come from the Mississippi Valley (Woods 1987), Huronia (Trigger 1969: 28), New York (Sykes 1980: 50, 52), southern New England (H. Russell 1980: 140; Warner 1972: 41), the Carolinas (Merrell 1982: 15–16), Florida (Covington 1978: 10), and the Southeast in general (Hudson 1976: 290; Murphy and Hudson 1968: 28).

Were these claims not enough, and in a very real sense they are not, there exists solid evidence that slash-and-burn shifting cultivation was neither widespread nor a common practice. This evidence includes the difficulty with which forests were felled and therefore the likelihood that pre-European people would have maintained such a system; that stumps were removed from fields; that formerly forested areas can be continually cultivated for long periods without additives; that abandoned plots took exceptionally long periods of time to become reforested; that fallowing and field rotation, when documented involved removal of grasses not trees; and that archaeological evidence of clearing is of such a magnitude that vast areas rather than small field-size parcels were deforested.

As for the difficulty with which forests were cleared, five of the previously cited documentary accounts mention that the felling of trees was done only with great labor (e.g., J. Adair 1968: 405; J. Smith 1910*c*: 357). Three specifically state that lack of suitable land was the principal reason (Champlain

1907: 327; Kalm 1972: 270; Sagard-Théodat 1964: 8). There exist an additional three accounts which indicate that native tools were inadequate to clear forests. Describing conditions in late 16th-century Georgia, a Jesuit said that indigenous farmers 'do not have that with which to clear trees' (Sedeño 1946: 424; see also L. Larson 1980: 208). Nearly a century later on the Carolina coast, Hilton (see Swanton 1922: 63) said '*Indians* plant in the worst land because they cannot cut down the Timber in the best'. This observation was echoed in the early 1700s by another writer who noted that 'the Indians are not inclinable to settle the richest land, because the Timbers are too large for them to cut down' (J. Lawson 1937: 84).

Although a few scholars have recognized and appreciated the difficulty of felling trees with stone axes (e.g., Heidenreich 1971: 175), the problem seems to have eluded most. Doubtless this is a function of parallels drawn from swidden agriculture in the humid tropics today, a system often characterized by the 'ease' with which trees are felled. As has been recently articulated, however, slash-and-burn shifting cultivation in such environments may not be as ancient as long thought. Swidden agriculture in the Amazon, for example, may well be historic, owing its existence to the introduction of steel axes (Denevan 1992*b*, 1998). Old World farmers had a history of making superb tools for felling trees that extended back to the Stone Age (Iversen 1956), and Scandinavian settlers brought to North America a tradition of slash-and-burn agriculture (Jordan and Kaups 1989: 100–5, 127–9). The practice was carried west by descendants of European settlers as far as the Southwest and appears to have been sufficiently common to have gained a place in popular culture (cf., Grey 1925: 207–8, 217–19). Not all native North American farmers were adept at using steel axes, however, and continued to use them as they had the stone predecessors (e.g., Lafitau 1977: 70–1).

One of the hallmarks of swidden cultivation is the relatively short time fields are cultivated in comparison to the time they are left in fallow. Given that such fields are typically used only a few years, crops are planted between stumps which are left untouched. Of the documentary accounts of clearing mentioned previously, two do state that stumps were left in the field with crops planted between them (Champlain 1907: 92, 322). Two more, however, clearly state that roots were dug-up or removed eventually (J. Adair 1968: 405; Sagard-Théodat 1964: 8). One other document refers to root removal. Zeisberger's dictionary of the Delaware language includes a term for 'the rooting-out of trees' for farming (Mahr 1955: 210). In addition to documentary sources there are ethnographic descriptions of roots being removed from fields where trees once stood. Lincecum reported such from the Southeast (Campbell 1959: 16). G. Wilson (1917: 13) noted it for the Hidatsa, and Pennington (1963: 50) observed it among the Rarámuri.

That effort was exerted to remove stumps clearly indicates that aboriginal farmers were more prone to permanent cultivation than to shifting frequently. Digging up roots is hard work. It must, however, have been easier than chopping down trees, even if girdling was used. Accordingly, it appears that references to clearing by slashing and burning were observations of land intended to be cultivated for some time and not soon to be abandoned. When fields were abandoned, however, they were not intended to be recultivated either soon, or after the forest regrew. Evidence for this conclusion exists in the form of field tests of cultivation and observations that fields were not reforested quickly.

Perhaps the most graphic example of just how long a formerly forested area can be continuously cultivated comes from Mesa Verde. There, in 1918, a plot slightly smaller than 1 ha in area was cleared from the piñon–juniper forest. Crops were planted on this tract annually for seventeen consecutive years—the idea was to see how feasible rainfed agriculture would have been prehistorically. Neither fertilizer nor water was applied during the tests. Crop failures occurred in only two years (Franke and Watson 1936). The plot was then cultivated for another forty years, and during this time only one year experienced crop failure (Wenger 1980: 56). Long-term rainfed agriculture was not only possible (see also Ermigiotti 1989), it was also successful prehistorically (Cordell 1975; Cutler and Meyer 1965: 149–50) in this locale, and feasible in other forested upland areas of the Southwest as well (Bye and Shuster 1984; Van West 1994). Although those tests were not designed or intended to refute any notion of swidden or shifting cultivation, the conclusions certainly do. As will be demonstrated directly, they also are significant in one other unintended respect, reforestation after abandonment.

The rate of forest regrowth is certainly central to the issue of slash-and-burn shifting cultivation. After all, if formerly cultivated fields were to be reused after the forest has regrown, the frequency of cultivation, or the number of years in cultivation compared to the number of years fallow, would be contingent on the rate at which trees regrow. One excellent study of regrowth was undertaken in Michigan (Beckwith 1954). It made several interesting findings about plant succession, but the most important one in terms of prospects for swidden was that mature trees predominated only after twenty-five years (Beckwith 1954: 370). Two documentary accounts lend substance for historic times. Writing of New-Netherlands in the 1600s Adriaen Van der Donck (1841: 149) noted of his travels with natives:

> we have come to a piece of young woodland. When I have told them, in conversation, that they would do well to clear off such land, because it would bear good corn, that they said, 'it is but twenty years since we planted corn there, and now it is woods again'.

A century later in New Jersey, Kalm (1972: 245), as discussed earlier in the context of extensive tracts with abandoned fields, reported: 'large tracts covered with trees, only twenty, thirty, or forty years old' in which natives 'formerly had their little plantations in'.

Although this evidence is compelling, puzzling circumstances remain. In numerous places the reforestation rate was far longer than twenty to fifty years. As will be recalled, there are a remarkable number of documentary accounts that contain references to old fields or abandoned fields. Many of these were not cultivated during the decades immediately prior to being observed. For example, in the 1750s two separate writers in New England and Pennsylvania found old fields abandoned for at least eighty years and not yet fully reforested (H. Maxwell 1910: 85). So numerous were these long-cleared and little-regrown parcels that settlers of European ancestry sought them out for their own agricultural endeavors (e.g., Axtell 1992: 236; Jordan 1973: 251; Gabriel 1926: 32).

As hinted at earlier, the previously discussed experiments at Mesa Verde also have a bearing on the reforestation issue. Cultivation of the test plot at the national park terminated in 1975. By 1988, however, the 'natural' vegetation had barely begun to recover. Only a sparse stand of sagebrush existed where trees once stood and where crops were grown only thirteen years earlier (Stiger 1979: 136, 139).

Lending additional credence to the argument against slash-and-burn shifting cultivation are those documentary accounts of field preparation. Of the thirty or so reports discussed previously (see Table 5.5), *all*, explicitly mention the tillage and/or burning of brush or grass, not removal of trees!

There is absolutely no doubt that native North America slashed and burned forests, and shifted fields periodically. This does not, however, mean that they were practicing swidden agriculture. The length of the fallow period, when one existed, was relatively brief. Similarly, the amount of forest regrowth on abandoned or fallowed plots was quite limited. The bulk of evidence strongly suggests that once fields were cleared, the intent was to cultivate them permanently, or at least for very long periods of time, perhaps fallowing every other year or so. As more fields were needed, in response to either population increases or decreases in the productivity of existing fields, farmers cleared new fields from the remaining forests. Old fields were rarely reforested before being recultivated. As one prominent writer (Pyne 1982: 50) has said so eloquently, 'there is no reason to think that the process of converting forest to plain was not expanding'. The effect of this permanent cultivation with broad scale clearance—as evidenced by numerous fields (see Table 5.2, Fig. 5.1), extensively cultivated areas (see Table 5.3, Fig. 5.2), large fields (see Table 5.4, Fig. 5.4), and very few small fields (Fig. 5.4)—was obviously wholesale vegetation change. That conditions similar to those for which there is evidence at the time of contact existed prehistorically is confirmed by paleoenvironmental evidence, especially from the upland parts of the Southwest, and the Southeast.

The extent to which Mesa Verdeans cleared the woodlands for rainfed agriculture is an issue that has long been pondered (Rohn 1963*a*: 12–14). Studies conducted in the 1960s and 1970s addressed the matter and came to some rather definite conclusions. Changes evident in the pollen record were of sufficient magnitude to indicate that extensive tracts of mesa-top lands were cleared, cultivated, and then overgrown by secondary forests (Wyckoff 1977). Additional evidence extracted during coprolite, flotation, and faunal analyses was used to confirm that agriculture not only dominated the landscape (Stiger 1979) but it did so to such an extent that no lands remained fallow or uncultivated, thereby ruling out any possibility that a swidden system of cultivation prevailed (Kohler and Matthews 1988: 539). Similar, but less detailed studies in southeastern Utah (Matson, Lipe, and Haase 1988: 258) suggest that the Mesa Verde findings may well apply to upland areas in other parts of the Southwest.

The paleoenvironmental studies carried out in the Four Corners area were paralleled by similar ones in various parts of the Southeast. Materials extracted from paleofecal deposits in Kentucky (Watson 1974: 238), and charges in the composition of charred wood and seed assemblages in archaeological sites in eastern Tennessee (Chapman *et al.* 1982: 118; Delcourt *et al.* 1986: 347) and western Alabama (Caddell 1982: 34) have been used to argue that forest clearing for rainfed agriculture was most extensive, involving both large areas in specific locales and being characteristic of the entire Southeast (Lopinot and Woods 1993; B. Smith 1986: 51).

In summary, the notion of slash-and-burn shifting cultivation or swidden agriculture has a long and checkered history. Numerous scholars have written, and millions of lay people believe, that such a food production system dominated the pre-European landscape of much of North America. However, some scholars have written, and some lay people know, differently. To be sure, there is some fragmentary evidence to demonstrate that at the very least some aboriginal farmers, in some places, at certain times did practice a type of swidden. However, the overwhelming corpus of evidence indicates clearly and convincingly that although native North Americans cleared the forest by slashing-and-burning, they did so for permanent clearance. Cultivation typically involved little, if any, field rotation; perhaps biennial fallowing, but rarely, if ever, regular shifting of fields to new locations. The notion of swidden should perhaps be more appropriately labeled a myth. Of course, old myths usually suffer long, excrutiatingly painful deaths, if they die at all. It may well be that the swidden myth will never die despite incontrovertible evidence to the contrary.

The Notion of Fertilizing with Fish

Probably every school child has been taught that the first European settlers in New England had a difficult time surviving, and would not have made it had the aboriginal farmers not taught them various skills. One of these skills involved the burying of a fish in each hole along with the maize seeds that were planted. These fish, as legend holds, were fertilizer that ensured the maturation of the crop, and, hence, the prosperity of the people.

The earliest scholarly study giving rise to this notion was published more than a century ago (Goode 1880). It was based on linguistic evidence, specifically that *menhaden*, a type of herring, is the Algonquian word for fertilizer; and documentary evidence by Europeans who state: (1) that maize was planted in the 'Indian' manner, by burying fish with seeds (e.g., Johnson 1910: 114; Morton 1947: 60), and (2) that one particular native, Squanto, showed them how to do it. Numerous writers (e.g., Bennett 1955: 375; Blackburn 1929; R. Brown 1948: 12; Carrier 1923: 98; Wissler 1922: 23) have accepted this assessment without question and in so doing have contributed to its becoming common knowledge by the middle of the 20th century. It was not until three-quarters of a century after it was proposed that someone questioned the notion. To geographer Erhard Rostlund (1957) goes credit for first challenging this popular idea. He revealed that: (i) early documents refer to European settlers, not aboriginals, burying fish while planting seeds (e.g., J. Smith 1910*d*: 952); (ii) no document referred exclusively to natives planting in this manner but that some documents state explicitly that they did not manure (e.g., Laudoniére 1975: 15); (iii) Squanto had spent the years 1614–19 living in Spain and England; and (iv) the linguistic evidence is spurious in that it is the same as people today equating dung with manure—for example, we often say manure when we refer to dung even though dung is not manure until it is applied to fields.

Rostland's article, for all its strengths, was not, however, comprehensive. Later, anthropologist Lynn Ceci (1975*a*) used documents not studied by Rostland, assessed Squanto's European travels and experiences, evaluated the ecology and economics of fertilizing with fish, and concluded that the practice did not have native North American origins. A few authors accepted Rostlund's and Ceci's findings (e.g., Axtell 1981: 294; Cronon 1983: 45). Some, however, did not (e.g., Fowler 1970; Hughes 1983: 66; H. Russell 1975; 1980: 166–7; Vaughan 1965: 31). In response to critics, Ceci (1975*b*) used data not presented earlier to demonstrate that the burying of fish would have been counterproductive and inefficient. Given the energy exerted in catching the fish and the nutritional values of both fish and maize, New England aboriginals would have been doing themselves a great service had they eaten the fish and not buried them.

In his detailed studies of corn hills in Ontario, Heidenreich (1974: 379) searched for remains of fish bones as indicators of fertilization. His results remain unpublished, but he found none (Conrad Heidenreich, personal communication 1996), thereby providing support for Ceci's claim. In contrast, of the 21 corn hills analyzed at the Sandy Point site, Cape Cod, three contained fish remains, specifically bones and scales (Mrozowski 1994: 57, 58). On one hand, this finding challenges Ceci's conclusions. On the other hand, however, if fertilizing with fish was a normal practice, then one should expect to find remains in all, not merely one-seventh, of the corn hills studied. Furthermore, remains of other types of organic matter were also found during analysis of the Sandy Point corn hills (Mrozowski 1994: 57), suggesting that fish remains may have been only one part of the normal household refuse that was hoed into the hills as fertilizer. The final verdict on burying fish is, of course, not yet in. Current evidence does not confirm that a single fish was buried in each hill as maize seeds were planted. It does, however, indicate that in certain coastal locales fish remains comprised at least part materials used as fertilizer.

For the most part, native North Americans whose agriculture depended entirely on direct rainfall located their fields in valley bottoms. These environments flooded periodically. The floodwaters did not irrigate the fields, but the sediment left on the surface after the waters receded helped to replenish soil fertility (Waselkov 1997: 180, 187). Also, the practice of hilling is the equivalent to deep mixing by a plow and can double the thickness of topsoil. The mixing-in of needed materials provides additional organic matter. Finally, the permanent intercropping, especially of beans and maize, retarded the depletion of soil minerals and nutrients.

Overview

In terms of the total number of hectares under cultivation, plain fields dependent on direct rainfall, in warm environments and where excess water posed no problem, was the dominant form of agriculture at the time of European contact. Documentary evidence, bolstered principally by ethnographic parallels and somewhat by archaeological data (e.g., Hudson *et al.* 1985), indicates that numerous, large, contiguous fields, frequently covering extensive areas were devoted to rainfed agriculture, particularly in the Eastern Woodlands. Fertile, well-watered bottomlands were the preferred locations, but these lands required the removal of large trees. Upland areas were also cultivated but to a lesser extent. Forest clearance involved slashing or felling, followed by burning. Fields were tilled with hoes and the grasses which sprouted after each year's harvest were typically burned prior to planting the following year. Planting was done with the traditional digging stick and usually involved, intercropping. Two crops per year were not unusual in southern lands where the growing season is quite long. In the cooler north, only one crop per year was harvested. That cool conditions toward the north affected the prospects of agriculture is evident in the number of seeds planted per hole. Northern farmers were clearly faced with problems not encountered by their counterparts in the South. They were, perhaps, more ingenious in the strategies they employed.

Mounding or hilling was a common practice. It not only eliminated weeds, but aided soil fertility. In some cases, particularly in northern settings, hills intended only as temporary (yearly) features have persisted to the present day.

Rainfed agriculture as practiced on fields without features intended as permanent landscape modification was not homogeneous or monolithic. In addition to those differences found between southern and northern locales, are differences between east and west. The principal difference, of course, is that it was much less common in the West than the East, a function of less frequent, and lower amounts of rain. This form of cultivation was restricted to two types of environments, valley bottoms and mountains. The former were cultivated on the central and, especially, northern plains, and in northern New Mexico. The latter were cultivated in the Four Corners region, and in the sierras of northwestern Mexico. Both environments are marginal for this type of farming, but they presented no other alternatives to native cultivators.

That rainfed agriculture was commonly practiced in areas covered prior to farming with dense forests, and in some places involved relatively poor soils, has given rise to a few strongly held beliefs about native cultivation practices. The fact that farmers both slashed and burned the forests has led many people to infer or accept the inference that native North Americans practiced on a broad scale some type of swidden agriculture. Evidence, however, shows otherwise. The notion that natives fertilized their crops by burying fish along with seeds during planting is also doubtful.

In summary, whenever possible, rainfed agriculture was practiced without the use of permanent field features. It was an easy form of agriculture, especially once the forest was cleared. The greatest difficulties were in marginal environments, uplands with less than ideal soils, in the Southwest and on the Plains where rainfall was less and erratic, and in the north under cooler conditions and a shorter growing season prevailed.

6

Ridged Fields in Cool Environments

The growing of crops in areas with ample, but not too much, moisture and a sufficiently long growing season was an activity that did not necessarily require the construction of permanent features on the North American landscape in pre-European times. As has just been illustrated, indigenous agriculture over much of the continent, and especially in the East, took this form. Of course, some landscape modifications associated with prehistoric rainfed agriculture do persist on the eastern landscape today, but these—the relict corn hills—were intended to be temporary when they were made; they were not constructed as permanent features. Their location is, nevertheless, curious. Surviving close to the northern limit of cultivation begs the question of an environmental correlate. Although record of their existence has been credited principally to careful observers, it seems highly likely that these particular corn hills might also have helped alleviate problems involving cool conditions and the short growing season. The previously discussed inverse correlation between the number of seeds planted per hole and the length of the growing season is corroborative evidence.

Cultivating crops on an artificially elevated surface can certainly offset some problems associated with low temperatures (Gartner 1997*a*). Stated simply, cold air is dense, heavy, and, therefore, descends. It behaves exactly the opposite of warm air, which, as anyone familiar with the principles of hot air ballooning knows, rises. Because cold air drops, temperatures on the tops of hills are, all other things being equal, higher than those on surrounding lowlands. This effect is especially pronounced at night when temperatures are not only lower than they are during the day, but are dropping. The coolest hour of the day, it must be remembered, is just before dawn. Known as 'cold air drainage', the effect is much like that of flowing water. Cold air responds to gravity, and, therefore, flows down hill. Cross a bridge over a river or a canyon during a summer night and one will easily detect not only the differences in temperature between the bridge and the land on either side, but also the cool breeze which blows down valley. On a much more localized scale, and one with close parallels to the problem of cold air and cultivation, all experienced campers know not to unroll their sleeping bags and spend the night in gullies or similar low spots; they are markedly cooler than surrounding high ground.

Hilling or mounding served a number of functions, and these undoubtedly varied from place to place given differences in local conditions. One of these functions may well have been to compensate for harmfully cool temperatures. However, when it comes to landscape modifications for the purposes of affecting temperatures, elevated planting surfaces separated by ditches are much more commonly considered. In no small way this is a function of the research conducted on prehistoric fields and present-day agricultural practices of indigenous peoples living in highland New Guinea. There, it was found that crops grown on raised fields were not damaged while those grown on the surrounding normal ground surface suffered serious frost damage on cold nights (Waddell 1972).

Be they modern, historic, or ancient, raised fields, defined as 'any prepared land involving the transfer and elevation of soil above the natural surface of the earth in order to improve cultivating conditions' (Denevan and Turner 1974: 24), are not, however, always associated with temperature modification. In fact, they have long been used for a variety of other purposes including the ease of cultivation (Bell 1984), increasing soil depth, texture, and fertility (Miracle 1967: 162), suppressing pathogens that affect crop roots (Thurston 1990), and particularly, the production of crops in seasonally or permanently inundated low-lying areas that otherwise could not be cultivated (Denevan 1970). Without doubt, the best-known examples of raised fields built for the purposes of cultivating wetlands are the ancient *chinampas* that contributed significantly to the food supply of the Aztecs who inhabited the Basin of Mexico (M. Coe 1964).

Raised fields built for whatever purposes were not very common in North America at the time of European contact. At least there is not as much evidence for the practice here as there is from Mesoamerica and South America. There does, however, exist a sufficient amount of evidence to indicate that some native farmers were elevating planting surfaces. As will be seen, the surfaces they raised tended to be narrow. They were not much, if any, wider than the ditches separating them. And, they were relatively low with shallow ditches. In a very real sense, they look like a giant washboard and, hence, are known as 'ridged fields'.

Exactly why these people built ridged fields remains unknown; surprisingly little systematic research has been conducted on the topic to date. It is highly probable that they were doing so for different reasons, in different places, at different times. Given that most of the evidence of ridged fields in North America comes from the upper Midwest, close to the limit of cultivation based on the length of the growing season (Moodie and Kaye 1969), and in proximity to the relict corn hills, it seems likely that many served to ameliorate low temperatures, especially those associated with well-drained upland areas. The evidence of ridged fields for cultivation of low-lying wetlands, as well as for other purposes is covered later (Chapter 12).

Documentary Evidence

Only two written accounts even hint at the ridging of fields for the possible purpose of temperature control. Not surprisingly, these are from the far northeastern corner of the continent, in an area characterized by both the existence of ancient corn hills and a large number of seeds planted per mound. The first is from Pierre Gaultier de Varennes (1980)—the Sieur de la Vérendrye—who, in the early 18th century, led an expedition from Montreal westward in search of information about the lands between there and the Pacific Ocean. The expedition never made it to the coast, but returned to Montreal. There, la Vérendrye told Kalm (1972: 419) that he saw:

> large tracts of land free from wood, but covered with a kind of very tall grass, for the space of some days journey. Many of these were every where covered with furrows, as if they had been ploughed and sown formerly.

On hearing this, Kalm began talking to others in Montreal who had trekked into the interior. He reported that:

> All of those who had made long journeys in *Canada* to the south, but chiefly westward, agreed that there were many great plains destitute of trees, where the land was furrowed, as if it had been ploughed. In what manner this happened no one knows, for the corn-fields of a great village, or town, of the *Indians* are scarce above four or six acres in extent; whereas those furrowed plains sometimes continue for several days journey (Kalm 1972: 421–2).

Kalm (1972: 419–20) also concluded 'that the nations, which now inhabit *North-America,* could not cultivate the land in this manner, because they never made use of horses, oxen, ploughs'.

It is clear that Kalm did not actually see these ridged fields himself, but heard of their existence from others. His comments, while made in the 18th century, do, however, seem to be accurate portrayals for three reasons. First, they are based on accounts he received directly and independently from more than one person who claimed to have seen the fields. Second, his identification of technological differences between pre- and post-European farming practices is correct. And, third, the comparison of large abandoned fields to then currently cultivated smaller fields fits the general picture of native North American agriculture for that time period. Subsequent research has concluded that the fields observed by la Vérendrye were probably in what is today Michigan (Fox 1959: 2, 4; Hubbard 1878: 1).

Kalm's method of assessing and portraying others' comments about ridged fields, appears to have been accepted by one 20th-century agricultural historian who replicated it, albeit without careful consideration. Writing in the 1930s, that scholar claimed that men who participated in the 1779 campaign against the Iroquois in:

western New York noted a peculiarity in Indian fields. In some . . . there were patches of ridged land. They soon found that these uneven surfaces were brought about by the Indian method of planting and tilling corn. The ridges came from planting the corn in continuous rows instead of hills (Hedrick 1933: 25).

Exactly where this scholar obtained his information is not at all clear. A thorough perusal of the *Journals of the Military Expedition of Major General John Sullivan* (Cook 1887), which includes the detailed diaries of several commissioned and non-commissioned officers, revealed not so much as a single mention of ridges. To be sure there were plenty of references to fields of various sizes and states of cultivation (see Tables 5.1–5.4), but not to ridges! If some other source of reference was used, it was not cited. If some other source was not used, the idea of ridges was either fabricated, or adapted unwittingly from an earlier publication of Kalm's book.

Although the documentary accounts of ridging are less than perfect, the two reported by Kalm, a highly reputable source, indicate that at least some agricultural fields were elevated by native North American farmers. That these ridged fields were actually used to compensate for cold conditions is not at all clear from the accounts themselves. It is possible that they could have been used to facilitate the cultivation of low-lying wetlands. If they were, however, one might expect that the observers would not have failed to mention the unusual environments in which the fields were located. One has to conclude, therefore, even if only tentatively, that ridged planting surfaces were made for other reasons. If environmental factors are critical, then temperature modification is a most probable explanation.

Ethnographic Evidence

Unfortunately, there exists no ethnographic evidence of indigenous people ridging fields for purposes of modifying temperatures to improve agriculture in either the 19th or the 20th centuries. It had been long-claimed that aboriginal farmers in the upper Midwest and Canada made corn hills exclusively during the historic period (e.g., Riley and Freimuth 1977: 4–5). Ridged fields, it was argued, were found only in prehistoric contexts (Gallagher 1992: 120–1). This pattern was first noted in the middle of the 19th century (Lapham 1855: 17–20; Schoolcraft 1851*a*: 57) and repeated in numerous studies since (Boszhardt, Bailey, and Gallagher 1985: 48; Fox 1959: 18; Moffat 1979: 244–5; T. Riley 1987; Riley and Freimuth 1979: 283; Riley, Moffat, and Freimuth 1981: 104, 112). The accepted shift from ridged fields to corn hills *c.* AD 1500 evoked the hypothesis, that climatic changes associated with the Neoboreal episode known as the 'Little Ice Age' may have played a major roll (King 1993; see also Gallagher 1992: 122–3). An alternative hypothesis submits that population declines due to the spread of infectious European-introduced diseases may have resulted in a disintensification of agriculture (Gallagher 1992: 124; Moffat 1979: 244).

Adherence to the party line that ridged fields are prehistoric and corn hills are historic has been rather strict despite some evidence that it probably should not be. Both types of features are often found together (Fig. 6.1). Archaeologist Robert F. Sasso (n.d.) has recently addressed this dilemma. He examined the Wisconsin state archaeological files of 272 sites with ridged fields and/or corn hills, and found that more corn hills seem to be in prehistoric rather than historic contexts, and that nearly one-quarter of the ridged fields appear to be in historic rather than prehistoric contexts. As good as this seems, however, Sasso's analysis should only be considered a first attempt. The site files he utilized are not without problems, most of which stem from the ways in which sites were classified, dated, and recorded early in the 20th century (Charles R. Moffat, personal communication 1996). Also, different types of features on the same site could reflect separate occupations of differing ages. Although an association between ridged fields and historic peoples has yet to be demonstrated conclusively, the case should not be considered closed.

As has already been mentioned, there remains no ethnographic evidence

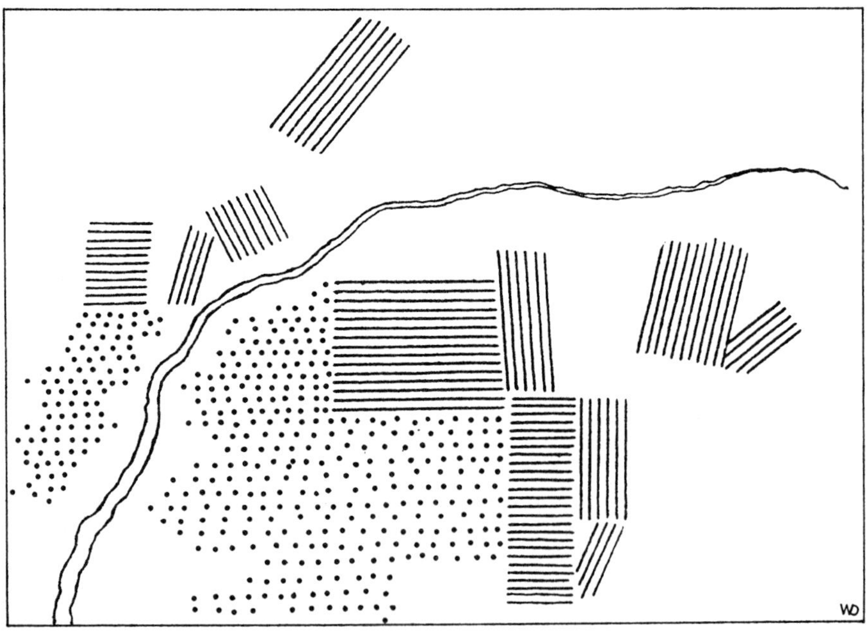

Fig. 6.1. Sketch map of the Red Banks site on Green Bay in Wisconsin. This agricultural field site contained remnants of both corn hills and ridges (after C. Brown (1909: 100, plate 2). No scale or North arrow). (Redrawn by author.)

that can be used to help understand the construction of ridged fields, much less how such features affected temperatures. Archaeological data do exist, however. And, some data derived recently from experimental ridged fields built by archaeologists is illuminating.

Archaeological Evidence

Remnants of ridged fields, often and long referred to as 'garden beds', came to the attention of North American archaeologists working around the Great Lakes in the middle of the 19th century. Henry R. Schoolcraft (1851*a*: 55) observed them on 'level and fertile prairie-lands' extending from far northern Michigan through northern Indiana. Subsequent research has confirmed the first, modified the second, and rejected the third of Schoolcraft's observations. Nearly all of the ridged fields were located on naturally well-drained upland prairies, most of which are today overgrown with dense stands of trees (Peters 1980: 34–6). Some existed in the eastern portion of the state (Fox 1959: 3–4). None have yet to be verified in the northwestern part of Michigan, and none seem to have ever existed in Indiana. Indeed, in speaking of ridged fields in Indiana, the archaeologist most familiar with the state, once quipped 'I do not know of a single specific instance! In view of what I do know, if I were writing on the subject, I would have to say there are none in Indiana' (Glenn A. Black, cited in Fox 1959: 1). Two sets of ridged fields have been found in Iowa (Billeck 1986; Lensink, Finney, and Fishell n.d.), but most, including the best preserved and the first identified archaeologically (Lapham 1855: 19, 57, 71–2), were located in Wisconsin.

A total of fifty-nine ridged field sites have been found and recorded in Michigan (Fig. 6.2), but published details concerning them are actually quite few. Schoolcraft (1851*a*: 55) claimed they were of various sizes, involving between 8 and 40 ha, in general, but some ranged upwards of 120 ha. Individual fields were composed of series of parallel ridges that were approximately 4 m wide and sometimes more than 40 m in length (Schoolcraft 1851*a*: 56, plates 6 and 7). Some ridges were nearly flat-topped, whereas others were convex. In some cases the ridges were adjacent to each other, in other cases they were separated by a 'path' that varied in width up to a distance equal to the width of the ridges. Fields were oriented in no particular cardinal direction, but some were aligned in groups. No height of ridges was provided by Schoolcraft, but, on the basis of the cortical rings counted on a tree which was felled atop one ridge in 1827, he concluded the ridged fields had to have been used prior to AD 1502.

The most detailed information pertaining to ridged fields in Michigan, particularly the southwest corner of the state, comes from the work of Bela Hubbard (1878). He made two contributions of great significance. First, he provided a typology of fields based largely on morphology, albeit one that has

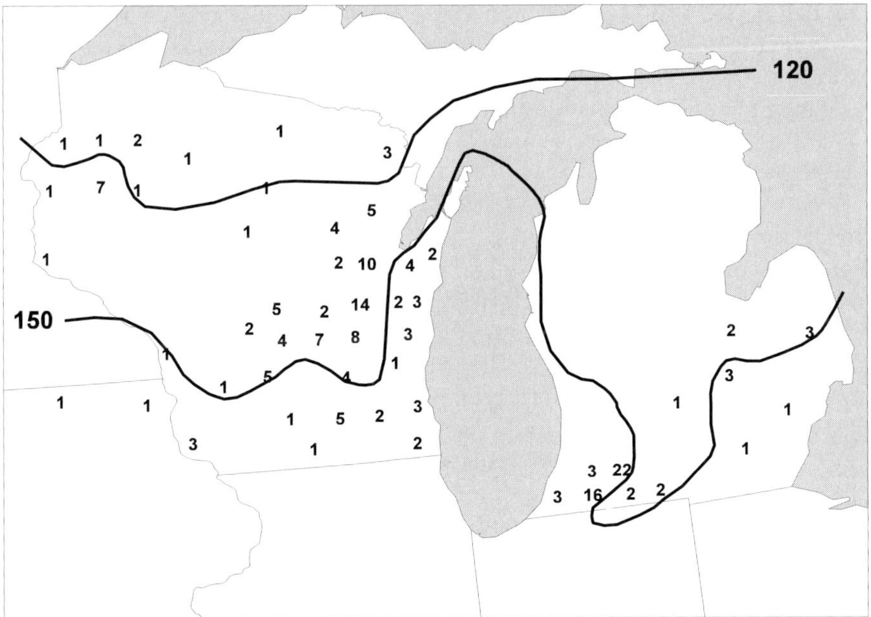

Fig. 6.2. Distribution of archaeological sites comprised of remnants of ancient ridged fields. The numbers refer to the number of such sites found in individual counties. Fox (1959: 3, 5), after whom the Wisconsin data were taken, claimed to have identified 121 sites but his map enumerated 132. Iowa data were taken from Billeck (1986; Lensink *et al.* n.d.). The lines indicate the length of the growing seasons by the number of frost-free days (after Riley and Freimuth (1979: 274, fig. 2).

not gone uncriticized (see C. Brown 1909: 105). Second, he presented some simply stunning diagrams showing the various configurations of raised agricultural platforms. Hubbard (1878: 3) identified eight types of ridged field complexes.

(1) Wide convex beds, in parallel rows, without paths, composing independent plats. Width of beds, 3.7 m; paths, none; length, 22.6–35 m.
(2) Wide convex beds, in parallel rows, separated by paths of same width, in independent plats. Width of beds, 3.7–4.9 m; paths, same; length, 22.6–40 m.
(3) Wide and parallel beds, separated by narrow paths, arranged in a series of plats longitudinal to each other. Width of beds, 4.3 m; paths, 60 cm; length, 30.5 m.
(4) Long and narrow beds, separated by narrow paths and arranged in a series of longitudinal plats, each plat divided from the next by semicircular heads. Width of beds, 1.5 m; paths, 46 cm; length, 30.5 m.
(5) Parallel beds, arranged in plats similar to class 4, but divided by circular heads. Width of beds, 1.8 m; paths, 1.2 m; length, 3.7–12.2 m; height, 46 cm.

(6) Parallel beds, of varying widths and lengths, separated by narrow paths, and arranged in plats of two or more at right angles (north–south and east–west) to the adjacent plats. Width of beds, 1.5–4.3 m; paths, 30–60 cm; length, 3.7–9.1 m; height, 20 cm (Fig. 6.3).

(7) Parallel beds, of uniform width and length, with narrow paths, arranged in plats or blocks, and single beds, at varying angles. Width of beds, 1.8 m; paths, 60 cm; length, approximately 9 m; height, 25–30 cm (Fig. 6.4).

(8) Wheel-shaped plats, consisting of a circular bed, with beds of uniform shape and size radiating therefrom, all separated by narrow paths. Width of beds, 1.8–6 m; paths, 30 cm; length, 4.3–6 m.

Curiously, and unlike Schoolcraft, Hubbard did not mention the cross-sectional shape for every type of field. He did, however, hint at this distinguishing characteristic in his illustrations. Ridged field types 4 and 5, like types 1 and 2, seem to have had convex planting surfaces. Types 6, 7, and 8 appear to have had flat surfaces (see Figs 6.3 and 6.4). And, type 3 looks as if it had a flat surface with beveled edges. Of course, since he did not mention this characteristic, it should not be expected that he would have offered an explanation for the variation. He did not, and for good reason; data probably did not exist to proffer any rationalization.

As for their spatial extent, Hubbard is vague, in large part because he fully recognized that he saw only remnants of features that sometime in the ancient past covered more area than they did in the late 19th century. In at least one place, however, he reported fields that extended for 800 m in one direction, and 550 m in another, covering approximately 40 ha (Hubbard 1878: 4).

As for their function Hubbard (1878: 6–7) was convinced they were agricultural, but he was unsure if they provided staple crops, specialty crops including medicinals, or even ornamentals as in the case of 'pleasure' or 'botanical' [*sic*] gardens. The existence of type 8, the wheel-shaped plats, seems to lend credibility to this explanation, at least in some cases. A more recent interpretation challenges the notion that these series of ridges were even agricultural. On the basis of an 1873 letter from one E. S. Moore to archaeologist Bela Hubbard, which stated that the ridges at one site were so prominent and extensive as to impede the speed of horseback travel to Kalamazoo, it has been proposed that:

> These groups of ridges may have been constructed and used for killing buffalo. The method would have been to stampede a herd in the hope that it would run across one of these artificial grids, which would be invisible in the grass of the small prairie. The speed of the herd would certainly be checked and in the resulting confusion some could be shot with bow and arrow. But more than this, it may be that those who reached the ridges first would stumble and fall, to be killed by those rushing from behind. The almost invariable disposition of the ridges in groups at angles to one another lends considerable support to this explanation for it would insure quickly successive changes of pace on the part of the herd with results disastrous to it after three or four patches had been crossed (Greenman 1957: 20).

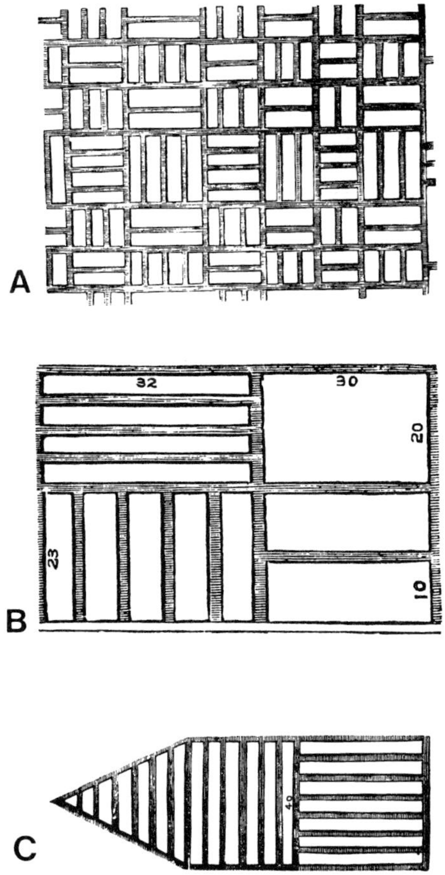

Fig. 6.3. Ridged fields arranged in plats with planting surfaces perpendicular to those in adjacent plats found in Kalamazoo (A and B), and near Galesburg (C), Michigan (Hubbard 1878: fig. 6).

Buffalo speed bumps? Moore's interpretation is almost absurd. It is, however, so novel, and even humorous, that it warrants inclusion here, if for no other reasons than to be fair and acknowledge that other non-agricultural alternative explanations do exist, and to demonstrate the enigmatic and fascinating qualities these features possess (see also Sasso n.d.).

Whereas the ridged fields in Michigan have received very little attention since the late 19th century, those in Wisconsin continue to be studied (Gartner 1999). The fields in both states were first reported by archaeologists within a few years of each other—Schoolcraft on Michigan in 1851, and

Fig. 6.4. Ridged fields arranged in plats at various angles found on Prairie Ronde, Michigan (Hubbard 1878: fig. 7).

Lapham on Wisconsin in 1855—but perhaps because they were not located so close to rapidly growing urban areas, were more numerous, and more widespread, at least some of those in Wisconsin survived and, hence, remained available for examination and interpretation. The Wisconsin fields also might have benefited by having a discoverer who was interested solely in ancient landforms and therefore set a precedent for archaeological studies in that state. Indeed, Lapham is considered the father of Wisconsin archaeology. Schoolcraft, in contrast, was more interested in the ethnography of surviving native North Americans throughout the Midwest and Great Plains (see e.g., Schoolcraft 1851*b*,*c*). To him, the relict fields in Michigan were more of a curiosity than a focus of attention. Whereas Schoolcraft proved not to be a catalyst for ridged field studies, Lapham did.

Lapham was the first person to provide detailed information about the ridged fields in Wisconsin. Near Milwaukee he found twenty-five parallel ridges within the space of 30 m. They were low, 15 cm high, broad, 1.2 m wide (Lapham 1855: 19), and on high ground, well above the Milwaukee River and low, marshy land nearby (Lapham 1855: plate 8). Further north, near Mayville, Lapham (1855: 57) recorded a second set of ridged fields. No areal extent was given, but they must have covered a sizable tract. They measured 30 m long, 1.8 m wide, 20 cm high, with a 38 cm path separating each ridge. Although details about the locale in which these fields were located was not spelled-out, it is reasonably clear from an accompanying map (Lapham 1855:

plate 38), that they were on an interfluve and, therefore, above standing water. The last set of ridged fields mentioned by Lapham was located near The Dells. In one sense he said very little about these fields—'there are extensive tracts of ground worked into garden-beds, or low flat ridges, as before described' (Lapham 1855: 72). In another sense, however, he said a great deal—(i) ancient raised fields were found in yet a third locale, (ii) they were all similar in form, and presumably in setting, and (iii) they covered reasonably large parcels of land.

In the hundred and fifty years since Lapham first started recording sites, no fewer than 132 sets of ridged fields have been recorded in Wisconsin (Fig. 6.2). Of course, some of these served purposes other than to modify temperatures, but many if not most of them did, and of those, several have been described in the literature. A major impetus for the burst of publications that included mention of ridged fields was the founding of the Wisconsin Archaeological Society in 1903. Numerous members scoured the state looking for evidence of prehistoric people (Moffat 1979: 224). Much of the information they gathered on ridged fields was summarized by Charles E. Brown (1909), who also reported, in detail, three more sites from naturally well-drained locales. Near the Red Banks on Green Bay he was shown no fewer than twelve plots, some with as few as five rows of ridges, some with as many as twenty-seven rows. The ridges were approximately 1 m wide and spaced 1 m apart. The longest ones measured 15 m in length, and the most prominent were 15 cm high. Brown (1909: 101) explicitly states: 'This planting ground occupies all of the higher ground between a small creek and two drains'.

Toward the center part of the state, Brown relayed the description of a second site. This one, between the Fox River and Lake Puckaway was adjacent to a permanent habitation site with several other types of mounds (Brown 1909: 103). The individual fields were 27 m long, 1.2 m wide, 15 cm high, and 30 cm apart. They were divided into two distinct, but adjacent, parts in a configuration not unlike the type 6 identified by Schoolcraft. One plat or set contained twenty-one ridges that were aligned north–south. Next to these were twenty ridges oriented east–west (Brown 1909: 99). Although Brown did not provide details about the environment in which these fields were situated, it seems highly probable they were on high ground; in broad, general terms they were between two low points, specifically, they were adjacent to a village that would not have been built in a wetland.

The third garden bed site reported by Brown was one he found not too far from that described by Lapham on the Milwaukee River. As was the case with that site, this one was also on high ground, and it was quite small in areal extent. As was the case with the site between the Fox River and Lake Puckaway, this one, too, contained other types of mounds indicative of a settlement (Brown 1909: 103). The individual ridges numbered only ten. They varied in length from 7 to 11 m, and width from 90 cm to 1.2 m. They were from

10 to 15 cm in height, and 80 to 90 cm apart. At the time he discovered them, they were, according to Brown (1909: 99) 'overgrown with trees and shrubs' and 'no longer very definite'.

This site has actually changed very little during the past century, perhaps because the land on which it is located became designated public space when the city of Milwaukee expanded. The ridged fields were rediscovered in the late 1970s in a heavily wooded part of what is today Kletzsch Park (Moffat 1979: 233, fig. 1; Riley, Moffat, and Freimuth 1981: 111, fig. 5), and reported initially in the form of some rather striking wintertime photographs in which the ridges were delineated by bare earth, while the paths, or spaces between ridges, retained snow which had not yet melted (Fig. 6.5). A more detailed study of the site in 1978 uncovered two important pieces of information. First, there were two sets of ridged fields—one with eleven ridges and one with seven—approximately 200 m apart, suggesting that some time in the past a total of at least 4 ha may have been cultivated (James and Benchley 1980: 16–17, 75). Second, the ridges varied in length between 11 and 16 m with the average distance between ridge tops being only 1.6 m (James and Benchley 1980: 77). Inspection of them in the summer of 1988 revealed that unlike previously discussed ridged fields, these are characterized by convex ridges

Fig. 6.5. Prehistoric ridged fields in Kletzsch Park, Milwaukee, Wisconsin, 1977. These features are visually enhanced by differential snow melt, illustrating that ridge tops are warmer than the paths between. Provided courtesy, and published by permission, of Charles R. Moffat.

tangent to concave paths or ditches. At best, the vertical distance between the high and low points is no more than 10 cm. Of course, it is most probable that material has eroded off the ridges and filled the paths as sediment; these ridged fields are barely evident even to trained observers.

Much more pronounced, and numerous, than extant ridged fields in the Milwaukee area are those in upland locales in the Lake Winnebago region. Some of these were reported initially by members of the Wisconsin Archaeological Society, but never by themselves; they were given only brief mention while much attention was paid to various other recently discovered antiquities. The importance of ridged fields in this area was first brought to attention by archaeologist George R. Fox (1922), whose graphic and detailed descriptions of ancient cultivation ridges on the Eulrich's family farm precipitated one follow-up study, and captured the attention of many archaeologists who continue to visit the site, and refer to it, although never publishing anything additional about it. Fox (1922: 50) reported that when he saw the ridges they covered approximately 8 ha, but he estimated that at some time in the past they must have encompassed twice that much land. Although Fox recognized that many ridges had been obliterated by more recent cultivation practices, he did not think that the remaining ridges had been reduced much in size. Indeed, he said: 'These beds are as distinct as the day they were made, in nearly all places rising about six inches [15 cm] above the dividing paths' (Fox 1922: 49). These ridges and paths showed great regularity in size. Most were 1.6 m wide, although some were as small as 1.3 m in width. The paths between the ridges were somewhat less regular than the ridges themselves, varying in width from 37 cm to 45 cm. In addition to being distinct and covering a sizable tract of land, perhaps the most fascinating characteristic of the Eulrich Farm ridged fields is the configuration of their layout. Unlike previously discussed ridges that were laid out in straight lines, these ridges were curved, and run in several, not simply one or two, directions (Fig. 6.6). Sets of ridges often begin parallel to each other and wind up perpendicular, in some cases even terminating at right angles to other ridges. A later study that actually involved archaeological excavations revealed that the ridges were made-up of dark topsoil devoid of large rocks which are present in the subsoil and unworked topsoil (Peske 1966: 193).

The follow-up study of the Eulrich Farm ridged fields was also conducted in conjunction with a second detailed restudy of some other pre-contact agricultural ridges nearby. At Lasley's Point, on reasonably high ground adjacent to a large prehistoric village site, approximately 6 ha of ridges were found and reported at the turn of the century (P. Lawson 1903). These were later subjected to two restudies. The first in the early 1940s revealed that the ridged fields were clustered in groups over a 24 ha area, and that the some ridges measured 90 cm from crest to crest, whereas others measured 1.5 m. Also, some were 15 cm high, while others were 22 cm (Bullock 1942: 34). During the second restudy, G. Richard Peske (1966: 191) excavated a 3.5 m by 75 cm

Fig. 6.6. Curvilinear ridged fields on the Eulrich Farm west of Lake Winnebago, Wisconsin. Features are visually enhanced by differential snow melt. Photograph by George R. Fox, 1915. (Published by permission of the State Historical Society of Wisconsin, Neg. No. WHi (X3) 48812.)

trench across one set of these ridged fields in order to see how they were constructed. He cross-sectioned three ridges and three paths. The distances between ridgetops were 1.25, 1.58, and 1.2 m, respectively, suggesting the fields were somewhat larger than stated in the earlier restudy. This discrepancy is probably attributable to improved measuring techniques. The vertical distances from the tops of the ridges to the surfaces of the paths were 7.5, 8.8, and 6.2 cm, respectively, or approximately half that reported in the first restudy. The depth of the dark topsoil was 2.5 cm on the ridgetops and between 7.5 and 12.5 cm in the paths. The difference in the depth of topsoil between the ridgetops and the paths is cited as a function of erosion and sedimentation. Adjusting for these processes, it was estimated that at the time the fields were originally constructed the ridges would have been 15 cm high (Peske 1966: 191), exactly as reported in the first restudy. Did erosion and sedimentation occur only during the past 50 years? Probably not. Given that the 1940s study erred in measuring the distances between ridges, it is likely that an error, perhaps due to an 'eyeball' estimation, was made in determining the height of ridges.

The Eulrich Farm and Lasley's Point sites are certainly important in their own rights, but they are also important in terms of the history of ridged field studies in Wisconsin. Both were not only subjected to restudy, but they also constitute two of only seven sites that were actually found to be extant in a late 1970s attempt to locate ridged field sites recorded in Wisconsin over the previous 120 years. In the summer of 1977 Charles R. Moffat (n.d.) carried out a two-phase study which can only be described as ambitious and admirable (see also Moffat 1979: 229–41). He began by gleaning information on site locations from data collected by the Wisconsin Archaeological Society. Armed with these materials, he then undertook a field survey of nineteen counties in eastern and central Wisconsin. Of the seventy-six ridged field sites for which he had locational and other infomation, Moffat could find remains of only seven. A total of forty-three sites are known to have been destroyed and twenty-six sites could not be checked for a variety of unspecified reasons (Moffat n.d.). Assuming that the proportion of extant to originally recorded sites is the same for the sites not visited as it is for those that Moffat checked, another three or four sites, or a total of ten or eleven sites, may remain visible today.

Moffat (n.d.) actually added little to what is known of the Eulrich Farm and Lasley's Point sites. For the former, he noted that many of the ridges that Fox observed have been destroyed, but that three groups of 'not very distinct' ridged fields remained at the time of his visit (Moffat n.d.). A field inspection of this site eleven years later revealed that the ridges seen by Moffat may have subsequently been obliterated. Although the area was overgrown with vegetation between ankle and knee high, a careful search for ridges was undertaken. None could be found. As for the Lasley's Point site, Moffat (n.d.) notes that the ridged fields are difficult to pick out, and, perhaps more importantly, he confirms that the ground they are on is fairly level and not marshy. In other words, the ridges were not intended to cultivate wetlands.

Another ridged field site rereported by Moffat lies just west of the village of St. Cloud. He relays that the site consists of two sets of ridges, both 'located on higher ground above the river' (Moffat n.d.). Ridges that comprise the smaller of the two sets average 30 m in length, and are 2.1 m apart, presumably measured from ridgetop to ridgetop. The rows of the larger field are said to be more 'irregular' (?) and reach lengths up to 45 m (Moffat n.d.).

A fourth ridged field site rediscovered by Moffat, the Leeman Site, was located in Outagamie County. The features were originally reported to cover between 4 and 8 ha, but were thought to encompass a much larger area when they were used. When he visited the site, Moffat (n.d.: 7) found it to be fairly distinct and consisting of a group of 'quite straight ridges and furrows'. The landowner told him that there were two sets of fields, but he and Moffat could find only one because of the presence of large trees and thick undergrowth.

Two of the ridged field sites that Moffat visited are today within the confines of public campgrounds. Each night several people park their recreational vehicles and walk their dogs among features the antiquity and function of which are beyond their knowledge. If visitors even notice the ridges, they probably attribute them to some seemingly unexplainable natural process. Most people, however, do not notice them as the ridges are not too distinct, and many have been damaged or destroyed as roads and parking spaces were cut and paved, and as electrical and water lines were laid. At Mirror Lake State Park in Sauk County, the ridges are best seen along the Echo Rock Trail at the far north end of the campground. They are reported to form a 'complex pattern', meaning that sets of ridged fields are oriented in various directions and often join each other at various angles. The widths of the fields vary slightly, from 1.35 to 1.4 m, and the spaces between them range from 1.1 to 1.7 m. No height was given but Moffat (n.d.: 8) claims they are not as prominent as those at the other campground site, in part because of the heavy tree litter on the surface.

The prehistoric ridged fields in the Rocky Arbor State Park in Juneau County are quite distinct, and, according to Moffat (n.d.), measure 30 cm or more in height. Moffat also made some detailed measurements of these fields' widths and spacings. He reported that these ridged fields vary from 1.1 to 1.4 m wide and are between 50 cm and 1.1 m apart. Moffat provides no information on the lengths of these fields, but does say they form complex patterns similar to those at the Mirror Lake State Park, and he notes that they are situated on an upland, approximately 30 m above the Wisconsin River (Moffat n.d.).

A visit to this site in 1988 confirmed everything reported by Moffat (Fig. 6.7), and more. These ridged fields are, indeed, well above the river and any ground that might be even broadly defined as a wetland. Located at the north end of the campground, they are, however, situated in a saddle between two small hills. Although a low point, the localized environment in which these ridged fields are set is one in which standing water was not a problem with which ancient farmers had to contend. Cold air, on the other hand, especially that draining off the two adjacent hills would have been problematic. Field inspection revealed that these ridged fields are aligned between the hills. Longitudinally, the ends of the ridges point toward the peaks. The ridges are not continuous, but instead are broken into sections, or have lateral openings. Cold air draining off the hills would, theoretically, flow through the paths parallel to the ridges, collect in the low point, at least temporarily, and then drain through the openings in the ridges toward the river far below. Were the ridged fields not utilized, the locale might not be suitable for farming. The ridges arguably put the planting surface above the accumulated cold air. The lateral openings in the ridges then facilitate drainage of the cold air from the area with the fields to the river.

The final site rediscovered by Moffat (n.d.) is 5 km west of Wisconsin Dells in Delton Township. Perhaps the same fields mentioned by Lapham (1855: 72), these were described by Moffat as appearing to be similar in size and form to the ones previously discussed at the two state parks, and very distinct. He also said that even though many of the fields had been destroyed due to widening and realigning of a road, 'scattered remnant patches mark its

Fig. 6.7. Prehistoric ridged fields in the Rocky Arbor State Park, Wisconsin, as they appeared in the summer of 1988. The person in the photograph, William M. Denevan, is standing on one elevated planting surface. The kinks in his shadow and those cast by the trees are a function of the curvature of the ridge tops and the paths between ridges. (Photograph by author.)

former extent' (Moffat n.d.). Field inspection in 1988 proved Moffat to be correct on some points and wrong on others. These ridged fields are morphologically similar to those at Rocky Arbor State Park and they are distinct. However, they are much more prominent than any other raised fields in the state, perhaps anywhere in North America, and they are spatially contiguous and still cover upwards of 8 ha. They are, however, overgrown with trees and a dense undergrowth.

A detailed study of this site was initiated in 1989 by William Gustav Gartner, a graduate student at the University of Wisconsin–Madison (Gartner 1991). With a crew of volunteers, he cleared much of the vegetation, did some mapping, carried out some excavations, and took soil and sediment samples for analyses. The ridged fields at what is now known as the Hulbert Creek site were 'sinuously shaped' and aligned roughly 10 degrees east of north (Gartner 1993: 115). They typically measure 10 m long, 2 m wide, 40 cm high, and are 1 m apart (Fig. 6.8). Archaeological excavations revealed several paleosols documenting ridged field construction and rebuilding (Gartner 1999: 673). Much of the relief is the result of ditching. Ancient farmers first

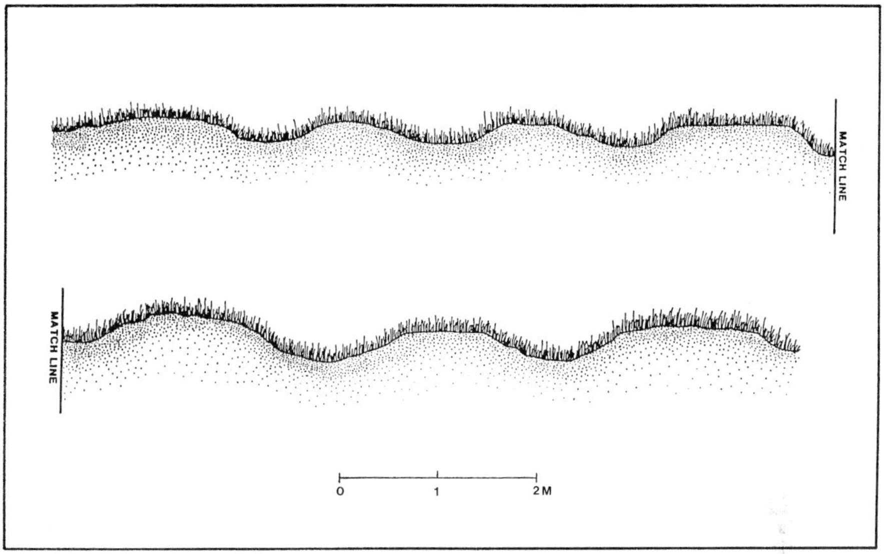

Fig. 6.8. A cross-section of prehistoric ridged fields at the Hulbert Creek site, Wisconsin. Based on the yet to be published 1990 field research of William Gustav Gartner and Richard L. Green. (Redrawn by author.)

cut and burned a large area. Fine materials then eroded off of upper slopes and accumulated downslope. Ditches were subsequently excavated into this newly deposited surface and materials were used to create raised planting areas. The agroecological functions of these ridges were multiple, including to affect hydrology, improve soil fertility, and, not inconsequentially, to facilitate frost drainage. Radiocarbon dating revealed that the fields were built and used *c.* AD 1000 (Gartner 1993: 115).

Much research remains to be done at this site, on the ridged fields of Wisconsin, and elsewhere in the upper Midwest. Given the large number of sites recorded, and the length of time they have been known, a complacency seems to have set in amongst scholars who should be more concerned than they are. Unfortunately, many people think that these features are well-understood, when, in fact they are not (Riley and Freimuth 1979: 273). Also, many people think that because of recent agricultural activities, forest clearing, and urbanization, ancient ridged fields have all been destroyed. Moffat's 1977 study demonstrates convincingly that while many sites no longer exist, others do survive, but are in the process of disappearing. This situation is no more evident than in the case of those features in Kletzsch Park which have been reported in the literature, and in the case of similar ridged field sites of unknown number which are not so widely known. For example, on a farm west of Larsen, Wisconsin, ridges and paths today are only barely visible to the trained observer as faint traces beneath a heavy growth of wood, but were easily discernible only forty years ago (James Clark Jr., personal communication 1988).

In addition to the sites he rediscovered, Moffat has recently been involved in the actual discovery of one new ridged field site (Moffat, Boszhardt, and Arzigian 1995: 133–51). This site is 60 km north of the one at Wisconsin Dells, on the west side of Petenwell Reservoir. It is located on high, well-drained land, and consists of a series of meandering ridges and ditches running parallel to and on a gentle slope (Moffat, Boszhardt, and Arzigian 1995: 134–5, fig. 52). The ridged field complex covers 1.3 ha and is similar to those at Eulrich Farm, Lasley's Point, and Rocky Arbor (Moffat, Boszhardt, and Arzigian 1995: 150–1). On the basis of location, shape, and similarity to other sites, these ridged fields are also interpreted as having facilitated cold air drainage. Not inconsequentially, this site also proves that one should never assume that previously unrecorded sites have all been destroyed.

Evidence that prehistoric ridged field sites may have existed, and may still exist, in places where they are not expected, exceeds the boundaries of Wisconsin. For example, it had long been thought that no ridged fields were known from Iowa. A review of some old site cards in the files of the state's historical society not long ago revealed information about one such site on the north side of the Cedar River (Billeck 1986). The ridged fields there were a series of parallel north–south trending ridges on a slightly sloping bank. They were 50 cm wide, 1.05 m apart, 7.5 to 12.5 cm high, and covering an area that measured 24 by 36 m (Billeck 1986: 6). A field survey revealed that these features had been destroyed since first being recorded in 1927. In 1994 a previously unknown ridged field site—the Litka site—was discovered in northwestern Iowa (Lensink, Finney, and Fishel n.d.). Excavations by Gartner (personal communication 1998) confirmed that these features were ancient fields and not natural, and unlike at the Hulbert Creek site (Gartner 1993: 115), were constructed solely by mounding without ditching.

All in all, research conducted on ridged fields during the past two decades has concluded that some sites remain to be studied, and that many of the features themselves served to modify atmospheric temperatures to facilitate cultivation. Moffat concluded from his survey and assessment of prehistoric ridged fields that: 'the vast majority are on relatively high, well-drained soils and not on marshy ground or on floodplains. . . . All of the extant sites described above are on well-drained ground' (Moffat n.d.); and, 'the ridged-fields of the Upper Midwest were not involved with the reclamation of wet, low-lying areas to any important degree. Indeed, low-lying, poorly drained places seem to have been avoided for garden bed construction' (Moffat 1979: 238). His documentation and findings are significant. They are also supported by some innovative experimental research.

Experimental Archaeology

During the field survey portion of his examination of prehistoric ridged field sites in eastern Wisconsin, Moffat was occasionally accompanied by a few different individuals. One of these persons was archaeologist Thomas J. Riley, who, along with Glen Freimuth, had only the year before begun to test the notion that fields were ridged primarily to drain cold air off the planting surfaces. Their research explored the possibility that ridged fields were adaptations to early frost conditions (Riley and Freimuth 1977: 7). It was precipitated by three facts. First, previous research had concluded that because all prehistoric archaeological sites yielding remains of cultivated plants were south of the line representing the 120-day frost-free season, maize agriculture in the upper Midwest during prehistoric times must have required a minimum of 120 frost-free days (Yarnell 1964: 149). Second, a number of known prehistoric ridged fields are located north of the 120 frost-free day line (Fig. 6.2). And, third, many ridged field sites are clustered around the 150 frost-free day line, a line that corresponds closely to places where the average date of the last freeze is 10 May—quite late (Riley and Freimuth 1977: 6). In other words, the region in which prehistoric ridged fields have been found is an area where agriculture is subject to a high risk of frost (Moffat 1979: 228).

Riley and Freimuth (1979: 274) hypothesized that ridged fields served as frost-avoidance mechanisms to protect a young crop from a late spring frost or a nearly mature crop from an early autumn cold snap. Defined as the condition that exists when the air temperature around plants falls to 0 °C, a frost is different from a freeze in that it is local and of short duration. Generally speaking, frosts involve localized cooling due to rapid radiation from plant and soil surfaces under clear skies and light wind or still conditions (Rogers and Swift 1970: 1). Ridged fields, then, should retard frosts on the planting surfaces by creating conditions in which the dense cold air is drained by gravity into the lower-lying paths or ditches where it settles and does little or no damage to the crop.

To test this hypothesis Riley and Freimuth constructed some experimental ridged fields at the Ornamental Horticulture Laboratory of the University of Illinois in the late spring of 1976 with the help of volunteers and students from a class on archaeological methods. On a nearly level plot measuring 36 m east–west by 23 m north–south, this group built two sets of fields. One set involved simple parallel ridges, much like Hubbard's (1878: 3) types 3 and 4, but without additional features at the ends of and separating groups of ridges. The other involved ridges laid out in a checkerboard pattern reminiscent of Hubbard's (1878: 3) type 6 (Fig. 6.3). It comprised six plots of three parallel ridges and paths, placed perpendicular to one another. The individual ridged fields varied between 1 and 1.5 m in width, and were 30, 61, and 93 cm high. Variations were intentional in order to represent differences reported in the archaeological literature. Thermographs were placed on the tops of 61 cm high, and 93 cm high ridges, and in the paths between. Data were recorded continuously, and collected at one-week intervals (Riley and Freimuth 1979: 275, 278, 1977: 7).

Riley's and Freimuth's data collection began in the fall of 1976, albeit on a limited basis. On the mornings of 16 and 17 October they recorded low temperatures of −1 °C in the paths and 1 °C on the ridged fields (Riley and Freimuth 1977: 8). These results were sufficiently promising that a more detailed study was undertaken the following year. Data collection in 1977 began on 4 October, before the first killing frost for the Champaign–Urbana area, and continued until the first snow, which occurred on 25 November. In addition to the collection of temperature data, wind speed and the degree of cloud cover were also noted. Conditions conducive to crop damage were recorded on many dates, but none more significant than 4, 13, and 20 October, and 13 November when there were no clouds and no wind. On each of these dates, the paths served to drain cold air off of and away from the ridges; the 61 cm high ridges averaged 2.6 °C warmer, and the 93 cm high ridges averaged 2.9 °C warmer than the paths. The date 13 October is particularly important in that had crops been growing on the ridged fields, they would have been spared a killing frost that morning (Riley and Freimuth 1979: 278–82).

According to these experimental data, ridged fields clearly facilitate cold air drainage. In the process of doing so, they also extend the effective length of the growing season by a considerable amount. The first killing frost recorded from the top of either a 61 cm high or a 93 cm high ridge occurred on 11 November, a full 29 days after the first frost was recorded in the paths, that is, at normal ground level (Riley and Freimuth 1979: 276–7). Given that they have the same affect on temperatures in the spring, at the beginning of the frost-free period, as they do in the fall, ridged fields can extend the effective agricultural growing season by two months. In places such as southern Wisconsin, where the frost-free season spans at least 150 days, ridged fields could extend the growing season to a full seven months. In so doing, two crops of varieties which mature in approximately 100 days could, theoretically, be grown sequentially in the same plot of land, whereas without ridged fields only one crop could be grown. In other places, such as northern Wisconsin, where the frost-free season lasts only 120 days, ridged fields would extend the agricultural season to six months. This is probably not long enough to double crop, but production of one crop without the risk of frost is almost guaranteed. To be sure, the difference between southern and northern Wisconsin in terms of the frost hazard to agriculture prior to AD 1500 was nearly erased by the use of ridged fields. The same can also be said of Michigan, Iowa, and any other northern locale where ancient ridged fields may have been used. Not without consequence, most of the known locations of prehistoric ridged fields in North America are in that area characterized by a frost-free season greater than 120 days (Fig. 6.2), but less that 150 days (Riley and Freimuth 1977: 6, 1979: 273).

As convincing as it is, the experimental work by Riley and Freimuth should not be considered the final word on the subject. More recent work has refined, at least tentatively, some of their findings (Gartner 1998). Four points merit mention. First, few prehistoric ridged fields have relief as great as Riley and Freimuth's experimental ridges. Second, and clearly a function of the first, experiments at the Hulbert Creek site (Gartner 1993) found the growing season to be lengthened only slightly. Third, these same experiments found that ridging does have a microclimatic function, but it is related to soil temperature rather than cold air drainage; ridged fields warm-up more quickly than flat surfaces in the spring if they are aligned north–south (Gartner 1999: 674, 677). Fourth, adjacent landforms and site location are critical factors in affecting temperatures, which helps explain why many sites with abundant evidence of maize agriculture have no ridged fields nearby (Gartner 1997*b*).

All in all, ridging does affect temperature and hence crop production. Much is known about the processes involved, but much awaits to be learned. One final point. On the basis of their spatially distinct distribution, ridged fields built for purposes of responding to cool conditions in North America (Figs 6.5 and 6.6) are most probably an autochthonous innovation (Gartner 1998; Riley, Moffat, and Freimuth 1981) and not a result of diffusion from Mesoamerica or South America as some scholars (e.g., J. Parsons 1985: 158–9) have speculated.

Overview

Data on the aboriginal use of ridged fields to compensate for conditions too cold at times for agriculture in parts of North America are, admittedly, less than perfect, and, hence, less than convincing. The two documentary accounts state only that ridged fields were used; the interpretation that they were for temperature, and not moisture, control is based on things other than what the early European explorers state themselves. Ethnographic data are non-existent. For reasons not yet understood (principally because to date no one has systematically addressed the issue), native farmers in North America seem not to have used ridged fields and corn hills exclusively of each other, around the time of contact with Europeans. The archaeological data are certainly the most complete, but they too await thorough analyses. Although there is a long history of scholarship involving prehistoric ridged fields in the upper Midwest, it deals mostly with recording sites, some with providing morphological details about sites and their features, and very little with understanding the mechanics and the ecologies of the features themselves. More research waits to be done. And, done it must be, before what few relict fields that remain are obliterated by modern activities. The best data that exist on how ridged fields in North America compensate for cool conditions come from controlled experiments. As with all experimental data, however, these resulted in the asking of more, rather than the answering of a few, questions. They certainly are provocative, and compelling nonetheless.

Part IV

Dryland Systems

7

Dry Farming

That agriculture, be it modern or ancient, requires water is a truism. It is similarly true that when we think of water in the context of farming we typically, and rightfully, place importance on rain. As the two previous chapters have illustrated, native agriculture in North America was not only dependent on rainfall wherever it was adequate, but rainfed agriculture was a widespread practice with a long and distinguished history. The principal reason for this rich tradition and broad distribution is the relative ease in which rainfed agriculture can be, and was, carried out. Besides various clearing, planting, tending, and harvesting activities, which are required of all types of agriculture, nothing else needs to be done in order to practice rainfed agriculture. When it comes to farming drylands, however, the picture changes dramatically.

Raising crops on lands lacking rain that falls in adequate amounts and at regular, dependable, and appropriate times can be a difficult task. It is not enough, however, that a certain amount of rain be received annually. Indeed, ways can be and have been devised to farm lands regardless of the rainfall regime. In terms of water, the critical issue for agriculture is not the rainfall itself, despite popular perceptions, but rather soil moisture. Plants absorb requisite water through their roots not their leaves, which transpire moisture. Accordingly, a minimum amount of soil moisture is a necessary condition for agriculture. But, this is not a sufficient condition. It is also important that soils be adequately moist at certain times of the year, when the crops are in particular stages of growth and development. For example, maize grows reasonably well and even produces a substantial yield in dry conditions if soils are sufficiently moist at two critical periods—at the time of germination and at the time of tassling. Studies have shown that maize grown under dry conditions, but having wetted soils when plants begin to sprout and again when they pollinate, produce yields approximately 80 percent as great as maize grown under ideal conditions, that is with adequate soil moisture throughout the growing season (Denmead and Shaw 1960; Russell and Danielson 1956).

A number of techniques have been devised to compensate for deficiencies in soil moisture at critical periods. In terms of aboriginal agriculture in North America prior to and at the time of European contact, at least four general types were employed. Two of these involve the collection of water from distant sources and its transport to fields. Two others involve conservation of this scarce and precious resource, one on relatively relief-free lands and the other on slopes. These four techniques are discussed in this and the three subsequent chapters, in order of their general technological complexity, labor input requirements, and population correlates.

The most basic form of arid lands agriculture is clearly dry farming in those locales needing little if any other form of landscape modification; that is, lands with relatively little or no relief. As was mentioned briefly in Chapter 5, 'dry farming' is a term often used by some scholars in describing rainfed agriculture. Typically, this confusion is made by arid land specialists (e.g., Lawton and Wilke 1979: 3–4), especially prehistorians (e.g., Doyel 1993: 10; Glassow 1980: 45; Winter 1976: 111) who distinguish the practice from irrigation, or the artificial application of water to fields. In a more general context, dry farming is considered to be any form of cultivation practiced in dry lands and involving any one of a number of soil moisture retention techniques (W. Moore 1974: 70–1).

Of the various ways in which farmers can practice dry farming, two very broad and general strategies can be identified. First, farmers could plant in those select locales with just the right natural conditions (Amsden 1929: 13). For example, certain locales are favored with sandy areas downslope of springs. Water from the springs is retained below the surface of the sand, but at a level sufficiently shallow that crops planted in the sand can utilize the subsurface water. On the Hopi Reservation fields of this type are referred to as 'seepage fields' (Hack 1942: 26, 34). In other settings, sands retain ground water, resulting from the proximity of lakes or streams, which percolates upward. Fields laid out in such favored locales have been called 'water-table fields' (Glassow 1980: 46–7) for obvious reasons. Cultivation in both types of settings has been referred to by some scholars (e.g., Wedel 1941: 21, 1961: 143, 1986: 116) as 'subirrigation', a term that actually describes only half of the agronomic functions. The sands that constitute fields of both these types function not only to transport water upward through capillary action, but also as a mulch to keep water from evaporating.

In addition to naturally sandy areas that receive water from below are sandy locales that receive water from above, in the form of either rainfall or floodwaters. Those places receiving direct rainfall might seem to be better classified as rainfed fields and discussed in Chapter 5. Here, however, the critical element is the sand and its water storage capacity (e.g., Ford 1985*d*: 32–4). Whereas rainfed fields receive moisture when the crops need it, dry farmed fields do not. Instead, they rely on water that fell sometime earlier, perhaps even as winter snow, that is retained even through long periods without rain. Fields that receive water through periodic floods are discussed in Chapter 11 because whereas sandy soils are advantageous for cultivation of flooded lands, the floods themselves are absolutely essential.

The second general form of dry farming involves the creation of microenvironments, typically by addition, whereby soils are given new properties that allow them to retain moisture during periods when they would normally be desiccated. The principal strategy here is mulching, and is not unlike the use of rock piles for the prehistoric cultivation of *agave* discussed in Chapter 3 (e.g., Corey and Kemper 1968; D. Lightfoot 1994). The principal difference, however, is that the mulching materials were used to insure the success of fully domesticated grains as opposed to the cultivation of large, and woody plants.

Dry farming can be carried out in certain locales without any form of landscape modification. Sandy locales are favored because of adequate, but not excessive, moisture just below the surface and can be farmed without leaving any physical remains. As such, there are problems identifying some long-abandoned dry farmed tracts in the archaeological record (Doyel 1993: 31; Glassow 1980: 88). In those places where mulching was practiced, especially with gravels, however, the likelihood of encountering vestiges of fields used in times past is greatly improved. Gravel mulched fields, even those long abandoned, can be quite conspicuous (D. Lightfoot 1994).

Methodological problems also exist with documentary sources or early explorers' accounts of dry farming. In many cases it is impossible to tell if chroniclers were describing dry farming or rainfed agriculture. For example, in 1598 the Oñate party passed 'farms, and planted fields on both banks of the' Rio Grande (Hammond and Rey 1953: 319). Were these fields dry farmed by taking advantage of residual soil moisture, or were they dependent on rainfall? The answer may never be known. In other cases it is quite clear. A few years later, in 1601, one Marcelo de Espinosa reported of this same general area that some fields 'depend on seasonal rains. They do their planting in May and harvest in August. During this time there is some rain, although not very much, for the land obtains its moisture from snowfalls, which are frequent from September to April' (Valverde 1953*a*: 634).

Perhaps not surprisingly, one of the keys to distinguishing between fields dependent on direct rainfall and those that were dry farmed according to documentary accounts lies in the nature of the soil itself. For example, as he accompanied Coronado through Zuni in 1540, Jaramillo (1940: 299) wrote 'They have sufficient provisions of maize, beans, and calabashes. . . . The soil is somewhat sandy'. In a recent study of Zuni agriculture Eric Perramond (1994:35) found that contemporary farmers at Zuni prefer to cultivate very sandy areas because such soils remain damp, below the surface, through the dry season. Perramond was hesitant to speculate on the antiquity of this form of soil agriculture, but noted southwestern archaeologist Keith W. Kintigh (1984: 223) concluded that 'it appears highly unlikely that aboriginal corn could have been grown *anywhere* in the Zuni area in the absence of some kind of effective water-control system'. On the basis of statements such as these, it is reasonably safe to infer that farming practiced on certain sandy soils in environments where the evapotranspiration rate is quite high can be considered a true form of dry farming, albeit an elementary one.

Dry farming, by definition, is restricted to arid or semi-arid lands. Accordingly, the practice, in all its various forms, was spatially limited around the time of European contact to the southwestern quadrant of the continent (Kirk 1973). It is not without consequence that this region was also the first to receive maize from southern Mexico, the hearth of this plant's domestication. Indeed, some of the earliest evidence of maize cultivation in North America goes hand-in-hand with dry farming as the earliest form of agriculture. W. H. Wills, who carried out extensive re-excavations of Bat Cave on his way to refining our understanding of early agriculture, found that arable land on the San Augustine Plains of highland western central New Mexico was confined to the margins of the Pleistocene lake bed. Whereas the former lake bed itself consists of impervious clays, unconsolidated sandy sediments constituted the beaches, which are not only near the cave, but in all likelihood were cultivated prehistorically (Wills 1988: 468). That these sands were dry farmed rather than rainfed can be inferred by the paucity of summertime precipitation and the abundance of winter snowfall. Also, on the basis of their proximity to the lake and according to one field typology (Glassow 1980: 46–7), these would have been 'water-table fields'.

Documentary Evidence

Early explorers' accounts of dry farming are, perhaps not surprisingly, few in number, and less than thorough. Indeed, they involve only references to either cultivation of sandy soils, or reliance on winter snow melt which was retained in the sands. They do not include mention of any deliberate mulching, suggesting one of three possibilities: the practice was not use in the 1500s and 1600s; it was in use but the explorers never saw it; or it was in use and seen by early Europeans who for whatever reasons did not report it. Insufficient evidence exists at this time to even proffer a guess as to which of these scenarios might be correct.

In addition to Jaramillo's account of sandy soils being cultivated at Zuni and Espinosa's report of fields in the upper Rio Grande Valley being dependent on snow melt, both cited above, there are only nine other explicit documentary references to aboriginal dry farming in North America. There are, however, several other accounts that hint at or support the notion of dry farming. All come from the Southwest, specifically northern New Mexico, and northern Arizona (Fig. 7.1). Four of these mention cultivating sandy soils exclusively, and four mention reliance on snow melt stored in soils that are presumably sandy. Jaramillo's description of farming at Zuni is echoed almost verbatim twice in the *Traslado de las Nuevas*, a document written by an unknown member of the Coronado party in 1541. In one

Fig. 7.1. Locations cited in the text for which there is documentary evidence of dry farming.

place it says: 'The country is sandy where they live near together and where they plant. They raise corn, but not very much, and beans and melons' (Anonymous 1922: 191). A little further on it says: 'They raise corn and beans and melons. . . . The land where they plant is entirely sandy' (Anonymous 1922: 191–2). Four decades later, Pérez de Luxán (1966: 185) reported during planting time at Zuni that 'most of the showers in this land come in the form of snow'.

The *Traslado de las Nuevas* also mentions cultivation of sandy soils at Tiwa pueblos. Specifically, it says: 'They raise as much corn as they need, and beans and melons. They raise cotton. The country is all sandy' (Anonymous 1922: 193). In his 1601 testimony which included mention of agriculture at a Tiwa pueblo, Captain Juan de Ortega does not mention sand, but he does discuss retention of snow melt, thereby not only confirming but improving on the description provided in the previous document. Ortega said: 'the land is short of water, even in summer, as it rains but little. The greatest relief from the lack of moisture comes from the snow, of which there is a great deal. It moistens the soil' (Valverde 1953*a*: 660).

Elsewhere along the upper Rio Grande are yet further accounts which connect sandy soils and snow melt retention for native agriculture at the time of contact. Espejo (1966: 221) noted in 1582 that: 'On both sides of the river there are sandy stretches extending for more than a league, naturally adapted for the production of abundant crops'. In 1601, Luis de Velasco (1953: 610) claimed that: 'The land is not all fertile because of the lack of rain, though the heavy snows help give it moisture'. Similarly, Francisco de Valverde (1953*a*: 653), wrote that: 'it rains very seldom, although there is much snow, which helps to moisten the ground so that they [Indians] may harvest what they plant'.

Without doubt, the single most complete and definitive documentary account of dry farming in the northern part of the Southwest comes from a Hopi pueblo in what is now northern Arizona. Writing in 1582, Pérez de Luxán (1966: 191) relayed that: 'The soil is very fertile for maize, cotton, and for everything sown in it as it is a temperate land. The natives cultivate sandy places without difficulty because they carefully guard the moisture from the snow'.

Together, the comments presented above give a rather clear picture that natives in the northern part of the Southwest at the time of European contact were dry farming, in the sense that they were planting in sandy places in order to take advantage of the moisture held below the surface and deposited there as the springtime melt of snow that fell in the previous winter. So compelling is the evidence that other documentary references to agriculture at these locations, but not mentioning either sandy soils or retention of snow melt (Espejo 1966: 227; and Gallegos 1966: 108 for Zuni; and Espejo 1966: 227; and Pérez de Luxán 1966: 192 for Hopi), may have been based on observations of dry farming.

Ethnographic Evidence

Nineteenth- and 20th-century descriptions of dry farming on relatively relief-free lands are as few and as varied as the documentary accounts. Some only mention the cultivation of sandy soils in New Mexico. Others provide much insight into the intriguing cultivation of sand dunes by the Hopi of northern Arizona. While yet others comment briefly on subirrigation but not necessarily of sandy soils in far western Texas, southern Arizona, and California. The single most important common denominator of these assessments of dry farming is that they all involve the Southwest (Fig. 7.2).

In his travels around the Southwest, the astute archaeologist–ethnographer Adolph F. Bandelier observed one locale where sandy and gravely lands were not only cultivated, but were quite extensive. This was the stretch of land along the east bank of the Rio Grande from Bernalillo to Albuquerque (Bandelier 1890: 222). Although he does not mention this, it is entirely possible, indeed probable, that this tract was the same one described by Espejo (1966: 221) three centuries earlier. Bandelier (1890: 31) also reported that the Piro cultivated the '*Médano*', or dune, while the Tano depended on winter

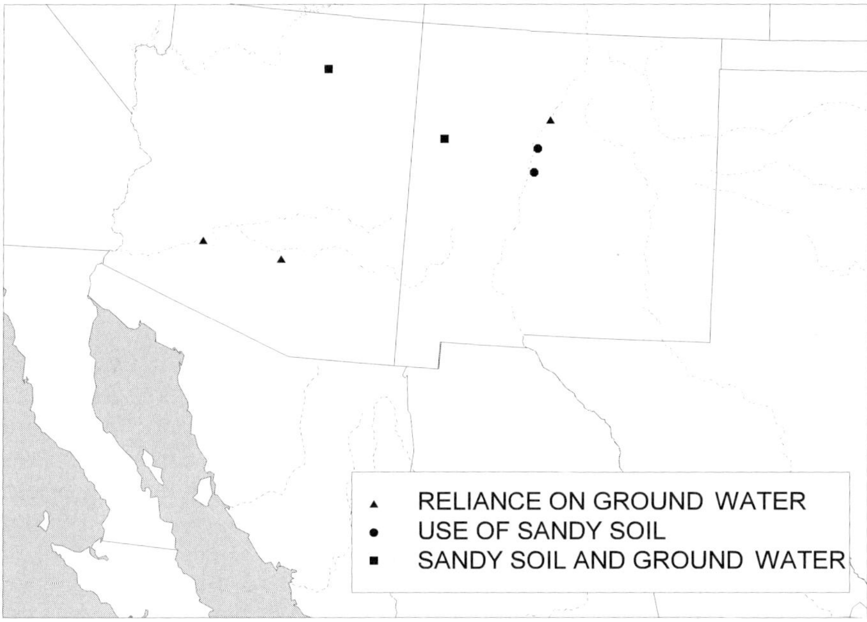

Fig. 7.2. Locations cited in the text for which there is ethnographic evidence describing and assessing dry farming.

snows to provide moisture for crop production a few months later. Although there is no mention of ground water in reference to the former group and no mention of sandy soils for the latter, reference to both peoples, their environments, and agricultural practices in the same paragraph suggests that dry farming was a common form of cultivation.

At about the same time that Bandelier was carrying out his field endeavors, an equally, and perhaps more, colorful character was doing likewise not too far away. Frank Hamilton Cushing (1974, 1979) is legendary for his exploits among the Zuni. Working initially with the Smithsonian Institution on its expeditions in the late 1800s, Cushing went native at Zuni. In the course of a little more than four years he became one of them and even advanced through certain social ranks. Fortunately, he recorded much of what he learned first-hand, thereby providing some of the best ethnographic observations ever made. Cushing not only mentioned the cultivation of sandy soils and the reliance on water trapped beneath them for agriculture, but he even described how the Zuni created dry farming fields. His description is so insightful, it is better reproduced than paraphrased.

Along the western sides of his field, as well as of such spots throughout it as are worn out or barren, he [a Zuni farmer] thickly plants rows of sagebrush. . . . As the

> prevailing winds . . . hail from the southwest, . . . these winds are laden . . . with fine dust and sand, behind each row of the sagebrush a long, level, deep deposit of soil is drifted. With the coming of the first (and as a rule the only) rainstorm of spring-time, the water . . . retained lower down by the 'earth bins', redistributes this 'soil sown by the winds' and fixes it with moisture to the surface it has usurped (Cushing 1974: 165–6).

Cushing's portrayal of agricultural conditions and practices is so detailed and authoritative that it has long been accepted as the best work on the subject. Also, and despite being engaging, a quality that seemingly should inspire additional research, Cushing's descriptions are accepted as the final word. Surprisingly little research has been done on Zuni agriculture during the past century (Woodbury n.d.*b*). The limited work that has been done was initiated in one case in response to litigation (Ford 1985*d*), and in another case out of the intellectual curiosity of one master's student (Perramond 1994). Both projects relied on secondary sources and some field observations. In reality, they do little more than provide simple descriptions of fields and state where in the general area such fields were located. *Soyalthlo deachi*, or 'sand on top of field' sites were identified by Ford (1985*d*: 32–5) as being 'fairly flat', on lands with 'no arroyos', and marked by wooden posts or cairns. Such fields were found north of the village of Nutria, below Pia Mesa. They were also located south of the village of Ojo Caliente and near the village of Pescado. Perramond (1994: 35) reported that the largest area containing the most dry farmed fields on sandy soils is just south of the main pueblo.

A visit to the Zuni Reservation in 1988 resulted in findings that parallel those just mentioned. Most of the dry farmed, sandy fields were quite large and located on relatively relief-free land only a short distance south of the pueblo with which it shares a name (Fig. 7.3). Fencing has replaced the traditional method of marking field boundaries, but it could well be that this is an adaptation that involved nothing more than stringing barbed wire between existing wooden posts. Another recent adaptation may be fallowing.

None of the documentary or other ethnographic accounts of dry farming at Zuni mention fallowing. The practice is, however, not uncommon today. In contrast, the planting of sagebrush in order to aid in the deposition of wind-borne sediment is not common today but was in times past. As Cushing noted (see above quote), some fields are known to have become 'worn out'. In addition to acting as a mulch to inhibit evaporation of soil moisture, the wind-borne sediment trapped by the planted sagebrush apparently contains greatly needed minerals and nutrients. Their deposition therefore helped to replenish soil fertility. According to local informants, sandy dry farmed fields at Zuni today are quickly overgrown with sagebrush once they are harvested and no longer tended. The fields are then abandoned and the sagebrush allowed to trap additional fertile sediment. When once again needed, these formerly worn out, but recently replenished, fields are cleared and replanted.

Fig. 7.3. A distant view of dry farmed fields south of Zuni pueblo. Note the size of the fields, their number and contiguity, and the relatively relief-free terrain on which they are situated. (Photograph by author.)

The Zuni are easily underappreciated as dry farmers who cultivate sandy soils. Exactly why this is the case remains unclear. One highly regarded scholar (Bohrer 1960: 181) suggested that the large number of dry farmed fields at Zuni is a relatively recent, and not an ancient phenomenon. What is clear, however, is that the Zuni's Western Pueblo counterparts, the Hopi, are very much recognized for their longevity in dry farming sandy areas. The Hopi have long been regarded for relying on a number of different types of farming for centuries, nearly all of which are carried out on sandy soils. Spanish explorers and early ethnographers from the eastern United States, alike, noted the diversity of Hopi agricultural practices. That the sands have certain water-retention properties, however, was not identified until the 1930s by anthropologist C. Daryll Forde (1931*b*: 361).

A few years later Ernest Beaglehole (1937: 36) took a closer look at the soils preferred by Hopi farmers. He not only reported the folk taxonomy of sands based on color, but made some interesting observations in terms of fallowing. Beaglehole noted that the Hopi have 'no system of crop rotation', but sometimes these farmers do leave their fields fallow for a year. More commonly they plant the same area year after year, but shift their fields slightly to one side. According to Beaglehole, the prevailing winds tend to blow sand away from the western sides of fields. Because 'it is necessary to have the subsoil covered with six to eight inches of top soil in order to retain adequate moisture in the ground', the custom has arisen of moving the boundaries of cultivated fields 5 to 10 m farther to the east each year (Beaglehole 1937: 36). When it is no longer feasible to shift eastward, fields become deflated and eventually are taken out of use. Fields no longer cultivated are then invaded by sagebrush, which collects blowing sand. After two or three years of abandonment, sands are sufficiently deep to retain soil moisture and, therefore, allow for the cultivation of a crop.

Although Beaglehole provided much insight on fallowing and field rotation, he did not mention the issue of decreased soil fertility as was the case with Cushing at Zuni. However, his predecessor at Hopi did. Forde (1931*b*: 389) mentioned, albeit briefly and without elaboration, that the Hopi planted several seeds per hole, as was common across the continent. Unlike other groups, however, who planted in the same spots year after year, the Hopi planted in between the clumps of the previous year's stubble. Indeed, the stubble from the previous year's crop was not removed after harvesting, but allowed to remain in place and act as a marker to show where not to plant the following year's seeds. Although this is not shifting cultivation or field rotation in the usual sense, it is a subtle form of fallowing that prolonged the life of fields. It is so subtle, in fact that some later researchers at Hopi (e.g., Titiev 1944: 184) did not notice it, although they did recognize the importance of the sands for retaining soil moisture in the face of high evaporation rates (Titiev 1944: 182, 183). When looked at in concert, and in the context of observations made by scholars working at Zuni, the pioneering ethnographic assessments of Forde and Beaglehole, indicate that the Hopi are master landscape domesticators. This fact was not lost on John T. Hack, a physiographer with the Peabody Museum's Awatovi Expedition in the late 1930s.

Hack's work is, to put it simply, legendary among southwestern ethnographers, archaeologists, and geographers. In no small way, this is because it covers so comprehensively the physical environments and their transformation by the Hopi for the purposes of food production.

The Hopi pueblos are situated along the southwestern edge of Black Mesa, a plateau that slopes to the southwest (Fig. 7.4). According to Hack (1942: 26), 27 percent of the land cultivated by the Hopi was devoted to the dry farming of sand dunes. The dunes on which these fields are found occur in three general locales; on the interfluves of streams below the mesa (Fig. 7.5), along the top of the mesa (Fig. 7.6), and at the foot of the mesa (Fig. 7.4) where sands are deposited by the prevailing southwesterly winds. Most of these dunes are very ancient and stabilized by vegetation which has to be removed prior to cultivation. Some dunes are active, but these are not cultivated (Hack 1942: 33). All dunes have great waterholding capacity (Hack 1942: 25). Water stored in these sands and used by crops comes from two sources. Some fields receive water only through precipitation (including winter snowfall), which amounts to only between 270 and 300 mm annually (Hack 1942: 25, 20).

Fig. 7.4. A view of Third Mesa, one lobe of a larger physiographic feature known as Black Mesa, and the pueblo of Oraibi (upper right) from the east. Note the stabilized dunes which have been cleared and planted along the base of the mesa. (Photograph by author.)

Others receive moisture both through precipitation and by means of seepage from springs (Hack 1942: 34).

Hack (1942: 32) is very explicit in his assessment of sand dune agriculture being dry farming in the true sense of the term. He notes that the loose surface of sand acts as a dry mulch which absorbs precipitation and then prevents evaporation. Although a thin layer of sand on the surface does dry out, the sand below remains saturated for a long time. Aiding in this process is a less pervious layer of soil at an even greater depth. This layer keeps the water from percolating below the root zone (Hack 1942: 32).

As advantageous as the sand might be for agriculture, it is not without at least one disadvantage. In order to conserve soil moisture, a large area around the field must be cleared of natural vegetation, thus decreasing the transpiration from non-crop plants. This clearing leaves the crop plants completely exposed to the winds, which can be quite violent in the spring when tender young sprouts of maize and beans are coming to the surface. Materials transported by these winds can literally sandblast a crop to shreds. To compensate for this hazard, the Hopi of today use split tin cans which they place around young plants (Fig. 7.7; Hack 1942: plate vd). Presumably they did something similar but with more ancient technology in times past. Indeed,

Fig. 7.5. A dry farmed sand dune field along Dinnebito Wash west of Third Mesa. (Photograph by author.)

Fig. 7.6. A dry farmed sand dune field on top of Third Mesa near the pueblo of Oraibi. The principal crop of this field is beans. The apricot trees are an introduction from the Old World and a favorite among the Hopi. (Photograph by author.)

Fig. 7.7. A split metal can placed to protect a young plant from blowing sand in a field near Oraibi. (Photograph by author.)

one writer noted that small windbreaks made of twigs were arranged around each plant, and that stones were sometimes placed behind crop plants for the same reason (Forde 1931*b*: 391). It has also been reported that planting several seeds in one hole afforded protection against damage caused by blowing sand. That is, as several plants began to sprout, those on the outer edges of the clump protected those toward the center. The outer plants would eventually die, but those in the middle would survive (Forde 1934: 230; G. Page n.d.). On a somewhat larger scale, the Hopi also have a long tradition of placing brush in rows 2 to 5 m apart perpendicular to the direction of the wind (Fig. 7.8; Hack 1942: plates ve and via). The brush is held in place by heavy stones which typically become buried by blowing sand, thereby leaving the brush projecting 30 cm to 1 m above the surface (Hack 1942: 33; Colton 1960: 95, fig. 35). In some respects these rows are not unlike those described by Cushing at Zuni (G. Stewart 1940*b*: 232, 233). Here, however, they are intended to protect against blowing sand, not to promote deposition.

Sand dune fields are most common near the base of the mesa where the cover of sand on colluvial soil is thin (Hack 1942: 33). Here, too, some fields benefit from water that seeps from the sides of the mesa and is retained beneath the dunes. Hack (1942: 34) referred to these as seepage fields, and did little other than to provide a rudimentary description. Details of how seepage fields actually worked awaited the investigations of Maitland Bradfield in the late 1960s.

Fig. 7.8. Rows of brush held in place by rocks across a mesa-top sand dune field near the pueblo of Hoteville. Note the natural vegetation on the stabilized area surround that which has been cleared for cultivation. (Photograph by author.)

Geologically speaking, Black Mesa can be envisioned as a 'layered cake' (Bradfield 1971: 7), albeit a sloping one. Although there are three major strata, only the top two are important in terms of agriculture, especially dry farming on the seepage fields. The uppermost strata is the permeable Toreva formation, a local representation of the Mesa Verde sandstone. It is underlain by the impervious Mancos shale. Precipitation that falls on top of the mesa, does not runoff so much as it percolates through the Toreva formation. When this moisture makes contact with the Mancos shale, it moves toward the southwest—downslope—along the interface of the upper two strata, as it cannot penetrate the lower one. When it reaches the edge of the plateau, this water seeps out in a number of springs (Fig. 7.9), flows downslope, and accumulates under the sand dunes (Fig. 7.10; Bradfield 1971: 8–9).

In addition to elaborating the geological basis for seepage fields, Bradfield also pointed out a climatological dimension not previously appreciated. Hack and others had long recognized that rainfall, in terms of quantity, timing, and dependability, was nearly insufficient for agriculture on Hopi lands. Some type of water control was nearly always necessary, including the dry farming of certain sand dunes. What previous scholars had not pointed out, however, was the importance of the size of Black Mesa as a catchment

Fig. 7.9. A seep between the permeable Toreva sandstone (top) and the impervious Mancos shale (bottom) at the southwestern edge of Third Mesa. Note the vertical face of the sandstone and the salts accumulated on their surface, and the vegetation benefiting from the water on the sloping face of the shale. (Photograph by author.)

Fig. 7.10. Dry farmed sand dune fields below Oraibi pueblo (on top of the mesa). A spring approximately half way up the cliff face provides water which flows downslope and accumulates under the sand. (Photograph by author.)

area or watershed. Whereas the chances of any field receiving rain when it was needed are not too good, the chances of it getting adequate moisture through seepage are inevitable. Furthermore, earlier researchers had not mentioned the significance of snow, which fell on the Mesa in the winter and melted in the spring before soaking into the sandstone and emerging later as seepage. Credit for bringing these attributes to our attention clearly belongs to Bradfield (1971: 2–5).

Although cultivation of seepage fields is most commonly identified with the Hopi, it is not a practice limited to them. Indeed, there are a few brief reports of subirrigation, although not necessarily involving sandy soils, from elsewhere in the Southwest. Along the lower Conchos River of Chihuahua, Mexico and along the Rio Grande in far western Texas, the Jumano practiced limited agriculture. Early Spanish explorers (e.g., Pérez de Luxán 1966: 158) mention fields and crops but provide no details. The late anthropologist J. Charles Kelley (1952: 358) who studied these people and their environments for many years referred to the practice as 'riverine' farming as it benefited from the constantly moist sandy soils. In certain places along the Gila River the water table is very near the surface, and crops can be raised without irrigation and despite the paucity of rainfall. What have been described as 'rock reefs' cross the channel of the Gila at depths just below the ground surface. These features impede the flow of underground water, forcing it toward the surface. Three such locations proved to be advantageous to Akimel O'odham farmers who raised crops west of Casa Grande monument in the early 20th century (Hackenberg 1983: 165), and a few others were used by Maricopa farmers living further downstream in the vicinity of Gila Bend (Spier 1933: 58–9). Finally, some elderly Cahuilla farmers who were interviewed earlier this century recalled that when they were young, in the mid 1850s, there were places where moisture near the surface was sufficient to raise crops (Lawton and Bean 1968: 21).

In sum, the ethnographic evidence supports the documentary evidence in regards to dry farming on relief-free lands. There is evidence that sandy soils were farmed because of their moisture retention capabilities. In some cases, reports mention only sandy soils and not subsurface water, but other environmental data indicate that agriculture would have been next to impossible in these locales without adequately maintained soil moisture. In other cases, there is evidence of perennially moist locales that were exploited by native farmers, but sandy soils were not explicitly mentioned. Finally, there are those places, most notably the Hopi Reservation, where evidence of water retained under sandy soils is essential for agriculture is overwhelming and compelling. The common denominator among all of the ethnographic evidence is that the southwestern quadrant of the continent is home to this practice. The picture changes very little in terms of the prehistoric data.

Archaeological Evidence

As mentioned earlier, one of the greatest problems for understanding dry farming prehistorically is the paucity (in fact, in most cases, the lack) of evidence which survives to the present day. Cultivation of sandy areas that retain soil moisture, and farming in locales that benefit from either seepage or water not too far below the surface do not often require landscape modifications that are detectable in the archaeological record. In some cases they do, but typically they do not, and this has led to more speculation than proof.

The mystique of Hopi agriculture combined with the insightful work of Hack have intrigued and inspired numerous southwestern archaeologists. This is so much so the case, that there probably is not a scholar working in the region today who has not looked at a sandy area near a prehistoric habitation site and wondered about it being cultivated in times past. To illustrate, writers (e.g., Glassow 1980: 46–7; Kintigh 1985: 101; Lindsay n.d.; Schwartz 1960: 62) have commented on the possibility of sandy locales having been dry farmed prehistorically, *à la* the Hopi, in the Cimarron District of northern New Mexico, at Zuni, in the Kayenta Region of Arizona, and in the Grand Canyon (Fig. 7.11). Similar speculation has been made about other locales, although not in the context of the Hopi, where water not too far below the ground surface could have made agriculture feasible prehistorically. Falling into this latter category are the previously discussed lakeshore sands at Bat Cave (Wills 1988: 468), places near the Salmon site in far northwestern New Mexico (Nials 1980: 19), and even low-lying areas as far east as the Great Plains (Wedel 1941: 21, 1961: 143).

Quite arguably, certain agricultural practices speculated for the Plains area might not fall under the category of dry farming. A case could be made that agriculture in this region was rainfed (Chapter 5) or even flood recessional (see Chapter 11). Doubtless, many prehistoric fields on the Great Plains were dependent entirely on the rain that fell directly on them, while others may have benefited from floodwaters. However, Wedel, in his most recent statement on the matter, made a most illuminating comment. Writing about the headwaters of the Republican River, the area surrounding the point where the states of Nebraska, Kansas, and Colorado meet, he said: 'Over much of this westerly region, the surface stream flow diminishes sharply or disappears entirely into the deep sandy beds during the heat of mid summer. Here water could often be found by the knowledgeable wayfarer with a bit of digging in the sand' (Wedel 1986: 32). Given that astute travelers could find such water and that fields are known to have been located in these general areas, it is more than probable that prehistoric farmers on the western Great Plains took advantage of subsurface waters that were protected from evaporation

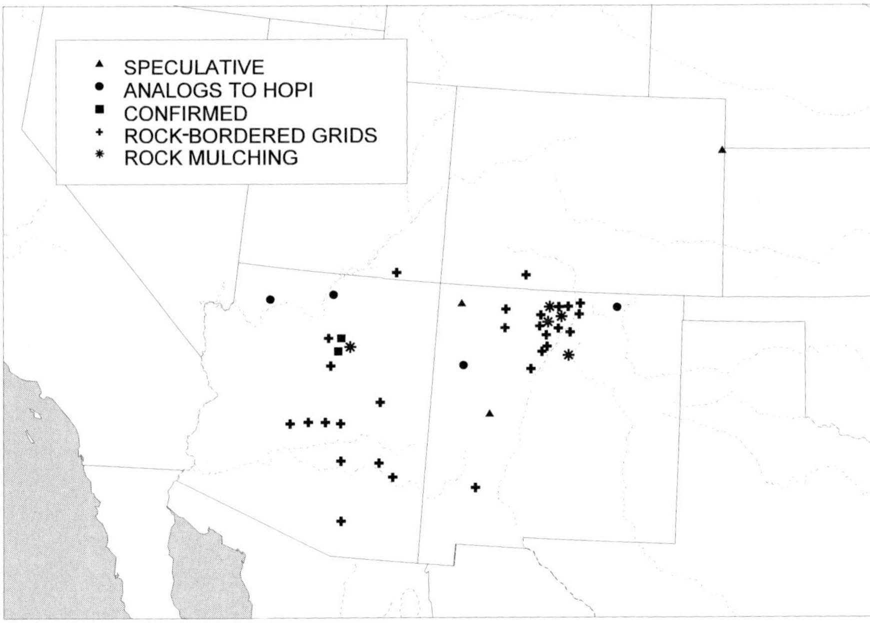

Fig. 7.11. Locations cited in the text and Table 7.1 for which there is archaeological evidence of prehistoric dry farming.

by the overlying sand. In other words, dry farming was in all likelihood practiced there.

Whereas speculation about prehistoric dry farming in North America has been more common than fact in many places, the situation begins to reverse itself in the area surrounding Sunset Crater and thanks to the pioneering work of Harold S. Colton (Pilles 1979: 459). Located on the Colorado Plateau in northern Arizona (Fig. 7.11), Sunset Crater is a volcano that erupted in AD 1064–5 and perhaps a second time in 1066–7 (D. Breternitz 1967). The eruption resulted in a cinder cone more than 300 m high, and spewed basaltic materials over an elliptical-shaped area measuring 58 by 48 km, or more than 2,200 sq km (Colton 1945: 12). The thickness of the layer and the size of the particles vary considerably, decreasing with distance from the crater. The area defined by a 10 km radius and immediately surrounding the crater is quite thick and comprised of lava and lapilli, or cinders, that range in size from 60 to 250 mm in diameter (Fig. 7.12). The remaining area is covered with sand or ash, and toward the limits of deposition, the layer is thin and made-up of dust-sized particles (Colton 1932: 584–7).

Single room surface structures, identified as field houses, and buried beneath the ash indicate that there was a small population living in the area before and at the time of the eruption. The exact nature of these people's

Fig. 7.12. A view of Sunset Crater. Note the thick layer lapilli around its flanks. (Photograph by author.)

agricultural base is unknown, but it has been suggested that it may have been similar to that of more recent Hopi (Ward 1978*b*). It is also unclear how many of these people died as a result of the eruption and how many emigrated (Hevly *et al.* 1979). However, it is quite clear that once the eruption(s) stopped and the land cooled down—*c.* AD 1071—and until approximately AD 1100, farmers recolonized the area (Colton 1945: 12). Settlement data, primarily in the form of field houses, indicate that farmers dispersed themselves over the landscape in individual farmsteads rather than agglomerating themselves into towns. These people were apparently exploring newly available lands (Pilles 1978: 131), and they were doing so under a social and political system that was more complex than previously thought (Travis 1990: 4.52–4.54; Sullivan and Downum 1991: 277–9).

Of the few factors that have been suggested as influencing this increase in the number of farms and farmers, the special qualities of the volcanic ash have long been seen as the most probable (Colton 1949: 24). The 'black sand' (Colton 1960: 9) of the Sunset Crater area is not at all unlike that of the island of Lanzarote in the Canaries, which has a recent but rich agricultural tradition (Dinkins 1969: 122–8, 138–9; J. Parsons 1985: 26, 28). While some people might think that volcanic eruptions may have a greater negative than positive affect on farming (see D. Breternitz 1967: 74), evidence exists to the contrary in certain circumstances. It has long been suggested that the ash functioned as a mulch, absorbing relatively large amounts of springtime snow melt and some summertime rainfall, retaining it through prolonged 'dry' periods, and inhibiting evaporation of soil moisture (Colton 1932: 589, 1945: 12, 1949: 24). Surprisingly, no systematic study of these properties has ever been undertaken. Similarly, no one has examined, and only one scholar before now (Pilles 1979: 468) has considered the possibility that the ash, because of its dark color, also retained heat thereby extending the growing season, and aiding germination. Controlled experiments have, however, demonstrated that the ash layer was not only beneficial, but crucial to the production of a crop.

An experimental plot was planted in July 1961 not too far from Wupatki ruin (Maule 1963). It was planted this late in the season in part to test the possibility of getting two crops per year. Four rows of maize were planted in soil containing no cinders at varying depths. The rows were 9 m long, 1 m apart, and oriented east–west. Sixty seeds were planted in each row, three to a hole, 450 cm apart. Of the four rows, the southernmost was left in soil. The second row was covered with 25 mm, the third with 50 mm, and the northernmost with 75 mm of cinders. No water other than rainfall was applied to the plot.

Interestingly, the best producing row was the one with the 25 mm layer of cinders. Production decreased in direct proportion to the depth of the cinders, but all of the rows covered with volcanic materials produced. In contrast, not one of the seeds planted in the row without cinders so much as germinated! This test not only demonstrated that the volcanic material was essential to the production of a crop, but that it also facilitated double cropping. The subsequent publication of the findings of a series of separate studies undertaken several years earlier confirmed these findings (Colton 1965). The earlier experiments included a large number of plots over a broad area and included several environmental zones. Although details have not been made public, two conclusions are clear. Maize planted in soils covered with a thin layer of ash or cinders outperformed maize planted in soils without such a cover, and maize planted in soils covered with ash more than 25 cm deep did not grow well at all. Maize planted in a deep ash layer apparently experienced too much of a good thing; its roots suffered from not adequately reaching the underlying soil (Colton 1965: 78).

Although the volcanic material was clearly a boon to prehistoric agriculture on this part of the Colorado Plateau, it proved to be a short-lived one. By the middle of the 12th century AD the population began to decline and by AD 1250 was back to its pre-eruption level. Five factors have been suggested as explanations: (i) disease caused by overpopulation and crowed living conditions; (ii) decreased rainfall; (iii) deflation of the ash layer which would have diminished the mulching properties (Colton 1949: 24–5); (iv) a cooling of the atmospheric temperature; and (v) systematic exhaustion of soil nutrients through overcropping (Pilles 1978: 131). The overcrowding explanation seems implausible as there were certainly larger and more dense populations elsewhere in the Southwest who did not experience such problems and suffer accordingly. Changes in both rainfall and temperature at the time the area was farmed are well-recorded (Pilles 1979; Hevly *et al.* 1979). Arguably, the mulching property of the ash and cinders were tapped for their ameliorating affects in an attempt to forestall the inevitable. There is also evidence of both deflation and diminished soil fertility.

Soil deflation is the oldest of the explanations proffered for the abandonment of agriculture in the Sunset Crater area. Beginning with his earliest professional publication about the eruption and its impacts, Colton (1932: 589) suggested that westerly winds, which blow quite strongly in the driest months of May and June, gradually removed the ash. This deflation, it was argued, was assisted by continual cultivation, which brought to the surface finer-grained, easily air-eroded particles (Colton 1960: 118). In an attempt to mitigate this problem, farmers are envisaged as having built rock-anchored brush windbreaks (Colton 1932: 588, 1960: 13, 95–6). Numerous sites containing parallel rows of rocks 1 to 3 m apart, almost all of which are aligned perpendicular to the prevailing wind, have been found throughout the ash-covered area (Marozas 1983: 55–64, 69–71, 100–6; Ritterbush 1984: 53–5; Travis 1990: 4.13–4.25), and are currently thought to be remnants of such windbreaks.

Does the deflation explanation sound like something developed on the basis of ethnographic parallels with the Hopi? It should, because it was. Colton was a contemporary of both Forde and Beaglehole, and he was in contact with them while they were conducting their investigations at the Hopi pueblos. He was also a contemporary of Hack, and, of course, more recent archaeologists working in the area knew of the latter's ethnographic work. The contributions of Hack have, of course, been discussed already. Not yet mentioned, however, was the archaeological work that Hack carried out at Hopi (Fig. 7.11); work that Colton and other archaeologists found significant and germane to Sunset Crater.

During the course of his ethnographic studies, Hack (1942: 33) noticed that when fields containing windbreaks were abandoned, the brush gradually rotted or was cut away by moving sand. Eventually the entire field was blown clear of sand, leaving visible on the surface only the lines of stones that once anchored the brush. Observing this process in action helped Hack to understand lines of stone that he found in other locales near long-abandoned pueblos. Along the southern foot of Antelope Mesa in Jeddito Valley, Hack (1942: 70–3, plate vf) found no fewer than forty-seven sites where lines of stones indicate sand dune cultivation in times past (see also Colton 1960: 96, fig. 36). Hack (1942: 70) estimated that such fields encompassed only about 20 percent of the available farmland and that many more areas of stone lines may once have existed but were destroyed by natural processes. One of these sites, the one nearest to Jeddito Trading Post was most spectacular (Fig. 7.13).

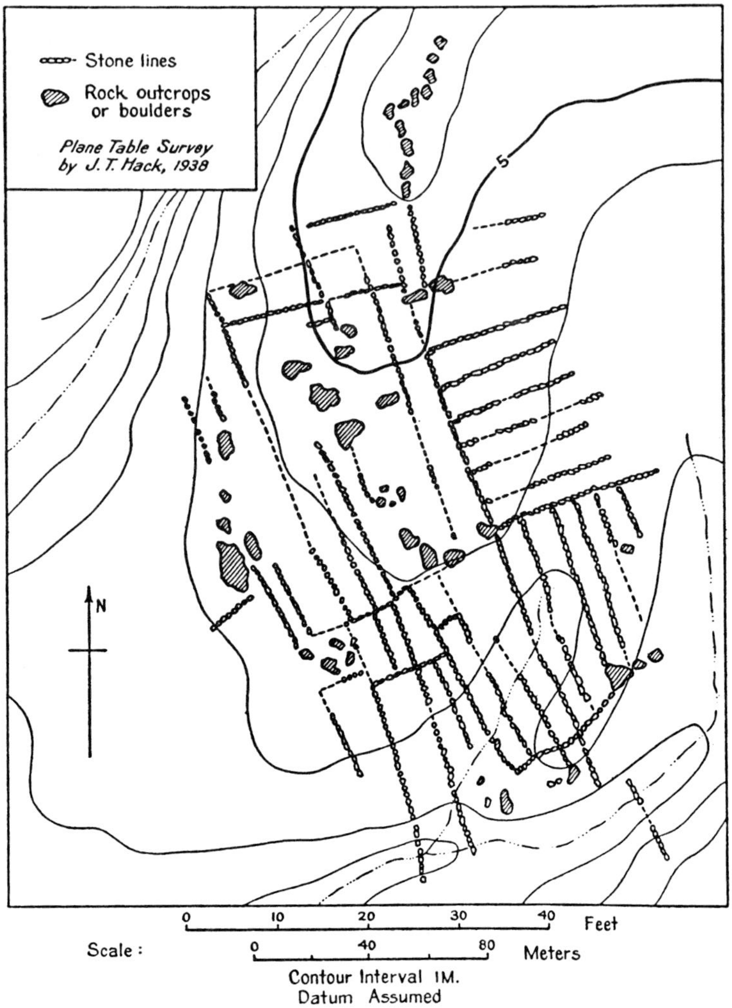

Fig. 7.13. Area of stone lines near Jeddito Trading Post on the Hopi Reservation. These stones have been interpreted as the anchors of brush windbreaks on previously dry farmed sand dunes (see Fig. 7.8). The sand has been removed by deflation. Appeared originally in Hack (1942: 72, fig. 49). (Reprinted by permission of the Peabody Museum of American Archaeology and Ethnology, Harvard University.)

Given that some stone lines were oriented at right angles to others, and that this was not a perfectly flat area, Hack (1942: 70) suggested that a small amount of water was undoubtedly furnished by runoff and underground seepage from the outcrops and sand dunes surrounding the area. By and large, however, this was once a sand dune field that was dry farmed. Most of the stones, certainly those aligned north–south, anchored brush windbreaks sometime in the past. The sands that once covered these prehistoric Hopi fields clearly blew away long ago (see also Brew 1979: 515). Accordingly, the notion of deflation provided by Hack for Hopi was quickly adopted by Colton for explaining the existence of stone lines at Sunset Crater.

In addition to the rock alignments found at Sunset Crater, there are other field features that have been described as 'linear cinder dunes' (Marozas 1983: 71–2), 'cinder berms' (Travis 1990: 4.12–4.13), and 'ash ridges' (Berlin, Salas, and Geib 1990). These features were first identified in the mid 1960s as a result of newly developed technologies (Martin and Schoenwetter 1960). Thermal infrared scanning images taken from an airplane revealed the presence of linear features in three locales of the ash-covered area. Each locale, thought to be a separate ancient agricultural field, was subjected to independent investigation. The first study involved analysis of a set of parallel lines of unspecified length and total area. Ground truth verification revealed that these lines were the remains of gray-black ash ridges separated by troughs of buff-colored soil. The ridges and the troughs vary from 3 to 4 m in width with a relief of 5 to 30 cm. Preliminary analysis of soils concluded that either the ridges were the result of deposition on a brush barrier and the troughs were the result of deflation, or that the ridges and troughs were the result of farmers removing ash from one place to another (Schraber and Gumerman 1969). Later studies concluded that the latter explanation is more correct.

The second study focused on the 'most geometric field area' (Berlin *et al.* 1977: 589). It measured 265 by 67 m and is composed of 10 alternating ridges and swales varying in width from 3 to 4.5 m. The ridges are capped by 10 to 30 cm of ash, with a mean grain size of 0.7 mm, overlying brown-buff soil which is 5 to 20 cm deep. The swales are characterized by large areas of exposed buff-colored soil and red-brown cinders and smaller areas where the soil is veneered by 2 to 25 cm of ash. Intrafield relief varies from 5.5 to 20 cm (Berlin *et al.* 1977: 590).

Palynological examination revealed that maize and squash were grown on this plot, with maize having been grown on the ridges, and squash in the swales (Berlin *et al.* 1977: 592–3). Soil samples collected beneath the ash layer were subjected to various physical and chemical analyses. The soils proved to be sandy and depleted of fine textured materials thereby suggestive of deflation. They were also low in soluble salts suggestive of continuous cropping. Ridge soils were depleted of nitrogen, phosphorous, and potassium which is expected in maize cultivation (Berlin *et al.* 1977: 594–5). This second study concluded that the area was really marginal for agriculture and that continuous cultivation in concert with high winds led to loss of soil fertility and indeed the soil itself. It also suggested that because the ash cover in this locale was originally too thin to act as an effective mulch, farmers collected ash from one place and piled it into thicker accumulations elsewhere. The swales are the result of ash being scraped from there and piled into ridges.

The third study of ash ridges stemming from the discoveries made by aerial photography of the Sunset Crater area focused on nineteen very large ridges that did not occur in series. The largest and smallest of these ridges measured 150 by 20 m and 25 by 9 m, respectively, with an ash depth varying between 17 and 42 cm (Berlin, Salas, and Geib 1990: 2). The size of these ridges resulted in some distinct vegetation patterns with shrubs found on the ridges and non-ridged areas being covered predominantly by grasses. The conclusion drawn here is that mulching properties of the ash layer promoted the growth of the former vegetation over that of the latter (Berlin, Salas, and Geib 1990: 5–9, figs 6, 7, and 8). Curiously, evidence of neither maize nor squash was found in the palynological record of these ridges. In their place, and quite unexpectedly, was found evidence of the family Chenopodiaceae and the genus *Amaranthus*—better known as Cheno-Am—suggesting that goosefoot and amaranth were the principal crops (Berlin, Salas, and Geib 1990: 11–13).

Findings of soil analyses mirrored almost exactly those of the second study, however, some additional evidence for the scraping of ash into ridges was provided (Berlin, Salas, and Geib 1990: 10). Numerous small, thin fragments of sandstone slabs were found in association with the ridges. These were derived from the Moenkopi formation which outcrops within 1 km of the study area. Because the ash ridges are situated on a basalt flow that caps and postdates the sandstone, the slab fragments have to be attributed to human agency. They are, therefore, interpreted as hoe blades or ash scraping tools (Berlin, Salas, and Geib 1990: 10).

In summary, the evidence from the Sunset Crater area is clear and compelling. Prehistoric farmers did not take long to identify the mulching properties of the volcanic ash. They moved in quickly and expanded cultivation of a variety of crops over a broad area. They did not cultivate areas where the ash was too deep and in places were it was too thin, they scraped it into planting beds. In effect, they employed a rudimentary form of gravel mulching (Fig. 7.11). Continuous cultivation and high winds combined with deteriorating climatic conditions led to depletion of soil nutrients, deflation, and overall environmental degradation. The ash mulching system of farming did not last long, but it worked exceptionally well while it did. It may have worked so well, in fact, that it was inspirational to farmers elsewhere in the Southwest who did not have ash, but had an abundance of gravel (Doolittle 1998).

Archaeologists working in various parts of the Southwest, but especially in northern New Mexico (Cordell 1984: 203–7), have been fascinated since the late 1800s by features on the ground that can best be described as rock-bordered grids. From the air many of these features look like giant waffle irons (Fig. 7.14), but these are not 'waffle gardens' as described in Chapter 4.

Fig. 7.14. An aerial view of a set of rock-bordered grids on a Pleistocene terrace overlooking the Gila River, 10 km west of Safford, Arizona. (Provided courtesy, and published by permission, of James A. Neely.)

They cover much greater areas and the individual grids are considerably larger than the plots at Zuni which gave rise to that particular moniker (Herhahn 1995: 78–80).

As striking as these features are, and as long as they have been known, surprisingly few studies have actually focused on how they functioned. Instead, many prehistorians have been all too quick to assume a single particular explanation in spite of some obvious differences that suggest otherwise. At least three types of rock-bordered grids can be identified. Some grids are comprised of individual rocks set into the ground in single rows one course high and are difficult to distinguish from remains of dwellings or other structures (Bandelier 1892: 51; J. Moore 1995*a*: 13–15). Others are made of rows of multiple rocks stacked two or more courses high on the surface. Some rock-bordered grids are clear of stones in the middle (Fig. 7.15), while others are filled with gravel (Fig. 7.16; Anschuetz 1995: 26, 29–32; Bugé 1984: 29–30; Maxwell 1995*a*: 117, 1995*b*: 7; Maxwell and Anschuetz 1992: 56–9). Of this latter type, some are divided into smaller 1 to 2 m grids, reminiscent of 'waffle gardens' (see Figs 4.5–4.7). All of them date between AD 1200 (Cordell 1984: 207) and the arrival of Europeans (Maxwell 1995*a*: 117), with most being used in the 1400s (D. Lightfoot 1993*a*: 116).

Fig. 7.15. A rock-bordered grid comprised of multiple rows and multiple courses of rocks stacked on the surface near Safford, Arizona. (Photograph by author.)

Fig. 7.16. A rock-bordered grid on a Pleistocene terrace overlooking the Rio Chama. Note the layer of dark pebbles inside the grid in contrast to the lighter-colored soil outside the grid. (Photograph by author.)

Assuming that each of these three types of grids functioned in a similar manner appears to be inappropriate. About all that is really known at this time of the former two types is their locations, which are dispersed throughout the Southwest (Table 7.1, Fig. 7.11), and their appearances. More research, much more, needs to be done on these fields. The latter type, in contrast, has been subjected to substantial systematic study recently and the results have confirmed a sophisticated form of dry farming in the truest sense of the term. Curiously, they seem to be concentrated in the valleys of the Rio Chama and its tributaries (Cordell 1984: 205) and in the Galisteo Basin (D. Lightfoot 1990; Fig. 7.11).

The first archaeologist to identify the possible mulching affect of small stones covering the ground surface within these rock-bordered grids was Florence Hawley Ellis (n.d.*a*), in an unpublished paper she presented in a symposium on prehistoric water control systems held as part of the 1970 Pecos Conference. Although she did not offer any evidence to support this interpretation, her claim was made—using such definitive language as 'garden plots edged with stones and mulched with gravel to hold what water fell upon them', 'masses of stone-lined and gravel-mulched garden plots', and 'dry farming on garden plots'. In a publication that appeared a few years later she elaborated on her earlier comments, noting that near Sapawe ruins overlooking El Rito Creek there:

> are acres of dry-farm gardens mulched with gravel from an open pit. Each plot was edged with a row of stones which helped to hold that mulch in place. Gravel mulch may sound strange to one of our own farmers, but it is an efficient method for holding water and retarding evaporation (Ellis: 1978: 63).

Mulching with stone apparently did not sound as strange as Ellis must have thought. Many, perhaps even most, southwestern prehistorians accepted the notion of gravel mulching even when they should not have. To illustrate, two syntheses of ancient food production in the region identify features as 'gravel-mulch bordered gardens' on the basis of aerial photographs which clearly show no gravels covering the ground within the grid borders (Cordell 1984: 206, fig. 6.4; Masse 1991: 210; and Fig. 7.15 above). In no small way, the widespread acceptance of the mulching notion can be traced to Ellis's persuasiveness and influence, and studies that have followed up on her lead—studies that investigated grids with gravel layers, ignoring those without gravel fill.

Ideas concerning the benefits of covering the field surface with gravel have been nothing if not plentiful. The three most obvious advantages that have been purported (e.g., Tjaden 1979: 26–7) are that the gravel would have inhibited evaporation of soil moisture (Lamb and Chapman 1943: 577), it would have increased the infiltration of moisture (Alderfer and Merkle 1943), and it would have facilitated the production of specialized crops such as cotton (Dean 1995: 354–5). As evident as these processes might be, they are not

Table 7.1. Information pertaining to simple rock-bordered grids gleaned from archaeological literature

Llaves Area, New Mexico
'field plots demarkated [*sic*] by lines of stone' (Dick 1976: 20)
'rock-bordered gardens are numerous in the area' (Cordell 1979: 143)

Llegua Canyon, New Mexico
'striking similarity [to] ancient stone lines near the Jeddito Trading Post' (Green, Danfelser, and Vivian 1958: 52)
'brush-lined garden plots . . . stones lying on the terraces' (Pattison 1968: 129)

Abiquiu Mesa, New Mexico
'squares and rectangles . . . single rows of boulders laid upon the surface . . . cross each other at right angles to form . . . checkerboard pattern . . . there are at least 2,000 of the squares' (Hibben 1937: 16)
'Grid borders', 'mulch-border fields [on]mesa tops, benches' (Cordell 1984: 205, 206)
'outlines of . . . cobblestones and small lava blocks . . . vary from 8 to 10 feet wide by from 15 to 20 feet long . . . follow the trend of the mesa for over 900 yards'—mistaken for a large pueblo (Hewitt 1906: 36)

Southeast of Abiquiu, New Mexico
'rows of stones forming little rectangles in one place, as if a few garden plots had extended' (Bandelier 1892: 56)
'stones arranged in a square . . . There were thousands of such arrangements'—mistaken for shrines (Jeancon 1923: 71; see Ellis n.d.*a*.)
'Grids' at site LA 11832 cover 1,825.5 sq m (Cordell 1984: 205)

Northeast side of Rio Chama Valley, New Mexico
'Three clearly outlined cobble stone grids' (Enloe, Smith, and Peckham 1974: 28)
'Extensive stone gridded areas, some on the lava-topped mesa' (Cordell 1979: 145)
'several hundred more of these remarkable [grid] structures' (Hibben 1937: 16–17)
'large ruin of the "foundation stone" type . . . all over the mesa top'—mistaken for a large pueblo (Greenlee 1933: 13)

El Rito Colorado Creek, New Mexico
Near Sapawe ruins are 'plots incased by stones laid on the ground, the outlines of which are marked by heavier pieces than the compartments. The quadrangles are quite large' (Bandelier 1892: 52)
Mistaken for pueblos and field houses (Skinner 1965)
'masses of stone-outlined . . . garden plots' (Ellis n.d.*a*.)
'Acres of small stone-outlined fields were cleared for farming' (Ellis 1975: 20)
'Grid borders' covering 2,164 sq m near Sapawe ruins (Cordell 1984: 205; Cordell, Earls, and Binford 1984: 237)

Rio Ojo Caliente, New Mexico
Three sets comprised of 'rows of stones, intersecting each other so as to form rude parallelograms and squares, found on levels' (Bandelier 1892: 41–2, 46, 51)
'small rectangular plots with cobble borders. Hundreds . . . constructed along . . . terraces. They are of varying size and shape' (Bugé 1984: 29, 1978)
'Like waffle irons, the remnants of Anasazi farms stipple the land . . . rock and cobble borders' (Canby 1982: 576–7)

Rio Grande de Ranchos, New Mexico
'grid bordered gardens' (Cordell 1979: 144)
'4 sites of grid borders', 'Gridded fields, small borders' (Cordell 1984: 205; Cordell, Earls, and Binford 1984: 238)

Taos Valley, New Mexico
'stone-bordered farm plots', 'larger gridded field system' (Woosley 1986: 158)

Table 7.1. *Continued*

Picuris Pueblo Grant, New Mexico
'parallel lines of boulders . . . .developed grid pattern' (Woodbury n.d.*a*.)
Nine sites with 'stone grids & borders', 'waffle gardens', 'grid system', covering '10,000 sq m', '20,000 sq m', '200 acres', 'Extensive' areas (Cordell, Earls, and Binford 1984: 237)

South of Puye Cliff Dwellings, New Mexico
'on this land are traces of garden plots . . . encased by rows of stones' (Bandelier 1892: 72, 73)

Los Alamos, New Mexico
'a garden enclosed with lava rocks or in checkerboard form' (Ellis and Dodge 1989: 38)

Cochiti Pueblo, New Mexico
'Garden beds, enclosed by upright stones' (Bandelier 1892: 180)
'grids', 'rectangular stone alignments' (Cordell, Earls, and Binford 1984: 238)

Zia Pueblo, New Mexico
'many grid gardens . . . their overall length approximately a 3-mile strip . . . their width roughly one quarter of a mile' (Ellis and Dodge 1989: 39–40)

Mimbres Valley, New Mexico
'frequently enclosed spaces without any rubbish' 'The . . . largest . . . measured . . . 13 by 7 meters'
'[T]hey were, as the Pima tradition states of the Arizona ruins, garden beds' (Bandelier 1892: 357–8)

Pinelawn–Snowflake, Arizona
'grid units', 'grid systems' (K. Lightfoot 1981: 259–60)

Point of Pines, Arizona
'grid borders' (Woodbury 1961*b*: 13)

Gila Valley near Safford, Arizona
'rows of stones marking off the surface in rectangles' (Fewkes 1904: 177)
'stones have been picked to outline the borders of gardens' (Hough 1907: 33)
'detention structure . . . rectangular in shape' (G. Stewart 1939: 114)
'12 separate agricultural systems of the type known as . . . gridded gardens' (Gilman and Sherman 1975: 9)
'large gridded . . . fields' (Woosley 1980: 322, 323)
'bordered gardens (also referred to in the literature as "waffle gardens")' (Masse 1991: 210)

Gila Valley near Florence, Arizona
'rocks have been gathered in rows that enclose rectangular areas' (F. Russell 1908: 88)
'Bordered gardens' (Dart 1983: 410–13)
'field areas surrounded by stone borders . . . surface pebbles and gravels formed a grid' (Crown 1984*a*: 245)

Tonto National Forest, Arizona
'grid borders . . . waffle gardens' (Wood and McAllister 1984: 274)

Santa Cruz Valley north of Tucson, Arizona
'The rock alignments are arranged into a grid pattern that may represent a waffle garden' (Downum 1993: 35, see also 85–6, fig. 4.30)

Lower Verde, north of Phoenix, Arizona
'bordered gardens . . . gridded' (Kearnes, Polk, and Teague 1975: 251)

Cave Creek, Arizona
'grid borders' (Ayers 1967: 108, 110)

Table 7.1. *Continued*

Agua Fria Valley, Arizona
'rock debris raked into rectangles varying from one meter per side to twenty to thirty meters per side' (Dove 1970: 1)
'hundreds of acres of waffle garden type farmland have been spotted in our preliminary surveys in a light plane' (Gumerman and Johnson 1971: 96)
'Gridded alignments' (Rankin 1989*b*: 975–8)

Chaves Pass, Southeast of Flagstaff, Arizona
'[Rock] alignments were formed into a chain of squares or grids' (Henderson 1979: 10)

Wupatki National Forest, Arizona
'gridded-border fields . . . stone alignments are placed perpendicular to one another' (Marozas 1983: 64)
'perpendicular gridded [rock] alignments' (Travis 1990: 20)

Beaver Creek, Utah
'Grid borders . . . formed by short lines of stones' (Lindsay 1961: 183)

Pagosa Springs, Colorado
'remains of extensive cobble-lined plots on adjacent gravel terraces' (T. Stewart and Walker 1998: 32)

without some interesting nuances that have been given some attention by archaeologists (Maxwell 1995*a*: 118; Maxwell and Anschuetz 1992: 64). For example, gravel mulching not only helps retain soil moisture over short periods of time (Fairbourn 1975), but also from one season to the next (Fairbourn 1973: 926), and even from year to year (Choriki *et al.* 1964). As for improving infiltration of water, it does so not only by having large pores (Fairbourn 1973: 925), but also by preventing the formation of soil crusts (Corey and Kemper 1968: 14).

But water is not everything. Archaeologists working in northern New Mexico, and familiar with the region's short growing season, did not take long to explore the possible atmospheric-moderating, as well as edaphic-improving, qualities of the gravel layer. Indeed, Linda S. Cordell (1984: 206; Cordell, Earls, and Binford 1984: 236) was the first to suggest that the gravel might have stabilized soil temperatures by reducing air movement near the surface and warming the top few centimeters of soil. Because the gravel is dark and, therefore, absorbs solar radiation and dissipates heat more slowly than lighter-colored and less rocky soils (Aase, Kemper, and Danielson 1968), Cordell maintained that the use of a gravel veneer may have been employed prehistorically as a means to extend the growing season artificially. As with the case of soil moisture, there are also subtleties in the affect of gravel layers on temperature and, hence, crop growth that have not escaped the eclectic vision of archaeologists (Maxwell and Anschuetz 1992: 65; Maxwell 1995*a*: 119). For example, higher temperatures resulting from the dark stone covering hastens germination, and increases both root growth (Hakimi and Kachru 1978) and tassling (Fairbourn 1973: 927) by affecting the rate at which crops take up water and nutrients from the soil (Othiero and Ahn 1980).

That covering fields with a layer of gravel might have compensated for both cool and dry conditions in northern New Mexico during pre-European times was accepted on the basis of experimental studies conducted under ideal conditions outside the region. Analogs have been borrowed freely (Maxwell 1995*b*), but controlled experiments carried out on some actual prehistoric gravel-covered fields in the region remain all too few. Indeed, there have been only two such studies. Unlike experimental studies conducted at Sunset Crater, which only compared crop growth in mulched and unmulched plots (Maule 1963), the ones carried out in New Mexico focused explicitly on the edaphic and atmospheric impacts of the gravel layer.

The first detailed study of possible prehistoric mulching was conducted in the Rio Ojo Caliente. It examined a field measuring 13 m by 14 m, and divided into 14 distinct units by cobble borders (Lang 1980: 16). The uppermost stratum, the suspected mulch layer, was 4 cm thick and comprised of gravels that appear to have been scraped up from a nearby depression identified as a borrow pit. Beneath this veneer is a layer of fine sand and gravel extending to 7 cm below the surface, and beneath this is a layer of aeolian sand containing some gravel and cobbles to the limits of the test excavation at 15 cm (Lang 1980: 17).

Comparative tests involved measurements taken within the gravel-covered area and in a relatively undisturbed area outside of it. The tests were actually rather elementary, which is certainly more than acceptable given the infancy of any experimentation, relying on a moisture sensing electronic probe inserted into the soil at various depths twenty-four hours after a measured amount of water was poured onto test areas. Although they were at best preliminary, the tests demonstrated that soil covered with the layer of gravel retained more moisture and at greater depths than did the area not so covered (Lang 1980: 22–4).

The second systematic experimental studies of ancient gravel mulching were conducted by geographer Dale R. Lightfoot (1990). His investigations were not carried out in the Chama Valley or one of its tributaries, where archaeologists have focused their attention, but on some equally impressive, and previously underappreciated prehistoric fields farther east in the Galisteo Basin (see Fig. 7.11). Furthermore, his studies borrowed much more extensively from related bodies of literature than did previous New Mexican studies.

As was the case with the systematic studies into the mulching qualities of volcanic ash at Sunset Crater, Lightfoot's investigations first involved aerial photographs. In this case, they were color infrared images at a scale of 1:13,000 taken by NASA in 1987 (Lightfoot 1993*b*: 356). Analysis revealed no fewer than ninety-six separate fields in the basin. Ground-truth verification of these fields indicated that they appeared on the aerial photographs, not because of differences in soil or rock color, but instead because of differences in vegetative growth. Although crops were no longer raised on these gravel-covered plots, the mulching properties were still in place. Natural grass growth was more prolific on the plots then off (Lightfoot 1993*a*: 116). These fields covered a total of 41,000 sq m. The borders consisted of cobbles ranging in size from 10 to 30 cm and the gravel layer was comprised of materials measuring between 5 and 30 mm. The latter were both scraped up and excavated from nearby borrow pits (Lightfoot 1993*a*: 120–1, 122, 1993*b*: 350). The gravel layer itself was between 5 and 11 cm thick (Lightfoot and Eddy 1995: 465).

Lightfoot (1993*b*: 357–8) mapped these features and analyzed them in the context of their geomorphic and edaphic contexts. He also excavated select fields and collected moisture, temperature, and biomass data (Lightfoot and Eddy 1994: 429). In general, he tested a number of the hypotheses and ideas that were previously put forth by others.

Given that some scholars had argued that temperature control might have been an important factor in the use of gravel mulching, Lightfoot examined this aspect. He found that while 74 percent had a generally southern orientation, this might have been spatially autocorrelated with natural occurring gravel deposits (Lightfoot 1993*b*: 365). All of the ancient gravel mulch plots he examined in detail were located on 'flat or relatively level sites', none were on a slope greater than 4 percent (Lightfoot 1993*b*: 366). Furthermore, all were located on soils formed on alluvium and pebble–cobble conglomerates of glaciofluvial origin (Lightfoot 1993*b*: 360). In sum, one of Lightfoot's conclusions is that gravel mulch agriculture could only have been practiced where necessary materials were at hand.

Lightfoot's soil moisture, temperature, and biomass tests were as insightful as his remote sensing, cartographic, and geomorphic appraisals. He excavated trenches 2 m long, 50 cm wide, and 12 to 14 cm deep in each of three fields. Using a thermocouple soil thermometer and moisture reading meter, sediment samples were analyzed as they were excavated. Two biomass samples were collected from inside and outside of each of 20 plots. These were air-dried and weighed (Lightfoot and Eddy 1994: 429). Soil moisture data mirror, almost exactly, those collected by Lang (1980) in the Rio Ojo Caliente, thereby confirming that the gravel layer does act as a mulch (Lightfoot and Eddy 1994: 430–1, 1995: 468–9).

Analyses of temperature were not as definitive as those of moisture, and they did not compare with controlled agronomic experiments or ethnographic data from elsewhere. Lightfoot found that at the 5 and 10 cm levels, soils in the Galisteo gravel mulch fields averaged only 0.3 °C warmer than in adjacent non-mulched soils, and that at the 15 cm level the soil temperature averaged 0.3 °C cooler. He suggests that the fields may have lost their ability to increase temperature through the accumulation of aeolian sediment incorporated into the fields since they were abandoned nearly five centuries ago (Lightfoot and Eddy 1994: 431–2). Despite such differences, however, the gravel-mulched fields still have important qualities that favorably affect plant growth. Results of the biomass collections show an average of twice as much material today on the prehistoric plots as compared to adjacent areas. Not only do grasses grow taller and denser, but the roots are more numerous and penetrate deeper (Lightfoot and Eddy 1994: 433–4, 435).

On the whole, there are many benefits to gravel mulching, some of which have long been recognized. Others, however, were brought to archaeological attention only recently by Lightfoot, principally as offshoots of his studies which included extensive analogs from various bodies of literature as well as field tests. To illustrate, agronomic tests have shown that unmulched plots experience four times as much soil loss to erosion as do gravel-mulched plots (Mannering and Meyer 1963). Lightfoot argues that the gravel layer on the surface shields the soil from the impact of raindrops which would otherwise dislodge particles and put them in motion (Lightfoot and Eddy 1994: 436). He also points out that agronomic experiments demonstrate how gravel mulching helps control weed growth (Lamb and Chapman 1943), and inhibits salinization of soils (Unger 1971), facts previously overlooked by researchers (Lightfoot 1993*a*: 121, 122). Similarly, he brings to light little appreciated evidence that gravel mulching can have detrimental effects in terms of field longevity, and uses this to explain the very short period of time such fields were used prehistorically in the American Southwest.

Agronomic experiments (Alderfer and Merkle 1943) have shown that the organic content of soil can decrease by as much as 10 percent after one year of gravel mulching. At least one ethnographic study offers additional support. Carried out in China, that study found that gravel mulched fields began to play out in twenty to thirty years, and after fifty to eighty years became 'old' and less productive than unmulched fields (Tsiang 1948). According to Lightfoot (1993*a*: 122) a permanent layer of gravel mulch on the field surface discourages the return of crop wastes following harvest and thereby disrupts the nutrient cycles. Over the long term, the effects on soil nutrient recycling can pose a significant obstacle to the sustainability of gravel mulch agriculture (Lightfoot and Eddy 1994: 434–5; see also Lang 1995: 51–5).

Gravel mulching was practiced in New Mexico just prior to the arrival of European explorers. Archaeological evidence in the form of ceramics indicate that it was begun in the 1200s, but reached its zenith at approximately AD 1350. The dates of its inception and proliferation correlate well with a documented period of drought (Lightfoot 1993*a*: 129–32), and migration (Doolittle 1998). The cultural–ecological implications for it being an adaptation to dry conditions seem clear. However, given their relatively small size, limited number, and restricted spatial distribution, it is doubtful these plots were anything more than experimental attempts to deal with deteriorating environmental conditions (Lightfoot 1993*a*: 134; Lightfoot and Eddy 1994: 436). Although they might have eventually been a failure, they are significant in that they show the tenacity and innovativeness of their builders. They also are evidence that aboriginal farmers had a keen perception of the biophysical environment and all its nuances and subtleties.

Overview

Dry farming was certainly a common practice in the American Southwest at the time of European contact. It was, however, not a simple and straightforward practice and the evidence is at best checkered. The documentary data yields little more than insight into native farmers cultivating sandy soils, presumably to take advantage of what might be considered natural mulching properties of these coarse sediments. In many cases, this evidence indicates that snow melt was trapped under the sands and absorbed by crops later in the year when rainfall was deficient. The ethnographic data are considerably better, especially in regard to the Zuni and the Hopi. Indeed, two 'classic' studies of Hopi sand dune agriculture, those of Hack (1942) and Bradfield (1971) are among the best studies of indigenous agriculture from anywhere in North America, the New World, and the entire world, for that matter.

Without question, however, the archaeological studies of dry farming shed light on a number of the various aspects of native North American agriculture. As is typically the case with archaeological studies of prehistory, there is much speculation, especially based on ethnographic parallels. There is also much very reliable evidence in its own right, and some of this evidence can only be described as intriguing. Outstanding in this regard is the evidence from Sunset Crater on the reliance on volcanic ash. Late prehistoric farmers in northern Arizona not only took advantage of the natural mulching properties of the ash, but as the climate became drier and the soil depleted of nutrients, they formed the ash into distinctive planting beds in an attempt to perpetuate the system. Their attempts seem to have been copied to some degree by slightly later farmers in northern New Mexico. These cultivators did not have the advantage of the ash, but they did experience drought conditions and apparently learned of mulching with gravel. These farmers built some very elaborate and easily identifiable fields, quite in contrast to simply cultivating sandy locales.

As distinctive as rock-bordered fields are, they are not at all well-understood. To be sure, and thanks to the work of Lightfoot (e.g., 1990), we now know a great deal about those fields which contain a veneer of gravel and functioned as a mulch. However, the rock-bordered grids that do not contain a gravel layer are far more numerous and widespread than mulched fields, but absolutely nothing is known of them. Did they somehow serve a dry farming function? Who knows? A few scholars have claimed that they were mulched, but that is pure speculation at best, and erroneous at worst. Of course, it may well be that rock grids served a variety of purposes and that even some in proximity to gravel mulched fields might have served an entirely different purpose. Soil erosion, and water control come to mind. Such functions are usually associated with slope lands, but then slope is a relative concept. Flat is an absolute. Even relatively relief-free land can have some gradient.

8

Terracing

Dry farming by its very nature is typically associated with nearly level or relatively relief-free land. In no small way this is a function of the need for stability *vis-à-vis* the powerful forces of nature, and holds as true for rock veneers laid down by farmers as it does for naturally deposited sand. Although gravel mulch is a protective device in one respect, it is not without its weaknesses. Principal among these is its unsuitability on slope lands where runoff is heavy and its effects great. Runoff, of course, is the combined product of precipitation and the nature of the land on which it falls. (For economic assessments of cultivating level as opposed to slope lands see Plog n.d., and Plog and Garrett 1972.)

It is a truism that arid lands are characterized by a general paucity of rainfall. But there is more to this than may be initially apparent, especially to the casual observer. In some cases, it is not so much the amount of rainfall received in a particular locale that is important, but the timing of it. As was seen in the preceding chapter, certain places, specifically higher elevations in the northern part of the Southwest, receive an inadequate amount of rainfall for agriculture during the summer, but an abundance of precipitation in the form of winter snowfall. In other places, for instance lower elevations in the southern part of the Southwest, the problem is with the temporally and spatially erratic nature of rainfall during the growing season. Locales can receive abundant rainfall during the growing season one year and hardly any during the growing season of the next year. Also, one place might receive a great deal of rainfall at one time, while a neighboring place goes wanting. For example, the town of Baviácora, Sonora, Mexico receives in July an average 91 mm of rainfall. In 1958, it received only 24 mm during this month and in the following year it received 130 mm. Baviácora is located only 14 km from the town of Aconchi, at nearly the same elevation and topographic setting. However, in August of 1951 Baviácora received only 87 mm of rainfall while Aconchi was inundated with 241 mm (Hastings 1964: 1, 16). Coping with such erratic precipitation is clearly one of dry land farmers' greatest tasks.

Were rainfall, with all its subtleties and vagaries, not problem enough by itself, its effects on the land can be equally problematic. Arid lands have relatively little vegetation, either woody or herbaceous, in large part due to precipitation deficits over space and time. Much of the surface area remains without vegetative cover for most of the year. When rains do fall, they are typically downpours resulting from violent thunderstorms. In the southwestern part of North America, these rains occur in the months of July and August, following three months (April–June) when very little, *if any*, rain falls (Ives 1949). During this three month dry spell, trees and shrubs have hardly any leaves, and the ground has no protective grass cover (Shreve and Wiggins 1964). When the monsoonal rainy season does begin, its consequences can be severe. The high impact of raindrops dislodges soil particles which are then carried away by sheetwash. Rain typically falls so quickly, so hard, and for such short periods of time, that very little soaks in. Most water runs off, carrying sediment with it. Erosion is, therefore, a second problem with which dry land farmers have to contend. It is also the reason why gravel mulching is not used on slope lands; a layer of gravel placed on a field with anything other than a very slight gradient would quickly be washed away.

Although runoff does have its negative aspects, it can be beneficial. The natural movement of downslope flow allows for water to be concentrated in select areas. Whereas the probability of any particular parcel of land receiving a sufficient amount of rainfall at the exact time it is needed in the arid Southwest is quite low, the probability that parcels situated at the lower ends of large sloping areas and at the far downstream ends of drainage basins will receive runoff when water is needed is rather high. A good analog for this effect can be seen in the pouring of a liquid from one bottle to another through a funnel. One can never be sure exactly where the stream of liquid being poured will hit the inside of the funnel, but 100 percent of the liquid will pass through the neck (see e.g., Rankin and Katzer 1989: 988–94, especially fig. 31.1).

A second benefit of runoff is its ability to replenish soil fertility. Runoff is not clean water; it is muddy. Sediment put into suspension upslope or upstream is eroded downslope or downstream. It remains in suspension as long as runoff is moving rapidly. As the gradient becomes less steep, runoff slows, and sediment begins to drop out (Cannell and Weeks 1979: 240). Fields situated in comparatively flat areas at the lower ends of slopes and drainage basins can, therefore, be sites of sedimentation. In addition to these inorganic materials, organic matter in the form of floral and faunal detritus (e.g., driftwood, leaf litter, microorganisms, feces, and carcasses) is deposited on fields (Nabhan 1983: 155–93, 1984). These materials add minerals and nutrients to fields thereby keeping soil fertility levels high even if fields are cultivated intensively. The Rarámuri of northern Mexico have an intriguing system of agriculture that involves inducing erosion and sedimentation on concave slopes. Briefly, they clear two or three times as much land as they intend to cultivate (Robert Bye, personal communication 1990). The less steep downslope third or half is planted while the steeper upslope half or two-thirds is allowed to erode (Fig. 8.1). The materials washed off the upper portion are deposited on the lower portion, often behind low rock walls built for this purpose. Curiously, this interesting practice has been known about for nearly a

Fig. 8.1. A typical Rarámuri field near the village of Cerocahui in the Sierra Madre Occidental of northwestern Mexico. The lower portion of the cleared area is planted in crops and benefits from nutrient-rich sediments that are eroded off the upper, uncultivated portion. (Photograph by author.)

century (Lumholtz 1902: 152–5), but no one has studied it systematically. A valuable and important investigation remains to be carried out.

Scarce water and needed sediments may be concentrated in downslope and downstream areas, but farmers wanting to take advantage of both frequently have a problem. They have to slow the flow long enough for water to soak into the soil and sediment be deposited on their fields (Bale and Sheng 1988). The ways in which this is done are simultaneously simple and complex. They are simple in their appearance, yet complex in their actual functioning. Equally important, features used by many native peoples today, were developed by their ancestors several centuries ago, and were in service at the time of the great cultural contact 500 years ago.

The principal method employed by aboriginals to impede overland flow was to build devices of various sizes and configuration perpendicular to the slope or gradient (Treacy and Denevan 1994: 93). The most complex of these used before, after, and at the time of contact, in north peaples are the famed *trincheras* of Mexico's northern Sierra Madre. Some of these features were 4 m high and involved multiple-courses of stone (Fig. 8.2). The simplest runoff control structures were single-course rows of rocks. Some rather rudimentary brush structures (Fig. 8.3) are used today, but the technology they represent may be of great antiquity. In all probability, the very first water and soil control devices were patterned after logs that fell naturally across slopes (Fig. 8.4) or small drainages (Fig. 8.5). There exists both archaeological (Wright 1962) and ethnographic (Kunstadter 1978) evidence for this from other parts of the world so it is not an unlikely scenario for North America (Doolittle 1990*a*).

Features used to impede runoff have been given a number of names by various writers (Masse n.d.). In a very general sense, however, all such structures, regardless of specific details, can be considered terraces (Bale and Sheng 1988). In what has become a classic article among scholars interested in agricultural landforms, J. E. Spencer and G. A. Hale (1961: 3) defined a terrace as 'any artificially flattened surface on which crops are grown subsequent to the flattening, no matter how small, how crude, or how purposeful'. In another classic work, R. A. Donkin (1979: 2) said that 'terracing represents an attempt to overcome the problems inherent in cultivating slopes'. These authors defined terraces so broadly simply because almost all land slopes to some extent (albeit some nearly imperceptibly), because terraces are found in a multitude of environmental settings ranging from V-shaped stream channels to convex slopes (e.g., Córdova F. de A. and Vázquez C. 1991), because there is such a wide variety of terrace configurations, and because terracing can serve many purposes in addition to simply impeding runoff and trapping sediment. Indeed, there are some subtle nuances to these functions as well. Although differences in setting and appearance are significant, the most important thing to consider when attempting to understand terraces is their purpose or function.

At least five functions can be ascribed to terracing: (i) create nearly level planting surfaces; (ii) increase soil depth; (iii) control soil erosion; (iv) manage the flow of water; and (v) to prevent frosts. As discrete as these functions may be, they are not necessarily singular or mutually exclusive. That is, although some terraces may serve only one specific function, more often than not terraces serve a multitude of purposes simultaneously. For purposes of discussion, however, functions will be treated here individually.

Creating nearly level planting surfaces. The most obvious function terraces serve is the creation of nearly level planting surfaces. Cultivation of slope lands can be difficult. Although it is not impossible (Fig. 8.6), cultivating exceptionally steep slopes involves greater expenditures of time and labor than would be necessary if nearly level land was available. One should not, however, think that terraces always result in perfectly level land. In fact, the only terraces that are truly 'level' are those on which water is allowed to pool-up for prolonged periods of time, such as in the case of wet-rice cultivation where paddies are inundated at varying depths (e.g., Bray 1986: 28–42).

Fig. 8.2. A large prehistoric terrace with a rock riser in the Sierra Madre Occidental of northwestern Mexico. Known locally as *trincheras*, such features are located across streambeds in steeply sloping terrain. The cultivated plots resulting from the sediment trapped behind the rocks are actually quite small (see L. Herold 1965*b*). (Photograph by author.)

Fig. 8.3. Brush and stake water spreader built perpendicular to the flow of runoff over a field in northwestern Mexico. (Photograph by author.)

Fig. 8.4. A natural tree fall in the Apache National Forest of New Mexico. Note how the fallen tree has rolled downslope, coming to rest against two trees such that it is perpendicular to the gradient. Note also that sediment has accumulated upslope of the log. Natural occurrences such as this could have inspired ancient cultivators to begin intentionally modifying slopes. (Photograph by author.)

Fig. 8.5. Two natural tree falls across a small stream in western central New Mexico. Note how the fallen tree furthest upstream, toward the top of the photograph and in the rear, has trapped rocks which in turn trapped finer sediment. Such 'log steps' (Marston 1982) might well be the natural antecedents of cross-channel terraces. (Photograph by author.)

Fig. 8.6. Unterraced steeply sloping fields in the Sierra Madre Oriental of eastern Mexico (see Rainey 1991). Clearly, farmers will cultivate slope lands if that is all that is available. (Photograph by author.)

In most cases, terraces do have some gradient or slope however slight (e.g., Sandor 1987: 182). Terraces, however, can and do result in nearly level planting surfaces on otherwise steeply sloping terrain. They can, quite literally, result in the creation of fields on slopes otherwise unsuitable for cultivation (Fig. 8.7).

Increasing soil depth. Soil and sediment are not synonymous, although most people do use the terms interchangeably. Stated perhaps too simply, soil, by definition, forms in place (Buol, Hole, and McCracken 1980: 9–13, 92–104) whereas sediment is deposited (Pitty 1971: 14, 224–8). Gradient or slope has long been identified as one of the five factors affecting soil formation (Jenny 1958: 10–11), and it has a negative or inverse influence. That is, in general, the less steep the slope, the more developed or 'thicker' the soil; the steeper the slope, the less developed or 'thinner' the soil. Slope lands, especially those in arid lands where both moisture and vegetation are sparse, have notoriously thin soils that often make agriculture prohibitive. How do farmers deepen soils on slopes? In a very literal sense, of course, they cannot. They can, however, create a sufficiently deep planting medium by either bringing in sediment artificially (via baskets), or trapping materials eroding naturally off of upstream/upslope areas (Denevan 1980: 622). To do so, some type of

Fig. 8.7. View of hillslope terraces at the Citadel site at Wupatki National Monument, Arizona. These three gently sloping to nearly level planting surfaces range in length from 12 to 30 m and in width from 2 to 12 m. They are supported by rock risers and have been called 'boulder bench terraces' by some scholars (Stewart and Donnelly 1943*a*: 42–4). (Photograph by author.)

retaining wall, typically made of rock, is usually needed to catch and/or hold the sediment in place.

Controlling soil erosion. Being as thin as they are, slope soils are as fragile as they are valuable (Denevan 1989). Accordingly, not only is it crucial to deepen them for purposes of cultivation, it is essential that they not be allowed to wash away. Rock alignments constructed perpendicular to stream or slope gradients can retard erosion (Cannell and Weeks 1979: 249). This can be best seen in several present-day settings where terrace-like features have been, and continue to be, built as environmental protection measures. Indeed, the similarity of some modern, historic, and prehistoric rock alignments often makes determining their respective ages quite problematical (Doolittle, Neely, and Pool 1993; T. Wright 1993).

In some cases, the surfaces behind, that is upstream or upslope of, rock alignments were cultivated. In others, however, it was land further downstream or downslope that was farmed. In these later cases, the rock alignments were not terraces, *per se*, but merely erosion control devices, or check dams, that kept soil from eroding off the slopes only to be deposited further below on the fields themselves (Doolittle 1985; Geyik 1986). Now, preventing erosion and sedimentation may seem counterintuitive in the light of what was discussed earlier, but in some cases these processes can have disastrous impacts, such as burying crops with relatively infertile materials.

A few scholars have questioned the notion that terraces were built as erosion control devices by indigenous people in times past. Spencer and Hale (1961: 26–7) emphatically dismissed the idea arguing that ancient farmers lacked the acumen to take preventive action against the long-term hazard of soil loss. Donkin (1979: 34) states that erosion control was, at best, an epiphenomenal advantage of most terracing. These arguments make some sense in reference to slopes cultivated prior to terracing. However, simple terraces made of stone lines or brush (Fig. 8.3) may be responses to sheet erosion. Terraces constructed incrementally or accretionally, taking advantage of natural geomorphic processes, indicate that farmers have long been aware that soils may be highly subject to erosion, and then trapped as sediment (Doolittle 1984*a*; L. Williams 1990). Paradoxically, trapping sediment for one's advantage in specific locales can, cumulatively, have negative consequences that ripple throughout the entire hydrologic system (Bryan 1941: 233).

Managing the flow of water. The hazard of soil erosion is, of course, a function of runoff, which by definition involves water. Indeed, a fourth function of terraces is to manage the flow of water. Donkin (1979: 22) argued that prehistoric terraces throughout the New World were built primarily for moisture retention and control. Terraces aid in managing surface water in three ways. First, as nearly level plots with deep soils, terraces act as sponges of sorts and retain water that falls on them directly in the form of rain or snow (A. Wright 1962: 99). Were terraces not built, precipitation falling on the slopes with thin soils would quickly runoff and be lost for crop production. A corollary of moisture retention is adequate drainage. Terraces with permeable stone walls allow excess moisture to drain freely and thus soil waterlogging is avoided (Treacy and Denevan 1994: 95).

Second, terraces trap water as it runs off areas upslope, thereby 'conserving' it (Vivian 1974: 102). In so doing, they actually collect moisture from areas much greater in extent than the fields themselves. This practice has been labeled 'runoff farming' for obvious reasons (Lawton and Wilke 1979: 4). Third, on broad, gently sloping lands, low terraces facilitate the distribution of runoff over the fields (National Academy of Science 1974: 23–37). Farmers in such settings have long been known to construct one or more, sometimes several, single course walls on their fields to ensure that runoff reaches every part of their field in approximately equal amounts (National Academy of Science 1974: 27–9). Some scholars refer to this form of agriculture as one variant of 'floodwater farming' (e.g., Bryan 1929), and distinguish it from flood recession agriculture (see Chapter 11). In some cases, however, it is difficult to distinguish from runoff farming (Lawton and Wilke 1979: 4; Doyel 1993: 9–10).

Preventing frosts. The final function terraces serve is to prevent frosts. In the northern hemisphere, highland areas experience many of the same problems as do northern areas in terms of temperature regimes and their affects on crops. Cool temperatures and short growing seasons can be problematic, even in lower latitudes. Terraces can compensate for these conditions in one of three ways. First, terraces can correct for problems associated with aspect and sun angle. South-facing slopes are nearly always warmer than north-facing slopes. This is probably no more apparent than in a comparison of adjacent north- and south-facing slopes on a sunny winter day following a snowstorm. South-facing slopes will typically be in the sun and free of snow while north-facing slopes retain a layer of snow in the shade. Of course, this also means that south-facing slopes tend to be xeric whereas north-facing ones are more mesic (Fig. 8.8). In general, terraces built on south-facing slopes will experience fewer and less extreme frosts early and late in the growing season than will north-facing slopes (Earls 1986: 100–24).

Second, terraces can interrupt the flow of air, not unlike as in the case of the microclimatic affects of gravel mulching, creating turbulence which inhibits

Fig. 8.8. A view west of a chasm in western New Mexico in January. Note that the slope on the left, the north-facing slope, is in the shade and still has a covering of snow that fell the night before. The slope on the right, the south-facing slope, is in the sun and the snow has already melted. (Photograph by author.)

cold air from descending upslope areas (Patrick 1980: 763). Frosts occur in mountainous areas because of night-time radiational cooling under clear skies, and katabatic movements of cold air. Crops growing above the point where cold air collects at night and below the point where temperatures begin to fall according to the normal lapse rate, often avoid frosts since cold air glides over the hillslope. The effect of thermal zones was surely understood by indigenous farmers early on (E. Adams 1979). Induced wind turbulence may not constitute a prime function of terraces (Treacy and Denevan 1994: 94), but it is certainly an advantage that befell farmers who built such planting surfaces. The same probably holds true for the third way in which terraces inhibit frosts—reradiation of heat from the rock risers. In effect, heat absorbed by the rocks during the daytime is radiated back at night when the temperatures drop. This reradiated heat can compensate for the otherwise cool temperatures and frost conditions (C. Field 1966: 430).

Terraces are certainly a fascinating type of agricultural landform. The variations in form and function (which are not necessarily related), environmental setting, and degree of slope on which they are found, are, to state it mildly, bewildering. Accordingly, they have given rise to much confusion in the terminology used by scholars interested in aboriginal North America. To be sure, there are many similarities in the various features and the practices associated with them (e.g., Glassow 1980: 44–54; Hack 1942: 26–31; Masse 1991: 207–14; Winter 1976: 83–4). Differences in terms and definitions are, however, often more contrived than real (Merton 1968: 74–9). For example, some scholars' 'check dams' (Stewart and Donnelly 1943*a*: 34) are another's 'terraces' (Woodbury 1961*b*: 11; see also Cordell 1984: 192). Similarly, one's 'contour terraces' (Vivian 1974: 97) are another's 'rock alignments' (Rankin 1989*b*: 954–65), and yet another's 'linear borders' (Woodbury 1961*b*: 12). And, how many courses are required before linear borders are sufficiently high to be classified as 'boulder bench terraces' (Stewart and Donnelly 1943*a*: 34)? Were this not all confusing enough, some scholars have gone so far as to change their designations from one study to the next (e.g., G. Stewart 1939; Stewart and Donnelly 1943*a*).

Farmers the world over and from time immemorial have devised a number of means and technologies for cultivating slope lands. The indigenous inhabitants of North America at contact were no different. Although evidence in the form of first-hand accounts at the time are extremely limited and subject to interpretation, there is abundant archaeological and ethnographic evidence, especially from the Southwest.

Documentary Evidence

Early explorers actually recorded precious little in the way of direct accounts of terracing in any form. As Sauer (1971: 294) noted, '[such] special skills were not noted by the earlier Spaniards, who did see, however, that these people were most industrious farmers and their pueblos were large and many, with limited amount of cultivated land'. Two possibilities exist as to why this is the case. First, explorers might not have seen many terraces. As will be discussed later, many aboriginal terraces were not only abandoned before Europeans arrived on the scene, but were in remote areas, well off the routes these chroniclers traveled. Second, they might not have recognized some terraces, especially smaller ones, as loosely defined above. Indeed, this is probably the case.

There is only one case where terracing was mentioned explicitly by one of the first Europeans to encounter a particular group of people. Surprisingly enough, it comes from California, a land whose native inhabitants have long been thought of as strictly hunters and gathers. In 1776, Rivera y Moncada, while pursuing some Kumeyaay who were defying mission authority and fleeing into the mountains of far southern California, encountered numerous low rock walls paralleling the slope, behind which crops were cultivated (Shipek 1993: 384–5). In most other cases, terracing is implied by virtue of agriculture being next to impossible without it. For example, Juan Mateo Manje, who in the 1690s accompanied the famed Jesuit Eusebio Kino, noted at least five places in what is today southern Arizona and northern Sonora (Fig. 8.9) where natives planted '*de temporal*' (Manje 1971: 291, 310, 464).

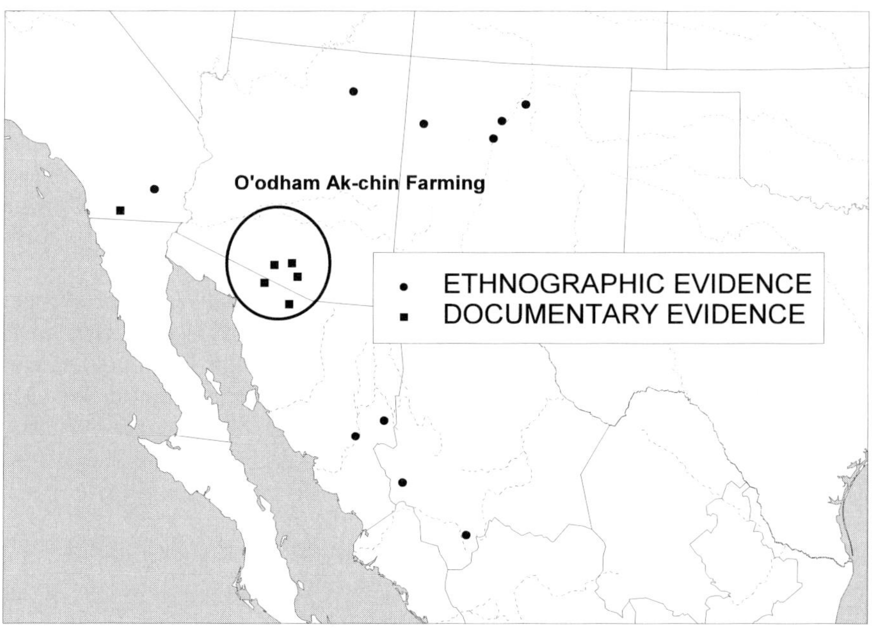

Fig. 8.9. Locations cited in the text where terracing was reported by early European explorers and more recent ethnographers.

Although some whose expertise lies principally in languages might think these are references to seasonal rainfed agriculture (e.g., Burrus 1971: 168, 191, 192, 253, 254), others more intimately familiar with agriculture in this part of the world recognize that these passages refer to farming at the mouths of arroyos which benefit from seasonal runoff and accompanying deposits of sediment and detritus (Bryan 1929: 444; Castetter and Underhill 1935: 29; J. Wilson 1985: 133). Observations made in later centuries indicated that this trapping of water, sediment, and organic matter involved ephemeral terracing, even though such was not explicitly mentioned.

Ethnographic Evidence

Quite in contrast to the early chroniclers who did not describe terraces and terracing, scholars working in the Southwest during the 19th and especially 20th centuries were captivated by the techniques indigenous people employed to produce a crop from an environment that seemed inhospitable, and, accordingly, they described them in great detail. The native people for whom terracing has been detailed include those living in a triangle with points in northern New Mexico, southern California, and northwestern Mexico (Fig. 8.9).

The people whose agriculture was referred to by Manje in the late 17th century as '*de temporal*', are today the Tohono O'odham. Together with their neighbors, the Akimel O'odham, these people's agriculture was assessed in detail first by Castetter and Bell (1942). These scholars wrote a meticulous study which was seminal to the work of later scholars. As good as it was, however, it fell victim to the exact same problem as that seen in the early explorers accounts—terraces were not discussed. Exactly why this is the case remains unknown. Perhaps, as was the case with Manje, ephemeral devices such as brush water spreaders, and low earthworks, were not seen as being important, although they did exist. Nevertheless, Castetter and Bell did provide some valuable information which later scholars built upon.

The sizes of these Tohono O'odham fields ranged from as little as 0.1 ha to as much as 2 ha, with the average being approximately 0.3 per person, with these fields providing 50 to 60 percent of the annual food supply (Castetter and Bell 1942: 35–6, 56–7). Only one crop per year was raised on these fields, and it was planted at the beginning of the rainy season in late June or early July (Castetter and Bell 1942: 145). Maize and tepary beans (*Phaseolus acutifolius* var. *latifolius*) were grown in mounds, 75 cm and 60 cm apart, respectively. No care was taken to locate mounds in the same place year after year, thereby facilitating a sort of rotation and maintaining soil fertility. Maize and beans were not, however, intercropped, but planted in separate fields (Castetter and Bell 1942: 148, 152–3, 155).

The first recognition of landscape features associated with the runoff fields, or *ak chin* agriculture, in this hot, dry corner of the continent was actually made several decades before the work of Castetter and Bell. An acting supervisor of what was to become the Tohono O'odham Reservation reported earthen embankments that retained runoff, and small stake and brush devices that checked the flow of water over field surfaces (Clotts 1915: 11; see also Nabhan 1981: 52; 1983: 34, 1984: 26, 1986*a*: 53). The Akimel O'odham are also known to have built earthen embankments around the lower sides of their fields in order to catch and retain runoff (F. Russell 1908: 88). Understanding how these embankments functioned was improved by parallels drawn to a successful system of agriculture practiced by non-aboriginals further south in northwestern Mexico. The *bolsa* system, as it's name implies in Spanish, involves the creation of a 'pocket' formed by an earthen embankment which impounds runoff. Water is trapped by this low wall and allowed to soak in. While the soil is still moist, crops are planted. Additional moisture may not be necessary, but runoff from later storms also accumulates and can be beneficial (Dobyns 1951; see also Hackenberg 1962: 190).

As for stake and brush water spreaders (Fig. 8.3), they are most often made from materials cleared from fields prior to planting or from vegetation growing near fields (e.g., Doolittle 1984*a*). Stakes 30 to 50 mm in diameter are cut into approximately 2 m lengths. These are driven into the ground half-way at 1 m intervals parallel to contours and across rivulets or small gullies. Smaller brush is then woven between the stakes. The primary purpose of these features is to dissipate the energy of flowing water, slowing it down and distributing it across the field surface (Nabhan 1979*b*: 248–9, 1983: 61, 1986*b*: 71).

Of course, earthen embankments and brush water spreaders located on lands with slopes of only one or two percent may not seem like terraces to some, but they do fulfill one terrace function—managing the flow of water—directly, and indirectly they can fulfill three others—control soil erosion, increase soil depth, and create more level planting surfaces, the final two resulting from deposition of sediment carried in suspension by the water. Similar features in similar settings have been observed by scholars working among native groups not too far from southern Arizona. In these cases, however, the connection to what everyone recognizes as true terraces is obvious in that more complex modifications of steeper slopes are also involved.

Two groups who practiced runoff farming in recent times live northwest of the O'odham people in California (Fig. 8.9). The Cahuilla practiced a variety not unlike *bolsa* agriculture. Although earthen embankments have not been described, these people did cultivate in 'small catchments' (Lawton and Bean 1968: 21) which undoubtedly were enhanced artificially. The other group for which there exists ethnographic descriptions of runoff farming, but this time with explicit references to terracing, are the Kumeyaay. According to their ethnographer, who during the early 1960s interviewed individuals between 80 and 110 years of age, these people described aligning rocks in parallel rows on alluvial fans where the desert meets the mountains. They built mounded rock alignments across small drainages in order both to spread the runoff over the surface and to slow it down, allowing more water to soak into the ground. Informants also relayed that these features facilitated the deposition of sediment. Crops were planted in the fertile, wet, leveled areas behind the alignments, or, literally, terrace walls (Shipek 1993: 379, 384). In many places, the Kumeyaay also built brush water spreaders, but rather than anchoring them with stakes as was done in Arizona, they weighed them down with rocks (Shipek 1993: 385). Among other things, these accounts lend first-hand support for the interpretation of features mentioned by Rivera y Moncada.

To the south and east of the O'odham reside a related group—the Névome—who are themselves divided into two groups, as well other indigenous people who also practice terracing (Fig. 8.9). The Névome who live in the foothills of the Sierra Madre Occidental cultivate small plots created behind rock-faced terraces. The terraces are between 3.5 and 4.5 m wide (Pennington 1980: 141–2), less than 1 m high and no greater than 30 m in length (Campbell W. Pennington, personal communication 1983). Their principal purpose appears to be to create level planting surfaces slightly downslope from a spring. Further to the east, those Névome living in the sierras proper also terrace, but here the practice is intended to halt erosion caused by cultivation of steep slopes. Rocks removed from fields during planting and cultivation are piled across gullies in rows which seem to be less than 1 m high, 6 m long, and several meters apart, depending on the degree of slope. Sediment accumulates behind these walls, thereby being conserved and resulting in a planting surface more level than was the case under natural circumstances (Laferrière and Van Asdall 1992: 159, 162, 163, fig. 3).

As has already been noted, the sierra's most famous inhabitants, the Rarámuri, also have to impede erosion on slopes. In these cases, however, it is erosion they initiate intentionally. The walls built for this purpose involve rocks collected from the fields themselves and nearby areas and, eventually, due to sedimentation, result in risers 1 to 3 m high. They vary considerably in length and spacing (Lumholtz 1902: 152, 155). Some are very much like those of the Mountain Névome. Others are much longer, in more regular series, and appear to be more well-planned and built (Lumholtz 1902: 152). In addition to the terraces that result from trapping sediment eroded by design, the Rarámuri also build small rock terraces across arroyos and on canyon slopes (Pennington 1963: 47, 1983*a*: 278). These are not numerous, however, and involve only crudely piled rocks which trap sediment creating level planting surfaces (Pennington 1963: 49). The Rarámuri's southern neighbors, the Odami, also build terraces. Indeed, these people have a long tradition of doing so. Terraces on slopes are described as having been constructed in an 'exceedingly crude arrangement of large stones' and in use for some period of time (Pennington 1969: 59). Relatively newer ones tend to be located across drainages, and built of small stones. They are also more numerous and found in at least five different locales. Details concerning Odami terrace agriculture are sparse, but it is known that their fields were 'not extensive in size' (Pennington 1969: 60).

Terracing appears not to have been a very important aspect of agriculture among the Puebloan people of the upper Rio Grande Valley of New Mexico during historic times. Of all the ethnographers to have ever worked there, only two provided even scant evidence. Informants at both Zia and Cochiti pueblos indicated that during the 19th century alluvial fans at certain arroyo mouths were farmed with the aid of log, rock, and brush structures that helped spread runoff (Euler 1954: 30; Lange 1959: 78). Another ethnographer did report arroyos being farmed at San Juan Pueblo, but does not so much as hint at the presence of terracing, although such is a distinct possibility (Ford 1968: 189).

Of all the New Mexico pueblos, the ones that have drawn the most attention to their features built for trapping runoff and sediment are those of Zuni (Ferguson and Hart 1985: 37). In no small way, the enthusiasm for studying Zuni agriculture can be traced to the pioneering work of the eccentric Cushing (1974, 1979). The technique he described for modifying arroyo mouths with earthen embankments to trap and conserve runoff is not unlike that known for the Tohono O'odham (Cushing 1974: 152–3; see also Bryan 1929: 449; Ford 1985*d*: 34–5; Ferguson 1985: 103, 106–7; Stewart and Donnelly 1943*b*: 141, 143). According to one investigator, the ridges which surrounded fields were 45 to 60 cm high and 90 cm wide at the base (G. Stewart 1940*b*: 335). These figures were repeated in a later study (Ford 1985*d*: 37), although a contemporary who also provided additional details, claimed they were 75 cm high, 75 cm wide, and used by at least eleven different farmers (Ferguson 1985: 105). Regardless of their size, these ridges collected runoff, thereby augmenting scarce rainfall, and they collected sediment and organic debris.

In addition to earthen embankments, Zuni farmers modify their agricultural landscape with three other types of terrace materials, rocks, stakes and brush, and logs (Winter 1976: 85). On slopes with gradients up to 4 percent they build small boulder bench terraces which effectively retain soil (Stewart and Donnelly 1943*b*: 134). On some fields the Zuni use simple rock alignments, ranging in length from a few to several meters, to spread runoff uniformly over the surface (Ferguson 1985: 107, 108, fig. 3*b*).

More common than rocks, however, in their efforts to control runoff and build up soil, is the use of stake and brush devices (Ferguson 1985: 110). Juniper seems to be the preferred material for the stakes, undoubtedly a function of its strength and durability, while juniper and willow branches and sagebrush are the preferred materials for weaving between the stakes (Ferguson 1985: 106). Some stake and brush water spreaders are upwards of 40 m long (Ferguson 1985: 108, fig. 3*a*, 111). Typically, though, they are only a few meters long. One clear advantage of small, less permanent features is the ability to construct them quickly depending on localized changes as they occur from one year to the next (Ford 1985*d*: 36).

Finally, the Zuni appear to be unique in their use of logs for terracing. Cottonwood and Ponderosa pine are preferred (Ford 1985*d*: 36), perhaps because of their great length, availability, and relatively light weight. They are used at the edges of fields to retain runoff and trap sediment and detritus (Ford 1985*d*: 36), and within fields themselves to intercept and divert runoff. As with stake and brush devices, log terraces can be built quickly, indeed if the logs are cut and at hand they can be moved into position at the onset of a runoff event and placed exactly where they will do the most good (Ferguson 1985: 106).

Runoff fields at Zuni have one other landscape modification that, although not terracing, *per se*, merits mention. In the 19th century, and presumably earlier as well, farmers here stuck tall cedar poles in the ground throughout their fields and strung cords made of yucca fiber between them. On these lines they hung tattered rags, decaying animal parts, and streamers of moss (Fig. 8.10). The purpose of this network was identical to that described by Bannister

Fig. 8.10. A Zuni maize field with crow traps. Appeared originally in Cushing (1974: plate IV; see also Cushing 1979: 253, fig. 57).

(Rountree 1989: 49) of the 17th-century Powhatan (see Chapter 5); to scare away or ensnare crows that would otherwise eat recently planted seeds (Cushing 1974: plate IV). Today, Zuni farmers have scaled-back this practice and now only occasionally hang a dead crow from a single pole.

Terracing at the Zuni pueblos is obviously related to that of the Tohono O'odham. Both groups rely heavily on altering the terrain of arroyo mouths. A third group does likewise, one that is related culturally to Zuni but is located geographically—both in terms of distance and environment—in between. This group, the Hopi (Fig. 8.9), also are responsible for some of the most impressive terraces in North America (Winter 1976: 85). Terracing at the Hopi pueblos can be dichotomized according to environmental setting and features (Hack 1942: 26). First, there are arroyo mouth locales, not unlike those discussed above (Brew 1979: 515), where small and rudimentary devices are employed (G. Stewart 1940*a*: 215, 1940*b*: 331). Second, there are steep slopes with elaborate and complex structures (G. Stewart 1940*b*: 333).

In terms of total hectarage, the preponderance of Hopi fields fall into the first category. One study estimated that 73 percent of Hopi fields are watered by runoff trapped by low terraces (Hack 1942: 26). Another noted that 372 ha of such fields were located near Third Mesa alone (Bradfield 1971: 19). Stake and brush devices are used as water spreaders by the Hopi (Beaglehole 1937: 37), but they are not all that common, and are used only on fields that are quite broad. The opposite holds true for earthen embankments which are found at the lower edge of nearly every runoff field (Hack 1942: 29, fig. 15; G. Page n.d.). Indeed, these ridges are so common and last for such extended periods of time that long abandoned fields can be identified by their presence (Hack 1942: 28). Simple rock alignments (Fig. 8.11) are also quite common and usually used in narrow arroyos. In such settings rocks are piled in rows completely across the drainages, at right angles to the direction of flow, and at regular intervals depending on changes in gradient (Fig. 8.12; see also Colton 1960: 93, fig. 33; Hack 1942: plate VC). Sometimes clumps of sagebrush (Page n.d.) or rabbit brush (Forde 1931*b*: 364) are stuck between the rocks on top. Exactly why rock terraces are used more frequently than stake and brush devices is not clear. In part, it may be a function of the abundance of rock and the paucity of vegetation (Page n.d.: 62). It also may be partially a result of scarce vegetation being used in making windbreaks as discussed in Chapter 7. Regardless, the fact remains that Hopi use both types of features to slow and distribute runoff and trap sediment (Beaglehole 1937: 37; Colton 1960: 92), thereby resulting in more level, fertile planting surfaces.

The more complex terraces located on steeper slopes near Hopi pueblos are visually quite striking (Connelly 1979: 543, fig. 10; Woodbury 1963: 67). Indeed, a photograph taken of one set near Wepo Springs in the 1930s (G. Stewart 1940*b*: 333) is so impressive that it was reproduced four decades later

Fig. 8.11. An upstream view of a tributary of Wepo Wash, just north of and below the village of Hano, visible on top of the mesa. Note the remnants of three abandoned rock terraces that have been breached. This arroyo and these terraces were discussed and mapped by Hack (1942: 30, fig. 16). See Fig. 8.12. (Photograph by author.)

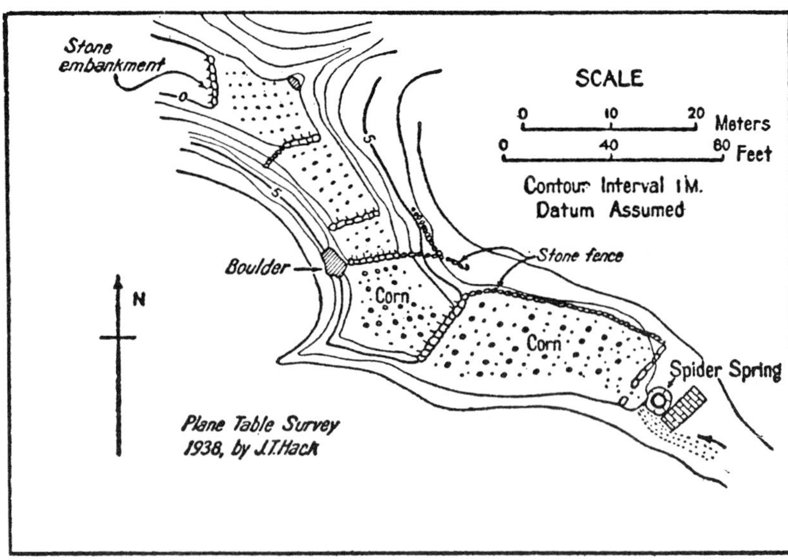

Fig. 8.12. A terraced arroyo below the village of Hano. See Fig. 8.11. Appeared originally in Hack (1942: 30, fig. 16). (Reprinted by permission of the Peabody Museum of American Archaeology and Ethnology, Harvard University.)

in the now classic volume *Agricultural Terracing in the Aboriginal New World* (Donkin 1979: 60, fig. 3.20). A comparison of that photograph (Fig. 8.13) with one taken in 1993 (Fig. 8.14) indicates that some things have changed, most notably only approximately 10 percent of the terraced fields are in use today.

These terraces vary considerably in size, depending on localized conditions. Risers are typically made of sandstone and range from a low of 60 cm (Forde 1931*b*: 364) to a high of 3 m (Hack 1942: 37), with most being approximately 1 to 1.5 m tall (Fig. 8.15). They are several courses high and taller ones tend to trapezoidal in cross-sectional shape and at least two rows of rock thick (Forde 1931*b*: plate XLIV, figs 1 and 2). Repair of these walls is the most laborious aspect of these terrace's maintenance. Forde (1934: 232) claimed, however, that it was minimal and done only occasionally as the need arose. A local informant in 1993 disagreed. He said that risers need constant maintenance, that the work was difficult, and that it was the greatest reason why so few people farm these terraces today. This informant did agree, however, that the rock walls trap sandy sediment which retains moisture (Forde 1931*b*: 364), thereby resulting in a nearly level planting surface (Hack 1942: 37). The narrowest terraces are 1.2 m wide (Clark 1928: 240) while the widest measure approximately 5 m. They can vary in length up to 30 m.

These terraces receive the benefit of direct precipitation and runoff from upslope areas. Lower terraces also receive surplus moisture from their higher counterparts (Clark 1928: 242). However, these terraces are primarily watered from springs and adjacent storage tanks. These reservoirs are either circular or roughly rectangular and made of rock. According to Forde (1931*b*: 365, 1934: 231), they are either built on a large rock or have a plastered bottom. Chinks in the sides are filled with a mixture of plaster and dried plant fiber. An outlet at the bottom is plugged with a bundle of cloth or hide and grasses.

In years past, water flowed from the cistern, trough a series of small canals to each terrace (Figs 8.16 and 8.17). These canals were rarely more than 30 cm deep and 60 cm wide (Forde 1931*b*: 392, 1934: 232, Hack 1942: 35, fig. 20, 36, fig. 21). On the individual planting surfaces themselves, the distribution of water was regulated by the construction of small earthen ridges a few meters apart (Fig. 8.13). Some terraced fields were gridded and irrigated manually with water carried in ceramic jars (Forde 1931*b*: 366, 1934: 231). Today, farmers plant in narrow enclosed furrows and use commercially available garden hoses (Fig. 8.18) in place of the canal systems (Soleri 1989: 13).

Exactly how far into the past the terracing tradition extends at Hopi is anyone's guess. One scholar (Soleri 1989: 12) has noted that certain terraces have been used for only about a century. The village associated with that particular set, Hotevilla (see Kennard 1979: 554, fig. 1), was established in

Fig. 8.13. Terraces at Wepo Springs, First Mesa, Arizona as they appeared in the 1930s. Appeared originally in Stewart (1940*b*: 333), and reprinted in Donkin (1979: 60, fig. 3.20). (Provided courtesy of Robin A. Donkin, and reprinted by permission of the Wenner–Gren Foundation for Anthropological Research, Inc., New York.)

Fig. 8.14. Terraces at Wepo Springs as they appeared in 1993. Compare with Fig. 8.13. (Photograph by author.)

Fig. 8.15. Close-up view of a rock-faced terrace at Wepo Springs. (Photograph by author.)

1906 by people displaced after flooding and gullying destroyed farmland at Oraibi. Given that Oraibi is known to be the oldest continually occupied town in the United States, dating back at least 600 years, it seems likely that terracing was practiced at various Hopi pueblos at the time of contact. To be sure, the archaeological record for terracing in the region goes back much further in time.

Archaeological Evidence

While ethnographers spent the past century doing an admirable job of reporting terracing among native North Americans in the Southwest, their archaeological counterparts were doing them one better in recording relicts of prehistoric terracing. The study of prehistoric terracing in the region has a long and rich tradition, but one that is almost paradoxical. Archaeologists have recorded literally thousands of terraces of all types and related field systems. Site files in places such as the Arizona State Museum in Tucson, and the Laboratory of Anthropology in Santa Fe are full of field notes with references to terracing of one form or another. Remnants of ancient terracing are so commonplace, so well-recorded, and so well-known, that many archaeological publications simply mention the practice and provide little in the way

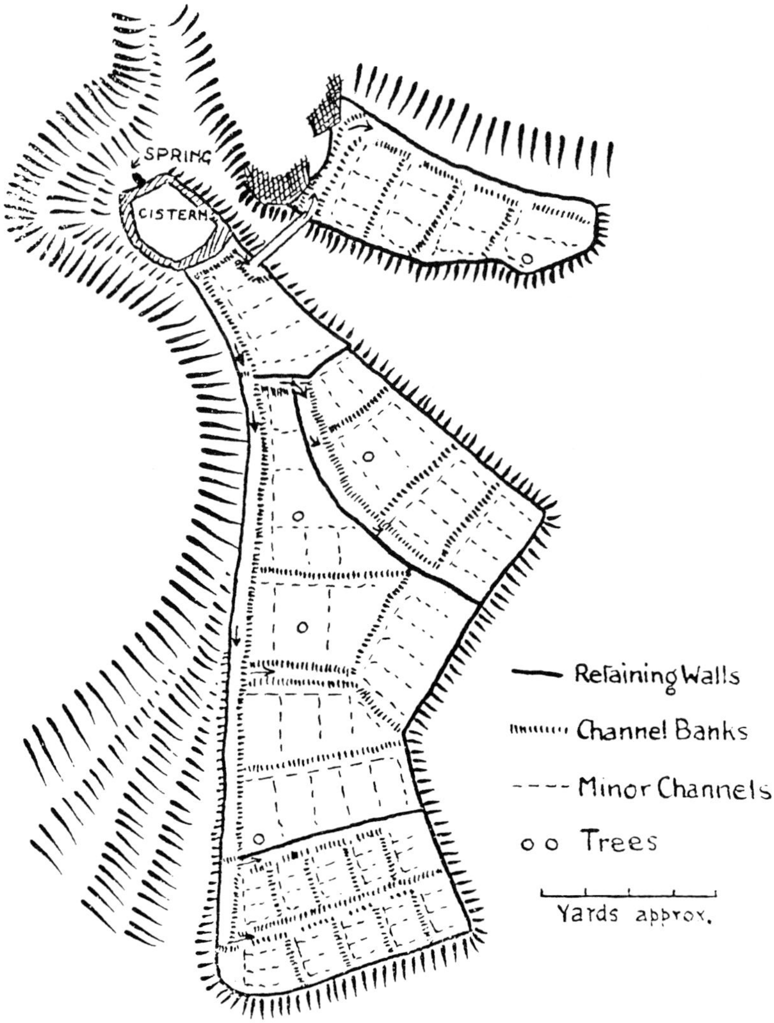

Fig. 8.16. Sketch plan of irrigated terraces near Sikyaki village. Appeared originally in Forde (1931*b*: 392, fig. 9) and reprinted in Forde (1934: 232, fig. 20). (Reprinted by permission of the Royal Anthropological Institute of Great Britain and Ireland.)

of detailed data. Relict terraces are so ubiquitous that they are often seen to merit little discussion. A few works should suffice as examples.

The Tonto National Forest in eastern central Arizona may well have more prehistoric terraces than any other part of the larger region. Its archaeologist, J. Scott Wood, has masterfully crafted a reconstruction of agricultural

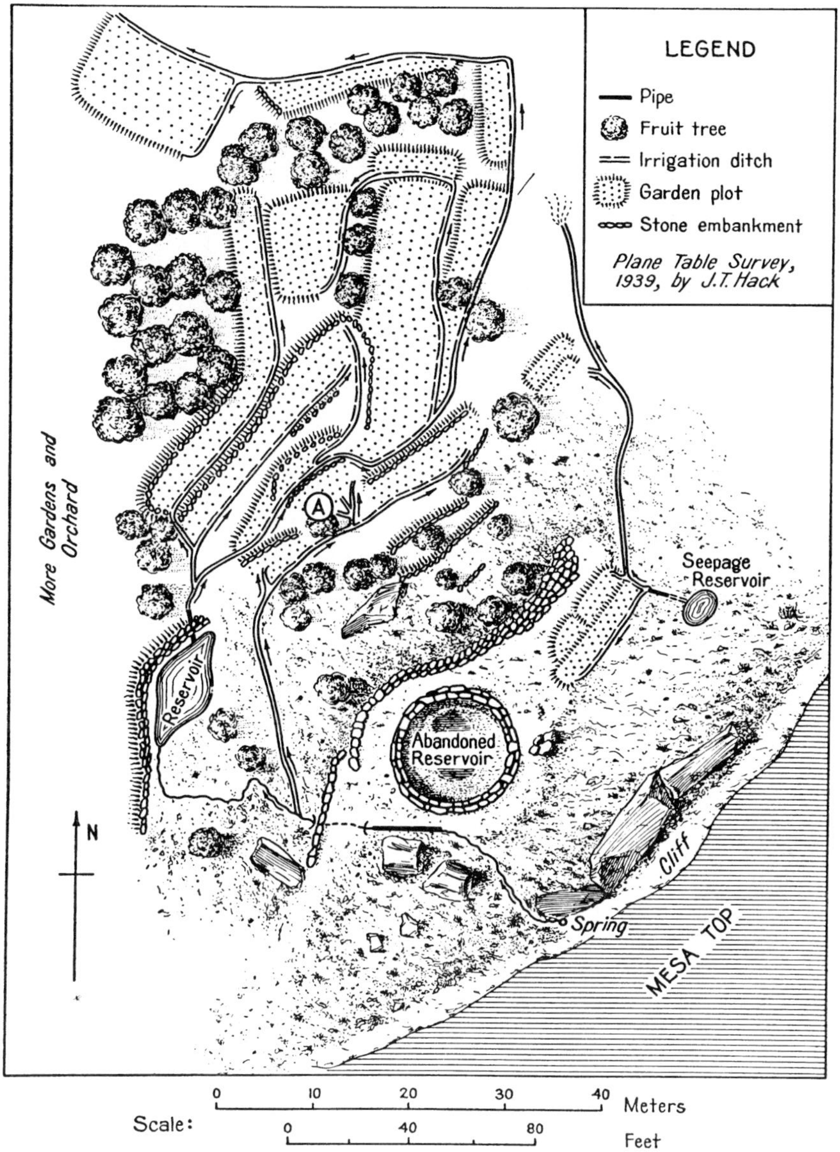

Fig. 8.17. Sketch map of a portion of the irrigated terraces in Tallahogan Canyon, north of Awatovi ruin. Appeared originally in Hack (1942: 35, fig. 20). (Reprinted by permission of the Peabody Museum of American Archaeology and Ethnology, Harvard University.)

Fig. 8.18. Small enclosed furrows now being watered with a garden hose. In years past water would either be provided from canals and would wind its way through small earthen embankments visible in Fig. 8.13 or it would have been hand-carried in ceramic jars. (Photograph by author.)

practices and land use based on his in-depth knowledge of the region, its archaeological sites, particularly relict farming features (Wood and McAllister 1984). His reconstruction is so compelling that it was later incorporated into the best brief overview of native landscape modifications throughout the whole of North America (Butzer 1990*a*: 40–1, fig. 2.8). Wood's work, however, contains no detailed data descriptions (see also Wood *et al.* 1987: 7, 16).

In a vein somewhat similar to that of Wood, James B. Rodgers (1985) provides a general description of the kinds of agricultural features, including various forms of terraces, found in the Hohokam Northern Periphery of central Arizona, and a listing of sites containing these features. Donkin (1979: 19, 58–61, 137–8, 140) and Winter (1974: 139–43) do likewise, but they cover the entire southwestern quadrant of the continent (see also Woodbury 1963: 62–5; and Lawton and Wilke 1979: 29–36). Despite their shortcomings, all of these are valuable contributions and should be regarded as initial readings for anyone wishing to pursue further study of terracing.

Two other contributions merit mention. The first is by James L. Moore (1981). It is, without doubt, one of the finest master's theses ever written on prehistoric agricultural features. Chapter 2 is an overview of soil and water conservation sites throughout the Southwest. The chapter suffers only by being somewhat dated now. The second noteworthy contribution, by David Doyel (1993), is more recent, but limited to Arizona. Like nearly all studies of ancient terracing in the Southwest (e.g., Woodbury 1961*b*: 9–10; Winter 1978: 85), its typology of features and fields is based on ethnographic parallels (Doyel 1993: 31–7), particularly the work of Hack (1942). This study does not provide a comprehensive and complete catalog of terrace sites, but it does list sixteen sites, at least ten of which contain terraces, that are valuable cultural resources deserving of preservation (Doyel 1993: 56–8). Furthermore, Doyel notes most explicitly that of the various types of non-irrigated agricultural sites in the Southwest, only check dams and terraces, 'features made of stone . . . survive in the archaeological record' (1993: 31), and that prehistorians frequently have to rely on 'indirect evidence' (1993: 31, 33, table 3–3).

In some cases, evidence of terracing exists, but remains unpublished. In others, data are simply non-existent. This is particularly true in those cases where slopes are not very great, even slight, where runoff is substantial and powerful, where sedimentation is great, or some combination of the three. The channels of arroyo mouths, the floodplains of ephemeral streams, alluvial fans, and highland basins are particularly sparse in terms of relict terraces. Scholars working in such locales have often found themselves determining lands that were most likely cultivated in times past on the basis of soils, vegetation, presence of stone hoes, proximity to storage and roasting pits and field houses, or other types of sites. They recognize that ancient features such as rock alignments, earthen embankments, or stake and brush devices were more than likely used, but have long since been destroyed. In some cases, scholars rely on ethnographic parallels; assuming that if natives farm an area today, their ancestors must have as well.

Studies relying on indirect data have been conducted in every part of the Greater Southwest where terracing of one form or another was probably employed (Fig. 8.19). Some specific locales include the Santa Cruz Valley of southern Arizona (Waters 1987: 215–16, 219; Waters and Field 1986: 341–4), two stretches of the Gila Valley (Lindauer 1984: 73; Welch 1994: 225–7) in central Arizona, near Grasshopper Ruin in eastern Arizona (Tuggle, Reid, and Cole 1984: 106–8), an alluvial fan in southwestern Utah (Marwitt 1986: 166), alluvium-filled canyons on the Colorado Plateau of southeastern Utah (Matson 1991: 60, 64, 83, 86, 113, 115, 212), along tributaries of the Dolores River in southwestern Colorado (Bye and Shuster 1984: 126, 129), in the vicinity of Salmon Ruin (Nials 1980: 19) and Escavada Wash (Cully *et al.* 1982: 157) in northwestern New Mexico, in the Cimarron District of northern New Mexico (Glassow 1980: 87–105), at Zia Pueblo (Hoebel 1979: 410), and Cochiti Pueblo (Lange 1979*a*: 368) in the upper Rio Grande, and in highland western New Mexico (Hogan 1987: 192).

Despite some problems with data not being available, not being published,

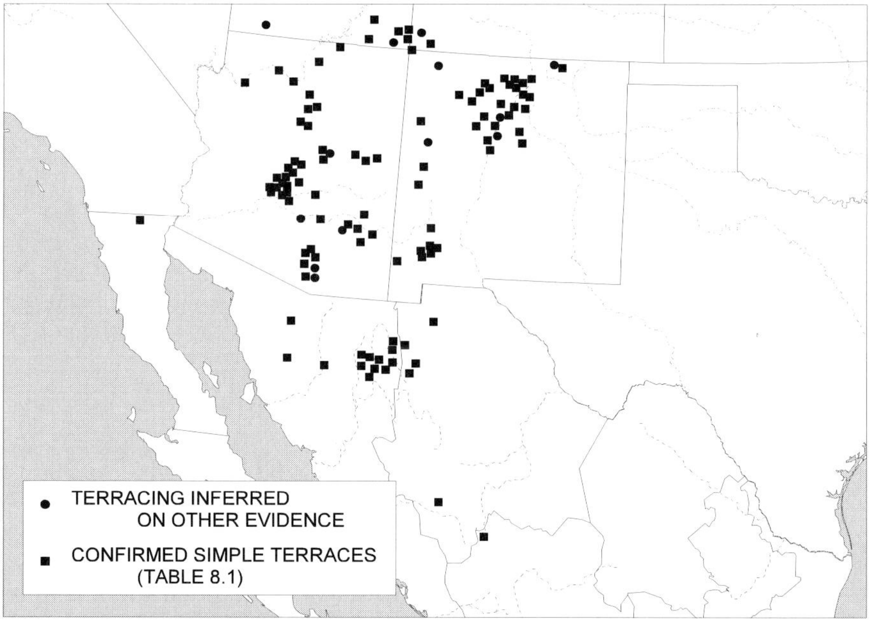

Fig. 8.19. Locations cited in the text and Table 8.1 for which there is archaeological evidence of terracing.

or treated only in a cursory fashion, there remains a wealth of pre-contact data on terracing in the Southwest. Indeed, there are so many data available in published form that space here allows for little more than a synthesis not unlike those criticized above. This holds particularly true for more rudimentary forms of terracing–rock alignments.

As mentioned in the introduction of this chapter, much confusion and disagreement exists over the names of certain features that generically can be described simply as 'rock alignments'. All too often, names reflect functions implied by form, despite the lack of supportive evidence. Complicating matters, many such features are often mentioned in publications with only a brief comment about function. Given the large number of such sources, much of the information available on ancient rock alignments with clear agricultural functions is presented here in tabular form. In some respects, this is a tabulation of data discussed by Moore (1981: 6–51), but is updated and expanded (Table 8.1).

A number of things should be obvious from this tabulation of 'simple' terracing. First, there is archaeological evidence confirming both the ethnographic and documentary reports of terracing in southern California. The tradition there, although localized, is very long. Second, evidence comes every corner of the Greater Southwest (Fig. 8.19). Third, a variety of terms

Table 8.1. Summary of archaeological reports of simple terracing

Locale	Features mentioned	Functions claimed: Level surface	Deepen soil	Control erosion	Control water	Source
Southern California						
	Parallel rock walls traversing slopes		x		x	Shipek (1993: 385)
Santa Cruz Valley, Arizona						
Tumamoc Hill	Check dams				x	Masse (1979: 167–9)
	Contour terraces			x	x	Masse (1979: 167)
	Terraces		x			P. Fish, S. Fish, Long, and Miksicek (1986: 564)
Los Morteros	Non-residential terraces	x	x	x	x	S. Fish, P. Fish, and Downum (1984: 61–5)
Rillito Peak	Linear trincheras		x			Wallace (1983: 196)
Marana	Contour terraces, check dams				x	S. Fish (1995: 105–6)
	Terrace alignments, check dams, earthern berms					P. Fish, S. Fish, and Madsen (1984: 86)
Los Robles	Check dams, rock alignments				x	Downum (1993: 36, 41, 42, 85–6)
Sulphur Springs Valley, Arizona						
	Terraced retaining walls	x			x	Caywood (1933: 23)
Gila Valley, Arizona						
East of Safford	Check dams			x		Seymour, Doak, and Ahlstrom (1994: 8.6)
	Terraces	x				Seymour, Doak, and Ahlstrom (1994: 8.8)
Mt Graham	Terraces				x	G. Stewart (1940*a*: 217)
Near Pima	Linear borders and terraces				x	Gilman and Sherman (1975: 7, 11, 13, 23)
West of Kearney	Rock alignments, check dams, and terraces			x	x	Debowski *et al.* (1976: 56–8, 90, 98, 108)
Near Florence	Contour terraces		x		x	Dart (1983: 414–16)
	Check dams			x		Dart (1983: 417–18)
Tonto National Forest, Arizona						
	Check dams, terraces			x	x	Wood and McAllister (1984: 272–4)
Lake Roosevelt	Rock alignments, check dams				x	Shelley (1994: 353–7, 363–4)

Table 8.1. *Continued*

Locale	Features mentioned	Functions claimed				Source
		Level surface	Deepen soil	Control erosion	Control water	
Vosberg locality	Check dams		x		x	Rodgers (n.d.)
	Ridge walls, linear borders, crescent walls, brush structures				x	Rodgers (n.d.)
Payson–Star	Check dams				x	Redman (1993: 106)
Lower Verde Valley, Arizona						
Fort McDowell	Check dams, contour terraces				x	Kearns, Polk, and Teague (1975: 250–1)
Bartlett Dam	Rock alignment					Homburg (1997: 105–6)
Horseshoe Dam	Rock alignments			x	x	Homburg and Ciolek-Torrello (1997: 69)
Lime Creek	Contour rock alignments			x	x	Homburg and Ciolek-Torrello (1997: 69)
	Cross-channel alignments			x	x	Homburg (1997: 112–15)
McCoy Mesa	Cross-channel alignments			x	x	Homburg (1997: 115–16)
Davenport Wash	Rock alignments					Homburg (1997: 116)
	Cross-channel alignments				x	Homburg (1997: 117)
Mullen Mesa	Contour alignments				x	Homburg (1997: 117–18)
	Rock alignments and terraces	x			x	Homburg (1997: 119–21)
	Cross-channel alignments				x	Homburg (1997: 122–4)
Cave Creek Valley, Arizona						
Site T:8:31	Series of terraces				x	Dittert, Fish, and Simonis (1969: 22, 23)
Site T:8:9	Extensive terraces				x	Kemrer, Schultz, and Dodge (1972: 32, 88)
Retention dam	Linear borders, terraces				x	Holiday (1974: 19, 41)
Cave Buttes	Check dams			x	x	Rodgers (1974: 40, 42, 44)
	Linear borders, terraces				x	Rodgers (1974: 40, 41, 42)
Fort Mountain	Contour terraces				x	Rodgers (1977: 55–63, 88)
				x	x	Rodgers (1978: 155)
	Check dams				x	Rodgers (1977: 55–9, 89)
				x	x	Rodgers (1978: 155)
Sites T:8:66, 91	Extensive check dams				x	Henderson and Rodgers (1979: 41–3, 59–63)

New River Valley, Arizona						
	Simple boulder terraces				x	G. Stewart (1939: 115)
	Rock alignments		x		x	Rankin (1989*b*: 954–65)
	Linear borders				x	Doyel and Elson (1985: 103)
Agua Fria Valley, Arizona						
	Simple boulder terraces				x	G. Stewart (1939: 115)
	Rock alignments		x		x	Rankin (1989*b*: 954–65)
Flagstaff area, Arizona						
Anderson Mesa	Terraces and check dams					Pilles (1978: 131)
Chavez Pass	Terraces				x	Stewart and Donnelly (1943*a*: 44)
	Terraces and check dams			x	x	Henderson (1979: 10)
Seligman	Boulder checks				x	G. Stewart (1940*a*: 217)
Walnut Canyon	Check dams and terraces				x	Downum, Harper, and Boston (1995: 4, 44)
Wupatki National Monument, Arizona						
Antelope Prairie	Linear borders				x	Marozas (1983: 54, 103)
	Check dams, rock alignments					Ritterbush (1984: 53–6)
Various sites	Rock alignments			x	x	Travis (1990: 11–12, 18)
	Terraces	x	x			Travis (1990: 12, 25–7)
	Check dams			x	x	Travis (1990: 12)
Grand Canyon area, Arizona						
Lee Canyon	Cobble alignments (aka check dams, terraces)	x	x		x	Whittlesey and Homburg (1992: 58, 67–73)
North Rim	Check dams and long terraces				x	Stewart and Donnelly (1943*a*: 38–42)
Shinumo Creek	Check dams and terraces					Schwartz (1960: 65–6)
Kayenta region	Terraces			x		Lindsay (n.d.)
	Cobble alignments				x	Lindsay (n.d.)
	Dams			x	x	Lindsay (n.d.)
Navajo Mountain	Stream checks, boulder field terraces					Stewart and Donnelly (1943*a*: 36–8)

Table 8.1. *Continued*

Locale	Features mentioned	Functions claimed				Source
		Level surface	Deepen soil	Control erosion	Control water	
Southeastern Utah						
Beaver Creek	Terraces		x		x	Lindsay (1961: 181)
	Linear borders			x		Lindsay (1961: 183)
Horse Flats	Check dams		x	x		Brooks (1974: 15–26)
	Hillside terrace					Brooks (1974: 15)
Hovenweep National Monument, Utah						
Site 42	Terraces, lines of rock			x	x	Schroeder (1967–8: 84–5)
Site 49	Terrace	x		x	x	Schroeder (1967–8: 85–6)
Various sites	Akchin terraces			x	x	Winter (1976: 91, 1978)
	Arroyo check dams	x		x	x	Winter (1976: 98, 1978)
	Slope wash and talus bottom terraces			x	x	Winter (1976: 99, 1978)
	Springside terraces				x	Winter (1976: 105; 1978)
	Rim dam terraces			x	x	Winter (1976: 105; 1978)
Four Corners						
	Rude stone dams					Prudden (1903: 239)
Mesa Verde National Park, Colorado						
	Check dams				x	J. Herold (1961: 103)
	Stone terraces		x		x	Woodbury (1963: 62–3)
	Terracing and temporary wood or brush checks		x	x	x	Osborne (1976: 43)
Cimarron District, New Mexico						
	Rock alignments				x	Glassow (1980: 117–33)
Upper Rio Grande Valley, New Mexico						
Various sites	Contoured terraces				x	Cordell (1984: 205)
	Terraces and check dams				x	Cordell, Earls, and Binford (1984: 236)

Taos	Stone borders				Woosley (n.d.*b*)
Fort Burgwin	Check dams		x	x	L. Herold (1968: 35–6)
	Stone terraces			x	Woosley (1980: 330–1)
	Contour terraces				Woosley (n.d.*a*)
Picuris Pueblo	Boulder terraces		x	x	Woodbury (n.d.*a*)
Chama Valley, New Mexico					
Various sites	Cobble-bordered contour terraces			x	Anschuetz (1995: 29–32)
	Check dams			x	Anschuetz (1995: 29–32)
Sapawe ruins	Terraces			x	Cordell (1984: 205)
Site LA 11830	Cobble alignments			x	Cordell (1984: 205)
Various washes	Check dams			x	Cordell (1984: 205)
Ojo Caliente Valley, New Mexico					
Various sites	Lines of stone, parallel terraces				Bandelier (1892: 46)
Site OC-7	Cobble alignments				Lang (1980: 5)
Middle Rio Grande Valley, New Mexico					
Nambé Pueblo	Agricultural terraces				Ellis (1978: 64)
Pojoaque area	Agricultural terraces				Ellis (1978: 64)
Los Alamos	Check dams		x	x	Neely (1993*b*: 233)
	Linear borders		x	x	Neely (1993*b*: 233–4)
La Bajada	Check dams				Herhahn (1995: 78)
Cochiti Pueblo	Terraces			x	Cordell (1984: 205)
Galisteo Basin	Linear borders			x	Lang (1995: 50)
	Linear retainers		x		Lang (1995: 51)
	Check dams	x	x	x	Lang (1995: 56)
	Terraced fields		x	x	Lang (1995: 56–7)
Gran Quivira	Terraces			x	Toulouse (1945: 366, 371)
Tijeras Canyon	Terraced gardens and check dams			x	Anschuetz (1984: 41, 153)
Zia Pueblo	Check dams, and cobble-step terraces			x	Anschuetz (n.d.)
Jemez Canyon	Terraces		x	x	Fliedner (1975: 371–4)
Northwestern New Mexico					
Puerco Valley	Contour terraces	x			J. Moore (1981: 108–12)
	Check dams		x	x	J. Moore (1981: 114)
Rio Gallina	Stairstep gardens				W. Douglas (1917)

Table 8.1. *Continued*

Locale	Features mentioned	Functions claimed: Level surface	Deepen soil	Control erosion	Control water	Source
Largo–Gallina	Garden terraces					Hibben (1940: 73–4)
	Rows of stone along the contour of the slope					Green, Danfelser, and Vivian (1958: 51)
Nogales Cliffs	Check dams			x		Pattison (1968: 129)
Llaves area	Lines of stone, terraces, and check dams				x	Dick (1976: 20)
Zuni Pueblo	Check dams				x	Kintigh (1984: 225)
Near Snowflake, Arizona						
Site P: 12: 74	Terrace wall			x	x	K. Lightfoot (1981: 260)
Site P: 12: 135	Terrace systems				x	K. Lightfoot (1981: 260–1)
Hay Hollow Valley, Arizona						
	Several terrace systems		x			K. Lightfoot (1981: 261–2)
Point of Pines, Arizona						
	Terraces		x		x	Woodbury (1961*a*: 10–12)
	Linear borders			x	x	Woodbury (1961*a*: 12–13)
Largo Creek, New Mexico						
	Rock alignments				x	Kayser (1973: 11, 17–19)
	Check dams and terraces				x	Kayser (1974: 31)
Devil's Park, New Mexico						
	Check dams			x	x	Peterson (1988: 123)
	Rock spreaders		x		x	Peterson (1988: 123)
	Linear borders		x	x	x	Neely (1995*b*: 246)
	Check dams				x	Neely (1995*b*: 246)
	L-shaped terraces	x			x	Neely (1995*b*: 249–51)

Sapillo Valley, New Mexico						
	Terraces		x			Sandor, Gersper, and Hawley (1986, 1990)
South of Mule Creek, New Mexico						
	Check dams			x		Doolittle (1985)
Arenas Valley, New Mexico						
	Check dams		x	x	x	Herrington (1979: 115)
	Terraces			x	x	Herrington (1979: 125)
Treasure Hill, New Mexico						
	Check dams, spreaders, and contour terraces			x	x	Herrington and Creel (1991: 58)
Cameron Creek, New Mexico						
	Check dams		x	x	x	Herrington (1982: 82, 83)
	Terraces			x	x	Herrington (1982: 82, 83)
Lampbright Draw, New Mexico						
	Check dams		x	x	x	Herrington (1982: 82, 85)
	Terraces			x	x	Herrington (1982: 82, 85)
Mimbres Valley, New Mexico						
	Check dams			x	x	LeBlanc (1977: 16–19)
	Terraces or dams			x	x	LeBlanc (1983: 153–4)
	Check dams or terraces					Minnis (1985: 76)
Rio Casas Grandes Valley, Chihuahua						
	Terraces, *trincheras*	x				Hard and Roney (1998: 1662)
	Stone wall, dike					Bandelier (1892: 550)
Sierra Madre, Chihuahua						
Arroyo de la Cuerda	Dams and dikes					Bandelier (1892: 558)
Various sites	Ledge, dike, dams	x	x	x	x	A. Leopold (1937: 120)
Gavilan Basin	Stone terraces or check dams	x	x			A. S. Leopold (1949: 11, 12)
	Rock terraces or check dams	x			x	A. Leopold (1966: 159)
	Stone dams or walls in series		x	x	x	McCabe (1955: 76–8)
	Trincheras				x	L. Herold (1965*a*)

Table 8.1. *Continued*

Locale	Features mentioned	Functions claimed				Source
		Level surface	Deepen soil	Control erosion	Control water	
Sierra Madre, Sonora						
Baserac	Artificial dikes					Bandelier (1892: 527)
Batesopa	Many dikes					Bandelier (1892: 519)
Quitamac	Dikes		x			Bandelier (1892: 518)
Los Matates	Dikes		x			Bandelier (1892: 512)
Vayuavavi	System of dikes					Bandelier (1892: 508)
Cerro de Nacori	Well-built *trincheras*					Lumholtz (1902: 44)
Cerro de Mochubabi	Four dams					Bandelier (1892: 504)
Arroyo de la Culebra	Dams, rows of boulders					Bandelier (1892: 502–4)
Various sites	Low stone breast-walls	x	x	x		Howarth (1895: 435)
Chihuahua–Durango Border						
	Masonry terrace walls					Kelley (1956: 132)
Durango, Mexico						
	Masonry terrace walls	x	x			Kelley (1956: 29)
Sonora, Mexico						
Valley of Sonora	Channel bottom weir terraces		x		x	Doolittle (1988: 48–9)
Cerro de Trincheras	Narrow terraces					McGuire *et al.* (1993: 23)
Altar Valley	*Pequenos represas*					McGuire and Villalpando (1991: 70)

have been used in describing the features. Some scholars prefer generic terms such as 'rock alignments' (e.g., Rankin 1989*b*: 954). Some prefer terms that imply function, such as 'check dam' (e.g., Neely 1993*b*: 233). Some prefer the name bestowed by local informants, '*trincheras*' (Herold 1965*a*). Finally, some scholars, particularly those from the 19th century who did not enjoy the benefit of decades of research by others, and did not have a clue what to call such strange features, came up with bizarre terms such as 'artificial dikes' (e.g., Bandelier 1892: 527). Fourth, there is great disparity in the functions various features are thought to have performed. Some scholars have attributed different functions to features with the same name (e.g., Rodgers 1974: 40, 1977: 55), whereas others have applied different labels to features that functioned alike (e.g., Kayser 1974: 31). Highlighting this confusion is not intended to denigrate the work of any scholar or scholars, but simply to point out the complexity of 'simple' features. Finally, there is a disparity in the functions simple features are purported to have performed.

The function most commonly attributed to simple terracing is water control. In nearly every reported case of terracing, something about slowing or stopping the flow of runoff is mentioned. Doubtless, this is a function of these features being located in arid environments, as well as parallels being drawn from present-day native farmers who currently employ similar devices for that reason. The next most commonly claimed function is erosion control. The link between runoff and erosion is tenacious. This is perhaps no more clear than to dwellers of arid lands, people who understand the fragility of their soil and the ferocity of runoff resulting from violent summertime convective thunderstorms. Increasing the thickness of the planting medium follows as being the third most commonly interpreted function of terracing. And, here a pattern emerges. Deepening of soil tends to be recognized by scholars working in the low desert, where soils are thin and often no more than a veneer overlying thick beds of cobbles. Also, scholars working in the higher elevations and on steeply sloping terrain recognize the problem of naturally thin soils in these environments. Finally, leveling of the planting surface is rarely recognized by researchers as a function performed by simple terraces. No doubt this is because in most cases, the settings in which the features are found have slopes of only a few degrees or percents. Not surprisingly, leveling is seen to be important on steep slopes, especially those in the higher elevations.

Two other functions have not been included in the tabulation, principally because southwestern researchers have rarely reported their occurrence. Presumably, this failure to report is a function of their minimal or locally restricted importance. These functions are improving soil fertility, and temperature control. Either very few scholars have found organic matter accumulated behind features oriented perpendicular to the flow, or they have failed to recognize that it was once detritus suspended in runoff. Adrianne G. Rankin (1989*b*: 964–5) is nearly unique for having identified such materials, and she recognized them as having improved soils in the New River and Agua Fria valleys of central Arizona.

In no place have terraces been interpreted as having served a temperature control function. However, terraces have been found in two environmentally favorable niches surrounded by an otherwise cool and less than optimal landscape for agriculture. Not surprisingly, these two locales are in the higher elevations and in the far northern part of the Southwest. At 2000–2200 m, Mesa Verde, Colorado is one of the cooler environments in the greater region. Winters are long and cold, with much snow. Summers are cool and short; the growing season barely sufficient. In addition to 'check dams' which trapped runoff, collected sediment and organic matter, thereby creating fertile and level planting surfaces on top of the mesas, ancient farmers built 'terraces' on slopes and across steep gradient streams. These features also trapped runoff, but their spatial distribution is not uniform. Approximately 75 percent of these terraces are located with eastern and northeastern exposures. Although not as warm as southern and western exposures, such locales benefit from having more mesic conditions (Erdman, Douglas, and Marr 1969: 58). Presumably, but this remains to be tested, increased moisture might well have ameliorated temperatures which surely are lower on slopes with such aspects. And, the wetter conditions might have mitigated damage from early and late frosts.

Evidence of terracing in the canyons of Mesa Verde is minimal. This condition has long been thought to be a function of cold air drainage, which would make agriculture there next to impossible. Not far from Mesa Verde, however, near Chimney Rock, Colorado, recent findings have resulted in this notion being challenged. There, among a series of check dams, archaeologists found high frequencies of panicoid grass phytoliths and low frequencies of festucoid grass phytoliths, suggesting warm and moist conditions in times past (Mobley-Tanaka and Eddy 1995:435–6). How could such seemingly cold locales support grasses requiring other conditions? Radiant heating off canyon walls. That is, dark gray shale absorbs warmth during the day and radiates it back at night thereby counteracting the otherwise damaging cold air drainage (Mobley-Tanaka and Eddy 1995: 439). Granted, the walls of the canyon and not terrace walls themselves resulted in temperatures favorable for agriculture. However, the terrace builders clearly had knowledge of climatic conditions and how to compensate for them.

Simple terraces are, it seems, not all that 'simple'. In addition to serving a multitude of possible functions, some not all that readily apparent, there exists a great deal of variation in their construction. Simple terracing can involve as little as a single row of rocks on the surface parallel to slope contours or perpendicular to the flow of runoff. Features of this configuration can be very short, 4 or 5 meters in length, or quite long, 30 to 40 meters in length. Regardless, they are always identified with very slight slopes. In some cases, rudimentary forms of terracing involve more than one course of stone, while in yet other cases the terraces look more like low linear piles of rock rubble (Fig. 8.20). These latter features have been described as 'sloping zones of boulders' (Woodbury 1963: 4), 'stepped-back terraces' (Neely 1995*b*: 249–50, fig. 16.6), 'banked' or 'gravel-banked terraces' (Dart 1983: 414–15, figs. IV.4.4 and IV.4.5), and 'broad rock alignments' (Homburg 1997: 120, plate 5.44). That terraces of this configuration appear as heaps of rubble (Fig. 8.21) has resulted in some writers (e.g., Fliedner 1975: 372) suggesting that they may be in part the result of clearing the ground of rocks prior to planting. Not inconsequentially, it is also possible that while serving as risers, retaining soil, sediment, and water for crops planted behind them, these lines of rubble might have been used as mulch beds for the cultivation of *agave*, not unlike the stone mounds discussed in Chapter 3.

Finally, it should be mentioned that simple terraces are found alone, as individuals, or in series of multiples (Fig. 8.22). In latter cases, the height of the riser and the distance between terraces is a function of the slope. Steeper slopes tend to have more, higher, and closer terraces than less steep slopes which are characterized by terraces that are fewer, lower, and further apart.

In addition to simple terracing, a few locales in the Southwest are known for more complex varieties. For example, near Snowflake, Arizona one terrace measured 100 m long, 50 cm high, and had a large field behind it (K. Lightfoot 1981: 260–1; Lightfoot and Plog 1984: 183). Another site, Kinnikinick ruin in north central Arizona, has 'a series of terraces with walls over 1 m high, the area behind them filled with trash to form rich farming soil' (Pilles 1978: 131). Similarly, at least four terraces have been reported from Chaco Canyon (Vivian 1970: 69, 71, fig. 3). Located just below the canyon rim, and dating to the 11th century AD, these terraces are comprised of dry masonry risers ranging in height from 50 cm to more than 3 m (Windes 1987*a*: 120, 123, plates 5.15 and 5.16). They range, as best as can be determined by their fragmentary nature, from as short (!) as 80 m to more than 280 m. Filled by alluvial sands, the 'cultivation strips' range in width from 11 to 25 m wide (Windes 1987*b*: 555–9, plates 7.8, 7.9, and 7.10). Building risers this high requires some masonry skill, and much labor. Filling the spaces behind the risers also required great energy inputs, and clearly involved deliberation. The building of such terraces prehistorically involved a combination of abundant labor, experienced craftsman, and the right physical environments. Three places stand out in this regard, Mesa Verde in Colorado, the Sierra Madre of northern Mexico, and the *cerros de trincheras* of the lower Sonoran Desert.

In some respects the terraces at Mesa Verde are not all that impressive. They are considerably smaller than some of those in both the mountains and deserts of Mexico. There are, however, a lot of them. Furthermore, they are well-studied and they are easily accessible, both points resulting from being in a national park. Gustav Nordenskiöld (1893: 72) was the first archaeologist

Fig. 8.20. A downslope view of a series of low, linear piles of rock rubble that functioned as terraces in prehistoric times. These features are part of site AZ O:14:66/01-1048 on Mullen Mesa, in the Lower Verde Valley, Arizona. This section of the site was cleared of vegetation for detailed soil studies by Jeffrey A. Homburg (1997). Appeared originally in Van West and Altschul (1997: 348, plate 9.3). (Reprinted by permission of Statistical Research, Inc., Tucson, Arizona.)

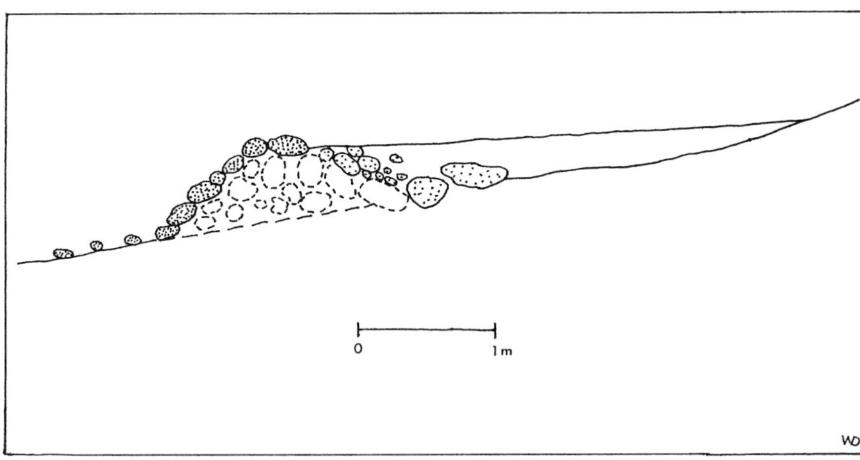

Fig. 8.21. A cross-section of a typical rock rubble terrace (after P. Fish *et al.* (1986: fig. 1)). (Redrawn by author.)

Fig. 8.22. A birds-eye view of a series of prehistoric terraces on a low gradient slope at the far western end of the Zuni Reservation, New Mexico. Discussed in general terms by Keith Kintigh (1984: 225–6), this site is similar to one described in detail at Point of Pines, Arizona by Richard B. Woodbury (1961*b*: 27, fig. 12). (Photograph by author.)

to note 'low walls' across depressions on the tops of the several small mesas which comprise Mesa Verde, and to conclude they served a water control function for agriculture. His work was followed up, albeit several years later, by a number of scholars, most notably Guy R. Stewart and Maurice Donnelly of the Soil Conservation Service, and Arthur H. Rohn and Alden C. Hayes of the national park. Perhaps thousands of individual terraces once combined to make hundreds of sites in the area. On Chapin Mesa no fewer than thirty-nine terrace sites involving more than 900 features have been reported (Rohn 1963*b*: 442). Immediately to the west, on Wetherill mesa, 136 sites with 943 features are known to exist (Hayes 1964: 76). Many terraces undoubtedly have been lost to erosion, and some to burial by sedimentation (Rohn 1977: 91), and many remain unreported on neighboring mesas.

From studies that have been conducted, it is known that three types of environments were subjected to agricultural terracing prehistorically at Mesa Verde, hill slopes, low gradient stream beds on the mesa tops, and steep gradient stream beds near the rims of the mesas. Frequently, terraces extended from one environmental setting to another (Hayes 1964: 79).

Terracing on hill slopes is known from only a few specific locales. Two such sites exist on Chapin Mesa. One is comprised of three 'long' rock alignments that parallel contours. Details concerning the length, height, and spacing of these terraces are, unfortunately unreported. However, it is thought they stabilized the sloping soil surfaces against erosion (Rohn 1963*b*: 445; 1977: 91). On Wetherill Mesa, only 'a few' terrace sites were found along the contours of hill slopes (Hayes 1964: 76). One of these though included approximately 100 terraces 4.5 m long in a series extending up and down slope 400 m (Hayes 1964: 79), for an average spacing of 4 m. Individual terraces were made of unworked boulders stacked one to three courses, resulting in a riser approximately 60 cm high which held 'considerable soil' (Hayes 1964: 77, fig. 65, 81, fig. 73).

Terraces built across shallow drainages on mesa tops are much more numerous than hill slope terraces, with Wetherill Mesa again containing the largest number of such sites, seventy-one, and individual terraces, 412 (Hayes 1964: 76). With the exceptions of their length and setting, these features are not unlike hill slope terraces. They tend to be made of unworked rocks stacked roughly one to three courses high. Given that they lie across drainages, these terraces are higher in the middle than at either end where they blend into the natural ground surface. The streams in which they are located have gradients of 1.5 to 4 percent, and obviously the flow of water was affected by the presence of these terraces, which might have impeded erosion as well. Indeed, the soils in which these terraces are found are relatively thick and adequate for crop production (Stewart and Donnelly 1943*a*: 32). Despite their small size, but because they are often located in open settings, terraces of this type are quite conspicuous and may well have been the ones which caught the attention of Nordenskiöld.

The ephemeral streams on the mesa tops flow over the edges of the mesas and as they approach the rim their gradients steepen, to 10, 30, and even 55 percent (Rohn 1963*b*: 443–4). Their channels also deepen and widen. As a result, terraces built in such settings tend to become longer, higher, and closer together, as they progress downstream. Depending on the specifics of their locales, terraces in channels near the rim of Chapin Mesa vary in length from 2.6 to 13 m (Stewart 1940*a*: 209), with most averaging 6 m (Stewart and Donnelly 1943*a*: 34). On Wetherill Mesa, they are larger, ranging from 9 to 15 m long (Stewart and Donnelly 1943*a*: 34).

The horizontal distances between terraces near the mesa edges vary considerably, both within and between sites. One Chapin Mesa site has a relatively uniform spacing with terraces 3.5 to 5.2 m apart (Rohn 1963*b*: 442). Nearby, another site has some terraces as close as 2.1 m apart and some as far as 20.3 m apart (Rohn 1963*b*: 443). Terrace spacing at one Wetherill Mesa site averages 2.6 m (Hayes 1964: 77). Spacing, of course, is related directly to the gradient of the stream and the height of terraces.

The risers of terraces in steep gradient streams are considerably different from those of terraces found in other locales. Here, they are made of flat sandstone blocks laid horizontally (G. Stewart 1939: 115). Rocks at the base of each terrace tend to be bigger and more irregular in shape than those at the top (Rohn 1963*b*: 442). In the upstream segments of the channels, terraces are two to three courses, or 30 to 60 cm high in the center. Downstream they involve several courses and measure more than 1.5 m in height (Hayes 1964: 77). Most of these terraces have been badly eroded, and are discernible today only by a few rocks firmly abutted in the walls of the stream channels and by blocks lying in the channel below (Rohn 1963*b*: 442). In one locale, sufficient material remained to reconstruct the terraces to their original appearance (Fig. 8.23).

The nature of the soil or sediment behind these terrace walls, and the function of the terrace systems, have not been without controversy. Obviously agricultural, these terraces provided level planting surfaces where few existed otherwise. The relatively limited amount of farm land at Mesa Verde stands in stark contrast to the numerous, large, and elaborate cliff dwellings for which the area is famous. The large number of people who lived at Mesa Verde in prehistoric times, clearly had to utilize every square meter of land, and then they had to make some land. The terraces located in the steep

Fig. 8.23. Reconstructed terraces across a deeply incised stream channel with a steep gradient near the edge of Chapin Mesa, Mesa Verde National Park, Colorado. Note the flat block construction. These terraces are part of site 395 (see Rohn 1963*b*: 441, 442, fig. 2). Appeared originally in Doolittle (1992*c*: 391, fig. 3). (Reprinted by permission of the Association of American Geographers.)

canyons at the edges of the mesas were, quite literally, human-made farm land (Hayes 1964: 79; G. Stewart 1939: 34).

Soils behind terraces that remain intact today are deep, fertile, clay loams or fine sandy loams that are high in organic matter (G. Stewart 1940*a*: 210). It was once thought that these soils were the combined result of both deliberate filling by people and natural sedimentation (Rohn 1963*b*: 442). The idea of filling seemed natural enough given the depth of the soil and the proximity of terraces to large habitation sites. It was soon discovered, however, that sedimentation was involved. One terrace which was thought to be only a single course of rocks proved to be a riser 1.8 m high after it was excavated archaeologically. Furthermore, it went through three or four stages of rebuilding (Hayes 1964: 79). Differences in rocks between the bottoms and the tops of terraces are, therefore, not so much a function of design as they are timing and periods of reconstruction. Profiles showed several strata of water-laid deposits and no signs of having been filled manually (Hayes 1964: 79; Martin and Byers 1965: 128–9). The conclusion to be drawn then, is that these terraces were built as level planting surfaces. They were, however, subject to sedimentation, and periodically the risers had to be raised. This was a very dynamic agricultural landscape both during occupation and after abandonment (R. Fletcher 1985: 668).

Although the ages of prehistoric agricultural features are notoriously difficult to determine, most southwestern archaeologists agree that terraces on the mesa tops were built prior to those further downstream (Rohn 1977: 262; Stiger 1979: 136–7), perhaps as early as AD 900 and as late as AD 1200, respectively. Indeed, such makes sense from an economic point of view (Boserup 1965) as the lands on the mesa tops were easier to cultivate and there terraces were easier to build than those further downstream. Apparently, however, erosion on the mesa tops was too great for the terraces built there to control. Presumably, the abandonment of Mesa Verde was at least in part linked to the difficulties of maintaining their environment and not simply with feeding the population.

Of course, feeding the population was a problem. As numerous as the terraces and the terrace sites are at Mesa Verde, individually and together they account for a very small amount of agricultural land. For example, the planting surfaces behind one of the smaller terraces measures only 1.5 by 3.5 m (G. Stewart 1939: 131), which is less than the area delimited by the lengths of the risers and the distances between them. Two factors explain why all of the area behind each terrace was not cultivated, both of which are a function of stream gradients and the V-shaped cross-section of the natural channels. First, the terrace fill is thinner along the sides and the back of each planting surface than it is immediately behind each riser. In some cases the soils along the sides and back are too thin for root development and, hence, crop production. Second, the shape of the cultivated surface was not rectangular, but trapezoidal, wider behind the top of the riser than at the foot of the next riser upstream.

When the average cultivable surface area of the 900 terraces on Chapin Mesa is multiplied by their number, a figure of 8.5 to 12.5 ha of terraced land results. If these figures are doubled on the assumption that evidence has been found for only one-half of all the terraces that might have been used, then terraces account for no more than 2.5 percent of the 970 ha of farmland on the mesa (Rohn 1963*b*: 446, 1977: 91). Wetherill Mesa is smaller than Chapin Mesa, and it has more and larger terraces. Accordingly, one can conclude that terracing was relatively more important on the former mesa than the latter. However, terracing still would have accounted for only a small portion of the farmland. Although figures are not available, it seems probable that terracing would have involved no more than 5 or 6 percent of the agricultural land on that mesa. Together, terracing probably constituted no more than 4 or 5 percent of farmland on Mesa Verde. Given that it was a relatively late practice, and one that was evolving right up until the time of abandonment, terracing seems to have been a relatively labor-intensive, or socially expensive, activity that was doomed. Ancient farmers at Mesa Verde were not getting much 'bang for their buck', to use military parlance. This condition was quite different from that experienced by their counterparts further south in the Sierra Madre.

Bandelier was the first archaeologist to note the presence of terracing in the Sierra Madre. In addition to numerous examples of simple terraces (see Table 8.1), some of which occurred in series measuring more than 1.5 km long (Bandelier 1892: 502–3), he noted several sets of very large terraces. For example, he reported the channel of the Cajon de Mechapa, just south of the town of Baserac, Sonora, as being 'traversed by a number of artificial dikes' 1.5 m high and 0.5 m thick. These were 'regular walls of broken stones laid in adobe mud'. Behind, or upstream of, each were 'triangular or quadrangular patches of soil' (Bandelier 1892: 523). Not too far away, in the Valle de las Escobas, he found similar terraces but made of 'dry walls constructed with boulders piled neatly' (Bandelier 1892: 526). Finally, on the eastern side of the Continental Divide, in the state of Chihuahua, he found some terraces similar to those observed in Sonora, that were 'very well preserved' perhaps in part because the walls were 1.5 m thick (Bandelier 1892: 564). In each of these cases, Bandelier interpreted the functions of these terraces to be the creation of level and fertile planting surfaces, with the medium being sediment that accumulated over time (Bandelier 1890: 62).

Bandelier was followed a few years later by Carl Lumholtz who observed 'traces of ancient . . . stone terraces, which had evidently served agricultural purposes' all the way from Nacori, Sonora to Casas Grandes, Chihuahua (Lumholtz 1902: x). Lumholtz was the first to refer to them as *trincheras*. He also noted: (i) that they did not appear to be very ancient,

(ii) that stones were not cut or faced, (iii) that risers were not vertical but rather inclined slightly upslope, (iv) that they resembled terraces built by the Hopi, (v) that some risers were 4.5 m high and 10 m long, (vi) that great labor went into their construction, and (vii) that the small size of the resulting cultivated plots made them seem hardly worth the effort (Lumholtz 1902: 20–1, 74).

The famed naturalist Aldo Leopold and his young son, Starker, visited the sierras in the 1930s and wrote about these terraces, albeit providing only personal impressions of those in the Gavilan Basin (A. S. Leopold 1937; 1966). The younger Leopold returned several years later with a larger party of scholars, principally to study the region's fauna (A. S. Leopold 1949). One member of his party took a special interest in the terraces, especially as works of prehistoric engineering. Although his one article on the subject did not provide much in the way of new information, it did point the way to detailed systematic studies of the terraces (McCabe 1955).

Everyone interested in prehistoric agriculture in the American Southwest recognizes the contributions made by geographer Laurance C. Herold and his colleagues in improving understanding of the *trincheras* of the Sierra Madre. Beginning in 1962, his team from the University of Denver began investigating these features in great detail. Coming from a geographical, rather than an archaeological or naturalist background, Herold and his associates looked at every aspect of the biophysical environment, and from a decidedly scientific perspective. Almost immediately, he hypothesized that the principal function of these terraces might have been to serve as moisture reservoirs (Herold 1965*a*: 83).

Two major monographs were produced after conclusion of the field work. One, by Herold (1965*b*) himself, provided details about the terraces and the environments in which they are situated. Significant findings reported in this monograph are many. Herold identified four types of terraces: those described by Lumholtz which were truly impressive (see Fig. 8.2); low rock alignments that followed the contours of low gradient slopes, and which did not accumulate soil but impeded erosion; and multiple course rock alignments on steep slopes which did accumulate sediment. He also discovered, along the Rio Piedras Verde, other features known locally as *trincheras*, which remain enigmatic but were probably not agricultural (Herold 1965*b*: 102–10).

Herold discovered through excavations that terrace walls are of four types. Rock alignments and linear piles of rubble were common and not unlike those found elsewhere and discussed above. In addition, there were two types of taller, multiple-course terraces. One was a rock-faced riser with rubble backing. The second involved two parallel rock walls. The outer one, or the riser was up to 1 m thick and sloped upstream. The second was made up of smaller rocks and was only 30 to 50 cm thick but stood vertically. Between these walls was a 30 to 50 cm space that was filled with rock rubble (Herold

1965*b*: 110–16). Terraces of this last variety were clearly the most architecturally sophisticated in all of North America.

Finally, Herold performed detailed analyses of both the material immediately behind the risers and the terrain surrounding them. He found no indications that the terrace fill accumulated by any process other than that normally associated with water deposition and mass wasting, that filling occurred over a long period of time and in observable sequences, that a sufficient amount of time had passed since abandonment for soils to begin to form, and that risers were extended periodically to accommodate increased sedimentation (Herold 1965*b*: 119–30). An attempt was made to evaluate terrace sediments for their moisture storage properties, but results proved, for a variety of reasons beyond the control of investigators at the time, to be inconclusive (Herold 1965*b*: 130–4).

Herold focused much of his attention in Gavilán Basin, the area featured by Aldo Leopold, and the refuge of the notorious 19th-century Apache rebel, Geronimo. Indeed, it was here that Geronimo was convinced by the US cavalry to surrender (Bourke 1958).

The second monograph produced by personal from the University of Denver looked at the greater distribution of terraces throughout the sierras. It reported the findings of several reconnaissance trips made during the course of one year, using mainly wheeled vehicles and traveling over the very rugged logging roads that criss-cross the region. Terraces were found in varying numbers and densities along a 900 km by 50–200 km strip straddling the Continental Divide from just south of the US–Mexico border to the Rio Urique which cut the famous Copper Canyon (Howard and Griffiths 1966: 8, fig. 3). This is an area that is characterized by pine and oak woodlands, that receives highly seasonal precipitation, is very rugged, and has poor soils. As Leopold's interest indicates, it was until quite recently very rich in wildlife. The region was very attractive to early prehistoric people as is evidenced by the numerous cliff dwellings (Fig. 8.24). Perhaps not surprisingly, terraces tend to be located near these ancient habitation sites (Howard and Griffiths 1966: 81–6), more so than to any environmental factor. These places were occupied even before people began farming. Once crops such as maize were diffused into the area from Mesoamerica, the only way they could be produced was on terraces.

As important as these two monographs were, and are, they precipitated little in the way of follow-up investigations. Two master's theses were produced shortly thereafter (Griffin 1966; Dennis 1967). Both focused on present-day terraces built by a logging company, not on ancient terraces; in effect, they were exercises in experimental archaeology. These studies demonstrated what Herold first suspected, that terraces impound water on their upslope side and allow it to pass slowly from one terrace to the next. As a result, the hydrologic regime was changed from one in which streams were ephemeral to one in which they were quasi-perennial (Griffin and Dennis

Fig. 8.24. Olla Cave in Cave Valley, Chihuahua, Mexico. This is perhaps the most famous cave in the Sierra Madre, largely because of its regionally unique and intact prehistoric granary. See Lumholtz (1902: 64), Sayles (1936: 19, plate VIIIa), and Lister (1958: 31, plale 5b). (Photograph by author.)

1969). The significance of these changes was not lost on the most prominent archaeologist working in northern Mexico.

The late Charles C. DiPeso served as the director of the Amerind Foundation, Dragoon, Arizona for several years. In that capacity, he oversaw the investigations of many significant archaeological sites in southern Arizona, southern New Mexico, and northern Chihuahua. He is most well-known for his long-term studies at the ancient city of Paquimé, near the present-day town of Casas Grandes, which has also lent its name to the archaeological site. Never one to think small, DiPeso concluded that the terracing systems in the sierras immediately to the west of his site not only provided food for people who lived near them, but that the hydrologic regime must have been sufficiently altered to affect farmland scores of kilometers away, specifically along the Rio Casas Grandes. He developed an elaborate model in which hill slope terraces diverted runoff onto cross-channel terraces, which sequentially absorbed and seeped moisture which eventually flowed into the lower valley where it was carried to broad floodplain fields through irrigation canals (DiPeso 1984: especially fig. 3).

There are only two things wrong with DiPeso's model. First, DiPeso made a huge leap of scale and assumed that if several drainage basins were altered locally, then collectively their impacts would be magnified the further one goes downstream. To a certain extent, of course, he was correct. His logic is appropriate, but he overestimated the degree and the extent of landscape modifications. Second, most of the terraces in the sierras are on the western side of the Continental Divide, exactly opposite of where they should be in order to have an impact on the Casas Grandes valley. Both of these problems are the result of DiPeso having developed his model without ever having conducted much work in the sierras. Most of his investigations involved excavations at Paquimé; he had little first-hand familiarity with the agricultural features and the environments of which he wrote.

One scholar not lacking first-hand familiarity with the region is Robert H. Schmidt, Jr. He has probably traveled through the Sierra Madre longer and farther than anyone, having devoted nearly thirty years of field study to the area. Trained as a climatologist, but one with geomorphological understanding, Schmidt, and the late archaeologist Rex E. Gerald, collaborated in a regional geoarchaeological assessment of terraces in the sierras after DiPeso presented his model. They looked at the spatial distribution of terraces, synoptic rainfall patterns, and the regional geology in concert (Schmidt and Gerald 1988). They found that terraces tend to be concentrated in areas with andesite surface rock which weathers rapidly and produces soils high in fertility. Terraces are much less frequently found on rhyolite-derived soils. Given that the former characterize the western side of the sierra which receives more moisture than the eastern side which is dominated by the latter type of geology, Schmidt and Gerald provide convincing evidence that the terraces were built by local groups to increase production on steeply sloping lands, and not as part of an elaborate water control scheme.

Further evidence for the localized nature of terracing in the Sierra Madre was provided about this same time by a small team of scholars, one of whom was Herold. The Elvino Whetten pueblo is located near the banks of the Rio Gavilán at the foot of a steep slope dissected by stream channels across which there are 233 terraces creating 3.6 ha of planting surface (Luebben, Andelson, and Herold 1986: 167, fig. 2, 168). There is also 2.5 ha of gently sloping land along the river, bringing the total farmland near the pueblo to slightly more than 6 ha. The eight- to ten-room pueblo is estimated to have been home to no more than twenty persons (Luebben, Andelson, and Herold 1986: 177). Accordingly, each person would have been sustained, at least theoretically, by 0.3 ha. This figure certainly does not allow for a surplus to have been transported to Paquimé, and, given that the Rio Gavilán flows westward, the terraces did not improve the drainage basin of the Rio Casas Grandes either (Herold and Miller 1994: 102).

While certainly not an abundance, the 0.3 ha of farmland per person at Elvino Whetten pueblo was barely adequate. The principal factor that ensured success was something Herold had suspected all along. In the late 1980s and early 1990s, Herold returned to the sierras to retest his original idea—to conduct new tests where previous ones had failed—on the moisture retention properties of the terrace soils. He collected sediment samples from different terrace terrain environments both before and during the summertime rainy season. Samples were collected at 10 cm intervals from alluvium along the Rio Gavilán, from behind rock alignments on mesa tops, from behind terraces on slopes, and from behind cross-channel terraces (Herold and Miller 1995: 146–7). From this, field capacities and wilting points were determined (Herold and Miller 1993: 185–6).

These tests revealed that terraced plots of all types retained more moisture to greater depths during the dry season than unterraced land, and that cross-channel terraced plots retained the most moisture the deepest. It was concluded that water acquired from late summer and winter precipitation soaked in to great depths. It then moved downslope to seep out at the base of the riser. From here it was absorbed by the fill behind the next terrace only to issue out at its base (Herold and Miller 1995: 149–52). In effect, cross-channel terraces retained moisture for use by crops during the late spring and early summer dry season. However, each terrace did not retain water for crops growing on its own surface. Instead, successful crop production depended on the hydrologic ecology of the entire terrace series.

If soil moisture deficits posed a problem to ancient farmers in the Sierra Madre, they must have made life nearly impossible for counterparts living in the Sonoran Desert. The operative word, however, is 'nearly'. Agriculture was practiced in the region and some of it involved terraces which are also known locally as *trincheras*. Here, in northwestern Mexico and southwestern Arizona, are found a number of individual, often conical shaped, hills that are characterized by series of rock-faced terraces on one or more sides (Fig. 8.25). These *cerros de trincheras* have been a mystery of sorts for more than a century. Their exact number remains unknown as every so often new ones are discovered (e.g., Bowen n.d.: 32–3; Hard and Roney 1998; McGuire and Villalpando 1991: 67). Their range, however, extends from the Valley of Sonora in the southeast (Doolittle 1988: 34–5), to the Rio Casas Grandes in the east (Crisell 1998), to the Agua Fria Valley in the northwest (Dove 1970: 16, 17, 21).

The earliest assessment of these features concluded that they must have been defensive (Shumacher 1882), places of refuge in times of siege. Later, the famed environmental determinist, Ellsworth Huntington (1913, 1914: 68–70) interpreted them as agricultural features, in large part to confirm his notion that until recently the Southwest was much more humid than it is now. His idea did not go unchallenged. In the early 1930s Sauer and his student Donald D. Brand, carried out an extensive survey of the region looking at numerous *cerros de trincheras*. They interpreted these hills and terraces as unsuitable for agriculture, and concluded that a defensive purpose, especially as lookouts and for signaling was a more likely explanation (Sauer and Brand 1931: 118).

Fig. 8.25. The Cerro de Trincheras in northern Sonora, Mexico. This is the largest and most famous site of its type. (Provided courtesy, and published by permission, of Randall H. McGuire.)

Sauer and Brand (1931: 119–21) were less than kind to Huntington, but they did contribute greatly to the archaeology of the region. Among other things, they discussed how the *cerros de trincheras* and their builders, who had their own cultural tradition—the Trincheras Culture—complete with distinctive ceramics, fit into the overall cultural mosaic of the Southwest. Their work gave rise to much enthusiasm among prehistorians, but not for the search of the true function of the *trincheras* themselves.

Until the late 1970s scholars pretty much accepted the *cerros de trincheras* as defensive. Since then, research has revealed that these hills served a multitude of purposes; they contain all types of sites and features, including terraces that were agricultural. Sauer and Brand are now having their conclusions refuted.

In northern Sonora, recent investigations under the direction of Randall H. McGuire have focused on *the* Cerro de Trincheras (Figure 8.25). Two separate archaeological field seasons were devoted exclusively to the terraces on this hill, one involved mapping, the other involved excavations. The Cerro de Trincheras is now known to have been 'a town with a complex activity structure', which served a number of functions (McGuire *et al.* 1993: 67). Features on the hill include walls, trails, reservoirs, house foundations, ballcourts, and a variety of smaller, less definitive structures. The most important in terms of agricultural possibilities, are, of course, the terraces.

From afar, and from the base of the hill, Cerro de Trincheras appears to be covered by long terraces that stretch the entire length of the north face. This, however, is an illusion created by distance and vegetation. What appear as unbroken terraces, are, in fact, numerous small terraces each built as a distinct unit (Fig. 8.25). Mapping revealed that no fewer than 879 terraces comprise the site. Some of these terraces are up to 100 m long, 5 m wide, with risers made of dry laid unworked rock stacked 3 m high. Others had similar dimensions except they were not as high and in the place of formal risers had loose rocks piled no more than 1.5 m high. Terraces fitting these descriptions were found on the steep upper and middle slopes of the hill. Although it is possible that agriculture may have been practiced on some of these terraces, other evidence suggests that non-agricultural functions were the principal reason for their existence (McGuire *et al.* 1993: 23–6, 69–70).

Terraces identified as having served purely agricultural purposes are those with single or double rows of rocks that measure less than 20 m long and 20 cm high, and are in series no more than 1.5 m apart. Terraces fitting this description were found on the lowest elevations and on the gentlest slopes of north, west, and south sides of the hill (McGuire *et al.* 1993: 23,70–1). These can be considered simple terraces, as they are really no different than those discussed above (see also Table 8.1).

In contrast to the well-known hill studied by McGuire, another *cerro de trincheras* was only recently discovered. Located in northwestern Chihuahua, this hill is not unlike its Sonoran counterpart in almost every regard. It is comprised of hundreds of short, disconnected terraces, most of which are of a simple nature (Hard and Roney 1998: 1663, fig. 4), and cumulatively result in only a few hectares of level land. That any of these terraces were actually cultivated has not been established, but remains of maize and squash were found, and they date to an astonishingly early 1100 BC (Crisell 1998; Hard and Roney 1998).

Much different from the simple terraces on the *cerros de trincheras* of Sonora and Chihuaua are some of the terraces on Cerro Prieto, 60 km north of Tucson. This site has been the focus of much attention for the past few years (Downum, Douglas, and Craig 1985), especially by Christian E. Downum. The hill is probably the largest *cerro de trinchera* north of the Mexican border and dates *c.* AD 1200–1300 (Downum 1993: 53). In addition to a number of simple terraces (Table 8.1) and small terraces that served as habitation sites with dooryard gardens (see Chapter 4, especially Fig. 4.9), this hill has several very prominent terraces which are now known to have been used exclusively for agricultural purposes on its lower northeast side (Fig. 8.26). Although the terraces appear from the base of the hill to be long, contiguous, and arranged in a stair step fashion, mapping has revealed that most are short

Fig. 8.26. Large terraces on the northeast side of Cerro Prieto (Arizona State Museum negative 86827). Appeared originally in Downum (1993: 80, fig. 4.28), and reprinted in Downum, Fish, and Fish (1994: 284, fig. 5). (Provided courtesy of Christian E. Downum and reprinted by permission of the University of Arizona Press.)

and discontinuous, displaying great variety in size and shape and having trails that link them together (Downum 1993: 61, fig. 4.7). The largest ones, however, measure more than 60 m long, with risers 4 m high, and 12 to 18 m apart. Details of the risers themselves are not known as all are highly deteriorated and slumping, but it is known they were made of unworked locally available volcanic rocks and were very thick, perhaps as much as 3 m across the top and 5 to 6 m at the base. Given that the riser of each terrace was lower than the base of the riser immediately uphill, the fill, which was nearly level, did not extend upslope as far as the foot of each riser. A strip approximately 3 to 4 m wide at the foot of each riser was not suitable for cultivation due to thin soils (Fig. 8.27). The fill behind the risers is a combination of boulders, cobbles, gravel, with upper horizons of silty clay loam to silt loam, and ashy cultural deposits (Downum 1993: 79). The presence of the latter has been used to suggest that farmers might have intentionally amended the soil to enhance its fertility (Downum 1993: 81).

The northeast slope of Cerro Prieto, as was the case at Mesa Verde, is more mesic than other slopes. Because exposure to solar radiation and

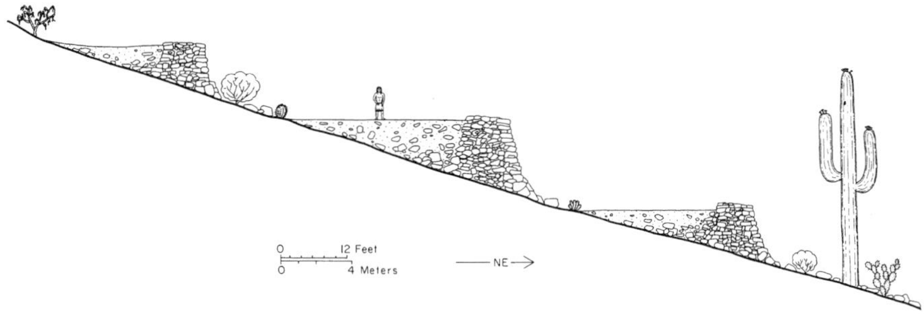

Fig. 8.27. Hypothesized reconstruction of large agricultural terraces on Cerro Prieto. Appeared originally in Downum (1993: 80, fig. 4.27). (Provided courtesy of Christian E. Downum and reprinted by permission of the University of Arizona Press.)

evaporation are minimal here, soil moisture derived from the winter rains is greater than on slopes with other aspects, and it is retained longer (Shreve 1934: 149). Also contributing the agricultural ecology of this, and undoubtedly other *cerros de trincheras*, are factors affecting temperature. Significant thermal inversions occur in the Tucson area. Because of cold-air drainage, winter freezes are fewer, minimum temperatures are higher and less prolonged, and the frost-free period is much longer on the hill sides than on the valley floors (Hastings and Turner 1965: 17). Night-time radiation of heat absorbed by the dark volcanic rocks during the daytime undoubtedly contributes to warming as well.

In sum, recent work at Cerro Prieto, combined with other archaeological investigations in the Tucson area and climatological data, indicate that the hillside environments of *cerros de trincheras* are not as hostile to agriculture as was long thought. Exactly what was grown on these terraces remains to be determined. It has been suggested that the usual maize, beans, and squash trilogy is most likely, and that certain specialty crops such as cotton and chilies might be possible, as well as certain cultivated 'wild' plants such as amaranth on the terraces and agaves on the risers (Downum, Fish, and Fish 1994: 283). Much additional research awaits to be done on the agricultural nature of these and other terraces on *cerros de trincheras* (P. Fish, S. Fish, and Downum 1992).

Overview

Terracing has always been a multifaceted phenomenon among aboriginal farmers of North America. Terraces have appeared in a myriad of forms, and they have served an almost unlimited variety of functions. The only thing that gives this aspect of agriculture any semblance of order is its regionalization. Terracing is entirely a southwestern practice.

In terms of data dealing with terracing, the overwhelming amount is provided by archaeologists. This is probably the case for two related reasons. First, some forms of terracing leave physical remains that are easily identifiable. Second, information about other aspects of prehistoric life are so fugitive that archaeologists simply cannot pass up the opportunity to study relicts so conspicuous as terraces. The same can probably be said for ethnographers. The simple fact that a practice foreign to the rest of North America could still be observed well into the 19th and 20th centuries, was something that cultural observers felt obligated to record. The early explorers however, perhaps because they came from foreign lands and were less provincial than more recent scholars, seem not to have to been so enamored of native terracing. Of course, they did not spend much time in some of those places where terraces built by earlier inhabitants remained conspicuous on the landscape, but they did encounter people who were building and using smaller terraces.

Although terracing was not well-reported by the early European explorers, it was in all likelihood being practiced in the Southwest at the time of contact. The persistence of terracing at the Hopi pueblos, and among other peoples, in concert with the abundant archaeological data, argue that the temporal lacunae in the documentary evidence does not adequately reflect conditions at that time.

The archaeological and ethnographic evidence of terracing is so abundant that it might leave one thinking that the practice of altering slope lands ensured a secure food source and hence existence. Nothing could be further from the truth. One thing should be clear from all the evidence. People who built terraces did so at great expense and for relatively little gain. Some of the very large and elaborate terraces resulted in the creation of paltry amounts of land. Furthermore, in no case was it exceptionally good farmland, and construction costs were not amortized over generations, as has been the case for terraces elsewhere in the world (Hanks 1972). Terraces in all parts of the Southwest, regardless of when they were first built, underwent constant maintenance and additions.

In most cases, terracing was something done to the most marginal of lands. Even well-intentioned terrace builders must have known they were fighting an uphill (no pun intended) battle. This is perhaps reflected in no better place than the archaeological literature where there have been several studies aimed directly at modeling risks taken by prehistoric southwestern farmers (see e.g., Hunter-Anderson 1986; Hegmon 1989; Wills 1995). Of course, just because they lived in the Southwest and farmed does not mean that they practiced terracing. Or does it?

One of the more curious aspects of terracing in the region is that it is found in proximity to some of the very earliest of agricultural sites—cave dwellings.

Also, terracing, as loosely defined and as broadly discussed here, is something that may well be associated with every form of agriculture, including some of the most rudimentary. Could it be that agriculture in the Southwest is intimately linked with terracing, and that the two phenomena have, since their very beginnings, been inextricably interwoven? Perhaps.

9

Water Harvesting

Native North American farmers certainly developed a number of techniques for conserving scarce moisture on fields. In some cases, however, conservation measures simply were not enough. All too often, either direct precipitation was insufficient, settings were not right for dry farming, or runoff was too limited even if terracing could have been employed. In order to ensure the successful production of a crop, therefore, farmers sometimes had to apply water artificially. Not surprisingly, these farmers lived in the Greater Southwest.

There are many ways to irrigate crops or the lands on which they are grown. As discussed in Chapter 4, for example, gardens were often irrigated manually with water carried in ceramic jars. This practice was, however, exceptionally labor-intensive and therefore not practical for larger fields where staples were cultivated. A much more economical way (in social and labor, if not monetary terms) of watering large tracts was by practicing 'water harvesting' (Lawton and Wilke 1979: 4). Unlike terracing which involves modifiying only the planting surfaces themselves, water harvesting involves altering the drainage basins upslope or upstream of fields to increase runoff and divert it to crops (National Academy of Science 1974: 9–22). For example, in the Negev Desert of what is today southern Israel, entire watersheds were cleared of moisture-retaining vegetation and rocks for the purposes of promoting runoff, and walls were built to direct the sheetflow to fields during Nabatean times (Evanari, Shanan, and Tadmor 1971). Aboriginal North Americans did not create water harvesting landscapes as elaborate and as well-known as those of the Negev, at least as far as currently available evidence indicates, but they did develop some interesting techniques for harvesting water, often, but not always, involving weirs and/or canals (Nabhan 1979*b*: 246, 251, fig. 4, 1981: 53).

In terms of technological complexity and development (Woodbury 1962: 302), water harvesting is transitional between runoff farming and canal irrigation, which involves diversion of water from permanent or perennial sources (see Chapter 10). Canals typically involve excavation along a very slight gradient so that water flows by the force of gravity from an upslope source to a downslope field. In some cases they may be lined to prevent water loss due to seepage, but most frequently canals are simply unlined with earthen sides and bottoms. In those cases where water harvesting involves canals, water is taken from ephemeral, intermittent, or seasonal streams, rather than constantly flowing rivers or springs (Masse 1980: 198, 201). In those cases where canals are not involved, weirs made of stakes and brush, rocks (Fig. 9.1), or a combination of the two, and not unlike some simple terraces (see Chapter 8), are placed diagonally across slopes or normally dry streambeds to divert runoff onto fields (Dobyns 1974: 325). Not infrequently, both weirs and canals are used together (Fig. 9.2).

As in the case of terracing, evidence for water harvesting is overwhelmingly ethnographic and archaeological. While it is quite detailed from some locales, it is sketchy from others. Water harvesting fields are notorious for being some of the most difficult to recognize. This is because of the small amount of rock used in the construction and because features are frequently destroyed by the very same forces of nature they are intended to harness (Stewart and Donnelly 1943*b*: 137). Weirs get washed away by runoff and canals can fill with sediment almost immediately (Fig. 9.3). In many cases, features built to harvest runoff are evident on the landscape for very brief periods of time.

Documentary Evidence

Early Europeans tended to either overlook or not report native water harvesting. Their failure in this regard probably had little to do with this type of agriculture being spatially limited to remote locales, and, hence, not seen.

Fig. 9.1. A rock diversion weir placed diagonally and partially across an arroyo streambed in Sonora, Mexico. (Photograph by author.)

Fig. 9.2. An earthen and stake and brush weir (right) which diverts periodic stream flow out of the arroyo channel (fore) and into the mouth of a canal (left) in Sonora, Mexico. (Photograph by author.)

Fig. 9.3. The same location as seen in Fig. 9.2. All evidence of the weir and the canal was obliterated with a single runoff event. (Photograph by author.)

Indeed, some areas from which there exists abundant ethnographic and archaeological evidence of water harvesting are the exact places where explorers and missionaries spent extended periods of time and reported other types of information, including much on agriculture. In all likelihood, therefore, early Europeans probably *looked at* water harvesting, but they never *saw* it. They failed to describe it in writing because they did not recognize it for what it was.

There exists only one case in which an early visitor actually described water harvesting (Fig. 9.4). Writing about the Névome of Sonora early in the 17th century, Pérez de Ribas reported: '*en algunos puestos acomodados, hacían sementeras de riego, sacando por acequias el agua de su arroyo para regarlas*' (1944, vol. 2: 150) or 'in some appropriate places, they plant irrigated fields, taking water out of the arroyo in canals to irrigate them' (Dobyns 1988: 50; see also Pérez de Ribas 1968: 137). That Pérez de Ribas mentioned both canals and arroyos together confirms that he witnessed water harvesting.

Ethnographic Evidence

Neighbors of the northern kin of whom Pérez de Ribas wrote, the O'odham, are undoubtedly the best-known water harvesters in North America. Indeed,

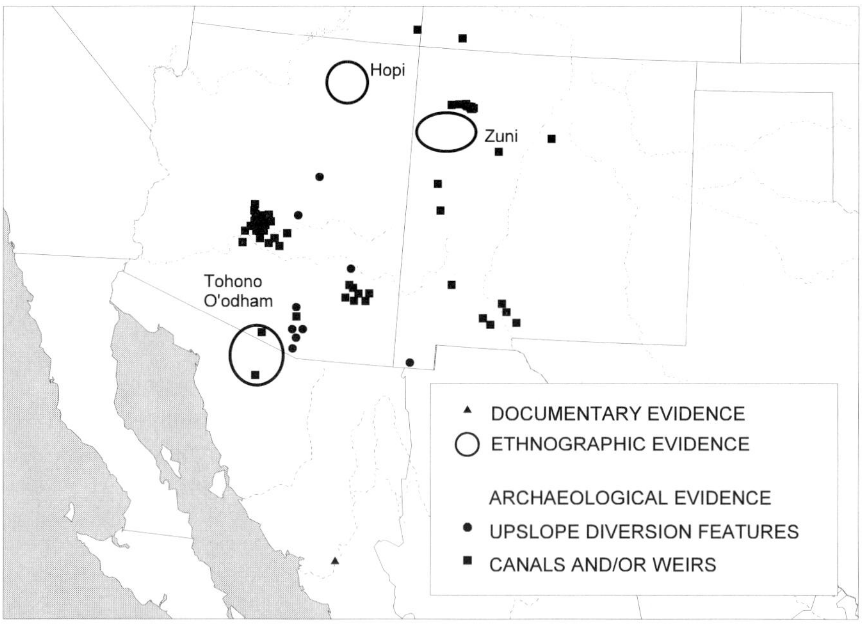

Fig. 9.4. Locations cited in the text for which there is evidence of water harvesting.

with the exception of the Hopi and the Zuni, they are the only known people who continue to rely on the practice (Fig. 9.4). The degree to which these people continue to both use and rely on water harvesting is as much a function of need (Bentley 1987) as it is available technology (Chiarella and Beck 1975). These people harvest water four different ways, one of which might be considered *passive*, and the other three *active*.

The passive system is, in some ways, the most fascinating of the four. Tohono O'odham continue to locate some of their fields in the far downstream ends of arroyos. These ephemeral streams have numerous tributaries in the mountains and an equal number of distributaries in the valleys (see e.g., Waters 1987: 209, fig. 14.2). The tributaries collect runoff from a large area—in some cases more than 1,200 sq km (Bahr 1983: 189)—and concentrate it into one channel which then bifurcates numerous times as the gradient becomes less steep. In order to ensure that runoff is concentrated in the one distributary which contains fields, and is not diffused into others, farmers build temporary dams of either loosely piled rock or stakes and brush (not unlike the feature shown in Fig. 8.3) across the mouths of all other distributaries (Fontana 1983: 131; Nabhan and Teiwes 1983: 16–17). These features are upstream of the fields and, therefore, are watershed-modification structures. In terms of harvesting water, this system is considered passive as it keeps water *from* flowing in directions where it would go naturally and where it is not needed. The other three systems are considered active as they involve the diversion of water *to* places it would not flow otherwise, and where it is needed.

The least complex of the active Tohono O'odham water harvesting systems does not involve arroyos, but rather the simple collection of slope runoff. According to one early 20th-century observer (Clotts 1917: 41), farmers built long dikes, which converged 'somewhat like the sides of a funnel'. The height, width, and materials used in construction were not described, but in all likelihood the features were no more than a meter or two wide, a half meter high, and made of locally available earth and rock. They were, however, said to be several kilometers in length, and oriented traversely across the valleys. Runoff was thereby collected from hundreds of hectares and diverted to pockets of only a few hectares where the soil was suitable for farming. A few writers (Morrisey 1949: 104; Nabhan 1986*b*: 67) have observed areas cleared of vegetation to facilitate runoff, but provide little in the way of details.

The other two active water harvesting systems used by the Tohono O'odham are much more common than either of the two just discussed, both in southern Arizona and among other groups in the Southwest. As in the case of the passive system, they involve seasonal flow from ephemeral streams. As in the case of the second system, they involve diversion structures. The third water harvesting system is very much like the second in that earthen dikes are utilized (Dobyns 1974: 325). In this case, however, dikes impound seasonal stream flow, raise the level of this water to a height greater than the channel banks, and divert it out of the channels and onto the fields (Gasser 1990: 4.12–4.15, figs 4.6–4.9; see also National Academy of Science 1974: 27, fig. 17).

The fourth, and final, water harvesting system relied on by the Tohono O'odham involves canals. The first ethnnographer to describe one of these systems was none other than the famed archaeologist Bandelier (1890: 254). He noted that near the base of steep mountains, native farmers dug small canals from arroyos to fields in order to lead the torrents of water rushing downstream after each shower to their crops. Bandelier (1892: 411, 446) made similar observations of the Maricopa. His descriptions were confirmed two decades later by Lumholtz (1912: 330) who provided a nearly identical account. Neither Bandelier nor Lumholtz mentioned diversion structures, but later observers did. One (Clark 1928: 246) reported the use of temporary dams or weirs across washes to divert water, but made no mention of the materials used in construction. Another (Dobyns 1951: 26), described these diversion structures as low earthen dikes, and yet another (Masse 1980: 201, 202, fig. 88) reported them to be stake and brush features. Variations in materials used in the construction of weirs, and indeed, the presence or absence of such features, is most likely a function of local settings. Some water harvesting systems with canals may not have needed weirs whereas others did. Similarly, where rock was available and brush in short supply, the former was probably used, and vice versa. Of the 18 Tohono O'odham fields subjected to one recent detailed investigation, four had weirs only, six had canals only, and five had both weirs and canals to harvest runoff (Nabhan 1983: 52–4, table 3.3, 1986*b*: 67–9, table 3).

In addition to using weirs and canals to capture and divert seasonal stream flow, some Tohono O'odam farmers also use one other type of water harvesting feature, reservoirs (Nabhan and Teiwes 1983: 16, 20). Some reservoirs are located in the arroyo channels above the canal intakes (Nabhan 1979*b*: 248, fig. 3), whereas others are located along canals between the arroyos and the fields. The dams which create the reservoirs are earthen embankments which appear from the air in the shape of a U. Known locally as *charcos*, these features store surplus harvested water to be used later in times of need. Reservoirs are small, rarely covering more than 1 ha, and the embankments are low, between 1 and 2 m high. Siltation is a problem, so most reservoirs are not only very shallow, but become increasingly more shallow over time.

Determining the ages and functions of reservoirs are not as simple and straightforward as they might seem. Some are clearly agriculturally related and date to the 19th century and earlier. Others seem to have been built in the 20th century and serve as tanks for livestock, and yet others seem to be prehistoric but were used for domestic water supplies as well as for agriculture (Crown 1987). Complicating matters further, many *charcos*, regardless of their age and function, appear very much like prehistoric ballcourts (Wilcox

1985: 650). Lesson to be learned? One must be very cautious when interpreting function on the basis of form (Ezell 1985).

The Hopi, in addition to being masters of dry farming and terracing, are also exceptional water harvesters, although remarkably little attention has been drawn to this facet of their agronomic expertise. In his classic study, Hack (1942: 29) said only that in some places along shallow arroyos water 'can be easily' diverted by earthen dikes. He went on to say that fields so watered are 'in reality' variations of arroyo mouth fields. Other writers, however, have provided more information than Hack, and when the pieces of insight each of them provides are examined together, it is clear that Hack uncharacteristically discounted this aspect of Hopi farming. One writer (Clark 1928: 235) reported that Hopi farmers go out when it is raining and construct small dikes that divert water into canals that terminate at fields. Another (G. Stewart 1940*b*: 329), made a similar observation but added that arroyos from which water is harvested must have very shallow channels, that the dikes are made of both brush and earth, and that they help impede erosion and gullying (see also Titiev 1944: 182). Yet another writer (G. Page n.d.: 61–2), reported that the diversion dikes are 45 cm high, 'as long as the diversion area demands', and have to be rebuilt after each runoff event. Although evidence is fragmentary, when combined, it is sufficiently compelling that Hopi farmers deserve more praise than they have received for their water harvesting knowhow.

One group that needs no more praise for their ability to harvest water is the Zuni. Much has been written about Zuni water harvesting, and it all stems from the writings of Cushing and one extraordinary illustration that he provided (Fig. 9.5). Water harvesting at Zuni involves first the selection of a gently sloping site which is bisected by a shallow arroyo. Then, at the far upstream end of the field, a dike is built across the channel. Construction of the dike involves the cutting of forked cedar branches which are then firmly embedded into the streambed in a line across its course, and extending a considerable distance beyond either bank. Against this row of stakes, Zuni farmers stack brush, rocks, and earth (Cushing 1974: 157). Similarly constructed, but smaller features are then built at varying intervals across the field surface, often forming grids (Cushing 1974: 160). Water flowing through the arroyo is diverted onto the field by the first diversion dike and is distributed evenly over the planting surface by the others. This water can be the result of springtime snow melt (Perramond 1994: 30) or summertime storms. In the latter case, it comes in foam-capped and sediment-laden torrents. By the time these summer floods recede, usually within a few hours, water has been carried to every portion of the field and it has deposited a fine layer of loam on top of the otherwise sandy soils (Cushing 1974: 164). Excess water flows off the field and back into the arroyo channel, with sediment being deposited there as well, thereby constituting an effective method of gully control (G. Stewart 1940*b*: 337). A second cross-channel dike located some distance downstream

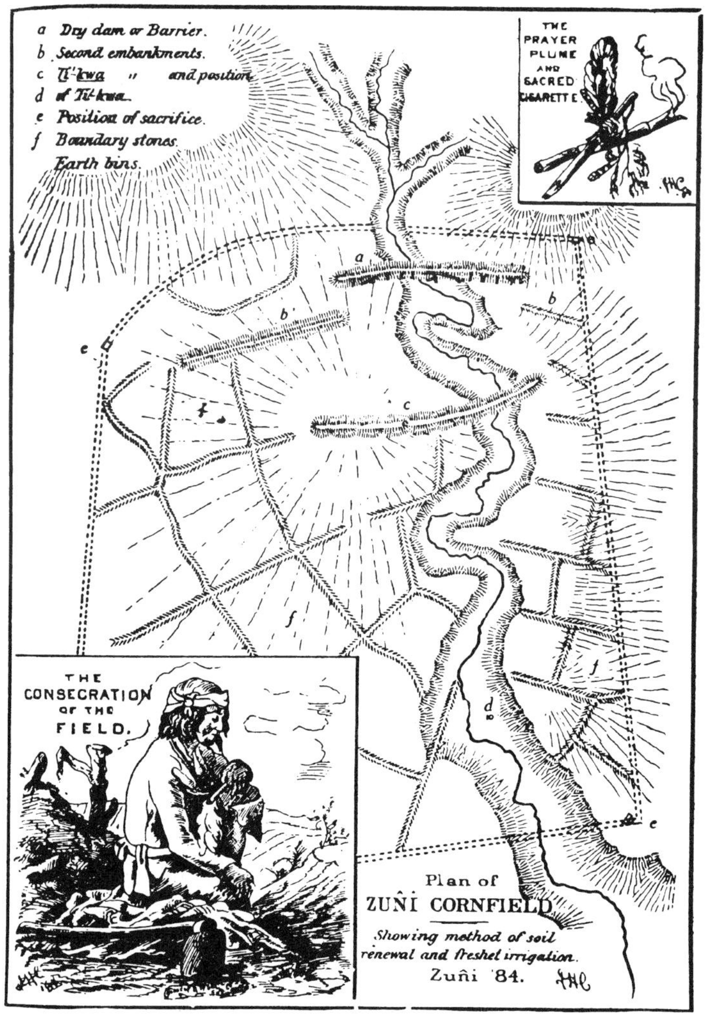

Fig. 9.5. Schematic diagram of a Zuni water harvesting system. Appeared originally in Cushing (1974: plate II; see also Cushing 1979: 265, fig. 59).

of the first one impounds water and collects sediment thereby contributing further to the water and erosion control (Stewart and Donnelly 1943*b*: 138). Accumulated sediment has recently been shown to be important in keeping fields from denegrating even after decades of continuous cultivation (Norton 1996: 55).

More than 10,000 ha of water harvesting fields may have been cultivated annually at Zuni in years past (Rhode 1995: 90). Such fields are found all over the reservation (Ferguson and Hart 1985: 13), but mainly on south-facing terraces above the valley floor where cold-air drainage was avoided (Brandt 1995: 296). They also are found where specific soils are available (Pawluk 1995). Fields vary in size, configuration, and construction details depending on environmental factors, readily available materials, and builder preferences. In one locale, a V-shaped device made of earth-covered stakes and brush and with the apex pointing upstream was built 50 m upstream of the diversion structure. This feature was intended to reduce the velocity and impact of runoff (Ford 1985*d*: 34). Near the village of Nutria, farmers used logs to dam an arroyo and divert runoff. This dam created a reservoir which retained water that was released later. Also, these farmers excavated a canal in order to transport water to their fields (Ferguson 1985: 94–6). A canal was also dug to carry water from an arroyo to a field near the village of Pescado, but in this case diversion was achieved by partially damming the stream channel with rocks (Ford 1985*d*: 37–8, fig. 3). In one case, Zuni farmers near the village of Bosson did not build dikes upstream of their fields, but, instead, built dams in the stream channels immediately adjacent to their fields. These dams impounded water and raised its level such that it flooded the fields laterally (Ford 1985*d*: 39). Although often characterized as monolithic—*à la* Cushing—water harvesting is a practice with numerous variations. Nowhere is its singleness more widely cited, yet differences more evident, than at Zuni.

Archaeological Evidence

There exists much in the way of remnant features on the southwestern landscape to suggest that water harvesting was widespread in pre-European times. Indeed, whereas there is documentary evidence from only a single locale, and ethnographic evidence from three areas, archaeological evidence comes from literally every corner of the region (Fig. 9.4). These data exist in two forms; sites which show some evidence of slopes having been modified upslope of fields, and sites with vestiges of canals that carried seasonal water diverted from the channels of arroyos. The former are far fewer than the latter.

When it came to modifying drainage basins, one practice stands out as not having been too important prehistorically in the Southwest, clearing slopes of rocks and vegetation to increase runoff. Archaeologists working in northern Nevada did find evidence of rocks being removed for such a purpose, but here it was to increase potable water supplies and not for agriculture (Irwin-Williams *et al.* n.d.). Another archaeologist, this one working in northern New Mexico, found evidence in one locale that trees were probably cleared to facilitate runoff to fields below (Anschuetz 1995: 33), but this finding stands out as being singular. At least two studies, both conducted in the Tucson

Basin, looked for, but found no evidence that clearing for such purposes was either known or practiced (Masse 1979: 172; S. Fish, P. Fish, and Madsen 1990: 203). Exactly why this was the case remains unknown. However, it might well have been because runoff was sufficiently great under natural conditions. This explanation is, in part, substantiated by the paucity of evidence for the diversion of slope runoff.

Archaeological evidence of diversion of runoff from slopes to fields below is reported from remarkably few places, all in southern and central Arizona. Four sites in the Tucson Basin have remains of such features; 'diversion walls', and 'channeling alignments' have been found at Marana (S. Fish, P. Fish, and Madsen 1985: 79, 83, fig. 3), 'stone walls to divert runoff' have been identified at Los Robles (Downum 1993: 38–9), 'channeling borders' exist at Tumamoc Hill (Masse 1979: 170), and 'boulder detention dams' have been recorded in the Cañada del Oro (G. Stewart 1939: 217). On the basis of its name alone, this last feature might not seem to be for water harvesting, but the accompanying description is much like that provided by Cushing for the dikes he described at Zuni (see Fig. 9.5).

Somewhat further north in Arizona, 'diversion dams' were identified but not described from a few sites in the Gila Valley (Debowski *et al.* 1976: 90, 108), and a 'rock diversion' was found on a slope above a small terraced field in Tonto Basin (Shelley and Ciolek-Torello 1994: 224, fig. 61, 225, 231). Finally, features identified as 'diversion walls', 'channeling walls', and 'brush structures', were seen to have been of local importance in harvesting water in the Vosberg Locality near the Tonto Rim (Rodgers n.d.).

Archaeological evidence for the use of weirs that diverted water out of ephemeral streams, canals that transported runoff to fields, or both is much more common and widespread in the Southwest than is evidence of harvesting slope wash. Three areas do, however, stand out as significant—the Hohokam Periphery, Mesa Verde, and Chaco Canyon.

The Hohokam are most well-known among ancient North American people for their development of and reliance on canal irrigation. Centered in southern and central Arizona, they irrigated vast tracts with water diverted from the perennial Gila and Salt Rivers (Chapter 10). They also cultivated small, individual parcels with water harvested from ephemeral streams surrounding the Hohokam Core area (Dove 1985: 66–7; Doyel and Elson 1985; S. Fish and Nabhan 1991: 32–5, fig. 2.2). Some of the archaeological evidence is quite old, indicating that the practice has long been recognized as important. Some of the evidence is also quite young, suggesting that the total importance of water harvesting is not yet known.

The oldest and most recently discovered evidence comes from Hohokam Southern Periphery, particularly southeastern Arizona. Writing more than a century ago, Bandelier (1892: 410, 414) found remnants of two water harvesting canals which originated at arroyos on the northern flanks of Mount Graham in the Pinaleños (see also Fewkes 1904: 178). One was 350 m long, varied in width from 1.25 to 3 m, and in depth from 15 to 20 cm. Forty years later, Sauer and Brand (1930: 421, 428, 429, 437) found more such canals on the northern, eastern, and southern flanks of this range. Futhermore, they, and a contemporary researcher, reported evidence of ancient canals which carried water from seasonally flowing streams on the eastern and western flanks of the Chiricahua Mountains (Caywood 1933: 22, 39; Sauer and Brand 1930: 434, 440). Exactly how many such canal systems there are in the area remain unknown, but James A. Neely (n.d.) and one of his students (Rinker 1998) found remnants of at least six in the vicinity of Lefthand Canyon, one of hundrēds of tributaries which flow into the Gila River. These canals are bordered on either side by large rocks set 1 to 2 m apart (Fig. 9.6). The longest canal was traceable for 4.6 km (Neely n.d.: 22). Gradients were quite steep, upward of 2.6 percent (Rinker 1998: 27). In some cases, canals bifurcated with certain branches being older than others (Rinker 1998: 17). Toward their downstream/downslope ends, the canals bifurcate more, forming networks of branch canals. These branches contain a number of openings, or gates, each of which allowed water to flow onto a small rock-bordered and rock-faced plot measuring no larger than 6 m square (Fig. 9.7). Once on a field, water flowed from one terrace to the next below, often assisted by gaps in the rocks

Fig. 9.6. Upstream view of the surface remains of a canal which was part of a water harvesting system on the northern flanks of the Pinaleño Mountains. (Photograph by author.)

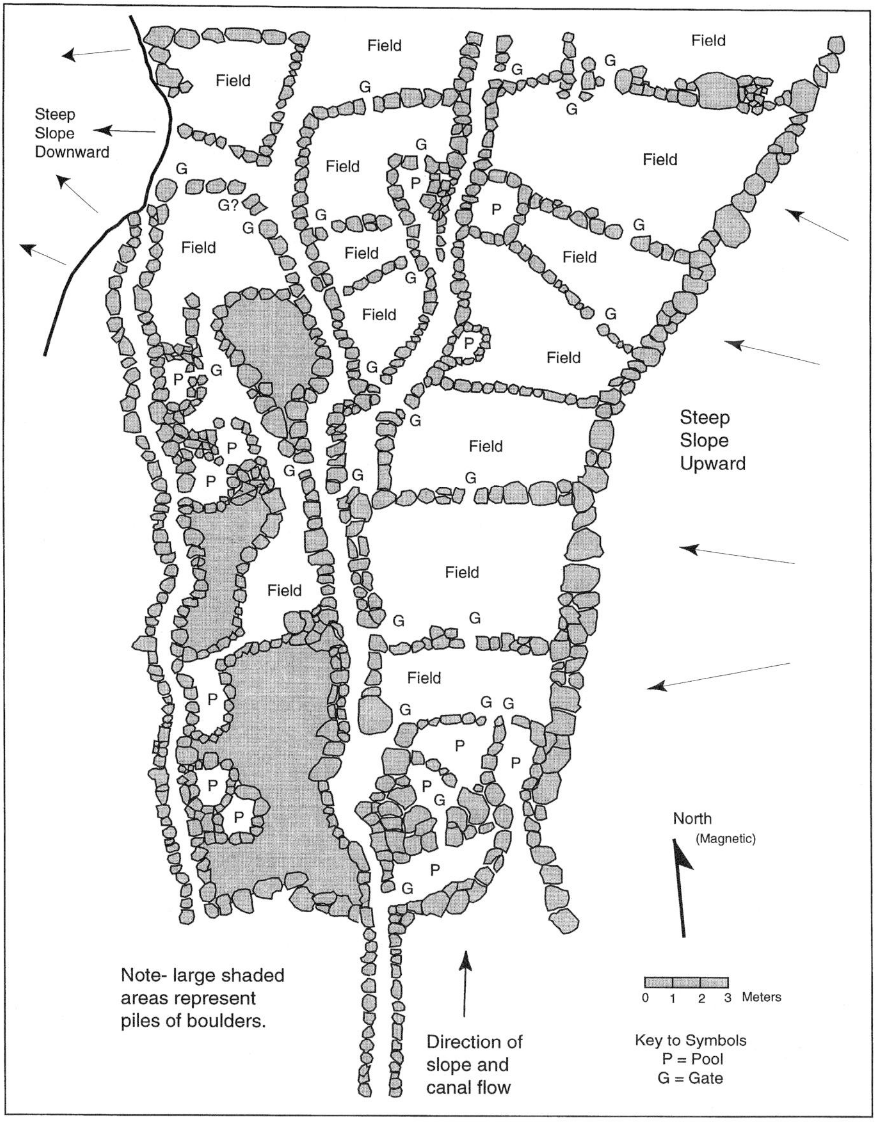

Fig. 9.7. Schematic diagram of rock-bordered canal and rock-faced terraces of water harvesting System 1 overlooking Lefthand Canyon, Arizona. Appeared originally in Neely (n.d.: fig. 4). (Provided courtesy, and published by permission of, James A. Neely.)

Fig. 9.8. Upstream view of a series of low terraces comprising part of water harvesting system shown in Figs 9.6 and 9.7. (Photograph by author.)

forming the risers (Fig. 9.8). Located in several places throughout the system are small circles of rock measuring 1 to 2 m in diameter. The actual function of these features remains unknown, but they appear to have been pools. Perhaps as much as 18 ha were cultivated in Lefthand Canyon (Rinker 1998: 45).

The water harvesting systems reported from around the Pinaleño Mountains all are located on bajadas or Pleistocene terraces. Further to the west, in the Tucson Basin a similar system was found toward the distal end of an alluvial fan. Details are not as abundant as those provided by Neely, but an unlined canal of unknown length, 1 m wide, 50 cm deep, and parabolic in cross-sectional shape, appears to have been rebuilt at least once because of sedimentation (S. Fish, P. Fish, and Madsen 1985: 81, fig. 3).

At another site, actually in Tucson, six segments of prehistoric canals were discovered (Mabry and Holmlund 1998). Three have similar alignments and profiles and were thought to be segments of one canal that measured 110 m long, was U-shaped in cross-section, 2.5 m wide and 1 m deep, narrowing to 1.6 m wide and 50 cm deep where it bifurcated into smaller canals. The three other canal segments ranged in width from 75 cm to 1.3 m and were 25 cm deep. Given the direction of flow, these canals were all interpreted as having carried seasonal runoff from surrounding slopes to fields below. The most important aspect of these, canals, however, is their age. They have been dated to as early as 800 BC (Mabry and Holmlund 1998: 284), making them some of the oldest known canals in North America, and lending credence to the notion that canal irrigation was developed independently in the Southwest and not a function of diffusion from Mesoamerica (Doolittle 1990*b*: 90, 1993: 147–8).

Further yet to the west, a prehistoric water harvesting canal of unreported size and morphology is known from the Tohono O'odham Reservation. This one is located in the foothills of the Baboquivari Mountains and dates to approximately AD 1400 (Joseph, Spicer, and Chesky 1949: 15). Finally, to the south, remnants of an ancient canal have been identified on the lower terrace of an arroyo in far northern Sonora. All that is known of this feature is that it was visible for only 200 m (Bowen n.d.).

Archaeological evidence of canals used for harvesting water is more abundant from the Hohokam Northern Periphery than it is from the southern counterpart, and comes principally from two areas, Cave Creek and the New River valleys (Elson, Doyel, and Hoffman 1985). In both areas fields were on low terraces overlooking the floodplains. Along Cave Creek, six canal sites were located on the west side of the stream channel (Henderson and Rodgers 1979: 14, fig. 3; Kemrer, Schultz, and Dodge 1972: 32, 101, fig. 16). Beginning near the southeast side of Fort Mountain, one (AZ T:8:6 ASM; 31 ASU) was reported more than thirty years before it was studied in detail (Rodgers 1977: 25–7, figs 5 and 6; G. Stewart 1940*a*: 219). Water was diverted out of Cave Creek and into the canal by means of a dike constructed exclusively of basalt boulders. The segment that actually crossed the stream channel diagonally long since eroded away, however, a section measuring 320 m long was found paralleling the canal on its downslope side as it circled Fort Mountain (Rodgers 1977: 31–4, figs 8 and 9). The dike/canal embankment measured 80 cm high, and 1.6 m wide (Rodgers 1978: 154). The canal itself was 750 m long, 6 m wide, between 50 and 70 cm deep, and with a 0.6 percent slope (Rodgers 1977: 28–31). In certain places rows of rocks reminiscent of check dams were found across the canal (Rodgers 1977: 29, 30, fig. 7.2). These features slowed the flow of water as is evident by the deposition of fine-grained materials (Henderson and Rodgers 1979: 33–5, figs 8 and 9). The canal bifurcated at least once (Kemrer, Schultz, and Dodge 1972: 85, 87, fig. 9, 105, fig. 27). Indeed, one branch is suspected as being earlier than the other, and perhaps part of a smaller system that was later enlarged (Rodgers 1977: 35–7). The total area watered by these canals is approximately 15 ha (Stewart 1939: 114) and lies south of the mountain (Holiday 1974: 17, 42). The field surface was characterized by a number of rock alignments which distributed water, facilitated the depostion of silt, and impeded erosion. The system dates to *c.* AD 1050 (Rodgers 1977: 34–5, 1978: 154).

The second canal site (AZ T:8:9 ASM; 35 ASU) had its head more than 1 km north of Fort Mountain. Although only fragments remained until recently, it appeared to be more than 1 km long, up to 1.5 m wide in places, and had cobble and earthen embankments (Rodgers 1977: 21, fig. 4, 66–7, 1978: 154). It was shallow, 25–27 cm deep, never narrower than 70 cm wide, parabolic in cross-sectional shape, unlined (L. Smith 1978: 23). The canal ran around the northern and western flanks of the mountain and irrigated a plot of undetermined size adjacent to and northwest of the field watered by the other canal (Kemrer, Schultz, and Dodge 1972: 88, 101, fig. 16). This field also included a number of rock water control features to aid in the distribution of water (Holiday 1974: 17, 19). The system was used between AD 1000 and AD 1100.

The third canal site associated with water harvesting in the Cave Creek Valley (AZ T:8:96 ASU) was found 4 km upstream from the second. It involved two rock weirs, one which blocked the mouth of one distributary funneling all the runoff into one channel much like the passive system of the present-day Tohono O'odham, and one which diverted water from that channel into the canal. The canal itself was 250 m long, 2 m wide, 50 cm deep, and lined by a series of rock borders and filled with fine-grained silts. As in the case of the first system, this one may have gone through two periods of construction. The fields watered by this canal contained the usual array of rock water spreaders. Unlike other canal and weir water harvesting systems though, this one also involved a rock wall 172 m long, one course high, and 3–5 courses wide, which paralleled the canal 50 m away along the side of a hill. This wall is thought to have trapped slope wash, filtered out silts, and directed runoff to the field. The system was dated *c.* AD 850–1050 (Henderson and Rodgers 1979: 68–9, fig. 20, 80–90, plates 18 and 19, fig. 24, table 12).

The final three Cave Creek canal sites are further north near its headwaters in an area known as Paradise Valley. Details about these water harvesting systems are not as abundant as the three downstream but some facts are known. One of the canals was approximately 1.6 km long. All three systems had fields marked by rows of rocks which were oriented at right angles to the flow. And, two of the systems were quite large, one had 20 ha of fields, the other encompassed 50 ha. The smaller third system involved a mere 4 ha of planting surface. All three postdate AD 1000 (Holiday 1974: 20–4, figs 3–5, table II, 41).

A number of impressive prehistoric water harvesting systems are also found west of Cave Creek in the New River Valley. Precipitating much of the work that has contributed to our understanding of these sites was Frank Midvale, who made an impressive photographic record of canals in Arizona from 1920 until 1971, and which is currently housed in the Department of Anthropology, Arizona State University, Tempe. Midvale did not publish much in the way of text, but he did take notes and make written comments. He also crafted some most informative maps. His work in the New River Valley was limited largely to that area south of the New River Gap, a pass between West Wing and East Wing mountains (Fig. 9.9). On his map of the

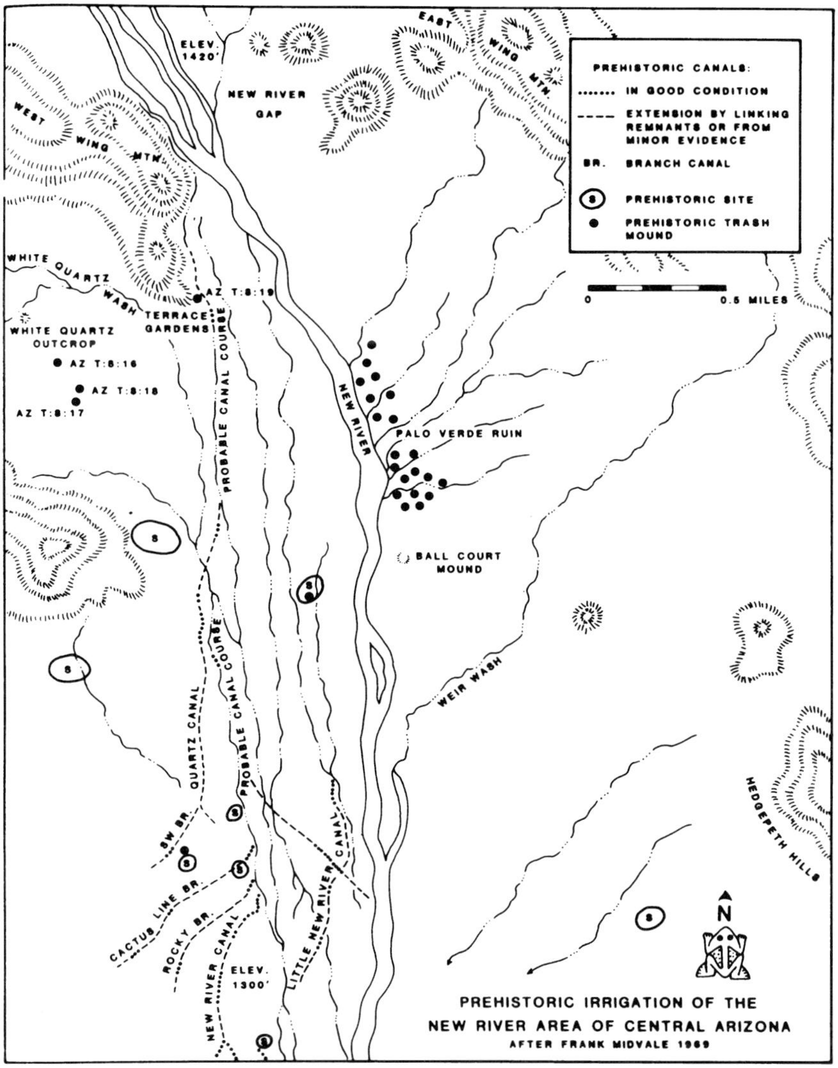

Fig. 9.9. Prehistoric canals south of the New River Gap, Arizona. This map is based on maps drawn by Frank Midvale. Appeared originally in Doyel (1985*b*: 695, fig. 15.3). (Reprinted by permission of Soil Systems, Inc., Phoenix, Arizona.)

area, Midvale (1969) provided an interesting description and reflection on both the environments and the ancient water harvesting features:

My first knowledge of these canals came from the late Isiac Gaylord a highway crew foreman who sketched out these canals in 1925. Some years later with the help of O. R. Higgins who was familiar with ancient canals and knew the area, most of these remains were traced out. The land west of New River at this time was in native desert clean and unspoiled. Surface features were plainly seen and many small ancient sites lay undisturbed. The canal segments found proved to be small and very widely scattered. Attempts to outline a unified system were futile. The sandy nature of the area lying wholly within the river flood plain left little to survive. Canal remains were mostly lines of river rocks left on high spots escaping the floods. New River is an erratic stream having no storage dams or flood control devices and recent floods have probably exceeded those of the prehistoric past.

In my earlier maps attempts were made to show the probable extensions of canals linked into a workable system. Although revised several times the map appeared sketchy and disjointed. During the last 3 years this area has been revisited, restudied and completely reworked.

As a result this edition of the map is conservative in treatment and scope. Only the canal remains that can be well substantiated are shown. A few in question are labelled as such. All others have been omitted. Dotted lines show linked extensions based on known evidence.

Four main canals appeared to be the nucleus of 2 large and 1 small irrigation systems. The total length of more than 30 canal segments proved to be less than 5000 feet. Although the total canal mileage is unknown, the map indicates that at least 8 miles were in use. Definite agricultural remains indicate probably twice that amount or roughly 15 miles. The entire area appears to have been abandoned by 1100 A. D. (Midvale 1969; see also Doyel 1985*b*: 693).

One canal seen and mapped but not studied in detail by Midvale was Feature 91 associated with the Terrace Garden Site, AZ T:8:19 ASM, (Doyel, Elson, and Hoffman 1985: 82, fig. 4.11). Because of its great length and prominence, this canal is often referred to as the New River Canal. This canal carried water diverted out of the stream channel very near the foot of West Wing Mountain, a few hundred meters north of the site. The diversion structure, if there was one, has not been visible in recent times. The main canal parallels the channel on the west and is more than 5 km long. Archaeological excavations revealed that in its upstream section the canal was 4.1 m wide and 82 cm deep. The cross-sectional shape is that of a stepped U, with two distinct lenses of fine sediment, suggesting that it may have been rebuilt at least once. The canal had a surprisingly steep gradient of 0.97 percent (Rodgers 1985: 283–91, figs 5.14 and 5.15). Midvale's map (see Fig. 9.9) indicates that this canal bifurcated perhaps as many as five times toward its downstream end, and that some of the branches may have bifurcated as well. The exact extent to which these canals watered farmland in prehistoric times remains unknown. Clearly, however, this was a large and complex water harvesting system.

Although relatively little is known of water harvesting systems south of the New River Gap, much is known of those north of it. Seven sites, individually and together, provide a wealth of information. Three kilometers up from the Gap is the Pinball site (AZ T:4:14 ASM), which is oriented north to south along a low terrace on the west side of the river (Doyel 1984: 43–7, figs 6 and

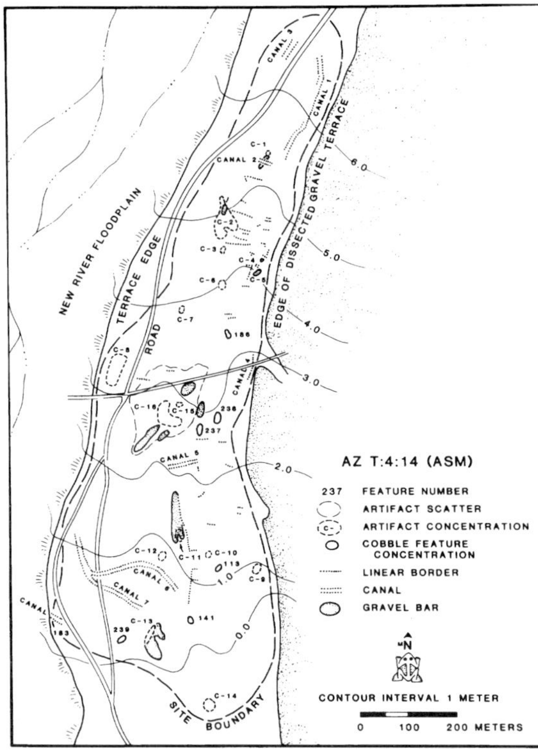

Fig. 9.10. Schematic diagram of features on site AZ T:4:14 (ASM), the Pinball site. Appeared originally in Doyel, Elson, and Hoffman (1985: 213, fig. 4.69). (Reprinted by permission of Soil Systems, Inc., Phoenix, Arizona.)

7). This is an impressive site complete with numerous features (Fig. 9.10). In contrast to the New River Canal, the intakes and upstream segments of what must be considered the main canal(s) for this site have long been obliterated. However, segments of at least eight canals remain visible in proximity to at least 60 other field features (Doyel 1986: 28, fig. 14).

Canal segments at this site are visible on the surface in the form parallel cobble alignments between 1 and 4 m apart (Doyel 1986: 29, fig. 16; Rodgers 1985: 278–83, figs 5.10 and 5.11, table 5.4). It appears that canal 1 (see Fig. 9.10) carried water from the New River or an eastern distributary of it. At 210 m long, 1 m deep, 2 m wide at the top, and trapezoidal in cross-sectional shape, it is the largest of the canals. Canal 3 might have been a branch of it. Canal 2, because it is oriented perpendicular to both canals 1 and 3, was probably a branch, sometimes known as a lateral canal. It was evident by only a 5 m long segment. Canal 4 appears to have been a downstream extension of canal 1, and canal 5 a downstream extension of it. That canal 5 curves back

Fig. 9.11. A row of rocks that probably anchored brush for the control of water across the field surface of the Pinball site. (Photograph by author.)

toward the main river channel suggests that it may have allowed surplus water to drain off the field. Indeed, it is the shallowest and narrowest of the canals studied. It was less than 30 cm deep and elliptical in cross-sectional shape (Rodgers 1985: 281–2).

Canal 6 appears to be a branch of canal 7 which had its intake off the New River just west of the site. These two canals, along with a small segment of an eighth which has an enigmatic location, watered the far southern end of the site (Rodgers 1985: 282–3, fig. 5.13). The field surfaces contain at least 60 rock alignments. Some of these were only a few meters long, involved crudely piled rocks, and probably functioned as water spreaders. Others involved lengthy single rows of rocks, and in all likelihood served to divert water across the site. Still others consisted of large rocks placed near, but not touching, each other (Fig. 9.11), which probably anchored brush which also helped distribute water (Rodgers 1985: 275–9, table 5.3, fig. 5.9). The site also contains some very large isolated rocks identified as field markers, and, along its margins, several piles of rocks. These were not for the cultivation of *agave*, as was the case elsewhere (see Chapter 3), but instead were refuse heaps resulting from field clearing (Doyel, Elson, and Hoffman 1985: 211–17, figs 4.67–4.73). The total cultivated field area measured 5.5 ha (Rodgers 1985: 275). However, given the number of canal segments evident, and that some of them

Fig. 9.12. Upstream view of remnants of the Morris Canal. Archaeologists Richard W. Effland, Jr., Margerie Green, Jerry B. Howard, Adrianne G. Rankin, and David E. Doyel (left to right) are walking in the canal. The rubble embankments are visible on either side under the creosote bushes. (Photograph by author.)

were not connected while others clearly were, it would seem that this water harvesting system might well have been built incrementally (e.g., Doolittle 1984*a*).

North of the Pinball site, and on the west side of the New River, is the Surly site, AZ T:4:13 ASM (Doyel, Elson, and Hoffman 1985: 193–5, figs 4.57 and 4.59). This site is a water harvesting system not unlike that just discussed, except it is larger, yet less complex. The total cultivated area covers 12.2 ha (Doyel, Elson, and Hoffman 1985: 269). A number of rocks, rock alignments, and rock piles were found on the surface and interpreted to function as brush anchors, water and soil control features, and spoil heaps from clearance, respectively (Doyel, Elson, and Hoffman 1985: 265, 271, table 5.2, 272). In addition to the size of the planting surface, the most impressive aspect concerning this site is the size of the canal that carried water to it. Known as the Morris Canal, it is evident on the surface by parallel cobble embankments up to 5 m apart and 460 m long (Fig. 9.12). No diversion structure was found, but it is clear that the canal transported water directly from the New River. Archaeological excavations revealed that it was 1 m deep, 3 m wide at the top, and trapezoidal in cross-sectional shape, and a gradient of 0.5 percent (Doyel, Elson, and Hoffman 1985: 266–70, figs 5.4 and 5.6).

Just to the west of the Surly site lies the Antelope Glyph site (AZ T:4:16 ASM). It, like the others, contains remnants of a canal and several field features (Doyel, Elson, and Hoffman 1985: 161–3, fig. 4.46). Two things, however, distinguish this water harvesting site from others in the area. First, it was the largest, covering 19.2 ha (Doyel 1985*a*: 297–301, figs 6.1 and 6.2). Second, and more importantly, the canal crossed an intervening drainage filled with soft sediments, thereby requiring reinforcement. The canal was traceable back to the point where it intersected the channel of the river. Here, the distance between spoil banks on either side measured nearly 5 m, making it nearly as large as the Morris Canal. It was 350 m long, from the channel to its terminus at the field (Doyel 1985*a*: 302). Archaeological excavations revealed that the canal was U-shaped in cross-section. Near the intake it was 4 m wide at the top and 1 m deep. Twenty meters further down its course, the canal measured 3 m wide and 75 cm deep. This configuration was maintained the remainder of its length (Doyel 1985*a*: 304–5, fig. 6.5). Exactly why the canal was wider and deeper near its intake than further along its course remains unexplained. It may be, however, that the wider intake facilitated diversion of stream flow, not unlike a funnel.

Fifty meters from its terminus, the canal crossed a small sediment-filled, but active, channel approximately 40 m wide. Archaeological excavations revealed that in order to get the canal across the channel, ancient water harvesters built reinforcing walls. They did so by excavating trenches on the downstream side of the canal near where it entered the channel, and on the upstream side of the canal near where it approached firm ground. In the center of the channel, trenches were dug on both sides of the canal. These trenches were then filled with large rocks. Most weighed between 20 and 25 kg, were oddly shaped and cumbersome, and required more than one person to carry. Furthermore, they had to be transported from an area which was more than 500 m away (Doyel 1985*a*: 306–10, figs 6.6 and 6.7).

A few other sites in the New River Valley have also been subjected to some investigation and contribute to knowledge of prehistoric water harvesting in the region. The Ridgetop site (AZ T:4:22 ASM) lies not too far to the east of the Pinball site. The most significant fact about this site is that near its edge, there are parallel cobble alignments, 3 m apart, indicative of a canal. On one hand this might not seem all that significant. On the other, however, this canal is on the floodplain and not on a lower terrace of the stream (Doyel, Elson, and Hoffman 1985: 229–31, figs 4.75 and 4.76). A diversity of locales were apparently used in water harvesting.

Between 10 and 15 km upstream of the New River Gap lie three other known ancient water harvesting sites. All are long and thin and situated on low terraces. Site AZ T:4:38 ASM has 4.2 ha of field area identified by 59 rock alignments, 17 rock piles, and fragments of two canals (Hackbarth 1989: 45, 47, fig. 3.1). Roughly parallel to each other and on opposite sides of the field, the canals' remnants are sufficiently offset to indicate that one might be an extension of the other. Cross-sectional differences add some support for this tentative finding. The section furthest downstream was the smaller of the two, measuring 10 m long, 80 cm wide at the top, and 25 cm deep. The upstream segment was slightly larger measuring 28 m long, 90 cm wide at the top, and 35 cm deep. It also had a rock alignment along one side which may have been an extension of a diversion dike (Hackbarth 1989: 52–3). This feature was 12 m long, along that portion of the canal furthest upstream—near the natural channel—and on the downstream side of the canal in relation to the channel from which water was taken. Although brush and stake weirs might have been used in some cases, the presence of this wall suggests that rock diversion structures were probably used at some sites (Rankin 1989*b*: 952–4).

Another important water harvesting site in the upstream reach of the New River Valley is AZ T:4:40 ASM. In some respects, there is nothing special about this site, but then in others, there is. It is certainly a large site with 142 rock piles, 88 rock alignments, 16.3 ha, and a 132 m segment of canal—evident by the usual parallel rows of rock on the surface—which bifurcates (Rankin 1989*a*: 69, 85, 91, 100–1, figs 4.1 and 4.3). The canal remnants, including the bifurcation, were at the very far upstream end of the site and extended for 132 m. Doubtless, two branches extended further down the site but have long since been obliterated. Archaeological excavations were made both upstream and downstream of the fork in the canal. Both sections were elliptical in cross-sectional shape, wide, and narrow. The upstream segment was 60 cm deep, 5 m wide at the top, and 1.25 m wide at the bottom. Downstream of the fork, one branch was 45 cm deep, 2.7 m wide at the top, and 1.5 m wide across the bottom (Rankin 1989*a*: 103–6, figs 4.4 and 4.5). Perhaps the most important characteristic of this site in terms of water harvesting technology is the presence of rocks piled and aligned at the point where the canals bifurcate. Small basalt boulders and river cobbles appear to have been fitted tightly together (Rankin 1989*a*: 103, 106). It is possible that these rocks were the abutments of headgates (Rankin 1989*a*: 163), but it is more likely they served as riprap, mitigating the erosional forces of the flowing water.

The final water harvesting site in the New River Valley has actually been subjected to very little study. Site AZ T:4:25 ASM has the usual array of rock piles and alignments on the field surface, as well as remnants of canals of various sizes. Two features, however, make this site interesting. First, it is far away from the channel of the New River, and harvested water from one of its tributaries. This is evidence that water harvesting was practiced on a variety of scales with systems of various sizes and varies often in proximity to each other. Second, several canal segments ran parallel to the natural contours of the field surface. In effect, they indicate that canals were used to distribute water across planting surfaces as well as transporting water to them (Rankin 1989*b*: 952).

Although water harvesting characterized agriculture in the Hohokam Periphery, it should not be thought of as having been absent from the Core.

Indeed, a few sites within the Salt–Gila Valley of central Arizona show signs of just how remarkably ingenious ancient farmers were at tapping ephemeral stream flow. Located south of the Gila River, across from the present-day town of Buckeye and not far from the confluence of the Agua Fria River, into which the New River flows, exists a canal more than 8 km long, 2 m wide, and 1.8 m deep. Running around the foot of the Buckeye Hills, this canal was once thought to be historic and to have carried water diverted out of the perennial Gila River. Archival records indicated that it was used early in the 20th century and that aqueducts had been used to maintain the flow across intervening arroyos. Not satisfied with written records and common knowledge, one archaeologist, James B. Rodgers, took a very close look at the canal and made some interesting discoveries.

Rodgers (1976: 47, 65) found one portion where the canal showed evidence of an earlier phase of construction and use. It involved the building of an earthen bank along the slope, opposite a natural bank resulting from a granite bluff. In places, the granite itself had been trenched to maintain a relatively smooth bottom to the canal and a constant gradient. Heavy axes would have been necessary for such work, and not surprisingly broken tool fragments were found in the spoil forming the adjacent bank. The canal proved to be smaller than its more recent counterpart. It was 2.2 m wide and 1.1 m deep.

This canal also turned out to be not one long canal, but several short ones. And, they each had separate ephemeral sources, not a common perennial one. Rodgers (1976: 48, 66) determined that seasonal stream flow in the arroyos draining northward from the Buckeye Hills was diverted into these canals. Each segment had its own source, and presumably diversion structure, but the latter no longer exist. From the canals, water was used to irrigate fields on the lower hill slopes. These water harvesting systems came into use AD 700–1000 and were abandoned AD 1100–1250.

Between the Salt and the Gila Rivers, approximately 15 km due west of the town of Florence Junction, Arizona, are a series of five very large water harvesting systems which depended on the seasonal flow of Queen Creek (Fig. 9.13). Information about these systems is due largely to the excellent field work and hydrologic expertise of archaeologist Allen Dart (1983; see also Crown 1984*a*: 208–13). Among other features, he found that each of the main canals appears to have had taken water directly from the creek channel. The three southernmost canals might, however, have shared one or more common heads, and the fourth might have been a branch of the third. The sections of these canals furthest upstream are no longer visible and therefore details concerning their points of origin remain unconfirmed. There is no question, though, that they carried water diverted out of Queen Creek. The northernmost canal was traced directly to the creek channel, but no evidence of a diversion has been identified.

As was the case in the New River Valley, canals in the Queen Creek area

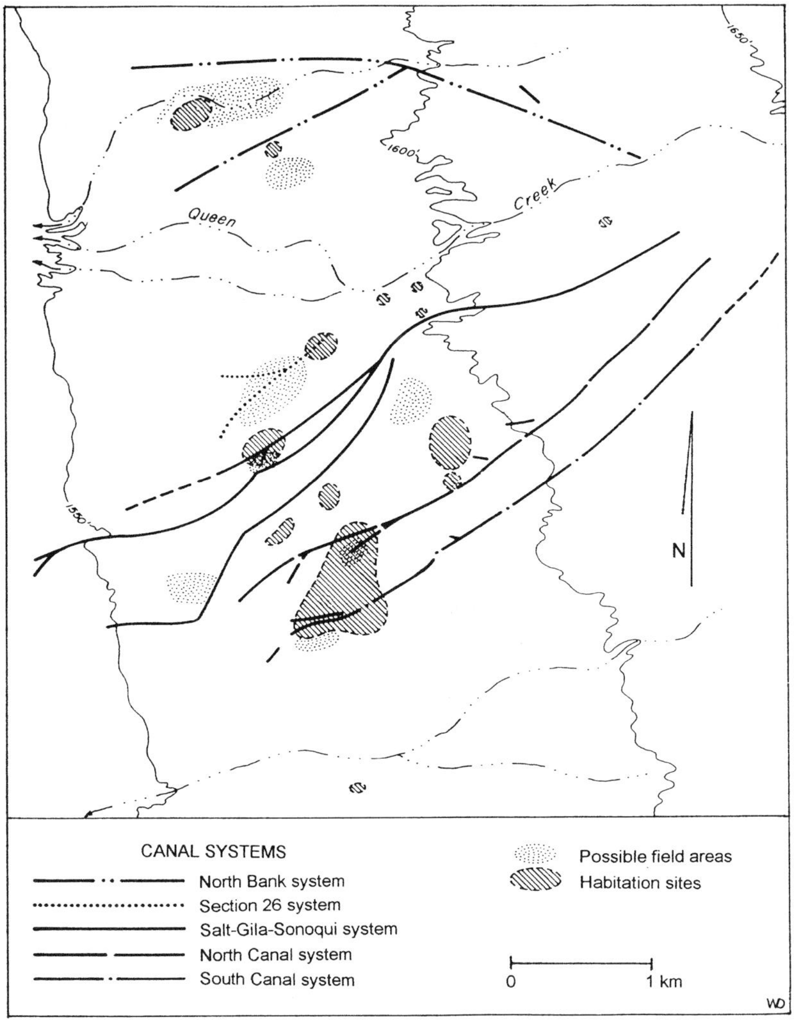

Fig. 9.13. Prehistoric canals and fields, and related habitation sites, comprising five large water harvesting systems in the Queen Creek area. After Dart (1983: fig. IV.3.1, 368). (Redrawn by author.)

were evident on the surface principally by parallel rows of rock and earth rubble, or the spoil removed during excavation (Dart 1983: 370, fig. IV.3.2). In some places, however, only slight rows of mounded earth were visible, and yet in other places canals were detected only by rows of vegetation more dense than usual. The most striking aspect of these canals has to be their lengths. The shortest main canal, that of Section 26 system (see Fig. 9.13), measured

920 m long, but was probably at least another 500 m longer (Dart 1983: 386–7). The longest, that associated with the Salt–Gila–Sonoqui system, was measured at 12.7 km long (extending well off the map in Fig. 9.13), but was undoubtedly longer (Dart 1983: 381). Each of these canals bifurcated at least once. In some cases, the known branch canals were quite short, 90–500 m, while in others they were very long, 1.6–4.3 km. The presence of long branches in all likelihood indicates that there were many more short branches than are evident from surface remains. A dry laid rock structure measuring 8 m long, 1.1 m wide, and 40 cm high was found at one bifurcation. Situated diagonally across the main canal, this feature clearly helped divert water into the branch canal (Dart 1983: 382–5, fig. IV.3.9).

Archaeological excavation of these canals revealed some interesting characteristics. To begin with, the gradients varied from 0.26 to 0.72 percent, with most being close to 0.50 percent (Dart 1983: 369, table IV.3.2). Next, the canals turned out to be surprisingly narrow, shallow, and variable in cross-sectional shape. The main canal of the southernmost system, for example, was roughly V-shaped in cross-section. It was 1.2 m deep, 1 m wide at a point 80 cm from the bottom, and 3 m wide at the top. Further down its course, the canal became elliptical in cross-section, 30 cm deep, and 1 m wide (Dart 1983: 371–3, figs IV.3.3 and IV.3.4). The canal just north of it, that associated with the North Canal system, is even more interesting. In one place along its length it was U-shaped in cross-section and 3 m wide at the top. The bottom of the canal was cut 70 cm below the ground surface (Dart 1983: 378–9, fig. IV.3.8). Near its downstream end, the canal maintained its cross-sectional shape, and narrowed to 2.2 m across the top. The bottom of the canal, however, was found to be *at* ground level (Fig. 9.14). In other words, all of the water transported here flowed entirely between the earthen embankments, which ranged in height from 70 cm to 1 m, and *above ground* (Dart 1983: 376–7, fig. IV.3.7).

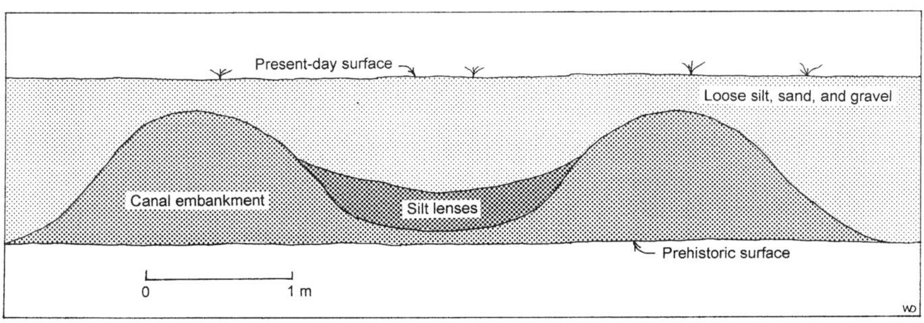

Fig. 9.14. Cross-sectional diagram of a downstream segment of a canal in the Queen Creek area. After Dart (1983: fig. IV. 3.7, 377). (Redrawn by author.)

Finding the bottom of the canal at ground level near its terminus should come as no real surprise. Water harvesting systems depend on gravity flow, which, by its very nature requires that water be applied to the field surface. Canals are usually thought of as features excavated *below* the ground surface. Such a condition is necessary at their upstream end of each canal, toward and at the intakes. However, in order to function, a canal's gradient must be less than that of the surrounding surface. With different gradients, a point will eventually be reached where the canal's bottom and natural ground level are identical. It is usually immediately downstream of such points that fields are located.

Canals as long as those in the Queen Creek area and with as many branches must have watered a substantial amount of land. Fields associated with these canals are, however, enigmatic. Rock alignments such as those associated with water harvesting systems elsewhere are conspicuous by their near absence here, perhaps in part due to much of the area being covered with a veneer of sediment. Several areas near the ends of canals have been identified as possible fields (Fig. 9.13) due to differences in soil and vegetation, or more precisely, a paucity of vegetation. The noted sparseness of vegetation on these cultivated areas has been interpreted to be a possible function of salinization resulting from evaporation of water harvested from Queen Creek (Dart 1983: 528–31).

The areas thought to be fields constitute a relatively small amount of land. Indeed, their total area roughly approximates that of habitation sites nearby. Farmland certainly takes up more space than habitational areas and this fact has led one scholar, Patricia L. Crown (1984*a*: 226–7, fig. III.4.6), to estimate the total area encompassing all the main and branch canals. Her estimate, which is admittedly imperfect, is 2,460 ha.

Now, was all of this land, and all the canals used contemporaneously? Evidence suggests not. First, estimates based on the morphology of canals and hydrologic principles reduce the size of the area which could have been watered at any one time. The canals themselves have limitations. At best, they could have carried only enough water for 1,490 ha. However, there was never enough water in Queen Creek to allow even this amount of land to be wetted at once. During peak flows, a maximum of 810 ha could be watered (Crown 1984*a*: 228–32, table III.4.3). Clearly, if 2,460 ha were cultivated, there would have been some type of water sharing and rotation system in place. Of course, this assumes that all of the canals were used during the same general time period. Again, evidence suggests otherwise.

Determining the dates at which these canals, and their various segments, were built and used was done on the basis of stratigraphic relationships with other features, artifacts found in association, and proximity to habitation sites. This last factor was most important in that several sites were located along each canal (e.g., Rice and Simonis 1977: 14–18, figs 6 and 7) and many date to different periods. When looked at spatially, construction on the upstream portion of the Salt–Gila–Sonoqui canal and its northernmost branch (see Fig. 9.13) began first, before AD 900. The main canal remained in use at least past AD 1300, but the branch seems to have been abandoned by AD 1175. The Southern Canal system and the Northern Canal system were both built next, before AD 1000. However, the former was abandoned before AD 1175, while the latter remain in use past this date. Construction and use of the southernmost branch of the Salt–Gila–Sonoqui system began AD 1000 but extended until perhaps as late as AD 1450. The central branch of this system and the entire North Bank system functioned AD 1000–1175. The Section 26 system was begun sometime during this same period, but was not completed until after AD 1300 (Crown 1984*a*: 213–16, fig. III.4.2).

When looked at temporally an equally interesting picture emerges. Approximately 14 percent of the canals were in use prior to AD 900. Between AD 875 and 1000 another 30 percent were constructed and 55 percent were in use. Forty percent were built and 84 percent were in use AD 1000–1175. Less than 1 percent were built AD 1175–1300 and only 40 percent remained in use. After AD 1300, 15 percent were built and 61 percent remained in use. Although many canal segments were taken out of service, the greatest building phase and period of use was between AD 1000 and AD 1175 (Crown 1984*a*: 248–51, 1984*b*: 6, 13, 15, 17, 1984*c*: 99–102).

Whereas the ancient systems in the Queen Creek involved the longest canals and the largest amount of land subjected to water harvesting, they are not the best known. Two other areas have water harvesting systems that have garnered far greater attention. These are at Chaco Canyon and Mesa Verde.

The Chaco Canyon system began to be studied formally early in the 1970s. However, what might be considered informal studies began much earlier. The person singularly responsible for initiating these studies was R. Gwinn Vivian. The son of Gordon Vivian, the National Park Service's chief archaeologist there for many years, Vivian literally grew up in Chaco Canyon (Frazier 1986: 100). His father first noted the presence of the systems but it was the younger Vivian who began to unravel their nuances and understand how they actually functioned. In no small way, the water harvesting systems at Chaco Canyon are a function of the physical environments in which they are located.

As a feature on the natural landscape, Chaco Canyon is the result of a westward flowing stream—Chaco Wash—which cut through extensive beds of sandstone. The main part of the canyon is approximately 15 km long, 30 to 50 m deep, and 1 km wide. The floor of the canyon is filled with alluvium suitable for agriculture (Hall 1985). It is here, especially along the north side, that most of the large pueblos and ancient field sites exist. Throughout human history, the areas overlooking the canyon have been characterized by limited soil development and sparse vegetation (Hall 1988). Much of these mesa top surfaces are pure bedrock. Rainfall in northwestern New Mexico, as throughout the Southwest, is slight and unpredictable. Although prehistoric farmers could never depend on rain falling directly on fields when needed, they could always depend on runoff from the surrounding mesa tops. The problem they then faced was harvesting this runoff.

Through intensive and complete surveys, Vivian (1972:3–4) found seventy-five features relating to prehistoric water control and agriculture in the canyon, many of which he then excavated archaeologically. Not inconsequentially, but certainly not germane to the present study, some of the features initially thought by Vivian to be ancient canals turned out to be roadways, and gave rise to another important line of research. The water control features that were confirmed also seemed to be associated with nearby pueblos. Detailed studies of several systems found that although they differed somewhat in size and configuration, all seemed to share certain commonalities. Indeed, Vivian (1972: 5) even went so far as to say that 'a uniform plan for water control' existed and that some features were so similar that they are 'monotonously alike'.

The principal method of water harvesting at Chaco Canyon involved first rain that fell on top of the mesa. Nearly all of this water then ran off the exposed bedrock surface, collecting in rivlets and small gullies before pouring-off the mesa and onto the canyon floor below (Fig. 9.15). In some

Fig. 9.15. View south down a drainage on the top of the mesa overlooking Chaco Canyon (distance) on the north. Note the paucity of vegetation and soil, the exposed bedrock surfaces, and the pour-off (arrow). (Photograph by author.)

places, such as near Pueblo Alto, masonry walls measuring several hundred meters long, 60 cm high, and 60 cm wide helped divert runoff (Vivian 1970: 73). Along the north side of the canyon there are twenty-eight drainages, including three very large ones—Cly, Mockingbird, and Gallo canyons. Water harvesting systems were found in fifteen of these drainages, including two of the major ones (Vivian 1974: 104). The watersheds or catchment areas for the twenty-five smaller ones ranged in size from 6.5 to 87 ha, whereas catchments for the large tributary side canyons ranged from 525 ha to 810 ha. There are on average eight small drainages between each of the major ones for a total drainage basin of approximately 4,250 ha (Vivian 1974: 104, 1990: 309).

Once it poured-off the mesa top and onto the canyon floor, the natural flow of water was impeded by dams built across the natural drainages (Vivian 1970: 74), close to the edge of the cliff. Dams across the large drainages were quite substantial. The one at Cly Canyon was built of earth and rock and measured more than 35 m long, 6 m wide, and 2 m high (Frazier 1986: 101). Similar ones existed in Gallo Canyon (Vivian 1972: 5), and at Rincon–4 North (Vivian 1990: 310). Some dams impounded water, thereby creating reservoirs, which seem to have been used as domestic water supplies as well as for agriculture (Vivian 1970: 73). Most, however, simply diverted water into canals which transported it to fields (Fig. 9.16). Canals averaged 2.7 m in width, and ranged in depth from 60 cm to 1.5 m (Frazier 1986: 101); they varied in size depending on locale. For example, the canal associated with the Rincon–4 North system was 230 m long, 4.5 m wide, and 1.4 m deep (Vivian 1974: 104, 1990: 310), the one associated with Pueblo Pintado was more than 5 km long and 15 m wide (Vivian 1970: 72). In most places canals were first excavated and then lined along the bottoms and sides with rock slabs. Where excavation was not possible, they were formed by slab and masonry walls (Vivian 1970: 72). In addition to transporting water diverted by the dams, the canals also collected water spilling over the escarpment along their lengths. This runoff, may, in fact, have been every bit as important to the operation of the systems as that from the dams (Lagasse, Gillespie, and Eggert 1984: 197–202), and it might well explain the great width of the Pueblo Pintado canal.

The canals terminated in features called 'headgates'. Made of rock slabs not unlike those used in the construction of canals, headgates had at least two intakes (Vivian 1972: 6, fig. 4), each approximately 2 m long and 30 cm wide (Vivian 1990: 311). These features received water not only from the main canal, but also from other canals that collected water from smaller drainages. Their purpose was to slow the water and channel it to the fields via smaller, distribution canals (Vivian 1974: 104). Restricting the flow of a 4 m canal into a 30 cm opening required substantial masonry wing walls which stretched several meters on both sides of the gate openings, and moveable rock slabs which could be used to close the gate (Vivian 1990: 311). The distribution canals between the headgates and the fields were no larger than the gates

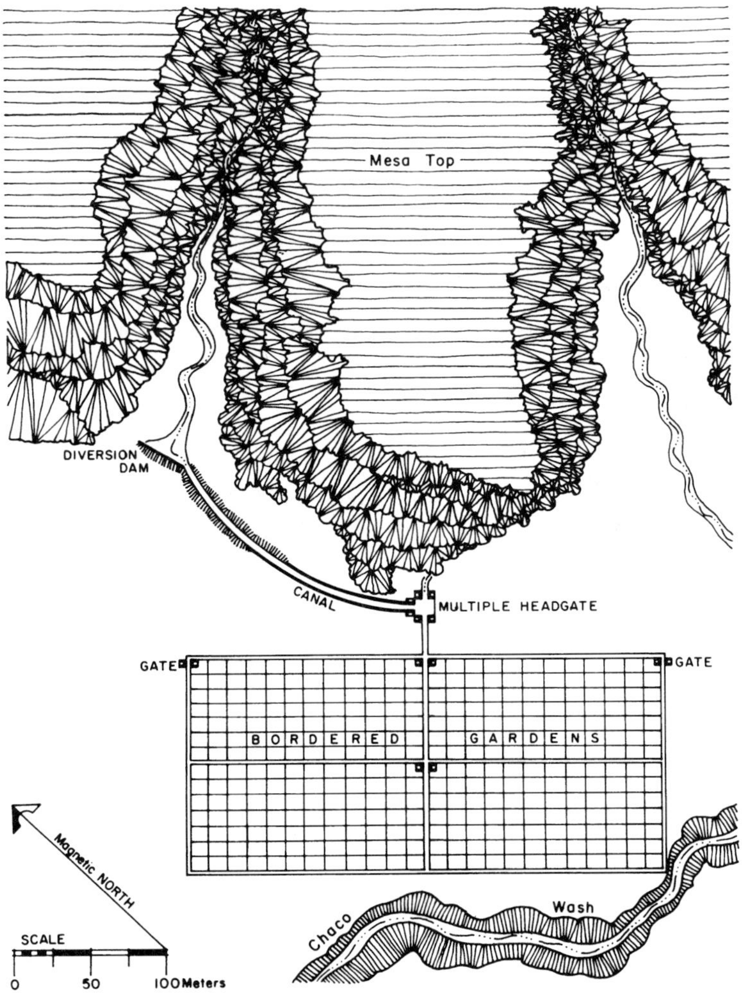

Fig. 9.16. Schematic diagram of the Rincon–4 North water harvesting system in Chaco Canyon. Appeared originally in Vivian (1974: 18, fig. 9.4). (Reprinted by permission of the University of Arizona Press.)

themselves (Vivian 1970: 5) and had gradients only slightly less than that of the natural surface (Loose and Lyons 1977: 143).

The fields at Chaco Canyon were first identified on aerial photographs taken by famed aviator Charles A. Lindbergh in 1929 (Loose and Lyons 1977). They appear as rectangular areas devoid of vegetation (Fig. 9.17) and show up on color photographs better than they do on either black-and-white or color infrared images (Gumerman and Lyons 1971: 128–9, figs 2

Fig. 9.17. A view east, up Chaco Canyon. Taken from the mesa top above Chetro Ketl ruins (fore). Ancient fields are clearly visible by the lack of vegetation in the rectangular area (arrow). (Photograph by author.)

and 3). Doubtless numerous ancient fields have been obliterated by erosion or sedimentation, and are no longer visible (Loose and Lyons 1977: 147). However, several fields remain visible on aerial photographs (see e.g., Lyons, Hitchcock, and Pouls 1977), with the one at Chetro Ketl ruins being perhaps the most conspicuous and well-studied. The Chetro Ketl field encompasses 4.8 ha divided into two rectangular plots separated by a distribution canal. Clearly visible on aerial photographs are individual grids within the plots. There are 42 such grids in each plot, averaging 14 m by 23 m. The borders forming the grids are made of ridged earth (Vivian 1974: 104–5; Loose and Lyons 1977: 134). Field measurements with a radiographic densiometer and a electron spin magnetometer revealed that the centers of the grids had higher magnetic readings than the borders, suggesting that grids contained iron-bearing clays deposited by irrigation waters (Loose and Lyons 1977: 137–9). Archaeological excavations bore this out (see also Lagasse, Gillespie, and Eggert 1984: 202, 205–7). They also revealed that the ridges were 40–50 cm high, and that significant field leveling was involved. Indeed, the cultivated surface had a gradient of 1.17 percent while the surface just off the field has a gradient of 1.98 percent (Loose and Lyons 1977: 139, 142). Apparently, individual grids were watered by temporary openings in the bordering ridges (Vivian 1990: 312). Experiments revealed that maize could certainly have been grown on these fields (Toll *et al.* 1985: 80, 89, table 3.3, 115).

Exactly how much land in Chaco Canyon was ever cultivated using these water harvesting systems, and whether or not everyone who lived there was fed exclusively by these fields, or relied in part on imports from surrounding areas (Reher 1979), has been the subject of debate for a few years. The principal point of contention has been over how much land was ever farmed. Vivian (1970: 5, 1991) has long argued that all the land along the north side of Chaco Wash was cultivated. Others (Loose and Lyons 1977: 147) could count no more than 30 ha of relict fields. Depending on how one wishes to: (i) account for post-use destruction of ancient fields, (ii) factor in fallowing versus fertilization through sedimentation, and (iii) estimate populations on the basis of ruins, almost any argument can be substantiated.

As in the case of the Chaco Canyon systems, features thought to comprise a water harvesting system at Mesa Verde have been the subject of much study and debate, albeit of a different nature. The alleged systems were first identified in the 1930s by James A. Lancaster, a member of the Mesa Verde National Park staff, who brought it to the attention of Guy Stewart (1939: 131, 1940*a*: 207). Located near Far View ruin on Chapin Mesa, it involved a reservoir known as Mummy Lake from which a canal was thought to have transported water to fields. Mummy Lake was originally recorded by Nordenskiöld (1893: 74) as being 30 m in diameter with surrounding walls 4 or 5 m thick. Nordenskiöld thought it might have been used for agriculture but found no canal or other evidence to support his claim. Park archaeologist Jesse W. Fewkes (1917: 464) thought, on the basis of its proximity to pueblos, that it served a domestic purpose. When seen in the 1930s, Mummy Lake contained 3 m of sediment, and had a history of being used by cattle ranchers as a stock tank (G. Stewart 1940*a*: 206–8).

The canal identified by Lancaster and Stewart as eminating from Mummy Lake was traced down mesa approximately 6.5 km. Thought initially by Fewkes (1917: 464) to be a trail, it varied in width from 6.8 m to 12.8 m, with an average of 8.5 m. Depth varied from 15 to 46 cm, with the average being 28 cm. It showed minimal signs of erosion and this was attributed to its broad, shallow cross-section and 'remarkably uniform gradient' throughout its entire length (G. Stewart 1940*a*: 207–8, fig. 3). A few follow-up investigations found that: (i) the canal actually extended upstream of Mummy Lake 1.2 km and 'would have intercepted an appreciable portion of the runoff from the upper part of Chapin Mesa' (Stewart and Donnelly 1943*a*: 31); (ii) rock alignments stabilized the bed at various points; (iii) along its course there was evidence of outtakes for watering fields and intakes for collecting additional runoff; and (iv) it is larger than any of the five other canals found in parts of Mesa Verde (J. Herold 1961: 103, 102, 216, fig. 23).

In the early 1960s, archaeologist Arthur H. Rohn (1963) undertook a detailed investigation of what had become known as Far View Ditch and

Mummy Lake. A trench excavated archaeologically in one place revealed that the canal was originally 1 m deep. In places it ran along the very edge of the mesa top and had to be supported by a row of rocks, and elsewhere its course was identical to a road built in the early 1900s (Rohn 1963*b*: 452). Rohn (1963*b*: 453) concluded that it was not associated with agriculture, but instead carried water for domestic purposes to pueblos at the southern end of the Chapin Mesa. His conclusion was based on the lack of any direct traces of off-takes heading toward known fields, of which there are plenty, and because most off-takes headed toward habitation sites (see also Rohn 1972: 217–19, 1977: 100, 105–7).

As for Mummy Lake, Rohn and one of his colleagues made some interesting discoveries. David A. Breternitz, conducted archaeological excavations which revealed that in addition to being formed by an earthen embankment, the exterior side on the east, and the entire inside was masonry lined (Fig. 9.18). Futhermore, there were a series of steps leading downward on the south side of the interior (Rohn 1977: 104). More importantly, however, the canals which were thought to carry water directly into and out of Mummy Lake were found to bypass it altogether on the west side. The intake for the reservoir was on the southwest corner. It was a rock-lined canal 1 m wide which made a 'hairpin' turn off the main canal and flowed in a northeasterly

Fig. 9.18. Mummy Lake at Mesa Verde National Park as it appeared in 1988. (Photograph by author.)

direction for approximately 10 m before emptying into Mummy Lake. This unusual and drastic change of direction has been interpreted as an attempt on the part of its builders to facilitate the removal of sediment. The curve slowed the flow markedly causing silt to be deposited in the intake rather than in the bottom of the reservoir. Given the differences in the depths of both features, sediment could be removed from the intake much easier than it could from the reservoir. Lending additional evidence for the reservoir not holding water to be discharged later into the Far View Ditch is the fact that the bottom of the canal lies at an elevation much higher than any water level in the reservoir. Mummy Lake was an appendage to, not an integral part of the system (Rohn 1963*b*: 449–50).

Following Stewart and Donnelly's lead, Rohn carried his studies further north, up what had been thought of as a 'feeder canal' to Mummy Lake. Not surprisingly, he made some interesting discoveries here as well. Over most of its course, the canal has a broad and shallow cross-sectional shape, approximately 1 m wide and 20 cm deep. In one place a rock alignment supported its downslope side (Rohn 1963*b*: 450, 452, fig. 13). Just up from Mummy Lake, it appeared to have a turn-out which headed westward (Rohn 1963*b*: 450). At the far upstream end of this canal, a double row of rocks measuring 35 m long, 1 m wide, and one course high, formed what Rohn (1963*b*: 450, fig. 10, 451, fig. 12) termed a 'gathering basin'. This feature marked the downslope/downstream end of a 10 ha 'catchment area' that contained at least six linear depressions thought to be 'collecting ditches' (Rohn 1963*b*: 450–1, fig. 11). According to Rohn, this system harvested water, part of which flowed into Mummy Lake, part of which was transported to reservoirs near pueblos further away, and part of which irrigated one small set of terraced fields. It did so between AD 900 and 1300 (Rohn 1972). The entire system was beautifully portrayed in an artist's rendering which appeared in the *National Geographic Magazine* (Osborne 1964: 192–3), and its operation is commonly accepted (e.g., Plog 1979: 126).

In some respects, this interpretation makes sense. On reflection, however, it is questionable, especially in regards to agriculture. Could a mere 10 ha generate enough water to fulfill all of these functions? Probably not. Given that the maximum distance from the upper end of the catchment area to the fields thought to have received water was only 1.8 km, was there that much difference in rainfall to merit such a complex system? On the basis that rainfed agriculture is known to have been practiced all over Chapin Mesa (see Chapter 5), the answer has to be no. Indeed, this point was questioned by others who worked at Mesa Verde (Stewart and Donnelly 1943*a*: 32; J. Herold 1961: 111). In sum, the suspected water harvesting system may have been for domestic purposes, but it certainly was not agricultural. One wonders if the 'collecting ditches' might not have been natural rivulets, and the 'feeder canal' and the Far View Ditch early wagon roads. The cautious Fewkes thought so.

Although the Hohokam Periphery, Chaco Canyon, and Mesa Verde are three places from which most of the archaeological evidence of water harvesting using canals, weirs, or both exists, they are not the only places. Indeed, there is some evidence, in many cases only scant, of the practice elsewhere in the Southwest. A canal lined with rock slabs appears to have carried water diverted from an ephemeral stream by means of a dam *c.* AD 1200 at Hovenweep (Schroeder 1967–8: 85, 92). A 'single, shallow, wide canal', dating AD 1370–1525, was found on the edge of a terrace of one arroyo in the Galisteo Basin (Lang 1995: 57). A 'diversion dam' built across a secondary tributary and remnants of a canal dating AD 1175–1300 have been recorded in the Rio Puerco Valley (J. Moore 1981: 115, 180). Similar features in similar settings were identified as having been used between the 11th and 13th centuries AD at two sites in Largo Canyon (Kayser 1973: 11, 19, 1974: 31). Dams and weirs, both made of boulders, and situated in normally dry stream channels have been located in association with canals excavated AD 1000–1150 in Devil's Park (Neely 1995: 247–8, 251, fig. 16.5b,c). An earthen dam measuring 9 km long and 7 m high was found across the valley of an ephemeral stream in far southwestern New Mexico (Gaillard 1896; Huntington 1914: 70). Although reported as being of great antiquity, not all scholars think it is prehistoric (Sauer and Brand 1930: 417). Local ranchers today claim it was built in the 19th century by 'Chinamen using baskets'. Given its exceptional length, the ranchers may well be right, but additional confirmation is needed.

A small, but interesting water harvesting system was found on the floodplain of Gavilan Canyon, a tributary of the Mimbres River (Creel and Adams 1986). It involved a narrow and shallow canal, a 400 m segment of which was discovered in the dry season by sighting a subtle line of green grass in an otherwise brown environment, two reservoirs, and a few rock alignments. Archaeological excavations revealed the canal to be 1 m wide, 30 cm deep, and elliptical in cross-sectional shape. The rock features were turn-outs to fields and a large rock found in one canal at the point of bifurcation (Fig. 9.19) was probably used to help control the flow of water. The system had two sources of water. One was the arroyo itself which the canal paralleled. The other was the uplands to the north which were drained by two arroyos that terminated in the two reservoirs near the canal. Apparently water collected in the reservoirs was released into the canal when needed. Only a few hectares of what was field surface remain today due to channel meandering. When the system operated *c.* AD 1400 (Creel 1990), the cultivated area may have encompassed 4 or 5 ha.

This system is probably typical of what may well have been hundreds, if not thousands, of small water harvesting efforts prehistorically across the Southwest. It is certainly a nicely documented system. Five locales containing a few not dissimilar systems, but markedly larger and involving diversion walls, have been identified on aerial photographs and ground-truth verified in southern central New Mexico (Hubbard 1987). None of these were

Fig. 9.19. An archaeologically excavated portion of the Gavilan Canyon canal showing a bifurcation and a large rock which was probably used in the control of water. (Photograph by author.)

archaeologically excavated. However, a canal with a cross-section reminiscent of that in the Mimbres Valley was later discovered and so examined in far west Texas (Hubbard 1992: 2, 5, 21–5, 27). Together, these studies indicate that water harvesting in late pre-European times was practiced further to the southeast than was long thought (Newcomb 1961: 221).

Overview

Water harvesting has long occupied an intriguing place in native North American agriculture. It may be uniquely southwestern, but there is nothing else unique about the practice; variety is its hallmark characteristic. Some water harvesting systems were very small and simple while others were exceptionally large and complex. In many cases the fields with which water harvesting was associated were characterized by features usually considered various forms of terracing. In some cases water harvesting involved canals characteristic of irrigation. If there is anything about water harvesting by means of canals that distinguishes it from pure canal irrigation, other than the arbitrary factor of ephemeral as opposed to perennial water sources, it is that water harvesting is less refined or polished. These systems tend to be more rustic. In no small way, this is probably a function of their being built to deal with things unpredictable and erratic, rather than phenomena that are very regular and only in need of management. Water harvesting systems are literally built to capture something that is non-existent but will arrive sometime and in unknown amounts. Precision in construction would, therefore, be uneconomical and perhaps even counterproductive. Canal irrigation systems, as will be seen in the following chapter, are built to redistribute a well-known, location-specific, and dependable resource. Precision in their construction is therefore essential (Plog n.d.). This is perhaps no better illustrated than in the canals themselves which are associated with the respective types of systems. Canals that constitute parts of water harvesting systems tend to have gradients much steeper—by a factor of two—than canals that form irrigation systems (e.g., Dart 1983: 369, 549–50).

Were the technological chararcteristics of water harvesting not sufficiently varied, the cultural factors surely are. To some native farmers the practice was a subsidiary activity to other forms of agriculture, while to others it was the principal form of their subsistence. There exists ethnographic evidence, from only a few places—the Zuni, Hopi, and Tohono O'odham reservations. In all probability, this is a function of native peoples being restricted to limited territories. Fortunately, these people provide some very detailed information. Ethnographic knowledge of water harvesting is attributable to the careful work of dedicated researchers and to the tenacity of native farmers who chose not to change even though the social and economic pressures to do so are great.

There is not much information on water harvesting as relayed by early explorers either from Europe or of European ancestry. In all likelihood this is a function of cultural bias. This is not to say that these chroniclers deliberately failed to report indigenous people harvesting water because they saw the practice as something backward or not worthy of mention. Instead, they probably had what they thought was a rather firm image of what agriculture entailed and did not recognize water harvesting for what it was—something very ingenius and special.

In contrast, the importance of water harvesting has never been lost on archaeologists. These professionals enjoy the benefits of working in the 20th century and therefore knowing the literature, being culturally sensitive, and focusing their eyes on the ground, literally not figuratively. With such perspectives it should come as no surprise that these scholars have discovered a plethora of evidence for ancient water harvesting. Relicts of ancient water harvesting are by their very nature quite fugitive, but they have not escaped the scrutiny of diligent observers.

When looked at together, both spatially (see Fig. 9.4) and in the contexts of technology and culture, the evidence for water harvesting on the part of native North Americans of the Southwest is most enlightening. Nearly everywhere there were people and suitable soils, native southwestern farmers manipulated the physical environment to their advantage. These people developed a multiplicity of techniques to milk this dry land for every drop of its precious water.

10

Canal Irrigation

There may well be no other form of agriculture that has garnered as much attention as irrigation. In a very basic and literal sense, irrigation is defined as the artificial application and distribution of water to otherwise dry lands in order to facilitate cultivation (Monkhouse 1965: 173; W. Moore 1974: 117). With a definition this broad, irrigation technology can vary in complexity from being as simple as carrying water by hand in containers, a practice that must be very old, to as complex as computerized and motorized sprinkler systems, a practice which is obviously quite new and beyond the nature and scope of this volume. Most frequently, however, irrigation is envisaged as involving canals through which water flows under the force of gravity (Rawitz 1973), and it is usually associated with permanent or perennial rather than seasonal or ephemeral streams (Lawton and Wilke 1979: 4–5; Masse 1980: 198).

Irrigation is both ancient and has tenacious cultural linkages, at least in academic veins. The domestication of the world's most prominent cereals—the oft-called 'origins of agriculture' itself—occurred in arid lands (Lawton and Wilke 1979: 5–9). The dependence of these crops on water long ago led a prominent anthropologist, Herbert Spinden (1915: 270, 1917: 182), to note that irrigation may predate agriculture, and, therefore, account for its very existence. Spinden's remarks were made in a general and a global sense, but he did go on to note that this agrotechnological sequence applied to the New World, including North America (1915: 273, 1917: 186). His ideas were seen as having merit nearly a half century later (Woodbury 1960: 301) and still maintain some validity today (Doolittle 1990*b*: 136–43). Indeed, as discussed in Chapter 2, Steward (1929) not only made a compelling argument for the development of water control through the use of canals prior to the cultivation of domesticates, but he provided ethnographic evidence which has been supported by later studies (Lawton, Wilke, DeDecker, and Mason 1976; Patch 1951). As will be seen in the following pages, archaeological confirmation of canals that predate the introduction of domesticates has yet to be found in North America. It may, however, be just a matter of time before such evidence is uncovered.

Arid lands were not only the hearths of certain domesticates, and the places where irrigation was developed, they were also the places where ancient civilizations arose (see e.g., Adams 1965; Butzer 1976; Kosok 1965).

It should come as no surprise, therefore, that in addition to the connection between irrigation and agriculture, there is also a connection between irrigation and civilization (e.g., Wittfogel 1957). The complexities of this relationship have long been the subject of debate (Leach 1959), and arguments show no sign of abating in the near future (e.g., Downing and Gibson 1974). Certainly, this is not the place to engage in or further that line of discussion. Suffice it to say that large populations living in dry lands have always relied on substantial canal irrigation systems, and that the sociopolitical mechanisms employed or developed to build and operate such systems were varied to the point of being bewildering, especially in terms of which came first and how one factor influenced the development of the other (e.g., R. Hunt 1988).

Native North Americans were responsible for building and using canal systems of every size and configuration imaginable. They did so long before Europeans appeared in the hemisphere and they continue to do so today. Many of the canal systems built in prehistoric times were abandoned prior to contact, but others have been well-maintained and continue in service. An abundance of all types of data provide insight into these indigenous peoples' hydraulic expertise. Not unexpectedly, it all comes from the arid southwestern quadrant of the continent. Indeed, water has long been seen as the lifeblood of the Southwest (e.g., Baker *et al.* 1973; I. Clark 1987; J. Clark 1978; S. Fish and P. Fish 1994: 101, fig. 2), and canals have been considered the arteries (Doolittle 1991).

Documentary Evidence

Unlike with terracing and water harvesting, the early European explorers in the Southwest made great mention of canal irrigation. This point, however, should not be taken to mean that they recorded irrigation every time they saw it, or that every locale being irrigated was mentioned as such in writing. Early Spanish chroniclers in the Southwest are particularly troublesome in regard to what they did and did not report. The documentation they provide is less than complete. Anthropologist Carroll L. Riley (1987: 61) could not have stated it better when he said: 'Spanish writers for this period are notorious for leaving out the obvious or, perhaps more confusing, mentioning it prominently in one place and ignoring it in the next'. To use a metaphor to further illustrate the point, early writers rarely mentioned the sky being blue. Does this mean that it was blue only in those few places where it was so mentioned, and some other color or colors elsewhere? Of course not. Conversely, just because early reporters mentioned canal irrigation in some places, one should not think that is was practiced everywhere. These two perspectives on how to infer spatial distributions on the basis of fragmentary documentary data are certainly academic. They do, however, have an applied dimension as well. Indeed, such contentions lie at the heart of on-going litigation over water rights between descendants of the first Americans and Spanish-descended farmers in northern New Mexico. Those defending the indigenous position favor the former interpretation, while those favoring the Hispanic position take the latter.

The documentary sources mentioning canal irrigation are so numerous and controversial that they should be looked at in two separate groupings. The first set of documents are those provided by chroniclers who appeared on the scene early, prior to any sustained Spanish activities in the area. These first-hand reports of canals and/or irrigation are literally explorers' accounts and are unquestionable references to such features and practices being aboriginal. The second set of documents are those provided by persons who were active in the region somewhat later. Principally missionary accounts, these references to canal irrigation are somewhat equivocal. In many cases, they report natives irrigating, but claim that these people learned the practice from earlier missionaries (e.g., Benavides 1954: 25, 1965: 24). In some cases, they may well be correct, but in others they are probably wrong.

The earliest European accounts of canal irrigation were provided by Fray Marcos de Niza after his travels of 1538. Writing about aboriginal people living in the town of Vacapa, probably on the Rio Fuerte, he noted '*tierra que se riega*' (Niza 1987*a*: 18), or 'their land is irrigated' (Niza 1987*b*: 18). He described land near present-day Mátape, Sonora simply as '*de regadio*' (Niza 1987*a*: lx), or 'irrigated' (Niza 1987*b*: 23), saving glowing praise for the Valley of Sonora (Fig. 10.1). According to him: '*riégase todo y es como un vergel*' (Niza 1987*a*: lxii), or 'it is all irrigated and is like a garden' (Niza 1987*b*: 25). Two years later, Juan Jaramillo (1940: 296), one of Coronado's chroniclers, reported 'irrigation' near present-day Ures, Sonora and that the Valley of Sonora was 'also irrigated' (Jaramillo 1940: 296, 297; see also Winship 1904: 224). According to Obregón, Ibarra reported of Oera, Sonora in 1565 that his expedition marched through '*un valle de dos leguas a lo largo, poblado de labores de maíz, frijol, y calabazas y melones, todo de regadío*' (Obregón 1924: 145) or 'a valley two leagues long planted with corn fields, beans, calabashes, and melons, all irrigated' (Obregón 1928: 159). He went on to say of these people, that '*sus labores de much concierto de acequias con que las riegan*' (Obregón 1924: 146) or '[t]heir plantations are well provided with canals used for irrigating them' (Obregón 1928: 160). Similarly, at Guarispi Francisco de Ibarra reportedly saw '*acequias de regadío para sus labores*' (Obregón 1924: 157) or 'irrigation canals for their plantations' (Obregón 1928: 173). And, finally, at Cumupa, he saw '*un valle de una legua todo de labores de regadío*' (Obregón 1924: 158), or 'a valley one league in extent entirely given over to irrigated fields' (Obregón 1928: 175).

These nine observations involving seven separate localities and three different writers in a matter of a few years, were sufficient to convince some scholars (Doolittle 1984*b*; C. Riley 1987: 59) that canal irrigation was

Fig. 10.1. A portion of the Valley of Sonora as it appears today. It remains an important agricultural area, criss-crossed by a network of irrigation canals, one of which appears in the lower right (arrow). (Photograph by author.)

widespread in eastern Sonora and northern Sinaloa, Mexico when the first Europeans arrived. Similar evidence exists for other parts of the Southwest (Fig. 10.2). For example, near the headwaters of the Rio Conchos in southern Chihuahua, Ibarra found a valley 'inhabited by a great number of people, with much food, and as skillful in the cultivation of their fields and in the irrigation of them as one can find in the world' (see Hammond and Rey 1929: 50, fn. 124; Mecham 1927: 81).

The earliest descriptions of canal irrigation in New Mexico are slightly later than those from northwestern Mexico. Nevertheless, they are great enough in number and thorough enough in detail to illustrate that in many locales aboriginal farmers were using technology developed by their ancestors long before people from the Old World arrived (C. Riley 1987: 233). No fewer than six accounts mention irrigation canals explicitly. Another six mention irrigation but not canals, although the latter seem to be implied. In 1582 Espejo (1966: 220; see also Bolton 1908: 178) reported fields 'irrigated by means of good ditches' at a Piro pueblo on the Rio Grande (see also Bandelier 1892: 237), and he inferred the same for Tiwa pueblos (Greiser and Moore 1995: 192). He also noted that the people of Acoma 'irrigate their farms by little streams of water diverted from a marsh near the river' (Espejo 1966: 224; see also Bolton 1908: 183), an observation supported by Pérez de

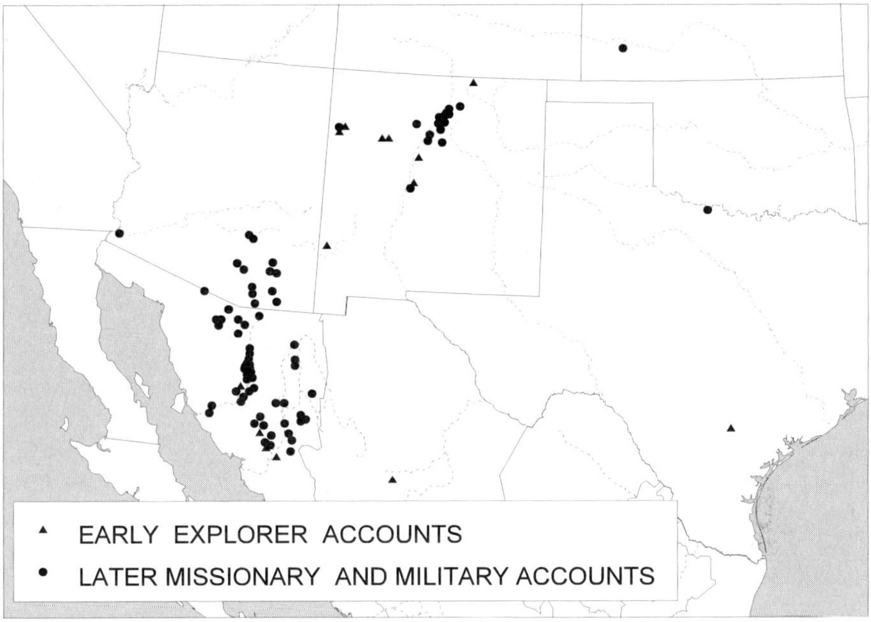

Fig. 10.2. Locations cited in the text for which there is documentary evidence of canal irrigation.

Luxán (1966: 182) when he said: 'We found many irrigated corn fields with canals and dams, built as if by Spaniards'. Near the Zuni pueblo of Hawikuh, Pérez de Luxán (1966: 186) also noted that 'there is a large marsh with many waterholes which enables the natives to irrigate some fields of corn. There are two canals' (Fig. 10.3). In 1591 Gaspar Castaño de Sosa (1966: 282) visited six pueblos—the Tewa pueblos of Tesuque, Cuyamungue, Nambé, Pojoaque, Jacona, and one unidentified pueblo (Wozniak 1998:11), but which may have been Pecos (Schroeder 1979*b*: 434; Schroeder and Matson 1965: 117)—all of which 'had canals for irrigation, which would be incredible to anyone who had not seen them with his own eyes'. A decade later one Joseph Bronata, who testified before Francisco de Valverde (1953*a*: 626), reported of the Rio Grande pueblos that: 'Some of their fields are irrigated by means of ditches'.

Simple accounts of either irrigated fields or irrigation, albeit without mentioning canals, include those by Martín de Pedrosa (1966: 118) of the 1582 Chamuscado–Rodriquez expedition near San Felipe Pueblo (Wozniak 1998:10), Castaño de Sosa (1966: 283) in 1591 near San Ildefonso Pueblo (Woosley n.d.*a*), and Marcelo de Espinosa before the Valverde (1953*a*: 634) investigation of 1601 near an unidentified pueblo. Obregón (1928: 325, 327)

Fig. 10.3. Irrigated fields, and a canal which transports water from a spring (Ojo Caliente) near the ruins of Hawikuh pueblo on the Zuni Reservation. This is either the system, or a successor of the system seen by Spanish explorers. (Photograph by author.)

noted irrigation at both Acoma and Zuni in 1584, but his second-hand accounts may have been taken from others. In 1630 Fray Alonso de Benavides (1954: 39, 1965: 36) mentioned puebloan people of the upper Rio Grande irrigating, specifically at San Ildefonso Pueblo (Schroeder and Matson 1965: 120), but he also mentions a number of Old World crops, thereby casting at least a hint of doubt on the antiquity of water control.

Two final accounts, albeit very late ones are worthy of mention. In the diary of his 1719 campaign, Governor Antonio de Valverde (1935: 115) mentioned seeing canals and irrigation among the Jicarilla Indé. In 1775 Escalante reported that Miera found along the San Francisco River in western New Mexico, native farmers 'who cultivate the valley . . . with the aid of irrigation' (A. Thomas 1932: 115). These people were probably Gileño Indé (Ackerly 1997: 353). Both the Jicarilla and the Gileño arrived in the Southwest in early historic times. The degree to which they were irrigating prior to their arrival and whether or not canals were being used previously in the areas which would become their homes remains for the moment unresolved.

Although some evidence is equivocal, it is clear that many New Mexican farmers were indeed irrigating by means of canals prior to migrants of European ancestry appearing on the scene. Canal irrigation is claimed to have been practiced at specific pueblos but with no firm evidence offered in support. Among these are Santa Clara (Arnon and Hill 1979: 303), Isleta (Ellis 1979*a*: 355), Cochiti (Lange 1979*a*: 368), Santa Ana (Strong 1979: 403), Laguna (Ellis 1979*b*: 441), and Zia (Euler 1954: 27). Some writers have maintained simply that the practice was widespread and commonplace (e.g., Bandelier and Hewett 1937: 35; I. Clark 1987: 6–7; Schroeder 1979*a*: 236). Evidence is not sufficiently great, however, to conclude that irrigation was practiced at all of the pueblos (Ackerly 1996: 7; Cordell 1979: 144; Wozniak 1998: 12). Furthermore, where it was practiced, irrigation seems to have been limited to small perennial streams that fed into, and not the floodplain of, the Rio Grande. The latter was too densely vegetated and flood-prone to have been a suitable environment for such a practice (Cordell 1984: 204).

The first European chroniclers to mention canal irrigation arrived in New Mexico later than their counterparts did in Sonora, and they arrived later yet in Texas. Most people do not think of Texas when they think of aboriginal irrigation, and this is probably appropriate. However, there are two descriptions provided by early explorers who thought they saw canals built by indigenous people. In 1709 Fray Isidro de Espinosa, one of two priests who accompanied a small military party, reported that in what is today the city of San Antonio: '*dimos en una acequia de agua muy poblada de arboles que era suficiente para un pueblo, y toda llena de tomas de agua por estar alta la acequia y colgadas las tierras pusimosle por nombre el agua de S. Pedro*' (Foster 1995: 99), or 'we came to an irrigation ditch, bordered by many trees and with water enough to supply a town. It was full of taps or sluices of water, the earth being terraced. We named it San Pedro Spring' (Espinosa 1930: 5). Nine years later, as diarist for the Martín de Alarcón expedition, Fray Pedro Pérez de Muzquía reported in the vicinity of present-day Palmetto State Park that: '*la llanada cogerá como una legua y está debajo de riego*' (Foster 1995: 133), or 'The plain covers about a league and is under irrigation' (Pérez de Muzquía 1938: 320).

Both of these accounts have been used as evidence that aboriginal Texans practiced canal irrigation (Weber 1992: 163; Foster 1995: 99). On one hand, the documentary evidence seems clear enough. On the other, however, it is not unequivocal. The Pérez de Muzquía account is certainly worthy of additional study, in part because it is provocative and in part because it has not been subjected to any follow-up scrutiny. The Espinosa account is simply wrong. What Espinosa thought was an irrigation canal is nothing more than the natural stream that heads at San Pedro Springs. The stream must have looked like a canal to Espinosa, much as it did to a later Spaniard, one Joseph de Urrutia, who portrayed it as such on a map he drew as part of the expedition of the Marquis de Rubí in 1767 (Moorhead 1975: plate 21, 156–7). In fact, however, this 'canal' is actually San Pedro Creek (J. Francisco de la Teja, personal communication 1996), a prominent, well-known, and important physical feature on the historical and present landscapes of San Antonio.

At approximately the same time as a few explorers were reporting what they thought was aboriginal canal irrigation in south Texas, a number of missionaries were making similar observations in Sonora and Arizona. In this latter area, however, their comments concerning canals, irrigation, and native farmers were valid. If there is anything about these documentary sources that makes them suspect, it is that missionaries had been working not too far away at the time and some Old World crops had already been adopted by indigenous agriculturists. In all likelihood, however, canal irrigation was practiced in these locales well before the Spanish missionaries arrived on the scene.

The earliest of the missionary accounts in Sonora was made by Pedro Mendez in 1628. He noted of the Rio Sahuaripa that 'there are streams of fine water which the Indians employ with no little ingenuity for irrigating their fields' (Bannon 1955: 46). In 1630 Martín de Azpilcueta noted that inhabitants of the lower Moctezuma Valley were 'expert in the use of irrigation . . . with numerous ditches' (Doolittle 1988: 46). Pérez de Ribas apparently borrowed these accounts (see Reff 1991: 58) when he noted '*dos valles muy fertiles de maíz y otras legumbres . . . con que riegan sus sementaras con notable artificio*' (Pérez de Ribas 1944 vol. 2: 178), or 'two fertile little valleys . . . irrigate their fields of corn and legumes with notable artifice' (Pérez de Ribas 1968: 146).

Further north in Sonora, and on into Arizona, Manje and Kino reported numerous places in the Sonoran Desert homeland of the Akimel O'odham and their immediate neighbors with canals, irrigation, or both. Their accounts are from the very late 1600s and very early 1700s, or the very beginnings of the mission era in that area. Although it might be tempting to infer that their reports of canal irrigation reflect Spanish and not native influence, there are at least two reasons to think that certain of their accounts refer to pure aboriginal practices. First, they frequently mention irrigation in places were there were no missions at that time. Second, they often refer to places where canal systems were non-existent, but could be built as part of mission activities. Theirs are literally scouting reports.

Near Pitiquín, they found '*acequias para su riego*' (Manje 1971: 293), or 'canals for irrigation' (Manje 1954: 11, 1971: 171). The site of Caborca '*tiene fertiles y feraces tierras, todas debajo de riego y acequias*' (Manje 1971: 293), or 'has fertile and fruitful lands, all under irrigation' (Manje 1954: 12, 1971: 172). West of Caborca they observed '*acequias y feraces tierras*' (Manje 1971: 304), or 'cultivated lands irrigated by canals' (Manje 1954: 31, 1971: 184). At Sonoyta they reported '*tierras pingües, con azequias para su riego*' (Manje 1971: 390), or 'rich lands with canals for irrigation' (Manje 1954: 110, 1971: 228), and '*pingües tierras de agricultura todas debajo de acequias y riego*' (Manje 1971: 473), or 'good agricultural lands here, all under irrigation' (Manje 1954: 158). According to them, at Quiburi in the San Pedro Valley '*todas debajo de riego*' (Manje 1971: 336), or '[a]ll the lands are under irrigation' (Manje 1954: 78, 1971: 361, 200). Indeed, they described the entire San

Pedro as a '*valle de pingües tierras de agricultura, con acequias para su riego*' (Manje 1971: 336), or 'valley of good agricultural lands with ditches for irrigation' (Manje 1954: 77, 1971: 360, 199), and '*ancho, largo, fertilísimo; y sus sementeras con acequias y riego*' (Manje 1971: 340), or 'wide and long and very fertile; and their fields are under irrigation' (Manje 1954: 83, 1971: 368, 205). The same held true for the Santa Cruz Valley (see Fontana 1983: 133; Manje 1954: 135, 1971: 461).

Toward its downstream end of this river there were '*buenos . . . tierras de agricultura y sementeras de muchas acequias para su riego*' (Manje 1971: 348), or 'good . . . agricultural lands with a canal for irrigation' (Manje 1954: 92, 1971: 377, 215). Near the mid-reaches there were '*muchas tierras de siembras, con muchas azequias*' (Manje 1971: 430), or 'rich lands all under irrigation' (Manje 1954: 125, 1971: 245–6). At Tumacacori, further upstream, '[*s*]*on tierras fertiles y de riego*' (Manje 1971: 349), or '[t]he lands here are fertile and irrigated' (Manje 1954: 94). Yet further upstream were more '*azequias para su riego*' (Manje 1971: 381), or irrigation canals (Manje 1971: 222). And, finally, near its headwaters, they described the Santa Cruz as '*un estendido valle y río, con muchas tierras de labransa, todas debajo de riego*' (Manje 1971: 461), or 'the extensive valley of cultivated and irrigated fields' (Manje 1954: 135, 1971: 251).

A few Spaniards who went beyond the range of Kino and Manje later into the 18th century also noticed indigenous farmers practicing canal irrigation. In mentioning activities along the Gila River, Jacobo Sedelmayr in 1744 (1955: 23) noted that '[t]hese Indians raise corn, beans, squash . . . by irrigation' (see also Ezell 1961: 36–7; Officer 1993: 291). In 1757, a Father Sanchez wrote: 'I took notice especially how they had conducted water by ditches to the land which served them for their crops' (Sauer and Brand 1930: 421). Further to the west, Sedelmayr (1939: 110) reported that in 1746 native farmers along the Colorado 'impound river water and lead it to their cultivated fields for irrigation'.

Spaniards continued to mention aboriginal canal irrigation well past the middle of the 18th century and in places where missionary activities were quite intensive (Radding 1995). In his description of Sonora and Arizona in 1764, Juan Nentvig reported canals, irrigation, or both at Oputo, Huásabas, Sátechi, Chamada, Onavas, Pónida, Sahuaripa, Movas, Tecoripa, Suaqui, Cumuripa, Mátape, San José de los Pimas, Batasaqui, Topisco, Arizpe, Tetoachi, Sinoquipe, Motepore, Banámichi, Huépac, Sonora, Aconchi, Baviácora, Concepción, Núñez, Ures, Gavilán, San José de Gracia, San Francisco, Pitic, Pópulo, Cocóspera, Imuris, Magdelana, Sáric, Caborca, the entire San Pedro Valley, the Florida Valley, and the Gila Valley (Nentvig 1980: 7–14; see also Pfefferkorn 1989: 46). Providing more detail, at Quitovac, Sonora, Anza (1966*b*: 150) reported that in 1774 water was 'collected in a ditch from five small springs . . . [to] irrigate some small bits of very sandy land, in which they plant about half a fanega of maize and some calabashes'. Writing of the San

Pedro Valley in 1767, Nicolas de Lafora (1958: 106) reported seeing '*acequias hondas*' or wide canals. In New Mexico in 1776 Fray Francisco Atanasio Dominguez mentioned canal irrigation at the pueblos of Quemado, Pojoaque, Tesuque, San Ildefonso, San Juan, Santa Clara, Abiquiu, Sandia, Cochiti, San Felipe, Jemez, Zuni, Isleta, Pecos, and Galisteo (Dominguez 1956: 41, 50–1, 59–60, 63, 71, 89–90, 118–19, 126, 143–4, 159, 181, 187, 196, 201, 207–8, 213, 216–17).

Accounts such as these are interesting, in no small way because they raise questions concerning the aboriginality of the canal irrigation systems mentioned. Clearly, some of the locales mentioned as having canals which carried water diverted out of perennial streams by native mission subjects in the late 18th century were places in which irrigation was being practiced when the very first Spaniards arrived (e.g., Mátape, Ures, Zuni, and Tesuque). Later Spaniards seem not to have been familiar with the writings of their predecessors, and they seem to have discounted the abilities of their subjects' ancestors. Nentvig, himself, provides what is perhaps the best example:

> Their irrigating canals, which bring the water from the river and some springs, were well constructed, the Indians having learned how to build them, no doubt, from Father Kino and other Jesuit missionaries who visited them from 1694 to 1751 (Nentvig 1980: 15).

Later Spaniards clearly attributed the presence of irrigation technology to their earlier colleagues (see also e.g., Benavides 1954: 25, 1965: 24); a fact which apparently contributed to later writers (e.g., Hutchins 1928: 263; Scott 1993: 38; Simmons 1972: 135; Sunseri 1973: 330) attributing certain aspects of irrigation to the Spaniards. To be sure, some of the missionaries did oversee native workers in the construction of canal systems (Ressler 1966, 1968). In 1638, for example, Gaspar de Contreras wrote of natives excavating canals for his mission near the headwaters of the Rio Conchos in Chihuahua (Sheridan and Naylor 1979: 11). Later and elsewhere, Manje reported that in 1694 with the help of native workers, Father Marcos Kappus built an irrigation canal at the Mission Dolores near the headwaters of the Rio San Miguel (Manje 1954: 67).

But, did these workers and overseers actually build entirely new irrigation systems? Or, perhaps, did they simply renovate earlier ones that had either fallen into disrepair or out of use? There are at least two cases in which there is firm evidence that canal systems thought to be historic in age and built by the Spaniards turned out to be much older, aboriginal, and only renovated by people of European ancestry. Both are in the upstream sections of the Santa Cruz Valley. The first of these involves the Jesuit mission of San Cayetano del Tumacacori, which was built on top of the Hohokam site known as Paloparado ruin. Archaeological excavations revealed three canal segments paralleling and proximal to each other. Two segments were definitely prehistoric and measured 1.8 m and 3.2 m wide, and 45 cm and 50 cm deep, respectively. The third segment, of smaller but unspecified dimensions, dated to the mission era (Di Peso 1956: 4, 202–3, plate 64). In concert, this evidence demonstrates that canal irrigation at the site was a prehistoric development and that later missionaries simply followed a technological precedent. A second study at the mission site of Guevavi, a little further upstream, was less detailed than the first, but it did find similar evidence. In this case, however, the mission era canal and its ancient forerunner were indeed one in the same (Robinson 1976: 139, 140). With such solid evidence in two places, there is good reason to think that the practice of Spaniards renovating prehistoric aboriginal canals was widespread (C. Riley 1987: 109–10).

There was a well-documented population decrease between the time the first Spaniards arrived in northwestern Mexico and the time when missionaries began working there (Reff 1991). Presumably there would have been a loss of farmland and a deterioration of agricultural technology commensurate with the population decline (Boserup 1965). The native population began to increase again during the mission era, and presumably this was paralleled by an expansion of farmland and greater need for agricultural technologies. Accordingly, it is entirely possible that what some 18th-century Spanish observers thought were new constructions, may well have been renovations. Most groups were quick to add Old World crops, tools, and techniques to their indigenously developed agricultural systems (Dunbier 1968:219; Morrisey 1949; Officer 1993; Spicer 1962: 541–2). In turn, it is not unreasonable to assume that Spanish missionaries relied on the talents and skills of native farmers, even if unwittingly. The contributions of aboriginal people to the mission lifestyle is certainly beyond the scope of this investigation. It is, however, a field of investigation (e.g., Radding 1979) begging for the study of agricultural technology transfers.

Perhaps the single greatest piece of evidence that confirms that aboriginal farmers gained little knowledge of canal irrigation from the Spaniards is that which is linguistic. Missionaries had as one of their tasks to teach natives Spanish (Polzer 1976: 56). If they taught their subjects about irrigation along with the language, there should be no native words for various aspects of irrigation. In fact, the reverse is true. Two dictionaries of native languages were compiled by missionaries in eastern Sonora in the 1600s, one for Névome and the other for Eudeve. Both dictionaries contain native words for irrigation canals (Pennington 1977: 4, 1980: 145, 1981: 100). A yet to be published dictionary also contains a word translated to mean 'canal of wood' thereby suggesting that split and hollowed tree trunks were used as aqueducts (Fig. 10.4) to bridge low places and maintain proper gradients in pre-European times (Reff 1991: 58, fn. 16; for an alternative perspective pertaining to New Mexico, see Condie 1989).

That indigenous irrigators might have influenced certain European-descended people is one thing. That they influenced other native peoples is another matter, and one more germane to the focus here. In his campaign

Fig. 10.4. A hollowed and split log aqueduct, known locally as a *canoa*, representing ancient irrigation technology in the Sierra Madres of northwestern Mexico. Appeared originally in Doolittle (1995: 318, fig. 5). (Reprinted by the permission of the American Geographical Society.)

against the Wichita on the Red River in 1759 Don Diego Ortiz Parrilla saw 'fields and crops of corn located on the south side of the river because irrigation was easier on that side (*como el mas facil para el riego*)' (Newcomb and Field 1967: 313). Forty years later José Cortés (1989: 85) reported for the same locale 'many springs and perennial fountains of fresh water, enough to irrigate and fertilize the extensive fields where they do their planting'. Now, the Great Plains are not normally thought of as having been irrigated by native peoples, and the account of Cortés is not without flaws (Elizabeth A. H. John, personal communication, 1995; see also John and Wheat 1991: 140–1, 172–3). However, nearly fifty years ago Wedel (1941: 12) entertained the possibility for a site in Kansas. He even went so far as to project this to be a function of puebloan influences. Less than a decade ago, Wedel's ideas crystallized from suspicion into fact. Although not providing a great deal of detail, he did write that in what is now Scott County, Plains peoples not only used canals to transport water from springs and perennial streams hundreds of meters to their fields, but that they received this technology from people who left Taos and/or Picuris pueblos as a result of Indéean contacts (Wedel 1986: 143, 148). Furthermore, he noted that later settlers from the East cleaned-out and reused these canals in the 19th century (Wedel 1986: 143). Documentary evidence of canal irrigation, while overwhelmingly from the Southwest, does exist for neighboring areas, and it shows that native southwestern irrigators had an impact on other people in other times.

Ethnographic Evidence

Details concerning aboriginal canal irrigation from the 19th and 20th centuries are less than perfect. Information pertaining to certain groups such as the Zuni and the Akimel O'odham are reasonably abundant and fairly useful as these people have remained in relatively well-defined territories, discrete from other peoples. In the cases of northern New Mexico and eastern Sonora, however, the picture is much different. To be sure, there is some information, but it is flawed in several ways. First, unless there is indisputable documentation, an extremely rare circumstance if ever there was one, it is notoriously difficult to determine with any accuracy the ages of canals, especially those in use at the time they are discussed. Second, many canal systems are undoubtedly quite new, and even though they are used by people who themselves, or whose ancestors, are indigenous, the antiquity and the cultural origin of the technology are not above question. Comments made by Spanish missionaries concerning the building and use of certain canal systems, as discussed above, should serve as ample illustration.

Third, many canals are undoubtedly very old, but have been abandoned and later renovated, often by non-native peoples. Perhaps the best example of this situation was provided by J. A. Jeancon. In writing of archaeological work done in the Chama Valley of northern New Mexico in the early 1920s, he noted that: 'No evidence of former irrigating ditches or other agricultural activities are to be found, unless we except the large "mother ditch" which is still used by Mexicans [*sic*] and is said to be of Indian origin' (Jeancon 1923: 2). The phrase '*said* to be of Indian origin' (emphasis added) should, in itself, be taken as a cautionary note concerning inferences about this canal, as well as others. Although he was referring to farmland in general, and not canal irrigation systems specifically, Woodbury's assessment of conditions is equally on the mark. According to him '[n]one of the New Mexican pueblos now has an economy that is sufficiently like the aboriginal situation to make their acreage under cultivation a safe basis for inferences about the past' (Woodbury 1961*b*: 37). In the case of eastern Sonora, native peoples have long ago been acculturated and assimilated into Mexican society (e.g., Hinton 1983:315–16; Pennington 1980: 1).

Finally, most of the ethnographic accounts of irrigation provide few if any details concerning the canals and related features that comprise the systems. Typically, ethnographers simply mention canals or irrigation and assume that readers need no additional information concerning the physical features themselves. In no small way, this is probably because most ethnographic data

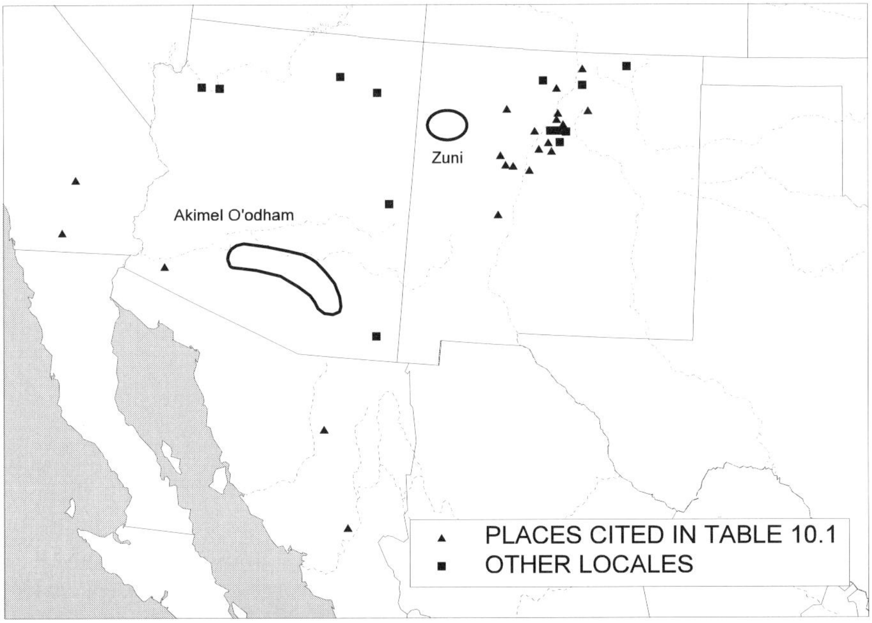

Fig. 10.5. Locations cited in the text and Table 10.1 for which there is ethnographic evidence of canal irrigation.

were collected by anthropologists interested principally in the cultural, social, economic, or political aspects of irrigation, and less so in technology and landscape modification. Given these problems, most of the information concerning native canal irrigation in northern New Mexico and eastern Sonora during recent times is relegated to tabular form (Table 10.1, Fig. 10.5).

There are, however, two exceptions to this, and those are the independent works of Ford and Ellis. Ford undertook a detailed study of canal irrigation at Picuris, one of the Northern Tiwa pueblos. Although the system had been modified some in the 20th century, the main canal was U-shaped, 90 cm wide, 90 cm deep, and originally 3.6 km long. A dam made of logs and boulders impounded and diverted water from the Rio Pueblo. At the downstream end of the system a series of lateral canals carried water to fields (Ford 1977: 143). The total irrigated area measured 40 ha divided by thirty families such that each had five to eight parcels. Each plot was watered three to five times per year as needed, each time with water standing 10 cm deep (Ford 1977: 148). The main canal was maintained communally, but each field was managed privately. Conflicts over water usage were not uncommon (Ford 1977: 144–6, 150).

Table 10.1. Observations of aboriginal canal irrigation gleaned from the ethnographic literature

Eastern Pueblos, New Mexico
'practice farming with an irrigation system of canals, dams, and ditches and gravity flow' (Jorgensen 1983: 692)

Taos Pueblo, New Mexico
'irrigation was easy' (Bandelier 1890: 157)
'is one of the best irrigated sections of the Southwest' (Bandelier 1892: 31)
'a series of stones placed across the small stream to turn water into ditches . . . only a foot wide . . . to provide water for irrigation' (Baker, Rae, Minor, and Connor 1973: 101)

Chamita Valley, New Mexico
'enjoys exceptional facilities for irrigation' (Bandelier 1892: 59, see also 88)

Tewa pueblos, New Mexico
'The Tewas irrigated on the Rio Grande' (Bandelier 1890: 31)

San Juan Pueblo, New Mexico
'Soil is improved by means of fertilization and new alluvium transported by irrigation waters' (Ford 1968: 188)

Santa Clara Pueblo, New Mexico
'It is fairly well off in irrigable land' (Bandelier and Hewitt 1937: 95)

Pojoaque Valley, New Mexico
'is not only fertile, but very well irrigated' (Bandelier 1892: 83, see also 88)

Nambé Pueblo, New Mexico
'Today . . . [there are] irrigated fields nearby' (Schroeder and Matson 1965: 115)

Pecos Pueblo, New Mexico
'irrigation was very easy' (Bandelier 1890: 157)

Cochiti Pueblo, New Mexico
'Cochiti ditches bring water to the various laterals supplying individual fields' (Lange 1959: 80–1)

Santa Domingo Pueblo, New Mexico
'It has ample land of great fertility and almost unfailing water for irrigation' (Bandelier and Hewitt 1937: 110)
'its fields, under irrigation, are . . . extensive' (Lange 1979*b*: 381)

Zia Pueblo, New Mexico
'In the 1970s they rely mostly on irrigated fields concentrated on both sides of the river immediately above town' (Hoebel 1979: 410)

Jemez Pueblo, New Mexico
'fertile when irrigated' (Bandelier 1892: 196)

Piro pueblos, New Mexico
'the Piros on the Rio Grande irrigated their lands' (Bandelier 1890: 31)

Tequima of Sonora, Mexico
'the lands . . . are rich and fertile. Irrigation was practiced where necessary' (Johnson 1950: 11)

Névome of Sonora, Mexico
'Pima Bajo who lived on the right bank of the Yaqui (Onavas, Movas, and near the Rio Chico) practiced some type of irrigation' (Pennington 1980: 355)

Cahitan peoples of Sonora, Mexico
'where water is available throughout much of the year canals have been constructed . . . to distribute water . . . weirs of stakes and brush are sometimes used to divert the water' (Dunbier 1968: 270)

More than a decade earlier than Ford did his research, Ellis undertook a rather detailed study of the history of Tewa water use. Focusing on the areas around the pueblos of Tesuque, San Ildefonso, Pojoaque, and Nambé, Ellis (n.d.*b*) interviewed numerous people, gleaned data from archival materials, and conducted archaeological surveys and excavations. She identified a total of forty-eight canal systems, thirteen at San Ildefonso, watering 833 ha, six at Pojoaque, irrigating 712 ha, twelve at Nambé, servicing 924 ha, and seventeen at Tesuque. Of these, five had always been used by Hispanics, thirteen were used by Hispanics but had previously been used by indigenous people, and thirty had been used by residents of the pueblos throughout the 19th and 20th centuries. Ellis determined that thirty canals were built originally in prehistoric times (to be discussed later), fifteen in historic times, two in modern times, and that the date of origin of one was indeterminable. Although there was some encroachment by non-aboriginals, the canal irrigation systems remained remarkably native.

As for details about these canals, she provided precious little. They all carried water diverted out of tributaries of the Rio Grande, some very near their heads at springs. Diversion dams were made of either brush (Ellis 1978: 61, 62) or logs and branches laid at an angle to the streams, and where the water flowed quite rapidly, they were held by piles of stone-filled baskets (Ellis n.d.*a*). The canals themselves varied considerably in size. Main canals ranged in length from 1.2 to 4 km, but most were described simply as either 'long' or 'short'. They were parabolic in cross-sectional shape, and varied between 1.8 and 3 m in width at the top and from 22 to 60 cm in depth. In several places these canals crossed by intervening arroyos by means of wooden aqueducts (Condie 1989, known locally as 'flumes' (Fig. 10.6). They branched several times into laterals which carried water to the fields.

The most detailed ethnographic information about native irrigation comes from the Zuni and the Akimel O'odham, with other groups in Arizona providing some shreds of evidence. It is possible that canal irrigation was practiced on riverine floodplains in the 19th century at Zuni, but firm evidence exists for only three spring-fed stream localities—Nutria, Pescado, and Ojo Caliente (Ferguson 1985: 51). Not surprisingly, it was Cushing who was the first anthropologist to describe irrigation at Zuni. In discussing Nutria Creek he noted that a dam, probably made of logs (Ford 1985*c*: 65), diverted its perennial waters into a 3.3 km long canal (see also Spier 1917: 216). This was probably no great feat as the stream channel here is less than 1 m deep (Ferguson 1985: 53). In a few places along its course, the canal crossed intervening drainages by means of split and hollowed log aqueducts. Fields were subdivided into plots measuring 3 by 3.5 m and separated by low earthen ridges. In no way should these gridded plots be confused with the waffle gardens discussed in Chapter 4, as these are five to ten times larger and not manually watered (Ford 1985*d*: 62, 63, fig. 8, 66, fig. 9). They extended up and down the valley for 2.4 km and across the valley nearly as far (Cushing 1974:

Fig. 10.6. A wooden aqueduct, known locally as a 'flume', typical of those used in northern New Mexico. (Photograph by author.)

355–67). More recently, the irrigated area encompassed only about 161 ha and was farmed by forty families for an average of 4 ha each (Ferguson 1985: 54). Bandelier, who visited the area a few years after Cushing, said only that Nutria was 'a very fertile and well irrigated valley' (Bandelier 1892: 340). Perhaps the irrigation was beginning to show signs of decline even then.

Stevenson (1904: 351) also noted the canal system at Nutria, and she also mentioned the one at Pescado. She did not say much about either, however, perhaps lending yet more speculation about the former's loss of importance. Instead, Stevenson focused her attention on Ojo Caliente. According to her, water from the spring at the latter locale was carried in two small canals measuring 75 cm wide and 50 cm deep (Fig. 10.3). Controlling the flow of water into one canal and not the other was accomplished by temporarily filling the opening of one canal with earth and brush. The longer of these two canals transported water to an area measuring 5.7 by 4.1 km, with the shorter canal watering a lesser area (Brandt 1995: 294; Kintigh 1985: 96). This is an exceptionally large amount of spring-irrigated land, and seemingly would have involved a larger and more complex canal system. It is not surprising, therefore, that the size of the area claimed to have been irrigated has not gone unquestioned (Jorgensen 1983: 687, 692; Woodbury n.d.*b*). Early in the 20th century, only 260 ha were irrigated at Ojo Caliente (Ferguson 1985: 56), suggesting further that any decreases in irrigated area that might have occurred since the late 1880s occurred shortly after Stevenson concluded her studies.

Not much is known of the canal irrigation system at Pescado. In addition to confirming Stevenson's observation that it existed, one early ethnographer (Spier 1917: 216) reported only that two springs there provided water for irrigation.

A far smaller amount of land remains irrigated at Zuni today (Perramond 1994: 72) than at any time in its past. Exactly why this is the case is not known and has not been systematically addressed. Two recent studies (Ferguson 1985: 55, 59; Ford 1985*d*: 42) have noted that irrigated land was never a large component of Zuni agriculture, although it was an important one. Its decline may be in part due to an overall decrease of local food production. Ford (1985*d*: 48–9) found, however, that siltation was a problem at both Nutria and Ojo Caliente, hinting that the value of the crops may have become diminished by the amount of work required to maintain the system. Evidence from elsewhere confirms that canal irrigation systems require high labor inputs.

Perhaps the best-known native irrigators in recent times are the Akimel O'odham. Whereas their neighbors, the Tohono O'odham are known for their singular reliance on runoff agriculture and water harvesting, the Akimel O'odham have long relied nearly exclusively on canal irrigation (Castetter and Bell 1942: 158, 161). Achieving icon status in this regard, at least among those interested in aboriginal water control, is one Ira Hayes. One of the more visible heroes of World War II, he helped raise the US flag on Iwo Jima. And, what was his ultimate reward, other than being immortalized in bronze? He was the subject of ballad and film, both of which portray his drowning in an irrigation canal!

The Akimel O'odham seem to have irrigated for as long as there is a written record, and, as will seen directly, earlier. Not only did Spanish explorers in the 1600s and missionaries in the 1700s see them irrigating (see above), but persons from the United States throughout the 19th century did as well (e.g., Bandelier 1890: 254; Bartlett 1854, vol. 2: 232; F. Russell 1908: 86–9; see also Castetter and Bell 1942: 157). And, of course, they continue to irrigate with canals today, in virtually every place where there is permanent water (Fig. 10.5).

Techniques used by these people are not that much different than those used by others. Dams were built to divert water out of rivers and into canals (Hackenberg 1983: 169, fig. 6). These consisted of two types, diversion dams or weirs, and retention dams designed to raise the water level. Weirs were built by first driving pilings of thick mesquite poles into the river bed in a zigzag arrangement, diagonally to the direction of stream flow. Mesquite branches were then stacked horizontally between the pilings. Retention dams were built perpendicular to the channel and involved pilings not unlike those for weirs. For this type of dam, however, arrowweed bushes were tied into bundles with mesquite or willow bark or rawhide and floated against the pilings. Trunks of larger trees, when available, were also used. The entire structures were then weighted down with rocks (Castetter and Bell 1942: 159).

Canals were excavated with digging sticks and wooden shovels (F. Russell 1908: 88, 97, fig. 10*a*,*b*). Details of canal size are sparse and somewhat conflicting. It has been suggested that the earliest canals were narrow and carried water to fields just above the normal flood level (Doelle 1981: 63). In more recent times, the main canals were much larger than appeared necessary, with much water being returned to the river (F. Russell 1908: 87). In fact, this partial use of water may have facilitated a flushing of toxic mineral salts which otherwise might have accumulated in the soil (Ezell 1961: 38; F. Russell 1908: 87). Also, these artificial water courses functioned as small oases that attracted all types of wildlife (Rea 1979) which must have made hunting much easier and more enjoyable than would have been the case were they non-existent.

The main canals and the branch canals, were considered as communally owned and were cleaned and kept by everyone who took water from them. Lateral or field canals were the responsibility of the individuals (Castetter and Bell 1942: 160). Fields ranged in size from 0.6 to 3 ha, were rectangular and arranged with reference to the canals (F. Russell 1908: 87–8, plate xia). They were often leveled in order to ensure uniform distribution of water. This was typically accomplished by carrying baskets of dirt scraped from high spots with shovels, to low spots and must have been a laborious and time-consuming task (Hackenberg 1962: 191). Small irregularities with field surfaces were often smoothed out during floods or with regular applications of irrigation waters (Castetter and Bell 1942: 161). Irrigation waters, as in the case of all forms of runoff is laden with sediment. These materials not only helped fill-in low spots, but they also acted as fertilizer, replenishing soil nutrients and eliminating the need to fallow (Castetter and Bell 1942: 172–3; Hackenberg 1962: 189).

Crops were not watered on any regular basis, but instead were irrigated whenever plant and soil conditions warranted and water was available. In general, three to five waterings were needed (Castetter and Bell 1942: 170).

Clearly, one of the greatest advantages of practicing canal irrigation in southern Arizona, was that it allowed for double cropping, the growing of two crops in the same field, the second one after the first was harvested. The low-lying Sonoran Desert is, as its name implies, a warm and dry environment. The frost-free or growing season is exceptionally long, 295 days on average (Jorgensen 1983: 689). It is the general paucity of water that limits cultivation. Where canal irrigation is feasible, however, neither of these normally limiting natural factors affects cultivation. In general, the first crop was planted immediately after the danger of frost had passed and the ground was sufficiently warm for seed germination. It was harvested early in the summer and the second crop was then planted. This planting always took place before mid August so as to avoid damage which could be caused later by early frosts (Castetter and Bell 1942: 144, 148). Canal irrigation as practiced by the Akimel O'odham was, according to the best authorities, 'on a lower technological level than that of the irrigated valleys of Sonora' (Castetter and Bell 1942: 59). However, it was more developed than that practiced by Yuman people who lived further downstream along the Gila River.

What might be considered 'conventional' canal irrigation was at no time practiced by Yuman farmers (Castetter and Bell 1951: 131; Jorgensen 1983: 692, 693). Instead, certain of these people had an agricultural base dependent on, and characterized by, flood recession (Chapter 11). Among the Cocopa, however, a 'modified' version of canal irrigation was practiced, and this is what Sedelmayr (1939: 110) first reported in 1746. The system involved fairly large swales which began near the edge of the river and just a little above the water level, and ran diagonally across the landscape. Some were reported by informants to extend for miles. During the winter months when the water level in the river was low, farmers would construct dams which were oriented diagonally across the channel. These were made of logs, supported by poles driven into the ground behind them. When the river rose in the spring, the dams diverted water into the swales (Castetter and Bell 1951: 133).

In addition to building the dams across the river channel, earthen dams were also built across the swales in series. As water entered the swales, it was impounded by the first of these dams and allowed to overflow the banks thereby irrigating fields on either side. Once the fields adjacent to the swales were sufficiently watered, the dam was broken and water allowed to flow to the next dam where the process was repeated (Castetter and Bell 1951: 134; Kelly 1977: 27).

This system could, arguably, be considered more a form of water harvesting than canal irrigation, especially because it involved the modification of natural features. However, the river was perennial and the dams built across the swales were eventually replaced by lateral canals as the dependence of seasonal fluctuations in the river level were lessened (Harwell and Kelly 1983: 75, 79; Kelly 1977: 27).

Elsewhere in the Greater Southwest there are only a few fragmentary accounts of canal irrigation practiced by aboriginal people in the 19th and 20th centuries. This is not to say that the particular agricultural systems have not been studied in detail by someone, or that the references are unreliable. On the contrary, the accounts seem to be reliable, but the publications resulting from in-depth ethnographic studies simply do not contain much about agriculture.

Certain lands in the Chino Valley near Palm Springs, California were not only cultivated since the first people entered the valley, but were irrigated by means of a canal as well. A half century ago, the old canal looked much like a small dry creek bed (Patencio 1943: 57–8). Diversion of water from streams and springs to fields by means of canals was also common in the mid 1800s among the Cahuilla of California (Lawton and Bean 1968: 20). In northern Arizona, the Hopi long irrigated terraces with canals that carried water from small springs (see Chapter 8). Reliance on water from larger sources, especially rivers, was neither common nor important (Brew 1979: 515; Forde 1931*b*: 361) until recently and then in only one locale. The floor of Moenkopi Canyon, near Tuba City, Arizona, is irrigated by means of a canal network that carries water from large springs and a permanent stream (Winter 1976: 110). As impressive as it is (Fig. 10.7), this is an entirely 20th-century native agricultural landscape. It is the result of people who migrated from Oraibi Pueblo after much of the lands there were destroyed by flooding in 1906 (Nagata 1970). Of course, it is possible that the technology evident here was indigenous, came from Oraibi, and was simply not recorded by earlier writers. It is also possible that it was not native, but something adopted from non-aboriginals.

Peoples living northwest of Moenkopi in Grand Canyon have been the subject of some notable ethnographic studies which mention irrigation. In the early part of this century the Havasupai had two diversion dams or weirs which diverted water from spring-fed streams. Although the canal systems

Fig. 10.7. View of Moenkopi Canyon from the edge of the north rim. Note the canal in the lower left, evident by more luxuriant vegetation in an otherwise dry environment. (Photograph by author.)

(Schwartz 1983: 13, 14) have been described as being small and 'not complex' (Spier 1928: 102), the weirs are truly impressive. They were made of heavy logs set into one bank and laid halfway across the streambeds. Stakes driven into the channel on the logs' downstream sides helped hold them in place. The banks of the streams near the dams were riprapped with stones to prevent erosion and keep the logs from washing out. Stakes were also driven into the channel bed in front of the weirs, but at an angle so that their tops rested against the weirs. Fascines of willow, approximately 1.5 m long and containing stones were laid end to end along the base of the weir with a second row placed on the channel bottom just in front. Willow brush was then piled thickly on the face of the weirs all the way to the tops (Spier 1928: 101).

The main canals were 60 to 90 cm deep and of unspecified width or cross-sectional shape. Lateral canals which carried water from the main canals to the fields were smaller (Spier 1928: 102). At least 71 ha were under irrigation in the 1940s (Breed 1948: 674). This agricultural landscape appeared in 1970 to look unchanged since pre-European times (Schwartz 1983: 15, 20–1, fig. 5). The Walapi, who have long occupied lands west of the Havasupai, built similar dams and canal systems along the Big Sandy and Bill Williams rivers (Dobyns and Euler 1976: 10–12; McGuire 1983: 33).

Finally, mention needs to be made of aboriginal irrigation in highland eastern Arizona and far northern New Mexico. These areas have been occupied throughout the historic era by Indéean peoples who rely on canals. As discussed earlier (Chapter 5 and above) these people were relatively late arrivals in the Southwest. One source (Opler 1983*a*: 370) argues that they were agriculturists upon arrival. Another (Basso 1983: 465) maintains that they learned farming from the western puebloan people. Regardless, at least three groups of them, the Jicarilla (Tiller 1983: 441), the Indé (Goodwin 1942: 150–1), and the Chiricahua (Opler 1941: 372, 1983*b*: 413), practiced canal irrigation. In doing so, water was typically taken out of rapidly flowing, steep gradient streams, often without benefit of a diversion device. It was carried through shallow, community-maintained canals that varied in length from a few hundred meters to 3.3 km to numerous individually owned fields measuring only 0.2 ha each (Buskirk 1949: 32, 59–60, 67–70; Goodwin 1942: 45, 156, 374). If these people were not practicing canal irrigation prior to either their arrival in the area or that of the European's, it is likely that their predecessors were (Castetter and Bell 1951: 134; Pool 1994: 93). There is certainly an abundance of data that demonstrate just how ancient and widespread the practice of canal irrigation was in the Greater Southwest.

Archaeological Evidence

Many of the canal irrigation systems used by aboriginal people in relatively recent times have undoubtedly been in use for a few or perhaps even several centuries (e.g., northern New Mexico). Indeed, for some (e.g., Zuni) there actually exists documentary data going back to the 1500s. Were this not sufficiently impressive in its own right, there is evidence that some systems claimed and thought to be historic in age were actually built prehistorically and simply renovated at a later date (e.g., Tumacacori). There is even evidence that canals built and abandoned prehistorically were brought back into use by Mormon settlers in the late 19th century (Halseth 1932: 166; Hodge 1907: 621; Judd 1931: 164; Midvale 1968: 28; Shetrone 1945: 384; Turney 1929: 21). Such evidence is, of course, archaeological. Although the accomplishments of native irrigators since the beginning of the 16th century are great, they pale in comparison to those of their ancestors and predecessors. Some of the systems developed by these earlier people were small and unimpressive, but others were on a par with some of the great waterworks from other parts of both the ancient and modern worlds. Indeed, this fact was not lost on the first non-natives to identify archaeological remains of ancient canal irrigation in the Southwest.

Interestingly enough, these first observers were not trained archaeologists, but rather a missionary priest and a soldier—Kino and Manje. Furthermore, Kino even noted the possibility of renovation. On seeing the ruins of Casa Grande in the Gila Valley in 1697, he noted:

> that it had a large aqueduct with a very great embankment, which must have been three *varas* high and six or seven wide—wider than the causeway of Guadalupe at Mexico. This very great aqueduct, as is still seen, not only conducted the water from the river to the Casa Grande, but at the same time, making a great turn, it watered and enclosed a champaign many leagues in length and breadth, and of very level and very rich land. With ease, also, one could now restore and roof the house and repair the great aqueduct (Kino 1948: 172; see also Nentvig 1980: 13–15 for a later but similar description).

Kino was, of course, describing a canal, not an aqueduct, and his measurements were of a spoil bank along one side. Converting to modern measurements, this bank was approximately 83 cm high, and 5 to 6 m wide. According to Manje (1954: 86, 1971: 342–3, 371, 208), the canal itself was 8.2 m wide and 3.3 m deep. By any standards, this was part of an impressive canal irrigation system (Fig. 10.8).

The magnitude of features such as this and the extensiveness of the area they watered in both the Gila and the Salt valleys has resulted in canal irrigation becoming a hallmark characteristic of the ancient Hohokam culture. According to one early archaeologist, Omar A. Turney (1924: 5), these are '[t]he greatest exploit of ancient man in America'. He went on to note that these people, the Hohokam, should be known as the 'Canal Builders' (Turney 1924: 9). These assessments have long been evident to professional archaeologists (e.g., Gumerman and Haury 1979; Hodge 1893), and are reflected in the popular literature as well (e.g., Doyel 1979; Haury 1945*a*). But there were also

Fig. 10.8. Downstream view of a modern irrigation canal south of, and which carries water from the Gila River. Houses on the left mark the northern edge of the present-day town of Coolidge, Arizona. The ruins of Casa Grande are on the right in the distance (arrow). Here, this canal parallels the prehistoric canal first seen by Kino and Manje (see Midvale 1965: 85). (Photograph by author.)

smaller systems outside this Hohokam Core area and these also warrant discussion (Greiser and Moore 1995: 189–92).

Archaeological remains of prehistoric canal irrigation systems have been found in a roughly square-shaped area (Fig. 10.9) with corner points in northern Sonora, southern Utah, northern New Mexico, and northern Chihuahua (Dart 1989: 8, fig. 4; Greiser and Moore 1995: 190, fig. 12.1, 191, fig. 12.2). Evidence ranges all the way from simple statements that remnants of ancient canals were seen, to detailed archaeological investigations.

Fragments of three canals were found near Caborca, Sonora and subjected to archaeological excavation. It could not be determined where and how water was diverted out of the river, and no dimensions have been reported. Profiles revealed superposition indicating that only a single canal functioned at any one time (Bowen n.d.: 28, 56). Whether the latest of these canals was one mentioned by Manje is unclear, but a distinct possibility. That canals were rebuilt strongly suggests that these farmers were constantly battling a very dynamic fluvial system.

The dynamic nature of rivers in the Sonoran Desert is perhaps no more evident than in the Santa Cruz Valley, and this has precipitated some debate

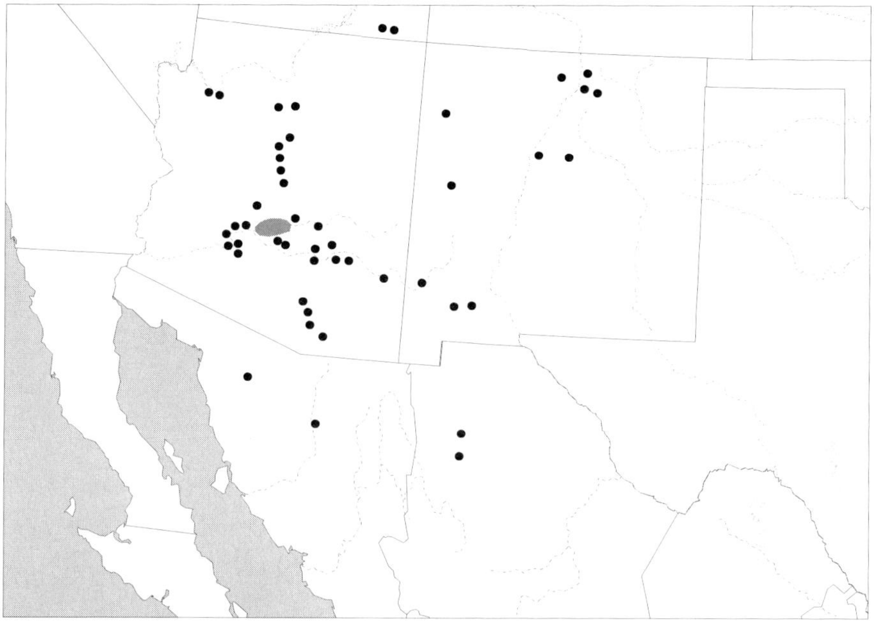

Fig. 10.9. Locations cited in the text for which there is archaeological evidence of canal irrigation.

on the role of irrigation in the overall subsistence landscape (Doyel 1977: 94–6; Grebinger 1976: 40–1). Indeed, there is some very good evidence that the building of canal irrigation systems in certain highly unstable locales contributed to environmental degradation which eventually led to the failure of the systems (Waters 1988: 480–1, 489). In other words, some canal systems might have carried the seeds of their own destruction. Several canals have been reported from the Santa Cruz, but few have been subjected to archaeological excavation (Dart 1989: 14). In addition to the one at Tumacacori, canals have been reported at a site close to downtown Tucson, another two sites a short distance north of town, and a fourth site further north near Marana.

Recent archaeological excavations at the Canal site resulted in the discovery of fragments of four canal alignments paralleling each other. The largest averaged 1.25 m wide, 75 cm deep, and was traced more than 370 m. In all likelihood it carried water diverted from the Santa Cruz and transported it to a terrace above the floodplain, between 1.2 and 1.8 km from the source (Mabry and Holmlund 1998: 285). A second canal was traced for 225 m and averaged 28 cm deep and 1.21 m wide. Prehistoric 're-engineering' created a series of canals within canals (Mabry 1997: 421, fig. 5.2). Each of these two segments consists of two or more anastomosing, superimposed canal alignments, each of these shows signs of multiple phases of cleaning and reconstruction (Mabry 1997: 422–4, figs 5.3–5.5; Mabry and Holmlund 1998: 289). The other two canal segments were shorter and smaller, and have been interpreted as 'experimental alignments' (Mabry and Holmlund 1998: 285).

Given the great lengths that the canals were exposed and excellent cross-sectional profiles, studies were undertaken to examine gradient and discharge. Gradients ranged between 0.07 and 0.12 percent, which is far less than typical for canals of this sort (Mabry and Holmlund 1998: 294). As a result, the velocity of the water flowing through the canals, and the amount available for crops would have been low. Not only would the irrigated fields have been small, but the amount of time and knowledge that went into the design, and the amount of labor that went into construction and maintenance would have been high. The canals were first used AD 1000–1100 and abandoned AD 1150–1450 (Mabry and Holmlund 1998: 283).

Remnants of two prehistoric canals were found when backhoe trenches were put across a site on the west side of the Santa Cruz near the confluence of Rillito Creek. Both canals had elliptical cross-sectional shapes and were filled with fine silts and sands. One was 1.4 m wide and 25 cm deep and buried 10 cm below the plowzone. The second was 1.8 m wide and 30 cm deep and was partially disturbed by historic plowing. Whether or not these two segments were used contemporaneously or at two different times is unclear. However, the former scenario is probable as the two segments are 33 m apart and oriented in different directions. It is suggested that given their small size, these were probably secondary canals which carried water to fields from a main canal (Kinkade and Fritz 1975: 29–31, fig. 5).

Perhaps the most intriguing canal site in the Santa Cruz Valley, and certainly the oldest, is the recently discovered Las Capas site (Muro 1998*a*). Here, archaeologist Jonathan B. Mabry discovered series of canals not unlike those at the Canal site. Details of this site and its canals remain to be published. Of the little that has appeared in print to date, however, the most important thing is the age of one canal. Radiocarbon dating of horizons immediately above and below the deepest one containing a canal remnant (Fig. 10.10) firmly place the date of that canal to between 1000 and 1100 BC (Muro 1998*b*: 15). This is contemporaneous with the earliest known canals in both Mesoamerica (Doolittle 1990*b*: 20–2) and the Andes (Zimmerer 1995). Not inconsequentially, a series of backhoe trenches across the canal and off the site, revealed that the canal actually carried water from the perennial Santa Cruz River (Jonathan B. Mabry, personal communication 1999). Doing so was a feat not matched anywhere else so early.

At the Los Morteros site in the north end of the Tucson Basin a segment of an irrigation canal was revealed through a backhoe trench as was the case further south (S. Fish, P. Fish, and Madsen 1985: 79–80, fig. 2). It was trapezoidal in cross-sectional shape, measured 2.5 m wide at the top, 1.3 m wide at the bottom, 75 cm deep, filled with fine silt and sand and 20 cm below the plowzone.

Fig. 10.10. An upchannel view of the recently uncovered oldest irrigation canal in North America, and discoverer Jonathan B. Mabry standing amidst his archaeological handiwork. (Photograph by author.)

The discovery of this canal led to further study which revealed portions of seven other canals forming two separate irrigation systems on the floodplain (Bernard-Shaw 1988: 156). On the basis of superposition and ceramic evidence it was concluded that these systems functioned at different times. No main canals were found, however, the one found in the earlier study may well have served this function. The two systems were each comprised of one secondary canal and a few lateral canals. The earliest system dated *c.* AD 900–1000. The largest canal associated with it had an average width of 2 m, and an average depth of 60 cm, and was traced for 72 m. The later system dated *c.* AD 950–1050. Its largest known canal averaged 3 m wide and 50 cm deep and was traced for 150 m. The canals were elliptical in cross-section and at 0.8–1.3 percent, had relatively steep gradients (Bernard-Shaw and Doelle 1991: 28, 32).

The steepness of their gradients, and the uniform nature of their fill, indicated that these canals conveyed large amounts of water fairly quickly. Hydraulic assessments concluded that the earlier canal had a discharge of 0.34 cu m per second, and the later one had a discharge of 0.55 cu m per second. A total of approximately 6 ha was irrigated by these systems at any one time (Bernard-Shaw 1988: 158).

No dams or weirs were found in association with these canal systems. In large part this is a function of the presence of thick alluvial deposits which comprise the floodplain and the incision and avulsion that has occurred during the 20th century. It is suspected, however, that some type of diversion structure must have been used as on the basis of geomorphic evidence, the Santa Cruz was a relatively low velocity stream (Bernard-Shaw and Doelle 1991: 32; S. Fish, P. Fish, and Madsen 1985: 91).

Although there is good archaeological evidence of canal irrigation from only a few sites in the Santa Cruz, it is probable that much of the valley was so farmed in prehistoric times. There are four reasons for making this conclusion. First, solid evidence comes from sites separated by great distance. Second, the documentary evidence indicates that much of the valley was irrigated in the late 1600s and early 1700s. Third, erosion and sedimentation are great, thereby obliterating or burying evidence of canals. And, fourth, there are numerous habitation sites situated in such a way that they look as though they had a canal orientation (S. Fish 1995: 104; S. Fish, P. Fish, and Madsen 1989: 256, 1992*a*: 51).

As is the case with the Santa Cruz, there are a number of known prehistoric irrigation canals in the Agua Fria Valley, but few have been described in any detail (Dart 1989: 13). The first of these was described, if that term applies in this case, by Midvale (1974). As was his style, Midvale portrayed much more in his maps than he said in the accompanying text. His map of the confluence of the Agua Fria with the Gila River shows a canal beginning on the west bank of the river, due east of the present-day town of Avondale. It ran for approximately 3 km before branching into at least four smaller canals, each approximately 1 km in length. There must have been a series of smaller canals branching off of these as Midvale's map reveals a total irrigated area of approximately 225 ha.

Twenty-seven miles up the Agua Fria, a second canal site was found during an archaeological survey and discussed, or at least mentioned (Dove 1970: 18–19). This canal extended for 1.5 km along the base of low terrace. It apparently functioned sometime between AD 700 and 1300 when the climate was drier than usual, irrigating approximately 40 ha of land which had been described earlier as containing 'traces of cultivation' (Turney 1929: 142).

The upper Salt River and Tonto Creek valleys comprise yet another area where there exists reports of canals but very little in the way of written descriptions (Dart 1989: 11). Bandelier (1892: 421) reported seeing an ancient canal on the north side of the Salt River under what is today Lake Roosevelt and another along Carrizo Creek, a tributary of Tonto Creek (Bandelier 1892: 400). The presence of Lake Roosevelt, a reservoir built early in the 20th century, has long led archaeologists to assume that numerous and large canal systems lie submerged (e.g., Turney 1924: 8; Wood and McAllister 1984: 277, 281, 285, 286, figs 5–7). Recent investigations into the region's geomorphology and hydrology by John R. Welch (1992, 1994) have concluded that canal irrigation systems were probably small and limited to only a few favored locales. Although the irrigated area undoubtedly increased as the population grew, there were probably never more than 610 ha watered by canals, and these were distributed in thirteen discrete locales (Ciolek-Torrello, Whittlesey, and Welch 1994: 455, fig. 15.5, 467).

Perhaps the most important area in terms of canal irrigation in the Hohokam Periphery is the Verde Valley. Not only have many sites been reported from there, but the details of some water control systems have been published (Dart 1989: 11–12). One study (Kearns, Polk, and Teague 1973) found no fewer than seven sites with confirmed remnants of prehistoric canals very near the confluence of the Verde with the Salt River. This figure probably does not reflect the actual number of canals, however, as there are many historic canals with ancient antecedents in the area, and these were not counted. Indeed, Bandelier (1892: 433) noted that ancient canals were quite common in the area. No archaeological excavations were conducted on the canals, but surface evidence, in the form of rock-bordered linear depressions, suggests they varied in width from 1 to 3 m. Surface depressions range in depth from 0.5 to 2 m, indicating that the canals might have been quite deep (Kearns, Polk, and Teague 1973: 251; see also Bandelier 1892: 434).

Exact lengths of the canals could not be determined as the active river and cross-cutting tributaries have resulted in only fragments remaining in place. However, in at least one locale, canal segments could be linked together by inference for more than 1.5 km. Bandelier (1892: 434) said that one canal was 30 km long, but that seems to be an exaggeration. Diversion structures, if any were used, have also long been obliterated. The fact that arroyos cut the canals in several places suggests that when the canals were in use, some type of aqueduct was needed (Kearns, Polk, and Teague 1973: 311–12, fig. 54). In all probability these were no different from others documented elsewhere (see Fig. 10.4). Finally, in at least a few places main canals on the lower terraces bifurcate into smaller branch canals that carried water to fields on the floodplain (Kearns, Polk, and Teague 1973: 252). These canals have been dated to AD 750–1150 (Kearns, Polk, and Teague 1973: 253).

The vicinity of Horseshoe Reservoir, 50 km up the Verde Valley, contains many remnants of ancient canal irrigation. Cosmos Mindeleff (1896) was the first archaeologist to make this discovery. He was particularly taken with the remnants of one canal that carried water to a tract of land that continues to be irrigated today (Fig. 10.11), albeit by a modern canal. The modern canal is necessary because of a great flood in 1891, which obliterated all but a few portions of the prehistoric canal, and resulted in heavy sedimentation which altered the entire geomorphology of the region. When Mindeleff saw the ancient canal, it was nothing more than a very shallow grass-covered trough bounded on either side by low ridge of earth and pebbles. On average, the remnants of the ancient canal were 3 m higher up the slope than the modern canal, but undoubtedly irrigated the same tracts of land (Mindeleff 1896: 238–9). No traces of a dam or weir were found, but Mindeleff (1896: 238)

Fig. 10.11. Irrigated floodplain fields downstream of Horseshoe Dam on the east side of the Verde River. Note the oasis quality of this locale. The prehistoric canal which once carried water to these same fields parallels the modern canal, running along the foot of the mesa out of view on the left.

suspected that the 3 km long canal began approximately 1 km downstream from where the modern Horseshoe Dam is located.

Although Mindeleff provided the first record of irrigation features in this part of the Verde, it was Midvale (1946) who provided the best documentation about ancient canals in the area. Just prior to construction of the dam and reservoir, Midvale rushed to the area and dutifully recorded canals and related features. Although he did not publish any descriptions, he did photograph the canals and the landscape, and he drew a map of the area which shows their locations. Midvale's map shows both the modern and the ancient canal discussed by Mindeleff. It also shows one that carried water diverted out of Lime Creek, a western tributary of the Verde, to fields near the confluence of the streams. Perhaps most importantly, however, it portrays two very long canals on the east side of the river.

One of these canals, the Little Limestone Canal, was actually a 1.5 km long branch off the other canal, and it watered lower floodplain fields, near a large ruin and north of Mullen Wash (Fig. 10.12). The other canal, the Limestone Canal, was at least 6 km long and crossed two large arroyos that empty into the Verde from the east. Where it crossed Mullen Wash, Midvale labeled a feature that he thought paralleled the canal on its downstream side, 'The

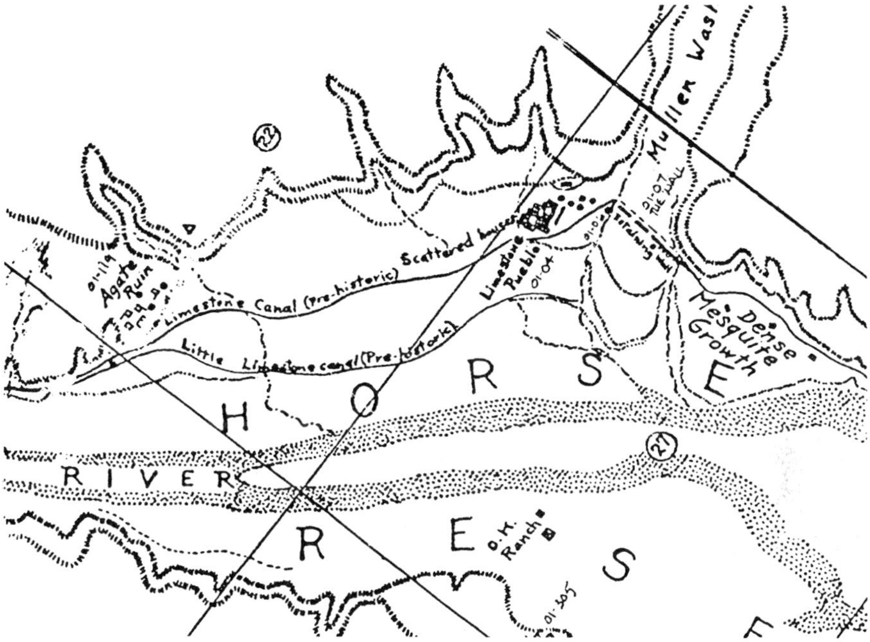

Fig. 10.12. An enlargement of one part of Midvale's 1946 map of the Horseshoe Reservoir area, including ancient irrigation canals and related features. (Published by permission of the Department of Anthropology, Arizona State University.)

Wall' in one place and a 'Retaining wall' in another (Fig. 10.12). Where it crossed Deadman's Creek, Midvale labeled a similar feature 'Stone was to Support Canal (Aqueduct Structure?)'. Both labels are correct and refer to identical features although different terms were used. In 1993 the level of Horseshoe Reservoir was lowered, and for the first time in a half century the feature at Mullen Wash was exposed. Both from the air (Fig. 10.13) and from the ground (Fig. 10.14) the feature looks like a wall. In reality, however, it was an aqueduct. All that remains are those sections on the floodplain, on either side of the arroyo channel. They are comprised of boulders which supported a canal that ran along the top, but has long since been destroyed. Exactly what the canal was made out of will never be known but it is likely that it was some type of plaster (for a parallel from elsewhere see Doolittle 1995: 316–20, fig. 10). Where it crossed the channel, the canal was probably split and hollowed logs not unlike those used in other places (Van West and Altschul 1997: 346, plate 9.4; see Fig. 10.4). By any standards, both the design and construction of this feature are remarkable. The existence of such an aqueduct is evidence of sophisticated planning and highly skilled labor, for getting water to flow above the ground surface is no easy feat.

Fig. 10.13. Aerial view of the remains of the prehistoric aqueduct where it crosses Mullen Wash. Note the break in the feature at the active channel. Appeared originally in Van West and Altschul (1997: 345, plate 9.2). (Provided courtesy of Jeffrey A. Homburg, and reprinted by permission of Statistical Research, Inc., Tucson, Arizona.)

Fig. 10.14. Ground view of the Mullen Wash aqueduct, with the author serving as scale. Appeared originally in Van West and Altschul (1997: 345, plate 9.3). (Provided courtesy of Jeffrey A. Homburg, and reprinted by permission of Statistical Research, Inc., Tucson, Arizona.)

Fig. 10.15. 'Part of an old irrigating ditch' in the Verde Valley (Mindeleff 1896: 241, fig. 300).

Further up river, 6 km south of present-day Camp Verde, Mindeleff (1896: 240–1, 243) found yet more traces of ancient canals. Details about these canals are scant, but a few things are known (see also Turney 1929: 156). First, they were very close to the river, only 1.25 m above the water. Second, they were buried by massive amounts of sediment as a result of post-use floods and they were exposed by more recent floods. Third, they were elliptical in cross-sectional shape (Fig. 10.15).

Remnants of canals have also been recorded but not described in detail along Beaver Creek, a tributary of the Verde (Schroeder 1960: 22, 27, 35). These carried water diverted from the creek. The most impressive of these systems has to be those located near Sacred Mountain. There, seven field complexes involving approximately 80 ha have been identified. From the air, these fields appear to be very similar to the rock-bordered grid fields near Safford (see Fig. 7.14). Three things make them different, however. First, there are remnants of canals that once carried water to the fields from Beaver Creek and Walker Wash, a tributary. Delimited by parallel rows of rocks, these canals branched into three different sizes, resulting in an elaborate water distribution system. Second, the grids, at only 2–3 m square, are

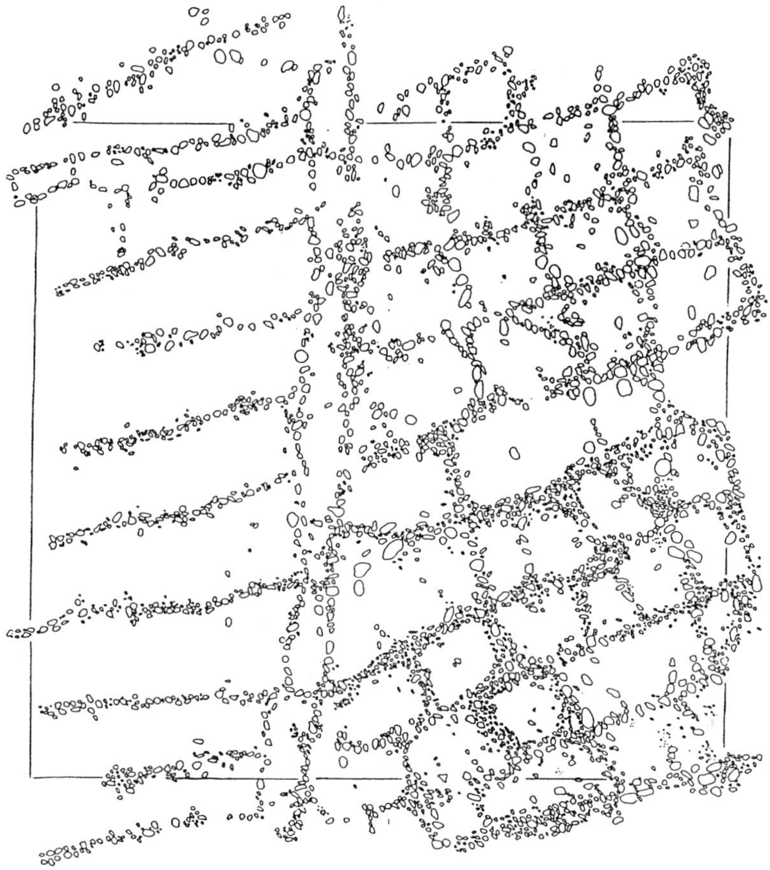

Fig. 10.16. Plan view of a canal and rock-bordered gridded fields near Sacred Mountain, Arizona. The grids average 2 m on each side. Appeared originally in Fish and Fish (1984: 154, fig. 4). (Reprinted by permission of the Department of Anthropology, Arizona State University.)

considerably smaller than those near Safford. And, third, the grids are not all totally enclosed by the rock borders. In almost every grid there appears to be an opening in the border that presumably would have allowed water to flow from one grid to the next (Fig. 10.16). These systems functioned *c.* AD 1300 (P. Fish and S. Fish 1984: 150–5, figs 2 4).

As unusual as these systems are, however, they are not the most unusual ones in the area. Indeed, a few kilometers downstream, in the southern unit of Montezuma Castle National Monument, is arguably the most unusual canal irrigation system in the Southwest. Montezuma Well is not a well, *per se*, but rather a sinkhole on the edge of Beaver Creek. It measures 120 m in diameter and 37 m deep. Springs producing 3,800 liters per minute (Baker *et al.*

Fig. 10.17. Montezuma Well. The fissure through which water flows from this spring-fed sinkhole is on the right, near the luxuriant vegetative growth. (Photograph by author.)

1973: 41), and a fissure on the side near the creek combine to keep it slightly less than half-full of water year round (Fig. 10.17). Two pueblos along the rim and three smaller ones just below it were home to perhaps 200 people between AD 1125 and 1400 (Schroeder and Hastings 1958: 29). These people did not let this much water go to waste. On the northern bank of Beaver Creek where the fissure debauches, they excavated a canal measuring approximately 1 m wide and deep which carried water to a minimum of 24 ha of fields 1.6 km away (Schroeder and Hastings 1958: 9). At least three branch canals have been identified at the fields (Schroeder 1960: 172a, fig. 6). More importantly, however, the waters flowing through the system were rich in calcium carbonates. Over the course of time these minerals resulted in the formation of travertine which fossilized the canal (Schroeder and Hastings 1958: 30), and, hence, created a most conspicuous feature on the local landscape (Fig. 10.18).

In addition to these canals there are numerous others that have been reported in the headwaters of the Verde (P. Fish 1974: 17). There were also more than 100 other small canals in use early in the 20th century that may have been prehistoric in origin, and only renovated in historic or recent times (Turney 1929: 155).

Irrigation was not only practiced below the Mogollon Rim in prehistoric

Fig. 10.18. The travertine encrusted canal which carried water from Montezuma Well to fields during prehistoric times. (Photograph by author.)

times, but on the Colorado Plateau as well. To be sure, evidence of such is not widespread, but it does exist. During excavations carried out near Flagstaff, Arizona in the 1950s two different canals were found, both unexpectedly as no evidence of such was visible on the surface. One of the canals was rectangular in cross-sectional shape, measuring 25 cm deep and varying in width between 35 and 75 cm. Approximately half of the 22 m length which was exposed, was found to be lined by volcanic rocks. On the basis of its location and direction it apparently carried water diverted out of Schultz Creek (D. Breternitz 1957: 28–30). The second canal was found while archaeologically excavating a pithouse on the east side of Eldon Mountain. No dimensions were reported but from a published photograph, the canal appears to be less than 1 m wide and U-shaped. It was also clay-lined (D. Breternitz 1957: 50,

52). Both canals were filled with alluvial gravels, unusually coarse canal fill, suggesting that they were abandoned due to heavy sedimentation. Ceramics in the fill indicate that both canals were used AD 700–900, a period of intense alluviation (D. Breternitz 1957: 29–30, 52–3).

There are reports of prehistoric irrigation in northeastern Arizona, but for the most part these are little more than statements that evidence of canals exists (Dart 1989: 17). Canals have been identified as having carried water diverted out of Show Low Creek and Hay Hollow Wash. Some ran for several hundred meters before branching into series of smaller canals (K. Lightfoot and Plog 1984: 184; Plog n.d.). A canal, claimed to be 32 km long, was reported along the upper reaches of the Little Colorado River (Turney 1929: 157). Others have been reported in the same area (Danson 1957: 63; Plog n.d.). Two canals further down the valley were described as being 'lined'. One was rectangular in cross-sectional shape and measured 76 cm wide and 25 cm deep. In all likelihood the lining was travertine not unlike that at Montezuma Well. The other was above ground, elliptical in cross-sectional shape, and was made from locally available rock slabs (Bandelier 1892: 388, 389).

There have long existed scattered reports of canal irrigation in locales along the Colorado River and its tributaries (e.g., Schwartz 1960: 64). For the most part, these systems were small yet complex. They often involved terracing of one form or another (E. Hall 1942; Lindsay *et al.* 1969: 174–87; Lindsay n.d.; Winter 1976: 81–122), and in this respect are similar to systems currently used by the Hopi (see Chapter 8). Two of the more intriguing sites yielding good evidence of prehistoric canal irrigation are in Glen Canyon. Although they differ in location and features, they are related in terms of technology and complexity (J. Jennings 1966: 44).

At the Creeping Dune site a dam was found associated with a spring. This feature was rectangular in shape and built against a slope. The three walls consisted of double rows of tabular rocks. The wall paralleling the slope contours was 8.2 m long and 2 m high. The side walls varied in height according to the slope and were 3.5 m long. The walls varied in thickness from 90 cm to 1.5 m with the bases being wider than the tops. The inside walls and the floor were plastered, and a drain was built into one corner. This consisted of a tunnel, measuring 10 cm wide and 15 cm high. A tabular rock measuring 32 cm square and 5 cm thick blocked this tunnel. However, a 5 cm diameter hole drilled through the center of this slab allowed water to pass. On the outside of the dam a canal measuring 30 cm wide, 15 cm deep was made of similar rock slabs. This canal had a gradient of 1.6 percent and all traces disappeared 9 m from the dam (Sharrock, Dibble, and Anderson 1961: 191–5, figs 3–9).

A number of canals not unlike the one just described were found at the Beaver Creek Community. They consisted of parallel rows of tabular rocks set vertically 20 to 40 cm apart. The canals had tabular rock bottoms and ranged in length from 10 to 30 m. They ran downslope, and although a source could not be identified, they probably headed at one or more springs. One of the more curious aspects of these canals are features which clearly aided in the control of flow. At irregular intervals along their courses these canals had tabular rocks set into their bottoms perpendicular to the flow. These rocks would have blocked the flow entirely were it not for semi-circular notches 10 to 15 cm in diameter cut into them. Adjacent to these notched rocks, small tapered ones (with tops wider than the bottoms) were set between larger ones forming the canal walls. These small rocks were not embedded, as were the large ones, but rather were loose so they could be easily removed and reinserted (Fig. 10.19). They have been interpreted as sluice gates that allowed for the control of water flowing to the adjacent fields (Lindsay 1961: 182–4, figs 6–10). Together with the dam at the Creeping Dune site, it is clear that prehistoric farmers along the San Juan River were sophisticated in their use of canals for irrigation.

Archaeological evidence of prehistoric canal irrigation in northern New Mexico is much more sparse than it is in Arizona, and this lies at the heart of many of the disputes over water rights between descendants of aboriginal people and descendants of settlers from Europe (see for comparison Plog 1978: 365–7; Wozniak 1998: 1–9). Evidence comes principally from three locales, El Rito Creek, Taos, and Tewa pueblos.

Traces of ancient canals have long been known on both sides of El Rito Creek near the ruins of Sapawe (Anschuetz 1995: 26; Bandelier 1892: 141; Ellis 1975: 20). According to Ellis, traces of two main canals were clearly evident in few road cuts, and could be traced on the surface. One was 2.5 km long, 2 m wide, and 45 cm deep. A spoil bank on its downslope side measured 60 cm high. It terminated above the Sapawe ruins. The other was 4 km long, 2.4 m wide, and 35 cm deep before narrowing below the ruins to 50 cm wide and 20 cm deep. Both were elliptical in cross-sectional shape, filled with silty sands, and, most importantly, contained only prehistoric ceramics. Smaller branch canals were found leading off the longer canal, downstream of the ruins. Large rocks found at the ends of these canals were interpreted as water spreading devices on the fields (Ellis n.d.*a*).

Archaeological evidence of canal irrigation on the lands near Taos Pueblo has long been known, but little studied. Remnants of what were thought to be prehistoric canals were first noted seventy years ago. One canal began in the foothills behind Taos Pueblo, presumably on the Rio Lucero, and carried water to fields 6.3 km away. Another began 4 km south of the pueblo and extended to the west 4.5 km. Although no ceramics were found to help in the dating of these canals, both ran past several prehistoric sites, and local informants knew nothing of their origins (Jeancon 1929: 3). The proximity to other prehistoric sites and the lack of legend both suggest great antiquity. Surprisingly, no other studies of these canals were undertaken for seventy years. A recent study has, however, begun to change this. It relied heavily on aerial photographs and surveys, but, unfortunately, very little on archaeological excavations.

This study found evidence of several canals originating on the Rios Lucero

Fig. 10.19. Stone-lined irrigation canal at the Beaver Creek Community, San Juan River, Utah. Note the notched tabular rocks in the canal and the rock alignments on the fields adjacent to them. Appeared originally in Lindsay (1961: 183, fig. 7). (Photograph taken by Alexander J. Lindsay, Jr., provided courtesy of the Museum of Northern Arizona, and reprinted by permission of the Society for American Archaeology.)

and Hondo. Canals were identified on aerial photographs largely by distinctive vegetation patterns. Ground surveys revealed narrow, shallow, linear depressions with dense grass and low shrubs in the bottom with sagebrush on the upper slopes and outside the depressions. Some canals were more evident on vertical aerial photographs while others appeared more clearly on oblique images. Some were easily visible on the ground, others were not. Some were more than 3.5 km long and branched several times. The land on which they were found has a gentle slope of less than 2 percent (Greiser and Greiser 1995: 222–7, figs 15.2–15.5). Within the study area, approximately 1,450 ha of land are thought to have been irrigated by these canals AD 1100–1250 (Greiser, Greiser, and Putnam n.d.). Only one of these canals has been archaeologically excavated. It proved to have a U-shaped cross-section, measuring 75 cm wide at the top and 70 cm deep (Greiser and Greiser 1995: 228–9, fig. 15.6). Clearly, more excavations are needed in order to understand prehistoric irrigation here.

South of Taos Pueblo, near the present-day town of Rancho de Taos are at least two canals, which although currently in use might have prehistoric origins. As is the case further north, these canals pass quite closely a number of archaeological sites. They carry water diverted out of the Rio Grande del Rancho for 7 km. Here, unlike at Taos proper, legend among farmers descended from Spanish settlers holds that these canals were in place prior to the arrival of their ancestors who merely renovated them (Woosley 1986: 159, 151, fig. 3).

Further up the Rio Grande del Rancho, near Southern Methodist University's Fort Burgwin Research Center are unequivocal remains of a prehistoric irrigation canal. This canal carried water from the Rito de la Olla, a tributary, to the floodplain of the Rio Grande del Rancho. It can be traced on the surface for more than 600 m (J. Moore 1995*a*: 16, 17, fig. 2.2). It is visible as a low swale, 10 to 20 cm deep, paralleled in places by a low embankment on the downslope side. Two archaeological trenches revealed that it was 1.2 m wide at the top, 30 to 50 cm deep, and U-shaped in cross-section. Archaeomagnetic dates from the bottom of the canal indicated that it was used prior to AD 1300. An area downslope from the end of the canal, and therefore thought to be a likely field site, revealed maize pollen (Moore 1995*b*: 212–17, figs 14.2–14.4).

As mentioned earlier, thirty of the forty-eight Tewa canal systems studied by Ellis (n.d.*a*,*b*; 1978) were interpreted as having been built and used in prehistoric times. As was the case with suspected ancient canals at Taos, Ellis based her conclusion on the proximity of canals to habitation sites, and ceramics found in various canal contexts. The Tewa irrigation systems were not all built at once. A review of the dates of initiation Ellis attributes to each canal reveals that an almost equal number of projects began in each of the 11th through 14th centuries. One canal, she claims, was begun *c.* AD 900. Although Ellis has done a masterful job of assessing historic Tewa irrigation and its possible antiquity, her conclusions are not unequivocal. This statement should not be seen as a denigration of her work—it is not—but rather a public recognition concerning the inexactness of dating canals. The case of the Caño Ditch, the uppermost of three canals used to irrigate the north side of the Pojoaque Valley, will serve as an illustration. Ellis (n.d.*b*) argued rather strongly that it was built in the 12th century. A more recent study (Tyler 1986: 25) concluded that '[t]he Caño might have been prehistoric. . . . That is was in use prior to 1859 seems clear, but exactly when it was built may never be known'. Somewhat cryptically, perhaps even disparagingly, the writer went on to say: 'In this kind of historical investigation, a good detective knows his limitations' (Tyler 1986: 25). He is, of course, correct. However, even if one fails on the safe side, he or she still fails. It is just as bad to be wrong for the right reasons as it is to be right for the wrong reasons.

The archaeological evidence for prehistoric canal irrigation in northern New Mexico is certainly in debate. Such is not the case, however, elsewhere in the state. At the Gran Quivira ruins in 1835–6 traces of a prehistoric canal were identified on the east side of the site (Woodward 1934: 96). Sixty years later Bandelier (1892: 286–7) visited the site, found the canal previously identified, measured it to be 2 to 3 m wide and 50 cm deep, and found a canal on the western side of the site as well. Each of these canals began at a circular reservoir which was spring fed. A later study revealed that each of these canals extended more than 350 m, and that the one on the east carried water to fields south of the site while the one on the west carried water to fields lying to the northwest (Toulouse 1945: 366, fig. 28, 368, 370–1).

Further to the east, near Fort Wingate, Bandelier (1892: 343) found remnants of another canal system that he thought was prehistoric. This one measured 2 m wide and carried water diverted out of the Rio Puerco to fields on its northern bank. This canal is figuratively and literally the closest thing to archaeological evidence of ancient canal irrigation at Zuni.

South of here, in the highlands of western New Mexico, Neely recently found evidence of a small canal that once carried water to fields from a tributary of Apache Creek. Traces of the canal remain barely visible on the surface as a shallow depression. In no small way this is because of the small size of the canal. It was traced more than 400 m, and watered 5 to 6 ha. Archaeological excavations bore out what was suspected from the surface; the canal was elliptical in cross-sectional shape, 2.2 m wide at the top and 15 cm deep (Neely 1995*b*: 249–54, figs 16.8, 16.9).

Evidence of canals used in prehistoric times comes from further south, along the Gila River. There, in 1756, Fray Bartolomé Sanchez reported seeing '*ruinas de edificios antiguous* . . . [and] *por el terreno qu conducion la acequia para la agua que serviria á sus semantaras*' (Sanchez 1856: 90), or 'ruins of ancient buildings . . .. [and] On the ground . . . they had brought an irrigation ditch which carried the water to their fields' (Ackerly 1997: 353). Although seemingly a documentary source, Sanchez's mention of ruins and the use of the preterit tense indicates that the canals were no longer in use when he saw them (Ackerly 1997: 354); his description is therefore archaeological.

The Mimbres River and its tributaries in southwestern New Mexico have long intrigued archaeologists in terms of irrigation (e.g., Shafer 1982). Bandelier (1892: 357), as was so often the case, was the first to mention evidence of ancient canals there, and these he found along the floodplain. A century later, however, not so much as a trace remains of a canal which can be definitively dated to prehistoric times (Shafer 1982: 9). This is not to say that canals do not exist; they do. However, they are all in the hands of people who settled the valley in the intervening period (Ackerly *et al.* 1993: 49–50). Of course, many of the canals currently in use might have prehistoric antecedents which were merely renovated. That is the contention of Paul E. Minnis (1985: 113), who did the most comprehensive assessment of prehistoric food production in the valley. Minnis substantiated his assumption by calculating the amount of irrigable land on the floodplain, estimating how many people could have been sustained by the canal-watered area, and comparing this latter figure with population estimates derived from the sizes of ruins of prehistoric habitation sites in the valley (Minnis 1985: 115–21). His figures concur.

The evidence lacking in the Mimbres Valley is not lacking in one of its tributaries, the Rio Arenas to the west. Here, archaeologist LaVerne Herrington did what others could not do. She, in the spirit of Bandelier, actually found traces of canals which could unequivocally be dated to prehistoric times. In no small way, this was due to tight land control during the past century, and the presence of archival materials which offer proof that the land on which the canals were found has not been irrigated in historic or modern times (Herrington 1979: 4, 21). The evidence of canals Herrington found involved linear depressions 10 to 20 cm deep with low earthen embankments along one side, and distinctive vegetation which was apparent from both the air and the ground (Herrington 1979: 127–33). At least eight such canals were found, varying in length from as short as 150 m to as long as 700 m. A few of these even bifurcated. All ended in fields divided into grids by low earthen borders (Herrington 1982: 83). None of the canals, however, could be traced to the points where water was diverted out of the river. However, Herrington (1979: 135–40) notes that these canals all line-up with outcrops of volcanic dikes which cross the channel. She concluded that a combination of natural and human-assisted diversion was employed. The total irrigated area measured 25 ha (Herrington 1979: 99).

Herrington (1979: 61–98, 147–81) backed-up her assessment of ancient canal irrigation with settlement and demographic parallels. At the same time she was conducting her study, another study employing almost identical strategies was being carried out independently further to the southwest in the Valley of Sonora (Doolittle 1988). This one, however, was strengthened even further by early Spanish documentary accounts, and a tantalizing petroglyph which is interpreted as a map (Doolittle 1984*b*; 1988).

East of the Sierra Madre, in what is today Chihuahua, Mexico, lie two of the more interesting remains of some ancient canal irrigation systems. One is interesting in that it also provided domestic water to the great trade center of Paquimé and it continues to provide water to fields. The second still awaits systematic investigation, but is important because of its size and the direction it points to where irrigation was first developed in North America.

The first of these begins at a large spring, Ojo de Vereleño north of the present-day town of Casas Grandes. In times past, water was carried 6 km through a canal that had a calcareous lining (Bandelier 1892: 555) to reservoirs in the heart of the city (Brand 1943: 141). Excess water and waste water were carried beneath the site through a series of drains which eventually emptied into the main canal of another system which irrigated fields on the floodplain (Di Peso 1974: 348–9; DiPeso, Renaldo, and Fenner 1974: 828–31). Along its way into the city, this canal crossed intervening arroyos by means of aqueducts which have long since been destroyed (Bandelier 1892: 555). Also, before reaching the reservoirs much of the water was turned out of the canal into a series of branch canals which watered floodplain fields north of the city (Doolittle 1993: 140, fig. 8.5).

As interesting and as important as this canal is, it is far less significant than the second one. Between 3 and 5 km south of Paquimé are the remains of a large canal that was elevated above the surface of the floodplain (Doolittle 1993: 135–7, figs. 8.1 and 8.2). The reasons for elevation are clear. These are the remains of the midsection or the downstream end of a very long canal system. As discussed in regards to the water harvesting systems at Queen Creek (Chapter 9), gravity-fed canals have to be higher than the field surfaces they irrigate. Whereas this canal had to be excavated at the place where it diverted water from the northward flowing Rio Casas Grandes, a point was eventually reached along its course where it had to be elevated above ground. Today only a single ridge of earth 80 cm high, 8 m wide, and a few hundred meters long exists. A small canal still carries water along its upstream end. One hundred years ago, it appeared much differently, less damaged by erosion and modern agricultural activities. When Bandelier (1892: 554) saw it, there were two parallel embankments 800 m long. One embankment was 1 m high, the other 1.75 m high, and they were separated by a 5.5 m wide canal. Some distance further downstream the banks were lower and the canal had narrowed to 4.3 m in width (Bandelier 1892: 555). This section was also visible in the 1930s (Brand 1943: 142) but cannot be found today.

As was also true for the spring-fed one, this canal dated to the 13th century AD and perhaps earlier (Doolittle 1993: 141). It also transported water to more than 1,000 ha of fields (Doolittle 1990*b*: 147, table 6.1). The sizes of this canal and the area it irrigated are important for one reason. Canal irrigation has long been thought of as something that originated in Mesoamerica and was introduced into the Southwest (e.g., Haury 1976: 50). However, up until the 1400s, there were no canals as large and as technologically complex and no irrigated tracts of land as expansive anywhere in Mesoamerica as those at Casas Grandes. There is simply no way that canal irrigation technology could have diffused from south to north. Where were there canals as large, as technologically complex, and irrigating as much land as at Casas Grandes earlier than the 1400s? The Salt–Gila Valley of southern Arizona. It is here, in southern Arizona, that the origins of canal irrigation lie (Woodbury 1961*b*: 554; Doolittle 1990*b*: 79–81, 1993: 147–8).

Vestiges of ancient canals along the Gila near the town of Solomon have long been known (Turney 1929: 157). Skirting the foot of a terrace on which a prehistoric settlement known as Pueblo Viejo, the Buena Vista Site, and the Curtis Site, is a canal that was used in relatively recent times (Hough 1907: 33, fig. 3.34). At least one writer (Fewkes 1904: 178), relying on accounts of late 19th-century farmers, concluded that these later irrigators simply renovated a prehistoric canal. Irrigation canals dating from the prehistoric era are thought to have been quite common in this stretch of the valley (Sauer and Brand 1930: 422). During the 20th century, however, the floodplain was totally reworked by land-leveling machinery and all traces of canals, save perhaps a few remnants along the terrace edges, have been obliterated. Occasionally, parts of these canals are uncovered during the construction of modern buildings. For example, in Safford, a trace of one ancient canal was revealed when a lot was cleared for the building of the local office of the Bureau of Land Management (Botsford and Kinkade n.d.). Overwhelmingly, however, prehistoric canals have been destroyed. The same condition holds true, unfortunately, for the area downstream, near Casa Grande (see Fig. 10.8).

This area might well have been the most irrigated section of the Gila Valley in prehistoric times (Cummings 1926). More than 100 years ago, Bandelier (1892: 458) noted that the area was little changed since ancient times, except for canals having been abandoned. He found remains of several such canals. These paralleled the river, were very distinct, being 'quite deep' with 'raised banks', and extending upwards of 15 km. A more detailed assessment in the 1920s found remnants of five canals (Turney 1929: 149–50). On the south side of the river, one canal carried water diverted from 15 km northeast of the present-day city of Florence. It passed south of town, before bifurcating 4 km east of the ruins. One branch terminated 1.5 km north of the Casa Grande. The other passed close to the north and west sides of the site, for a total length of 32 km. The second canal south of the river began due north of the ruins and extended west for 15 km. Together, these canals irrigate approximately 4,350 ha. North of the river, the canal system furthest upstream began 5 km east of Poston Butte, passed the butte along its southern side, extending a total of 13 km. Near where this canal ended, another one began and extended westward 10 km. Six km downstream from where the second canal terminated, the third one began, and it extended another 6 km. Together, these three northern canals provided water to 1,700 ha. No branch canals were found (A. Larson 1926; Turney 1929: 149–50).

Re-examinations of the area by Midvale, which included several years of walking over the terrain and conducting numerous interviews with local residents, provided a great deal of insight (Andresen 1987). Midvale suggested that 140 km of canals were used prehistorically, but only 14 km of disconnected fragments survived to the 1960s. He found evidence of six canals, one more than Turney, with as many large branches. The longest canal did not begin where Turney thought, but only 5 km northeast of Florence. Furthermore, it flowed north of town, past the Adamsville Ruins and branched at least three times. Midvale named it the Adamsville–Casa Grande Canal. The sixth canal he labeled the Pinal Canal. It began where Turney said the first canal had its head and terminated south of Florence. Midvale suggested that diversion devices were made of brush and earth, and had to be rebuilt each year. In a few places canals were evident by parallel earthen embankments 8 to 12 m apart and separated by a slight depression. Elsewhere they were filled and the embankments eroded into a wide and low ridge (not unlike is the case south of Casas Grandes, Chihuahua). The first canal on the north bank, what Midvale dubbed the Poston Canal, was paralleled by a modern canal, and the second canal is now paralleled by railroad tracks (Midvale 1965: 82, 83).

In addition to providing some exquisite maps (Fig. 10.20), Midvale (1963, 1965: foldout, 84) also noted that the second canal on the south bank may have been a somewhat later, and upslope, response to the destruction of an earlier canal. A separate channel known locally as the 'Little Gila River' may

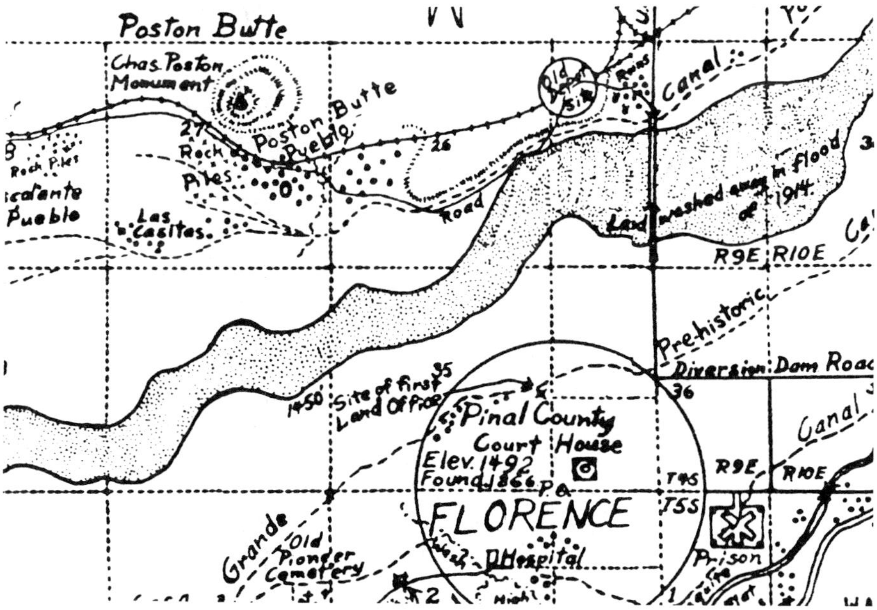

Fig. 10.20. An enlarged section of a map drawn by Frank Midvale (1963). Each dotted grid is the equivalent of 1.62 km square. Prehistoric canals are shown by dashed lines. The one extending from the right to the lower left and passing close to the north edge of the town of Florence is the Adamsville–Casa Grande Canal. The river channel is stippled. Note the 'MIDVALE' trademark among the stippling. (Published by permission of the Department of Anthropology, Arizona State University.)

actually have been a prehistoric canal that was pirated by exceptionally heavy flow, thereby becoming part of the river.

Two studies of this area undertaken in the early 1980s actually added little information to knowledge of ancient canal irrigation in the area, principally because modern agricultural activities continued to take their toll in altering the cultural landscape, but do show the nature of academic debates. One (Dart 1983: 393–7), essentially accepted Midvale's work. It concluded that nothing remains of the Adamsville–Casa Grande Canal, that some faint lines on aerial photographs were all that could be discerned of the Pinal Canal, and, only one segment of the Poston Canal—evident by parallel embankments 12 m apart with a central depression of 80 cm deep—could be found (Dart 1983: 396, fig. IV.3.11). It also found, on the basis of surface topography, that the Adamsville–Casa Grande Canal had a gradient of 0.17 percent, the Pinal Canal had a gradient of 0.05 percent, and the Poston Canal had a gradient of 0.15 percent (Dart 1983: 369, table IV.3.2, 399). The other study (Crown 1984*a*: 215–21) took into consideration a number of unpublished materials and concluded that Midvale's interpretations were not totally correct. Principally, the Poston Canal did not head as far upstream as he thought, and that the Pinal Canal was the upstream stretch of the Adamsville–Casa Grande Canal as Turney had noted (Crown 1984*a*: 219, fig. III.4.4). Exactly when these canals were used has not been determined, other than sometime after AD 1150. Together, however, they are thought to have irrigated 6,247 ha (Crown 1984*a*: 221, 226).

Although these studies differ considerably in their interpretations, three very important facts remain. First, there has been so much disturbance by historic and modern agriculture, both after and even *before* various scholars attempted to map these canals, that determining precise locations and relationships to each other has always been a rather futile exercise. This is compounded by the second fact; the dynamic hydrology and geomorphology of the Gila River which through avulsion, erosion, and sedimentation has altered the agricultural landscape after and even *during* the time when prehistoric people were irrigating. The locations of the canal intakes undoubtedly changed through time, and the canals themselves most likely were rebuilt, realigned, and connected and disconnected as conditions changed. And, third, and most importantly, these debates have been carried out without so much as a single archaeological excavation of a canal (Dart 1983: 398). We know the approximate locations of canals. Perhaps the next step is for someone to put in some backhoe trenches. Doing so will expand the information base by at least an order of magnitude.

The situation downstream to almost as far west as Gila Butte is worse than near Casa Grande. Ancient canals are known for this stretch of valley (Lindauer 1984: 79–81, fig. 1), but little research has been done. Starting near Gila Butte, however, things change greatly for the better. A short section of canal was long ago identified, near the terminus of the modern Santan Canal.

Given its location relative to a similar but much more recent feature, the late Emil W. Haury, long-time dean of Hohokam archaeologists, projected more than sixty years ago that its off-take might be approximately 8 km upstream (Haury 1937: 51, fig. 16, 52). On the surface, the canal appeared only as an irregular ridge of silty sand extending approximately 1.6 km. Archaeological excavations of this segment carried out by no fewer than three teams well after Haury's discovery revealed not one, but four canals. The first of these studies reported that the ridge was artificially created as part of the canal construction and that one canal, U-shaped in cross-section, ran along its length. It was dated post AD 1150 (Greenleaf and Vivian 1971: 13). A second study identified a second canal along the crest of the ridge paralleling the first, and a wide, shallow canal, elliptical in cross-section, under the ridge. This bottommost canal may date earlier than AD 750 (Brooks and Vivian 1978: 18–19). A third study concluded that the ridge was not cultural but a natural relict standing amidst an eroded landscape (Fig. 10.21). It also found a canal paralleling the ridge along one side (Rice 1978: 23–7).

Approximately 1.4 km north east of this segment, and, therefore, further away from the river, a second canal was identified (Antieau 1977: 22, 23, table 1). It appears on the surface as a slight depression 2 m wide and 1,750 m long, with vegetation more luxuriant than that of surrounding areas. Archaeological excavations revealed the canal to be 1.9 m wide and 30 cm deep (Brooks and Vivian 1978: 10–11; Rice 1978: 41). Given its location away from the river, this canal may have originated as far as 10 km further upstream than the other (Rice 1978: 7). One or more of these canals could have terminated at fields north of Gila Butte, or they may have continued on to the famed site of Snaketown (Antieau 1977: 22; Haury 1937: 52; Rice 1978: 22).

Snaketown was first excavated archaeologically in the 1930s (Gladwin *et al.* 1965). Along the southern edge of the site, at the edge of the highest

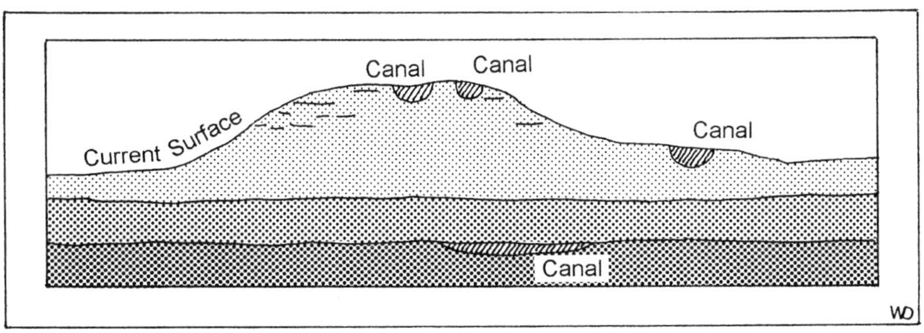

Fig. 10.21. Schematic cross-section of ridge and canals found west of Gila Butte on the north side of the Gila River. No scale (after Rice 1978: 27). (Redrawn by author.)

terrace, were preserved sections of a canal that was traced for approximately 5 km, and bifurcated just west of the site. One branch continued westward while the other trended toward the northwest. The canal was evident by a shallow depression flanked by low ridges (Haury 1937: plate XIIIb). The distance between these ridges averaged 10 m above the bifurcation and slightly less below. The gradient along the surface was approximately 0.06 percent (Haury 1937: 52). Excavations revealed three superimposed canals, all U-shaped in cross-section (Haury 1937: 54, fig. 17).

The bottom-most and first canal was excavated into the natural terrace sediments. It was 1.5 m wide and 1.8 m deep, but probably never carried water more than 1 m deep. It contained ceramics dating to AD 850–950, suggesting that it came into use at that time. There is, of course, the possibility that an earlier canal existed but was obliterated by this one. The canal gradually filled with sediments and ceased being effective sometime between AD 950 and 1150 (Haury 1936: 49). Sometime later, a second canal was excavated totally within the sediments filling the earlier canal. It measured 1 m wide, 1 m deep, and became lined with a thin layer of travertine (Haury 1937: plate XIIIa). Ultimately, it too filled with sediment, and the third canal measuring 1.2 m wide and 1 m deep was dug. It too became travertine lined and filled with sediment before being abandoned. Excavations downstream of the bifurcation revealed no traces of the lower two canals, suggesting that originally the canal did not extend below Snaketown (Haury 1936: 50). These canals were also U-shaped in cross-section. The one heading westward measured 1.5 m wide and deep. The one heading northward measured 2 m wide and 1.5 m deep (Haury 1937: 56, fig. 18).

Haury returned to Snaketown in the 1960s principally to search for evidence of yet earlier canals, and secondarily to learn more about irrigation at the site. He succeeded on both counts. He cut long backhoe trenches across the known canals in thirty places. He then focused on certain places that looked particularly promising (Haury 1976: 120–5). The point at which water was diverted from the Gila River, and the possible connection between the canal at the site and the segment near the end of the modern Santan Canal could not be confirmed. Haury (1976: 122–3) did suggest however that the point near Gila Butte might have been used at some early time, and later moved upstream, involving the other canal segment later when more land was under irrigation.

At one point along the main canal, archaeological excavations revealed a bifurcation which turned out to be the head of a branch canal. This later canal came off the main canal at nearly a 90-degree angle. It was U-shaped in cross-section varying between 1 and 1.5 m wide and 75 cm deep. It showed at least two periods of use and evidence of remodeling. Most interesting were a series of holes found in the canal floor near the juncture. These were interpreted as post holes for a sluice gate, probably made of woven mats. Also intriguing is evidence of the canal having been lined with a 'clayey material'. This was 5 cm thick and applied by hand at places that were particularly prone to erosion (Haury 1976: 125–8). This branch canal seems to have carried water to irrigate the 340 ha just below the site (Haury 1976: 144).

Lining was also used in places along the main canal where maintenance was a problem. In such places, a clay–loam mix was applied to a thickness of 8 to 10 cm (Haury 1976: 137). Fine cracks in several places indicate that the canals were periodically dry (Haury 1976: 144). There were also times when they carried too much water. In at least one place a 'turn-out' was found. It was really nothing more than a 20 cm wide groove cut in the side of the canal which allowed excess water to escape. This interpretation was based on the discovery of a small alluvial fan beginning at the groove, just beyond the canal (Haury 1976: 136).

Excavations at Snaketown did, indeed, reveal evidence of an earlier canal. It was not, however, an easy feature to identify simply because it was morphologically quite different from other superimposed canals. This one was elliptical rather than U-shaped in cross-section ranging in width from 3 to 5 m and being less than 1 m in depth. Ceramic evidence revealed that it dates prior to AD 750, perhaps as early as AD 300 (Haury 1976: 132–4). Excavations also revealed that the main canal had gone through several remodelings, at least five during one phase, and a canal dating to a later phase was also discovered. These rebuildings and remodelings show a continuous movement upslope, further from the edge of the terrace (Fig. 10.22). A separate study undertaken by Woodbury (Haury 1976: 141–2) found that the banks paralleling the known canals are the spoil of these activities.

Further down the Gila River, ancient canals are known and recorded, but little studied. At its confluence with the Agua Fria River, Midvale (n.d., 1974) found evidence of a canal he labeled the 'Liberty Canal'. His map shows that it paralleled the historic Buckeye Canal which was located slightly upslope for approximately 6 km. At one place along its upstream stretch, the canal

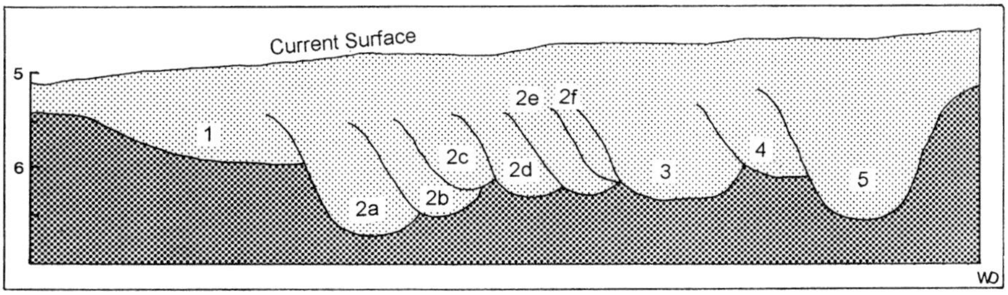

Fig. 10.22. Composite section of Snaketown canals showing sequential relationships. Number 1 dates to pre AD 750. Nos 2a–f date to *c.* AD 750–950. No. 3 dates to AD 950–1150. Nos 4 and 5 date to AD 1150–1500 (after Haury 1976: 150, fig. 8.48). (Redrawn by author.)

may have bifurcated. Evidence for this is tentative and exists principally in the form of a segment of ancient canal 2 km away and near the river. The canal definitely bifurcated at the 6 km mark. Here its 'Lower Branch' paralleled the South Branch of the Buckeye Canal and extended nearly 7 km. The main canal continued for more than another kilometer where the 'Liberty Branch' veered off for another 5.5 km. The main canal, known throughout the remainder of its length as the 'Van Liere Branch', terminated 5.5 km further on at a ruin which lent its name to the final stretch of canal (Midvale 1974). Midvale (n.d.) claimed these canals provided irrigation waters to more than 800 ha. He was correct. Canals this long and terminating at expanses of land the sizes he portrays on his map, probably irrigated three or four more times the land area that he suggests.

Further down the Gila, indeed west of the Maricopa Mountains and north of the present-day town of Gila Bend at least four other prehistoric canal irrigation systems have been reported. Traces of a main canal, evident on the surface in the form of segments of low parallel ridges 6.3 m to 8.6 m apart, have been found extending more than 20 km beginning at the Gillespie Gap. The modern Enterprise Canal parallels, and in many stretches obliterates, these traces. In other stretches, modern farming has erased signs of the ancient canal (Midvale n.d.). A second canal extending for 16 km was first observed by Midvale in the 1920s. Later visits showed progressive destruction as modern agriculture expanded (Midvale n.d.), and today probably nothing remains. The third canal in the Gila Buttes area was reported by Turney (1929: 69–70). He noted that it was 1.8 m wide, 90 cm deep, more than 1.5 km from the river and its bottom and sides were 'white from caliche-lime deposition'. The final canal received the greatest attention, having been subjected to archaeological excavation three decades later as part of a salvage project for the Painted Rocks Reservoir (Wasley and Johnson 1965: 24–5). This canal was traced for more than 8 km. It was U-shaped in cross-section, varied in depth from 1.3 to 1.8 m, and in width from 3 to 3.5 m. It was used between AD 900 and AD 1200, but given the low height of its banks, it was probably used only a short time (Doyel 1998: 227, 229–30, figs 2 and 3).

On the whole, evidence of prehistoric irrigation along the Gila River is spotty. There seems to be shreds of evidence everywhere there are expanses of fertile bottomland. Most areas have been visited by archaeologists, but surprisingly few areas have been subjected to systematic study. The area around Casa Grande is novel in that it has generated some debate, but with no excavations. The area around Snaketown has yielded the most information, but essentially this is attributable to one person. Things are much different in the Salt River Valley.

Much is known of ancient canal irrigation here for several reasons. First, there were numerous remains of large canals at the time settlers from the East began moving in—many of whom simply renovated the canals for their own agricultural pursuits (Fig. 10.23). Second, the valley has long been the

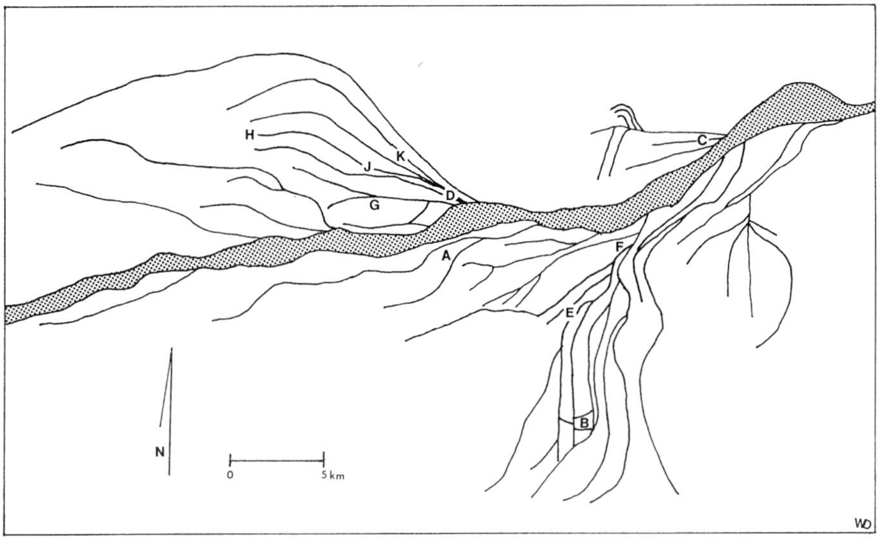

Fig. 10.23. Prehistoric irrigation canals in the Salt River Valley. Letters refer to specific sites discussed in the text. Nearly the entire area shown has been impacted by the growth of Phoenix and adjacent cities (after Turney 1929). (Redrawn by author.)

home to the densest concentration of people in Arizona. Many people saw many canals and wrote about them. Third, as the population increased, numerous federally funded projects required and funded much research. Fourth, and not inconsequentially, there have been some very dedicated archaeologists, professional and amateur, who distinguished themselves by recording canals before they were obliterated by development. The upshot of all this is that there is simply too much material to pay adequate justice to in an overview such as this.

Of course, one characteristic that is not much different in terms of the archaeology of irrigation in the Salt Valley is that Bandelier (1892: 436–7) was the first to mention canals in this valley. He noted that ancient canals here were numerous and large. He noted main canals paralleling the stream channel and branch canals running perpendicular to them. On the south side of the river, he reported a canal that measured 4.5 m wide, 32 km long, and was evident by low embankments on each side (Fig. 10.23, point A). Bandelier was followed shortly by a number of scholars who laid the foundations for future generations of archaeologists. Indeed, this group provided the basic knowledge of Hohokam canal irrigation, all within the first half of the 20th century. Their successors, in the second half of the century, refined interpretations and understanding.

The Hemenway Southwestern Archaeological Expedition (Hinsley and

Wilcox 1995) carried out some mapping and excavations further south (Fig. 10.23, point B) a few years later, but little was published as its director, Cushing, got sidetracked and went native at Zuni. F. W. Hodge (1893: 324–5), a member of the research team, did, however, report that some canals were lined with adobe which was baked by burning brush within the canals; that sluice gates were used as evident in the presence of post holes; that some canals were more than 2 m deep; and that one was 1.2 wide at the bottom, 9 m wide at the top, and had sloping sides with a step or bench 90 cm up from the bottom (Haury 1945*b*: 41, fig. 25); that nearly 250 km of main canals were involved throughout the valley; that the total area irrigated by the Hohokam was more than 100,000 ha; and that it was equally divided between the north and the south sides of the river. The lengths of canals, the extensive area which they involved, and their visibility prompted two early archaeologists to map them.

H. R. Patrick (1903) produced the first map of ancient Salt River canals. His was based almost entirely on his own surveys of the valley. In addition to his map, his field work resulted in several interesting findings. Although it was still rather early in the modern development of the valley, he noted that most of the ancient canals had already been destroyed or renovated. He also observed that remnants of diversion devices would never be found as the river bed had already become incised 2.5 to 4.5 m below the intakes of the prehistoric canals. Using ethnographic parallels, he speculated that the sluice gates would have been woven mats held into place by stakes or poles set in the canal bottoms. Patrick was the first to record an aqueduct where a canal crossed a wide swale. This feature was an embankment with the canal running the length of its top. Patrick also hypothesized a sequence of land use. According to him, small canals were first built in the downstream part of the valley. As more lands were needed, larger canals were built increasingly upstream (Patrick 1903: 3–8).

Although Patrick's map was an important contribution in its day, it was not perfect. It greatest flaw stemmed not from what it showed, but what it did not show. Although they existed, several canals were not portrayed including those south of Tempe (see Haury 1945*b*: 40, fig. 24). Correcting this deficiency was none other than Turney. He drew a map (actually several editions of a map) which incorporated a great deal of information provided by numerous sources (Turney 1924, 1929). Turney's map (Figs 10.23 and 10.24) has proven to be a classic that has been reproduced many times (e.g., Dart 1989: 9, fig. 5; Howard and Huckleberry 1991: 2.11, fig. 2.6). Indeed, it was so good that maps drawn by Midvale (1968) in the 1960s are small improvements (Fig. 10.25). Turney (1929: 12–52) also wrote an informative history of early canal studies in the area.

The early 20th century was a time of marked change, not only with the landscape of the Salt River Valley, but also with technology in general. Two

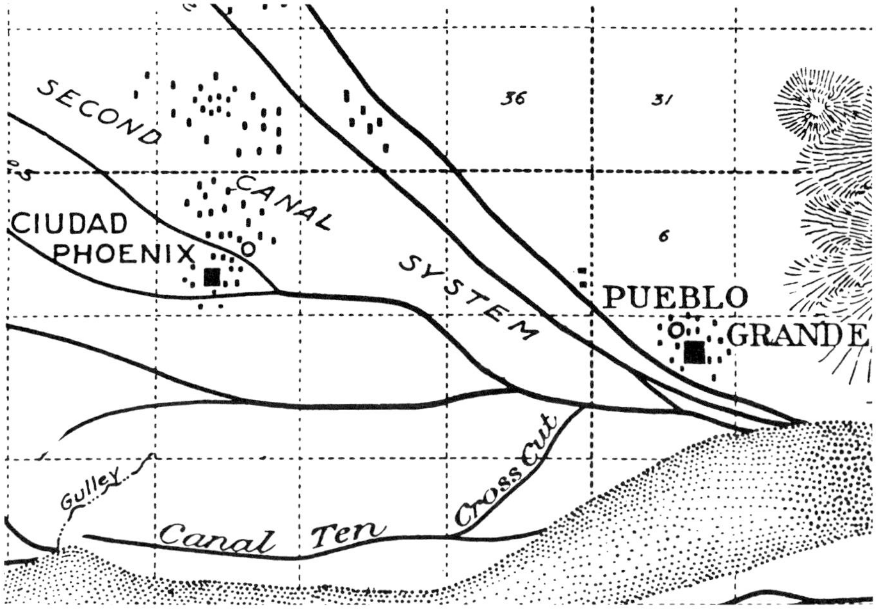

Fig. 10.24. An enlargement of that part of Turney's 1922 map showing prehistoric canals north of the Salt River. Each grid is the equivalent of 1.62 km square.

important developments occurred at this time with regard to understanding prehistoric irrigation. First, the birth of aviation and its marriage with photography resulted in some superb aerial images (Figs 10.26 and 10.23, point C) and a record of what the valley and its ancient and developing agricultural features looked like (Judd 1931). Second, archaeological treasures were being so rapidly and thoroughly destroyed that attempts were made to preserve some ruins and ancient canals (Judd 1930). The city of Phoenix established the Park of Four Waters, preserving a series of canals at the point where water was diverted out of the river (Figs 10.27 and 10.23, point D). It also created the position of City Archaeologist, with offices at the Pueblo Grande Ruins, and appointed Odd S. Halseth to the post (see Doyel 1987; Halseth 1932).

Halseth's early excavations bore some interesting fruit. He noted the parallels with irrigation systems in use at the time and concluded that waterlogging of soils would have been a problem without artificial drains like those in use during his day. Without a drainage network, and none has ever been found, the entire valley would have been unfit for farming within ten years, unless something else was done. And, what would that something be? According to Halseth (1936: 46–7), the only way the system could have

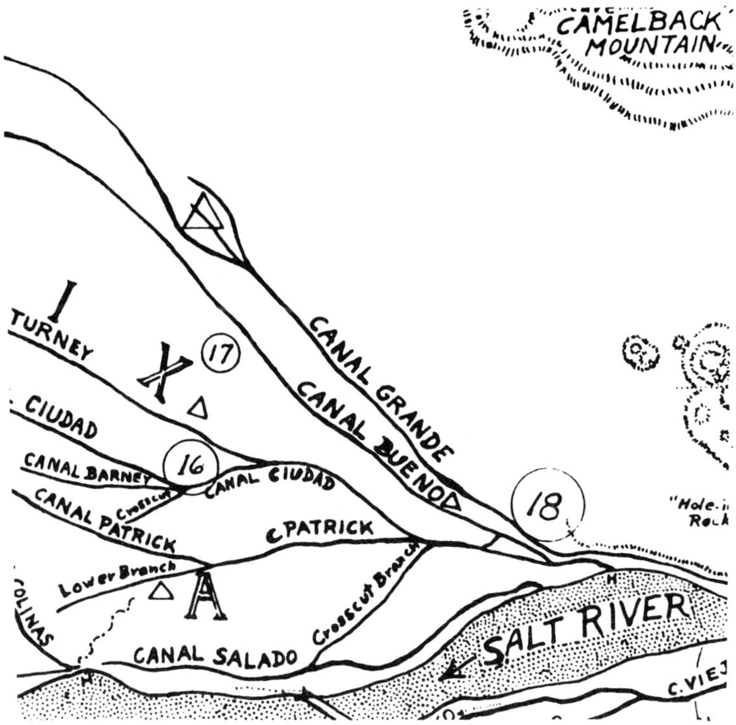

Fig. 10.25. An enlargement of one part of Midvale's 1966 map of ancient canals in the Salt River Valley. Compare with Fig. 10.24. (Published by permission of the Department of Anthropology, Arizona State University.)

been sustained was by all the known canals not having been used contemporaneously.

The issue of sustainability, using the term in its broadest sense and free of dogma that plagues the rhetoric of current development literature, was one that struck a chord with archaeologist Albert H. Schroeder (1943). He noted that 150 km of canal and 100,000 ha serving a population thought to have been 200,000 in prehistoric times was out of line with the 400 km of canal and 120,000 ha feeding 90,000 people in the middle of the 20th century, especially because the latter group had tools and techniques not available to the former. Furthermore, they were using water impounded in reservoirs and not relying on unregulated stream flow. He concluded that at any one time the Hohokam could not have had more than 9,000 ha under irrigation (Schroeder 1943: 380–1). Another archaeologist writing a short time later went so far as to hypothesize that not only did irrigation give rise to the Hohokam, but its excessive use might have led to the downfall of the culture (Shetrone 1945: 385).

Fig. 10.26. Oblique aerial photograph looking west, north of Mesa, Arizona in 1930. The Salt River is in the lower left. Remnants of a prehistoric canal and spoil banks are visible in the center. Modern fields are visible on the right. US Army Air Corps photograph. Appeared originally in Judd (1931: 161, fig. 138).

Fig. 10.27. Upstream view of a prehistoric canal and spoil banks in the Park of Four Waters. (Photograph by author.)

Schroeder pointed out three other features of interest concerning these canals. First, on the basis of appraisals made by local civil engineers, he concluded that canal construction was not as difficult as it might appear. In part, this was because canals were probably built incrementally (see Doolittle 1984*a*); small canals were built initially and then enlarged a short time later (Schroeder 1943: 381–2). Second, he concluded that the places thought to be canal intakes are probably wrong, and that lateral stream movement had long ago destroyed the first few hundred meters of canal. He argued that the intakes were further upstream (Schroeder 1943: 382–3). Acceptance of this idea is reflected in differences between Turney's and Midvale's maps (compare Figs 10.24 and 10.25, especially the lower right corners). Third, Schroeder (1943: 382) reminded everyone that with the exception of a few trenches cut by the Hemenway Expedition, no archaeological excavations had been made on canals up until this time. His call for such work apparently fell on deaf ears as the situation did not change for another two decades.

To Woodbury goes the credit for initiating excavations of ancient canals. As he stated it so eloquently: 'the body of information available has consisted largely of unsystematic comments on the surface appearance of these canal remnants' (Woodbury 1960: 267). He went on to note that even the maps appear 'to consist of about equal parts of observation and imagination' (Woodbury 1961*b*: 550). Woodbury carried out his excavations in the Park of Four Waters (Fig. 10.23, point D). In no small way, he was interested in determining how well the surface evidence of parallel ridges separated by swales (Fig. 10.27), reflects the actual morphology of buried canals.

Woodbury placed a 60 m long backhoe trench across the remains of three ridges and two intervening swales. From crest to crest, the ridges were 18 and 26 m apart. The top of the largest was approximately 2.5 m above the original ground surface. Erosion of these ridges filled the canals to such a depth that the central swales now are slightly higher than the original surface level. The two canals turned out to be quite different from each other and what is indicated by surficial remains (Fig. 10.28). The south canal, the one closest to the

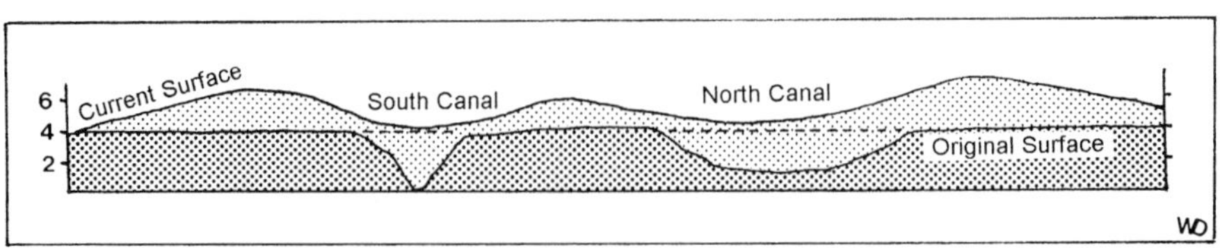

Fig. 10.28. Profile of an archaeological excavation across two prehistoric canals and three spoil banks in the Park of Four Waters. Note the cross-sectional difference between the original canals and the superficial evidence (after Woodbury 1960: 228, fig. 1). (Redrawn by author.)

river was V-shaped in cross-section, 4 m deep, and 6 m wide at the top. The north was trapezoidal in cross-section, 2.3 m deep, 4 m wide at the bottom, and 11 m wide at the top. It also had a plaster lining, but one that was put in place only after the canal had been used for a while. Both canals were filled with materials ranging from coarse sands to silts, much of it laminated indicating successive depositions Woodbury (1960: 267–8, fig. 1).

The ridges were much larger than the canals. Materials at their bases are gravely and identical to those in which the canals were excavated, indicating that the ridges were initially spoil banks. Materials on the tops of the ridges, however, are more like the canal fill, indicating that sediment was cleaned out of the canals and piled nearby. The position of smaller gravels high on the north side of the central ridge suggests that the north canal was dug later than the south canal. The bottom of the south canal is also lower than that of the north canal. It is approximately the same elevation as the channel bed of the Salt River, indicating that the canal intake was nearby. Having a higher bed means that the north canal had an intake further upstream, exactly as Schroeder had predicted (Woodbury 1960: 269). Ceramic evidence places the dates of these canals at AD 1150–1300.

Woodbury's archaeological excavation of Hohokam canals in the Salt River Valley was the first of many in recent times. Later studies can be seen in two categories, methodological and contractual. The methodological studies borrowed heavily from other sciences. Contractual studies are those carried out under federal and state law prior to certain types of construction which will impact archaeological sites. Both types of studies have contributed greatly to understanding Hohokam irrigation.

One of the earliest, most widely cited, and hence important, of the methodological studies pointed out how hydraulic principles can be applied in understanding how much water actually flowed through canals of various lengths, widths, depths, cross-sectional morphologies, and gradients (Busch, Raab, and Busch 1976). Another (Eighmy and Howard 1991) overcame one of the long-standing major problems—determining the ages of canals—by use of archaeomagnetism, measuring the directions of small ferromagnetic particles in canal sediments and analyzing their relation to movements of the earth's magnetic pole. Remote sensing was used in the Salt River Valley, much like it was in Chaco Canyon. It revealed that even though surface remains of most canals had been totally obliterated by modern urban developments, segments of many canals could still be identified on images exposed at various wave lengths of the photomagnetic spectrum (Ebert and Lyons 1980; Showalter 1993). Improved understanding of soils and sediments has facilitated interpretation of both environmental maps (Dart 1986) and archaeological contexts (Huckleberry 1992). Irrigation water and the canals that transport it, leave some not so obvious traces on the landscape; traces which, when appreciated, tell us something about cultural ecological relationships. Finally, parallels drawn from modern irrigation, its environmental impacts, and sustainability, principally by Neal W. Ackerly (1982, 1988, 1991), shed light on various aspects of canal operations.

Contractual studies are, of course, a mixed blessing. That they are done, means that archaeological evidence is being destroyed. On the positive side, studies that otherwise would not be funded are being conducted. Perhaps the greatest drawback to such studies is that they are limited to those areas impacted by construction; archaeologists are restricted to studying sites that might not be all that important and not those they might have chosen to study otherwise. However, sites selected for mitigation come from all parts of the valley, meaning that when looked at together a broader picture emerges. To be sure, this is tantamount to looking at a selection of disconnected pieces from various parts of a puzzle. Scattered pieces, however, tend to render a more accurate and less misleading picture than do connected pieces from one spot. The total picture is far from being perfectly clear, but it is coming into focus.

Contract projects in seven locales have contributed a wealth of detailed knowledge about Hohokam irrigation in the Salt River Valley. To one degree or another they all confirmed what was already known or suspected (e.g., superposition and rebuilding). However, they each have resulted in new discoveries which have added additional insight on the topic.

The first of these contract investigations, one conducted at site AZ U:9:46 ASM (Fig. 10.23, point E) in 1974, confirmed something most unusual. Aerial photographs had long shown a pattern of ancient canals oriented perpendicular to each other. Typically, branch canals take off at oblique angles so as to facilitate the flow of water and to prevent bank erosion at the point of bifurcation. Excavations at this site demonstrated that some Hohokam canal junctures did indeed contradict the norm. Furthermore, the bottoms of the branch canals were higher than the floor of the main canal, suggesting that the water level in the latter was fairly high (Herskovitz 1981).

Near the Las Acequias site (Fig. 10.23, point F) a later study revealed why it was possible for the Hohokam to have perpendicular canal junctures. The gradients of these canals were very slight. Indeed in some places they were flat, and in a few places appear to have been negative (uphill!). Furthermore, the bottoms and sides of some canals were not very smooth. The meaning of all this, of course, is that water normally flowed through the canals very slowly (Gardiner, Masse, and Halbirt 1987). It also flowed at a much earlier date than had previously been realized. The earliest canal in the valley was found at this location and dated to between 130 BC and AD 275 (Ackerly and Henderson 1989). The place at which it was found is known as La Cuenca del Sedimento, and is on the edge of the first Pleistocene terrace (Henderson 1989).

A project carried out on lands adjacent to the western edge of the Park of Four Waters (Fig. 10.23, point D) found a multitude of canals where only a few were thought to have existed. Whereas Woodbury found only two, this study found no fewer than eighteen prehistoric canals. These did not all operate contemporaneously, however. Evidence suggests very short life spans, and hence rebuilding and relocation. And, why were such renovations needed? Although the normal flow was quite slow, periodic floods of substantial magnitude left canals filled with sediment. Furthermore, most floods and canal rebuildings occurred between AD 950 and 1150 (Masse 1976, 1981, 1991).

Further west from the Park of Four Waters, near the Dutch Canal ruin, an opportunity was had to investigate the downstream segments of some prehistoric canals. This project found that not all canals had slight gradients, indeed some at this site were quite steep, especially the earlier ones. Water flowed through them sufficiently fast that erosion control features were built into some canals. Evidence of breaching and rechanneling were found and interpreted to be signs of poor engineering. Some canals appear to have been abandoned for prolonged periods, and then brought back into use. Also, a direct correlation between canal size and age was revealed. Small amounts of land were seen as having been irrigated early on near the river and larger tracts were watered at later dates away from the river. Clearly, there is a relationship here to the number of mouths to be fed and the number of workers able to invest in the construction and maintenance of the irrigation system (Greenwald n.d.; Greenwald and Ciolek-Torrello 1989).

The site of Las Colinas was subjected to intensive archaeological excavations (Fig. 10.23, point H). Its importance lies mainly in its location far downstream, near the ends of several known irrigation systems. Canals were investigated, but much more important were findings that pertained to the entire valley, its ancient agricultural environments. This aspect of the study relied heavily on reconstructions of stream flow through dendrochronological analyses of climatic conditions. It concluded that the stream flow was regular and predictable, but there was only about enough water to irrigate approximately 16,000 ha, and certainly no more than 40,000 ha, thereby proving Schroeder's estimate to be more correct than those of others, particularly Hodge. Furthermore, given the lengths of the known canals, irrigated fields would have been located within 400 m of the principal canals (Gregory 1991: 178). Branch canals would have been relatively few, and field canals almost non-existent. In all likelihood, border strip flooding and contour or bench border flooding were the most common methods of watering fields (Fig. 10.29). Maintenance was probably rather high as water-logging, salinization, and flooding posed major problems (Graybill *et al.* 1989).

Canals were also found and analyzed at the La Ciudad site (Fig. 10.23, point J). Especially significant findings involved water control structures. It was determined that a headgate was not used. Water from the Salt River flowed through the main canal constantly; variations in stream discharge were mirrored in canal flow. Sluice gates were, however, used to control the flow of water in branch canals. Indeed, this study involved a detailed

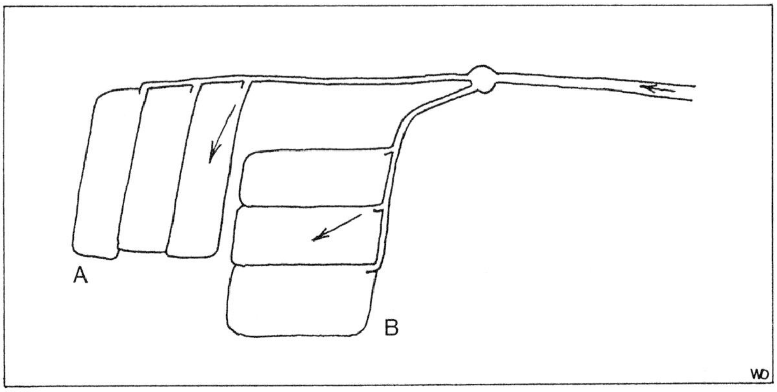

Fig. 10.29. Schematic diagram of Hohokam fields and branch canals in relation to a main canal. (A) Border strip flooded fields; (B) Contour or bench border flooded fields (after Graybill *et al.* 1989: 41–2, fig. 4.2; Masse 1991: 213, fig. 9.2; Howard n.d., fig. 5). (Drawn by author.)

assessment of post holes. This aspect of the study revealed that sluice gates involved three rows of posts across the canal mouths. Posts furthest upstream were larger than posts downstream, and those in the first two rows were set vertically while those in the third row were set at a 60 to 70 degree angle and functioned as a buttress. Small branches and twigs, probably willow, were woven between the posts in each of the first two rows and the space between was filled with earth and rock. Only a small section in the center of the canal was left open to allow water to pass when needed. When water was not needed, a sturdy mat of tightly woven reeds was placed across the opening (Ackerly, Howard, and McGuire 1987).

Finally, one contractual project in the Phoenix area investigated a series of sites containing a number of canals that comprised parts of different, unrelated systems. Conducted by Jerry B. Howard and Gary Huckleberry, this project had an advantage not enjoyed by others who were restricted to one site or locale with relatively few canal segments. Literally hundreds of archaeological excavations were carried out (Fig. 10.30). As a result, this study had good chronological control, was able to estimate hydraulic capacities of canals with some accuracy, and able to understand the structure of irrigation systems from river to fields. It also resulted in a map of all known prehistoric canals in the valley. This map was a great improvement over its predecessors (Fig. 10.31).

Equally as impressive as the map, are the project's conclusions, which are nothing short of monumental and revolutionary. For several years, it was thought that Hohokam canals were simply added to periodically with little labor investment, and that the irrigated area was constantly expanding. This

Fig. 10.30. Upstream view of a prehistoric canal that has been archaeologically excavated along one side. Archaeologist Jerry B. Howard is standing at the junction of the main canal with a branch canal. The other person is archaeologist John S. Cable. This canal is at the La Lomita site. A modern irrigation parallels this ancient one to the right, behind the row of trees. (Photograph by author.)

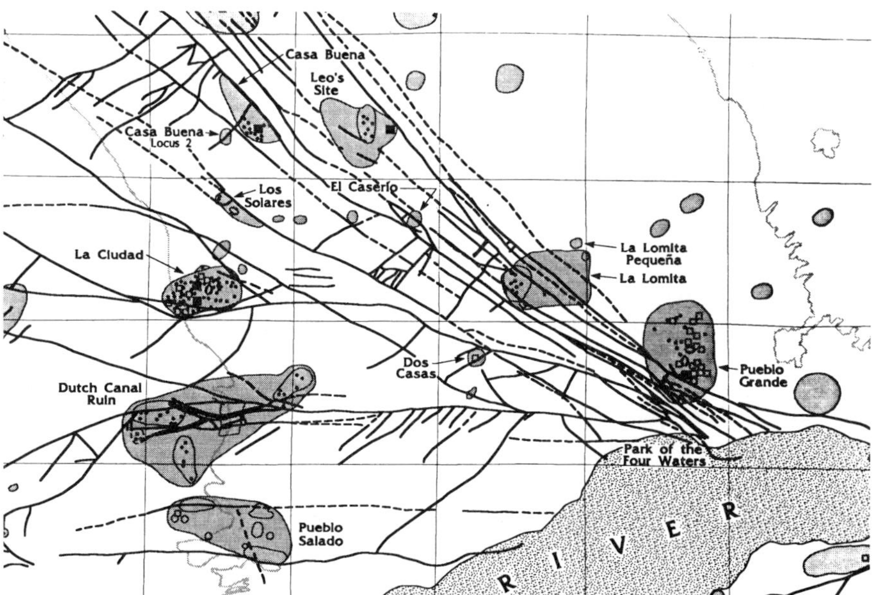

Fig. 10.31. An enlargement of that part of a map by Jerry B. Howard (Howard and Huckleberry 1991: fig. 1) showing the same area portrayed in Figs 10.24 and 10.25). (Reprinted by permission of Soil Systems, Inc., Phoenix, Arizona.)

view was based largely on the establishment and growth of habitation sites located along canals (Gregory and Nials 1985; Nicholas 1981; Nicholas and Feinman 1989; Nicholas and Neitzel 1984). Howard and Huckleberry found this not to be the case. Adding to what other contract projects had learned about the short lives and constant rebuilding of canals, they found that periodically entirely new canal systems were built with great labor expenditures and others abandoned. In other words, the entire Salt River Valley for the period extending from AD 0 to 1450 was a dynamic landscape of canals of various sizes and locations undergoing constant renovation and relocation, and with no small amount of social organization. To be sure, there were times when more area was under cultivation than at others, and that some canals were used longer than others. Overall, however, the valley can be best characterized in terms of the irrigated landscape as a constantly changing mosaic (Howard 1993).

In sum, the archaeological evidence of canal irrigation is abundant and varied in several respects. For some the evidence is quite tentative while for others it is irrefutable. Some prehistoric canal systems were very small and others were very large. In no small way, the sizes of the systems were spatially restricted by the amount of irrigable land available. Some canals were in use for very long periods of time; others were used only briefly. If there is any correlation, it might be that small canal irrigation systems were more sustainable than large ones. Those that are known to have begun prehistorically, were used in historic times, and continue in use today are all small. The very large prehistoric canals needed constant maintenance, and periodically were so damaged by floods that abandonment, and total reconstruction elsewhere was preferred. At the very least, this observation can be viewed as a hypothesis for future study.

Overview

Canal irrigation was associated with only a small corner of the cultivated landscape of native North America. It was, however, very important in that corner. Indeed, the cultures that arose in the Southwest prehistorically were linked tenaciously to that particular form of agriculture.

The extent to which canals were used to irrigate farmland *c.* AD 1500 will undoubtedly remain a point of theoretical and practical attention for some time to come. It is clear that the very large and technologically complex canal systems used in the Salt and Gila valleys were no longer in use at that time. However, there are still many questions concerning the origins, operations, and evolution of the systems for archaeologists to address. An unknown number of smaller systems were used in northern New Mexico shortly after that time, but which ones were in use when the Spaniards showed up will long be the center of debate amongst attorneys involved in settling water disputes.

The documentary record is certainly not without its pitfalls. Of course, it never is, but when lawyers are involved, the issue takes on an entirely new complexion, and a pretty one it is not. Suffice it to say, however, that canals were used at various northern New Mexico pueblos and throughout eastern Sonora. The ethnographic evidence is also flawed in that very good data come from only a few groups. Nevertheless, the insight provided by these people is both great, and applicable to others for which there are fewer data. In no small way, the characteristic that makes the data from both Zuni and the Akimel O'odham good is that it often agrees and rarely conflicts. What both peoples did, others probably did as well.

The archaeological evidence is at once abundant and sparse. There have been numerous studies from all over the Southwest, but on inspection it is evident that most of these were one shot deals. Rare is the case when there are two, three, or four studies done in any single area. There is still plenty of research to be done literally everywhere. The only areas that have been subjected to numerous investigations are the Santa Cruz and Salt River valleys. In these places, however, the abundance of work is really quite recent. And, what about the Gila Valley? Here is a place begging for more systematic archaeological excavations.

Many questions remain to be answered, and even asked, concerning aboriginal canal irrigation in North America. It should be clear, however, that native peoples in the Southwest were accomplished hydraulic engineers. Indeed, it is their ancient ancestors who were the originators of the technology. During the past three millennia, the landscape of this corner of the continent was certainly transformed from desert to cultivated oases.

Part V

Wetland Systems

11

Flood Recession Farming

Standing in stark contrast to drylands are wetlands. Whereas the former generally suffer from a paucity of moisture that inhibits agriculture, the latter have an abundance that also places extreme limits on cultivation. There are several types of wetlands including salt marshes, mangrove swamps, freshwater marshes, swamps, bogs, and lake shores. Although different in their own rights they all can be considered areas transitional between terrestrial and aquatic ecosystems where the water table is usually at or near the surface, or land is covered by shallow water and where at least one of the three following conditions exists: the land periodically, if not always, supports predominately hydrophytes; the substrate is undrained hydric soil; or the substrate is saturated non-soil (Cowardin *et al.* 1979).

Farmers have, since time immemorial, developed numerous specific strategies for practicing agriculture in wetlands. These, however, can be grouped into one of four general categories of agricultural ecosystems, each with its own level of technological elaboration or sophistication. The four strategies include reliance on hydrophytic crops, water recession farming, drainage, and the use of raised fields (Turner and Denevan 1985: 13–14).

The most elementary, or at least the most technologically simple in terms of landscape modification, of the wetland agroecosystems is the one involving hydrophytic crops. As the name implies, it is predicated on the use of water-loving cultigens. Hydrophytic systems were not used to any great extent in North America at or near the time of contact, although some notable examples do exist. The most conspicuous and important cases of the continent's native inhabitants raising hydrophytes were discussed previously in Chapter 2. These include, among others, the cultivation of cattails, and, especially, 'wild' rice in the lakes of the upper Midwest. Similarly, in his description of Virginia in 1612, Smith (1907: 93) noted that: '[i]n the low Marshes, growe plots of Onyons containing an acre of ground or more in many places . . .'. but this account is singular and unsubtantiated by other evidence. The cranberry (*Oxycoccus macrocarpus*), an historically and traditionally important food associated with the Thanksgiving holiday, would seem to have indigenous origins, but this hydrophyte, although gathered, was not cultivated until the middle of the 19th century, and then only by people of European ancestry (Phinney 1863; Schrambing 1988). All in all, and despite their limited technological sophistication and associated low labor input demands and landscape modifications, hydrophytic agricultural systems were developed in only a few places and only for limited food items.

At the other end of the spectrum, however, are the drained and, especially, raised fields. These will be discussed later (see Chapter 12). Here, emphasis is placed on the second type of wetland agricultural strategy outlined above, the cultivation of flooded areas after water recedes.

Water recessional agroecosytems are, as in the cases of hydrophytic systems, largely procedural, and include minimal technology and labor input. They involve the cultivation of seasonably inundated surfaces during the time of year in which annual and, therefore, predictable, floodwaters are not present. Planting usually occurs shortly after the floods recede, thereby taking advantage of the wet, fertile sediment. Harvesting takes place sometime before the next floods occur (Turner and Denevan 1985: 13).

Because these wetlands are in many cases dry for longer periods of time than they are wet, the term 'wetlands' is somewhat of a misnomer. Complicating the picture even further is the extent to which the floods and their recession are associated with farming. For example, the floodplain agriculture practiced on the Great Plains and in the Southeast (see Chapter 5) could be considered flood recession farming almost as easily but certainly not as accurately as it is considered rainfed agriculture. To be sure, the floodplains could not be farmed until the annual floods, which resulted in the deposition of nutrient-rich silt, receded (see e.g., Hudson 1976: 291; Wedel 1986: 38). However, a crop could not be grown on the floodplains of either the Southeast or the Plains without summer rainfall. Although floodplain agriculture in these areas was dependent on the recession of floods, it was not dependent on the actual floods themselves.

In a very real sense, flood recession farming is transitional between the irrigation of drylands and the drainage of wetlands. As in the cases of the floodplain agriculture on the Great Plains and in the Southeast, cultivation of certain floodplains in regions that are generally classified as drylands, would be impossible without the recession of the floodwaters. Unlike the former cases, however, the flooding itself is critical if agriculture is to succeed in recessional agroecosystems in the latter type of settings. Floodwaters saturate the soils to great depths, especially in those cases where soils are sandy and light. After the floods recede, soil moisture remains high for a prolonged period, even though high summertime temperatures can result in very high evaporation rates. This residual soil moisture is essential to agriculture in the absence of ample rainfall during the critical periods in the growing season. Due to the normally high temperatures and relatively brief periods of inundation and soil saturation, the natural vegetation of those wetlands transformed for flood recession farming is typically sparse grasses. Crops are commonly of quickly maturing species and varieties. Flood recession farming has been practiced in numerous places around the world for centuries, if not millennia (see e.g., Turner and Denevan 1985: 13). Without doubt, the best-known example involves the Nile River valley and the ancient Egyptians. Although this system gradually evolved into something technologically quite complex and involved extensive landscape modifications by means of a network of irrigation canals, it began, and continued to function principally, as a system of flood recession farming (Butzer 1976: 18–20).

Evidence of indigenous flood recession agriculture in North America is limited in number and spatial distribution. There are some rather detailed and graphic documentary accounts, and a few ethnographic descriptions that not only confirm the accuracy of the descriptions made by early Europeans, but also add considerable insight. There is only one study which provides archaeological evidence of flood recession farming. The practice, however, may have been more common prehistorically than the evidence suggests. The paucity of archaeological data is undoubtedly a function of this form of agriculture not requiring any permanent landscape modifications that might otherwise remain as relicts for study. All of these come from the southwestern quadrant of the continent (Fig. 11.1), a region that, in general, is considered dry, but within it has some places that are sometimes very wet. Although the data that do exist are not numerous, they are certainly not inconclusive. Explorers and scholars knew what they saw, and they described it well. Undoubtedly their diligence stemmed from a previous knowledge of

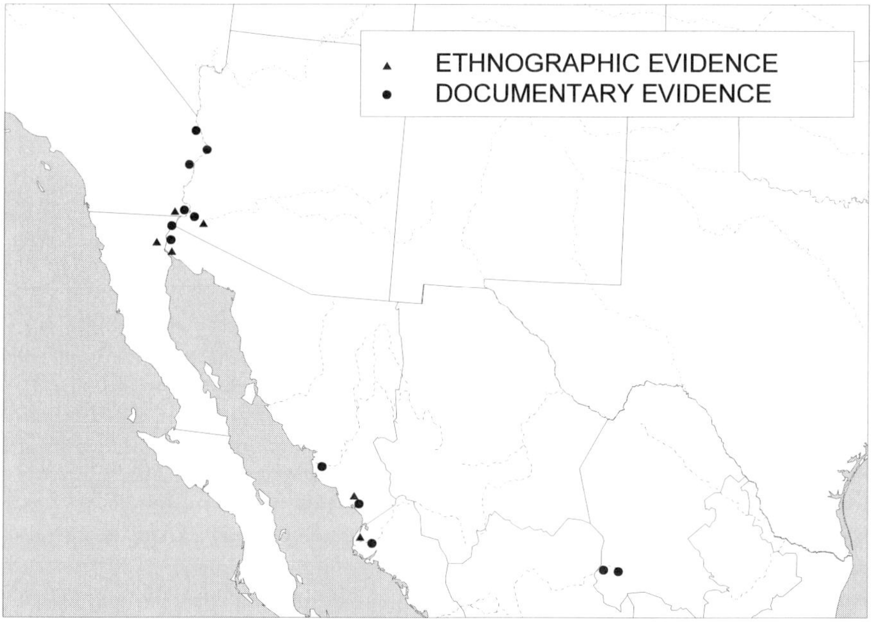

Fig. 11.1. Locations cited in the text for which there is evidence of flood recession agriculture.

Egyptian agriculture, its importance to ancient civilizations, and its singularly distinctive ecology.

Documentary Evidence

Some of the most detailed accounts of flood recession farming were provided by one Jesuit. First as a missionary, and later in his role as *Padre Provincial*, in charge of all Jesuit activities in New Spain, Pérez de Ribas (1944, 1968) traveled extensively throughout northwestern Mexico between 1604 and 1620, gathering much information about his domain. Although his predecessors and successors did likewise, Pérez de Ribas stands out as someone who was an especially astute observer, who recorded in detail what he saw and learned. His accounts, although somewhat repetitive, provide insight into this form of agriculture in four separate locales—the lower valleys of the Fuerte, Mayo, and Yaqui rivers, in far northern Sinaloa and far southern Sonora, and the margins of the Great Lagoon of San Pedro in Coahuila.

El río de Zuaque . . . [*e*]*s caudaloso y mayor en sus avenidas que el Guadalquivir en la Andalucía. Y aunque se ve salir de las serranías de Topia, no se sabe con certeza su nacimiento, porque debe de estar muy adentro de las sierra y las montañas . . . este río en espacio de treinta leguas a lo largo, por la cuales corre desde las faldas de la sierra hasta entrar en la mar. Fertiliza a trechos muy hermosas valles con sus crecientes y avenidas que suelen ser dos veces al año: el uno a tiempos de aguas y el otro por los meses del invierno. En éstos suele caer un agua menuda, que dura dos y tres días contínuos y con ella desatán dose y derritiéndose algunas nieves de sus montes, toma tan grande pujanza de aguas, que por algunas llanadas extiende su madre dos y tres leguas, con que deja regados los campos como se dice del Nilo en Egipto. A las naciones que alcanzan algo de sus valles les deja tierras regados y dispuestas para una, dos y tres sementeras al año; en la una de ellas no hay necesidad del rocío del cielo; porque sin lluvia la tierra con el riego que deja la corriente del río, da abundante cosecha de todos los frutos que usan sembrar los indios. Pero el año que faltan esta avenidas, padecen esterilidad estas naciones* (Pérez de Ribas 1944, vol. 1: 278).

The Fuerte River . . . is larger and has greater floods than the Guadalquivir River in Spain. And although it flows from the foothills of the Sierra Madre Occidental, its source is not known with certainty because it is well within the mountains. This river, over a distance of more than 120 km, flows from the foothills of the mountains until it enters the sea. Twice each year, in the summer and in the winter, it rises and floods fertilizing with sediment very beautiful segments of the valley. These floods last for two or three days, and those that result from the snow melt in the mountains are so great that in some places they extend from the stream bed 8 to 12 km watering fields as does the Nile in Egypt. The nations that occupy the valley can expect to make one, two or three plantings yearly. They are not dependent on rain as the floods provide ample water which results in abundant yields of all the crops that the Indians plant. But when these floods fail, the nations experience famine (see also Pérez de Ribas 1968: 51).

El río Mayo . . . el cual brota y sale como los demás de las montañas de Topia y corriendo por las llanadas y algunos valles por espacio de diez y seis leguas, desde que sale de entre montes, entra y desagua en el mar de Californias. Este no es río muy caudaloso y casi todo el año se puede vadear si no es al tiempo de sus crecientes que entonces recogiendo muchas aguas, ensancha su madre como los demás, y deja regados y fertilizados algunos de sus valles (Pérez de Ribas 1944, vol. 2: 9–10).

The Mayo River . . . flows forth, like the others, from the Sierra Madres across plains and through some valleys. From where it leaves the mountains and empties into the Sea of Cortes (Gulf of California) is a distance of 67 km. This is not a very large river and almost all year long one can wade across it, but not in times of flood. Then it has much water, overflowing its stream bed like the others, and irrigates and fertilizes some segments of the valley (see also Pérez de Ribas 1968: 87).

El Río Hiaqui, que es de los mayores que corren por la provincia . . . viene a ser casi tan caudaloso, como el de Guadalquivir . . . tiene su nacimiento . . . de la . . . cordillera. . . . Desde que sale de las serranías, corre por llanadas y entre algunas lomas, por espacio de treinta leguas, hasta desembarcar en el brazo de Californias. En las doce ultimas, a la mar . . . muchos valles, alamedas y tierras de sementeras, las cuales cuando el río trae sus avenidas y crecientes, que son ordinarias casi cada año, las deja regadas y humedecidas para poderse sembrar de verano, sin que tengan necesidad de lluvias para sazonarse y gozarse sus abundantes frutos. De suerte que antes que entren las aguas, que suelen comenzar a principio de julio, ya han alzado sus semillas los indios y esta es su principal cosecha; no obstante que por tiempo de aguas suelen algunos volver a sembrar, aunque desta cosecha poco caso hacen, porque su principal sustento les de la verano, que lo ordinario es abundante de maíz, frijol, calabaza, algodón y otras (Pérez de Ribas 1944, vol. 2: 64).

The Yaqui is one of the larger rivers that flows through the region . . . it comes to be almost as great as the Guadalquivir . . . having its source . . . in the high foothills of the mountains . . . From where it leaves the foothills, it flows through plains and between some hills for a distance of 120 km, until it empties in the gulf. In the last 50 km before the sea . . . there are many valley segments, groves, and fields that when the river rises and floods, a nearly annual occurrence, are watered such that the people can plant in the summer without being dependent on rain to ripen and produce abundant crops. Seeds of the principal crop sprout before the rains come, which is usually the first of July and result in an abundance of corn, beans, squash, cotton and others. Not withstanding some farmers return and plant after the summer rains (see also Pérez de Ribas 1968: 100–1).

These three accounts share many similarities, although each provides a bit of information lacking in the others. When read individually, they each say something. When looked at as an integrated unit, they say something more. There can be little doubt that Pérez de Ribas was describing flood recession farming. The parallel to the Nile Valley stands out as the most obvious indicator, and one repeated in 1764 by another Jesuit (Nentvig 1980: 9). That broad expanses of land were involved, and that floodwaters both irrigated fields and deposited nutrient-rich sediment thereby improving soil fertility is crystal clear. Although these rivers were claimed to be comparable to a specific large European river, they were mere trickles most of the year, becoming raging torrents only when they flooded. Indeed, one might change the old adage from 'When it rains, it pours', to 'When there is water in the streambed, there is water everywhere'. The northernmost of these three rivers, the Yaqui, flooded only once a year as the mountains where it heads are relatively low and receive little snow. Perhaps as a back-up strategy for when the flood failed, these farmers also planted a second, rainfall-dependent crop. Farmers in the Fuerte Valley, the southernmost of these rivers, did not plant a rainfall-dependent crop. They did, however, rely on two floods per year, snow melt from the high mountains contributing to one. Of course, if the floods failed, famine followed. From the accounts of both the Fuerte and the Yaqui rivers it is learned that more than one crop was produced each year, that yields were very high, and, most importantly, that farmers were dependent on both floods and their recession. By inference, the same is thought to hold true of the Mayo Valley (Crumrine 1983: 266).

Further to the east, inland from the Sonora–Sinaloa coast, Pérez de Ribas reported flood recession farming around the margins of a large lake in the center of an interior drainage basin. West of the present-day town of Parras and east of Torreon is a body of water, now typically dry, known as the Great Lagoon of San Pedro. In describing it and agriculture associated with it, Pérez de Ribas said:

> *La Laguna tiene de box y circuito, como cuarenta leguas, y en sus crecientes cincuenta y sesenta. Dije en sus crecientes, porque éstas las causa, por algunos tiempos del año, el río grande llaman de las Nazas, que entra en ella y la sustenta con sus aguas, que deja rebalsadas. Estas crecen y se explayan con las crecientes del río que es caudaloso, principalmente cuando lluve; y tiene su nacimiento de serranía de Topia .... La tierra jugosa, en que se da cualquier género de plantas con grande lozanía. Son las crecienetes y menguantes del río y su laguna, muy provechosas a sus moradores .... Y adémas de eso fertilizan muchos pedazos de tierras vecinas por donde se explayan. Las menguantes, retirándose, dan lugar a que en esas tierras ya regadas, pueden sembrar los indios sus semillas, y en ese mismo tiempo habiendo bajado el agua ... aunque siembran tambien algunas semillas, no cuidan tanto de ellas como las otras naciones* (Pérez de Ribas 1944: vol. 3: 246–7).

> The lake has a circumference of 160 km which increases to between 200 and 240 km. Its seasonal floods come from the large Nazas River, the waters of which, after filling the lake, spread over vast portions of the plain. The heavy flow occurs principally when it is raining near the river's source, in the Sierra Madre Occidental. Sediment deposited where the lake expands is very fertile. The soil produces any type of plant with great luxuriance. The floods, and receding waters of the river and the lake are, therefore, very advantageous to the area's inhabitants. As the waters recede, the Indians plant seeds in rich wet soils .... Although they plant a few seeds, they do not care for them as much as do other nations (see Pérez de Ribas 1968: 225, 226).

This description seems straightforward enough and is even corraborated by a separate account made several years earlier than that of Pérez de

Ribas. According to a 1604 relation made by a Jesuit named Francisco de Arista:

En este laguna, junto al pueblo de San Pedro, entre el río de las Nazas, que es el que la mantiene en ser, aunque en cierto tiempo del año se seca el río, por consumirse el agua en los arenales, corriendo debajo de la tierra . . . el río en la primera avenida . . . . De esta misma retirada de la laguna quedan también los prados y arenales con buenos húmedos, para sus sementeras de maíz, que sin más arado, ni más riego o cultivo, nace con tanta abundancia, que se han medido algunas mazorcas de más de media vara (Alegre 1958: 107; see also Alegre 1841 vol. 1: 417).

In this lagoon, next to the town of San Pedro, and along the Nazas River which is usually within its banks and at certain times of the year is even dry because what little water exists soaks through the sands and runs beneath the surface . . . [at other times] the river floods . . . . After the waters recede there remains also meadows and sandy ground good for the planting of maize, without plows, and with neither more irrigation nor more cultivation. It grows so abundantly that some ears are more than 40 cm long.

It has been argued, surprisingly enough, that these accounts do not refer to agriculture, but rather to the cultivation of non-domesticated plants as was discussed in Chapter 2 (see Driver and Massey 1957: 228; Martínez del Río 1954: 69–71). Exactly how such a conclusion can be made belies the evidence. Furthermore, two quite distinguished scholars with much expertise at translating and interpreting documentary relations of this general region and dating from the viceregal era accept these reports as descriptions of flood recession agriculture involving maize (Gerhard 1982: 220; Griffen 1969: 111–12).

The connection between the cultivation of non-domesticated plants and flood recession agriculture is, of course, documented. As was seen earlier (Chapter 2), a pair of priests involved with the Anza expedition between Sonora and California observed the sowing of a panic grass on the floodplain along the lower stretch of the Colorado River. These same two priests, along with one Fray Juan Díaz, and Anza himself, were the first to identify flood recession farming of domesticated crops in this area. To be sure, earlier explorers such as Alcaron in 1540 (Hammond and Rey 1940: 130) and Oñate (Bolton 1908: 245; Hammond and Rey 1953: 415) in 1598 found fields of maize, beans, and squash growing at the edges of the river. A century later, Manje (1954: 121) made a nearly identical observation. Manje's traveling companion, Kino (1948: 340–3), reported seeing both fertile bottom lands abundantly cultivated and soft muddy ground due to recent flooding of the river, but failed to make the connection between the two phenomena. Thus, it was not until Anza and his associates traveling in and through this area during the 1770s that the ecology of Yuman flood recession agriculture became known to people of European ancestry. Anza (1966*a*: 51–2) himself reported:

In the country which I have seen from the junction of the rivers [Colorado and Gila] to here, at intervals there are fields sown with wheat without irrigation, but so good and well sprouted that the best irrigated wheat in our country does not equal it. One sees also the places where they plant maize, beans, calabashes . . . . This fertility of the land, which they say is the same upstream and better downstream, is greatly aided by the annual overflow which the meadows receive after the coming of the spring, as a result of the melting snow in the interior, while it does not destroy or carry away the trees; and the meadows are silt covered for at least half a leagure [2 km] on each bank, and many persons tell me that the waters spread out to even a greater distance. This fertility is enjoyed by both rivers . . . (see also Anza 1966*b*: 174 for a nearly identical, but separate description).

According to Eixarch (1966: 321):

The land is so good that only with the bathing given it by the river during the time of its floods it conserves enough moisture so that it produces wheat, and also maize, beans, watermelons, calabashes, etc, and they say that of maize they harvest two crops. It is true that the seed is from what they call Apache maize, which matures very rapidly.

This observation was made in December, 1775. The following April he noted that:

The river is now formidable. It greatly overflows its channel. . . . The wheat of the Indians is now maturing and that which I have seen is very good. This is proof of the richness and fertility of this land, for without irrigation and solely with the moisture which the river affords at the season of its regular floods, these Indians get so important a harvest . . . (Eixarch 1966: 371–2).

In May, he continued:

The river is rising every day and now greatly overflows its channel, for this reason the Indians say that this year they will harvest plentiful maize, beans, etc. They have already cleared the land and are beginning to cut the wheat (Eixarch 1966: 377).

Font (1966: 98–9) echos many of Anza's and Eixarch's comments and adds some extra insight. He noted that:

The Yumas dwell on the bottom lands of the Colorado River . . . . The bottom lands extend on one side of the river and the other for about two leagues [8 km] . . . . Each year the river spreads out for a long distance through the bottom lands in the season of the floods, which come from the melting of the snow in the summer in the mountains to the north and far in the interior. For this reason it does not rise suddenly but gradually. Indeed it rises and falls nearly all the year, for it begins to rise in March and April and from that time each day it gets larger until July, when it begins to go down. . . . The lands which it waters are generally good, and since the water spreads over them so gently it does not injure them. On the contrary, from this irrigation they are greatly fertilized and have moisture for the crops which the Indians plant in them when the water recedes, and for the abundant harvests which they get . . . . The crops raised by the Indians are wheat, maize, which they call Apache maize and which matures in a very short time, orimuni beans, tepari beans, cantaloupes, watermelons and very large calabashes (Font 1966: 98–9).

Díaz (1966: 264) confirmed everything his contemporaries reported, but did so in a much more concise manner. In noting conditions near and downstream of the conflence of the Colorado and Gila rivers, he stated simply that:

the banks offer very great advantages for fields, so that these natives . . . get very good harvests of wheat, maize, beans, calabashes, melons, and watermelons, without the trouble of irrigating their crops, because with the moisture and the fertilization which the land receives during the great overflow of the rivers, as well as because of their proximity, no other irrigation is needed to raise any kind of crop (Díaz 1966: 264).

All four of these accounts make reference to Old World crops. These plants might have been introduced by earlier Europeans who passed through the region. It is more likely, however, that they diffused via exchange with other native groups. They do not, by any means, suggest that flood recession farming was a European introduction. Indeed, it is well-accepted that the practice was indigenous (Forde 1931*a*: 109). What these four accounts do demonstrate, rather convincingly, is that native North Americans in the far southwestern corner of the continent were practicing flood recession agriculture at the time of contact with Europeans.

Ethnographic Evidence

The lower valleys of the Fuerte, Mayo, and Yaqui rivers were inhabited at the time of European contact by Cahitan speakers who distinguished themselves among native North American peoples by their great resistance. Although many indigenous groups struggled to save their ethnic, racial, and cultural identies in the face of increased pressure from interlopers whose ancestry can be traced to the Old World, these people of northwestern Mexico did so with a vengence. With the exception of the Jesuits, from whom they adopted certain practices and technologies, they shunned virtually everyone else sometimes demonstrating great military might (Spicer 1962: 46–86). The Yoemen (Yaqui), who resided in the Yaqui Valley, were legendary in their efforts. Refusal to adopt European ways, especially in regards to abandoning flood recession farming, persisted well into the historic era (Spicer 1983: 254). The only factor leading to its demise, or, at least, its minimization, was the damming of rivers to provide irrigation waters to the flat, fertile, coastal plain lands distant from the rivers themselves (Dunbier 1968: 218–30). Indeed, this is the area where high-yielding varieties of wheat were developed and cultivated on a large scale as part of the 'Green Revolution'.

Although the creation of reservoirs upstream essentially put a halt to the floods and their recession downstream, as described by Pérez de Ribas, scholars conducting field research in the area during the 1920s and 1930s were certainly aware of what went on previously. To be sure, they knew of the documentary accounts. They also relied on elderly informants, and they saw some smaller, scaled-down versions of what they had read about and were told. Two cases stand out as exemplary in this regard, one being more detailed than the other.

In his study of these people and their many practices, one of which was agriculture, anthropologist Ralph Beals (1945: 5) found that '[o]nly to a very slight degree does modern agriculture follow aboriginal patterns'. He noted that the benches and bars along the rivers are still watered by floods, but that the rivers had been restricted somewhat by levees built of earth removed during the excavation of irrigation canals that carry water from the reservoirs to the fields. Beals (1945: 5) elaborated further, stating that these levees retained irrigation waters on fields in such a manner that they mimicked the effects of floods that occurred naturally, but not as regularly and as predictably, as before modern improvements were made. It is clear from Beals' description that true flood recession agriculture is not anywhere nearly as important to these people today as it was in times past, having been replaced by extensive canal irrigation. However, despite losing its importance, flood recession farming is seen to be characteristic of the valleys, and its legacy continues to remain great.

A second study of native peoples along the northwest coast of Mexico in the early 1930s focused specifically on one group, the Yoemen, and included an assessment of their agriculture. The report of this project is not as explicit about flood recession farming as was that of Beals, but it does refer to floods, agriculture along the river, canal irrigation projects, and times past as well as present (Holden *et al.* 1936: 116, 118). Anyone even remotely familiar with the topic can piece together the implicit allusion to flood recession farming in former times (Crumrine 1983: 266; Spicer 1983: 254).

Whereas only fragmentary ethnographic evidence of flood recession farming exists for those groups living along the northwest coast of Mexico, there is actually a substantial amount of information about the practice in recent times for native peoples living along the lower Colorado and Gila rivers, somewhat further to the northwest (see Fig. 11.1). Flood recession agriculture is mentioned explicitly in ethnographic overviews of Yuman people, in general (K. Stewart 1983*a*: 2), and, specifically, the Cocopa who live near the mouth of the Colorado River (Kniffen 1931: 52, 57; A. Williams 1983: 104), the Quechan who live at the confluence of the Colorado and Gila rivers (Bee 1983: 87), and the Maricopa who live along the lower reaches of the Gila (Jorgensen 1983: 689). It is also mentioned in reference to the Mojave (Fitzhugh 1926: 88; Kroeber 1925: 735; K. Stewart 1983*b*: 56, 57), and there are a few brief but fascinating assessments of this form of agriculture among two groups of the Ningwi (I. Kelly and Fowler 1986: 371) and among the Tipai (Drucker 1937: 5; Luomala 1978: 600). More importantly, however, there exist three detailed studies of indigenous agriculture in this region that provide a wealth of information about flood recession farming in recent times and, therefore, can be used as parallels for understanding its ecology in times past. One of these studies deals with all of the Yuman people (Castetter and Bell 1951), one deals with the Cocopa specifically (W. Kelly 1977), and the third deals with the Mojave (K. Stewart 1966). Given the similarities of the respective systems with which they deal, these studies are herein used in concert to portray a general picture of agriculture in relation to floods and their recession.

In general, the first thing involved with flood recession farming was the selection of a plot to be cultivated. Elderly informants reported that in the days before dams were built upstream farmers would begin by selecting a plot that had certain edaphic, hydologic, and vegetative properties. Soils with high salt contents—more common along the Gila than the Colorado—were avoided. These could be detected visually by either a 'frosty' appearance or by the presence of salt grass. They could also be detected by taste (Castetter and Bell 1951: 141). Preferred areas for planting were those over which the floodwaters moved rapidly enough to deposit coarse sediment. Areas flooded by still water were not highly regarded as the soil tended to be heavy, and upon drying subject to severe cracking (W. Kelly 1977: 28). One informant reported that he tested the ground by pushing a single arrowweed stick into it. If it penetrated easily when the ground was soaked, he regarded this as indication of a good place to plant (Castetter and Bell 1951: 140–1). In terms of vegetative cover, preferred areas were largely open, having few trees, and being covered with arrowweed (Castetter and Bell 1951: 140; K. Stewart 1966: 6–7).

Some individual farmers managed to obtain one relatively large parcel of land while others preferred to have a number of smaller ones. Separated fields were usually within easy access of each other, some even being nearly contiguous and in strips (Castetter and Bell 1951: 141). Regardless of the degree of fragmentation, fields varied in shape given local environmental conditions (W. Kelly 1977: 28). The size of each farmer's holdings varied with the number of people he had to feed and available to help with the work. It also varied from year to year and from one part of the region to another (W. Kelly 1977: 29). In general, however, holdings tended to be less than 1 ha (K. Stewart 1966).

In the spring, after plots were selected, farmers would chop down those trees that did exist and clear brush as well, using axes and knives (see also Forde 1931*a*: 110). This felled vegetation was then burned, after having been allowed to dry for a few weeks (Castetter and Bell 1951: 140; W. Kelly 1977: 30). New fields were selected each year, resulting in more than a little shifting of families and groups, but no formal field or crop rotation was practiced (K. Stewart 1966: 12; W. Kelly 1977: 26).

Fields chosen, farmers would then anticipate the floods, hoping they would be neither too great nor too small. Ordinarily, the floods would be of sufficient magnitude (K. Stewart 1966: 7). If they were not, or if they were too great and altered the conditions of the cleared plots significantly, farmers would identify just-flooded parcels fitting all the requisite criteria outlined above and quickly clear them for planting (W. Kelly 1977: 26). In some cases farmers anticipated larger than usual and potentially destructive floods, and built levees between the river channel and their fields (Castetter and Bell 1951: 135). Openings in the levees permitted floodwaters to reach the fields. When an ample, but not too great of an, amount covered the fields, the openings would be closed. The fields would then begin to dry out as the river continued to rise and then fall (W. Kelly 1977: 27). Although some scholars (e.g., W. Kelly 1977: 27) consider the use of levees to predate the arrival of Europeans, others (e.g., Jorgensen 1983: 692) do not. The argument against an old age for levees is based on their association with canals used for irrigation and known to have been introduced by Spanish missionaries (Castetter and Bell 1951: 131), rather than having been adopted from neighboring groups in earlier times (Jorgensen 1983: 693). The position supporting their antiquity is rooted in a claim by informants that these farmers 'had "always" used them' (W. Kelly 1977: 27). The weight of the evidence currently favors levees being recent, but the question is far from resolved and awaits future research. Whether or not pre-European farmers along the deltas of the Colorado and Gila rivers used levees, the next thing they did was plant.

Planting commenced as the floods began to recede. Sloughs, lagoons, and old river channels that intersected the wide fertile banks of the river were all planted. Native farmers followed the receding water; planting higher grounds first, and lower lands later, but always immediately after the flood subsided and in the freshly exposed mud (Castetter and Bell 1951: 137, 149; K. Stewart 1966: 7, 11). The entire planting process could take as long as a month, and ended when the moist soil ceased to cling to the feet of farmers (Castetter and Bell 1951: 148). Intercropping was not practiced, probably because its soil fertility-maintaining qualities were not needed given the regular deposition of fresh sediment. Because this is an area of very low precipitation, crops had to germinate and grow to maturity almost entirely on moisture furnished by the river during its periods of flood (Castetter and Bell 1951: 148). Accordingly, they were planted separately and in the order of their soil moisture needs and the lengths of their maturation periods (W. Kelly 1977: 31). Tepary beans were always planted first, on the high ground. Maize was planted next. And, squash was planted last, on the low ground (Castetter and Bell 1951: 149–50, 153).

Most informants reported that planting was done with the regular digging stick. In making the holes, the stick was thrust into the ground several times. Loosened earth was then scraped out by hand, forming a hole 10 to 15 cm deep and 15 to 22 cm in diameter. The person making the holes was followed by someone else, often a woman, who planted the seed. Four to eight maize seeds were dropped in each hole, and moist soil was then replaced by hand. Maize and beans were planted between one and one and one-half pace, or 1 m, apart and in rows, but not very uniformly (W. Kelly 1977: 28, 31; K. Stewart 1966: 12). Squash was planted 2 m apart (W. Kelly 1977: 31). Holes were made wherever digging was easiest. Seedlings were never thinned, but cultivation did involve two weedings, either by hoeing or by pulling weeds out by hand (W. Kelly 1977: 31). Hoeing resulted in the loosened soil around each plant being formed into small hills (Castetter and Bell 1951: 150–1, 153).

Growing crops received relatively little attention other than protection (K. Stewart 1966: 12). Men and boys were usually charged with the protection of crops and fields. Brush was often placed over freshly planted holes to protect the seeds from birds. Plants were sometimes sprinkled with water in which fish were boiled. Informants believed that the fish odor was objectionable to, and hence repelled, rabbits. Farmers frequently spent nights in their fields, protecting plants from mice and gophers. Scarecrows made of arrowweed, and children were used to drive away birds from fields with ripening crops (Castetter and Bell 1951: 153–4). In some cases rabbit entrails were spread over fields, and in other cases dead birds were strung on poles in the belief that their live counterparts would be scared away (W. Kelly 1977: 32).

Harvesting was done as soon as the crops matured, beginning in September and continuing into November (K. Stewart 1966: 12), and it involved everyone including women. Maize was harvested by pulling the stalks out of the ground, or breaking off at ground level, and piling them in heaps for husking in the fields soon afterward. Teparies were harvested by pulling up the vines and placing them in piles, where they were allowed to dry for two or three days. On a prepared spot of hardened ground they were trampled by foot and beaten with flails to separate beans from pods. Vines were then removed from the threshing area. Winnowing involved beans being sifted through the fingers to remove the coarser pods, and basket-loads of beans and pods being cast into the air, preferably when a light wind was blowing. Squashes were picked from the vines with the largest, best-developed, and blemish-free ones being reserved for next year's seed (Castetter and Bell 1951: 157–60).

All in all, flood recession farming, according to all ethnographic reports, was not very arduous. Furthermore, yields were dependable and abundant (see e.g., K. Stewart 1966: 5). In fact, this form of agriculture had so many advantages and so few disadvantages that even neighboring people who were known principally by their hunting and gathering lifestyles—having little need for agriculture and too few people to invest much effort—are known to have practiced it on occasion. Three groups fall into this category. The Chemehuevi branch of the Ningwi planted along rivers as floods receded, probably having learned how from the Mojave. Their related neighbors, the Pahranagat branch, planted along the edges of shallow lakes as the lakes dried (I. Kelly and C. Fowler 1986: 371). Finally, on at least a few occasions when floods were exceptionally great, groups of Tipai would come down to the mouth of the Colorado River, obtain seed from Yuman people, and practice flood recession farming (Drucker 1937: 5, 11; Luomala 1978: 600).

Archaeological Evidence

As seen in the previous two chapters, the earliest known Hohokam canals were on the edge of the first terrace overlooking the floodplains of the Santa Cruz and Salt rivers. In the latter setting, canal systems were later expanded over the first and higher terraces. Looking backward in time, especially in the light of expansion away from the river, begs the question about what was going on in terms of agriculture prior to the construction of the earliest canals on the terrace edges. To be sure, the answer may not be known for some time, if ever, but one study has suggested that the floodplain below the terrace may have been the site of flood recession farming.

In the early 1980s, habitation sites located on the terrace edge were archaeologically excavated. Remains of two types of impermanent structures identified as field houses, and segments of prehistoric canals were discovered. One type of field house was associated with the canals, situated back from the terrace edge, and dated to AD 800–1000. The other type of field house was located on the very edge of the terrace and dated to AD 500–700. A few structures of this latter type contained abundant fragments of mesquite (*Prosopis* sp.), trees that must have grown profusely on the floodplain, and cultigens. Putting two and two together, it was concluded that at this time and probably long before, flood recession farming was practiced along the Salt River (Cable and Doyel 1985). This interpretation is now widely accepted among Hohokam archaeologists. There is some recent evidence that a similar sequence might have occurred along the Santa Cruz River, farther to the south (Mabry and Holmlund 1998). In other words, flood recession agriculture might have been more widespread in times past that the current evidence suggests.

Overview

Although the data on flood recession agriculture are not very numerous, and deficient in terms of an archaeological dimension, they are certainly impressive. The documentary accounts provided principally by Jesuit priests, are some of the most thorough portrayals of any agricultural system anywhere. On one hand, this is highly peculiar given that the form of agriculture itself is so atypical. One might expect that something so unusual and observable only over a period of a few months and at a particular time of the year would be easily overlooked or not seen by explorers, especially those who were only passing through the area and had their thoughts on a more important mission and destination. On the other hand, the similarity of agriculture along the Colorado and Gila deltas to that long-known to have supported the ancient Egyptian civilizations along the Nile made it so conspicuous that these subjects of the Spanish crown were morally compelled to describe it in as much detail as possible.

Recognizing that flood recession agriculture dominated the domesticated landscapes along the lower courses of all the rivers from the Yaqui to the Colorado (see Fig. 11.1), and supported large indigeous populations, Spanish civil and religious authorities invested much time and energy in bringing these lands and their inhabitants into the colonial realm (Spicer 1962). What they did not realize, however, were the abundant water resources that could be utilized to cultivate even more extensive tracts of land beyond the floodplains of the rivers. The prospects for developing these resources to a much greater extent awaited the 20th century, the development of megascale water control technologies, capital inputs from the United States, and the existence of a global market economy (see Dunbier 1968: 176–90, 200–14).

To be sure, pre-Hispanic peoples in other parts of the Southwest, and the Spaniards themselves, had canal irrigation technology, sometimes on very grand scales. For reasons not identified previously, neither the indigenous people nor the European interlopers brought into cultivation, by means of developing an extensive canal system, lands distant from the rivers. In all likelihood, however, the respective valley's populations were more than sufficiently fed using flood recession farming, a type of agriculture that had the added advantage of not requiring great labor or technological inputs.

The determination that these exotic rivers, with their sources in the high mountains and mouths in one of the most arid deserts on earth, could be harnessed for development purposes coincided with a blossoming of interest in native peoples by ethnographers. As a result, we have today a small but impressive collection of detailed descriptions which articulate the ecology of native flood recession farming. Although archaeological data are lacking, they are really not needed to understand what was going on *c.* AD 1500 and earlier.

12

Draining and Ridging

Farming in seasonally inundated places after the water has receded is certainly a very basic approach to practicing agriculture in wetlands. It requires next to no technology, and, hence, little in the form of permanent landscape modifications. In some respects this is an ideal type of agriculture, especially in arid and semi-arid lands where successful farming is dependent on water other than that which falls as rain. Labor inputs are minimal and yields tend to be quite high in flood recessional farming systems. But this form of agriculture is predicated on the existence of a growing season during which the ground is neither too wet nor too dry. What about lake shores, freshwater marshes, and swamps—locales in which standing water or saturated soils are problems with which farmers have to deal during the growing season?

The latter two types of environments are quite different from each other, but they also share one important similarity. Freshwater marshes are perennially inundated, although the depth of the water may vary seasonally, and they are characterized botanically by herbaceous plants, particularly grasses (Horwitz 1978: 11). Swamps, in contrast, contain standing water in the winter and spring but dry out in the summer. They are identified botanically by woody vegetation (Horwitz 1978: 12). All three environments share an obvious abundance of water that must be dealt with in order to ensure successful cultivation.

Handling this water can be done in one of three basic ways. Surface water can be diverted away and kept from entering lowlands by means of levees, thereby not allowing them to become wet. It can be drained-off wetlands by means of ditches. Or, it can be replaced with earthen fill. It can also be, and indeed typically is, handled using various combinations of each strategy (Horwitz 1978: 34). Earth used to construct levees or raised fields, or to fill-in lowlands, has to come from somewhere. Rarely is it extracted from otherwise unused areas, resulting in the formation of borrow pits. Typically, it is excavated from what will become drainage ditches. Conversely, earth excavated during the construction of drainage ditches is usually piled up to form levees or raised fields, or spread over, thereby filling-in, the lowlands. It is rarely wasted by creating spoil banks. Sometimes, it is difficult to discern if a specific wetland field system involved ditching or raising (Jacob 1992).

The excavation of drainage ditches is a particularly interesting activity, and a very basic one at that. There is much evidence that drainage preceded irrigation, and that the technology developed for the former facilitated that developed for the latter (Blaut 1993: 86, 143). Curiously, however, although some very large and technologically complex canal irrigation systems were developed indigenously in North America during prehistoric times (Chapter 10; and Doolittle 1993), they seem not to have evolved out of drainage systems. Furthermore, when it came to cultivating lands that were inundated or soils that were saturated during the growing season, native North Americans, not unlike their counterparts in South America and Mesoamerica, relied principally but not exclusively on elevating planting surfaces.

Cultivating wetlands, especially by means of elevating fields, was, however, more important by several orders of magnitude in the latter regions than it was in the former. In no small way this was due to the great expanses of relatively relief-free land, exceptionally high rainfall associated with the low barometric pressure characteristic of the tropics, and large, agglomerated populations that needed to be fed. Prehistoric raised fields built principally for the purpose of cultivating wetlands in South America have been found in such diverse locations as the Lerma Valley of Argentina, Lake Titicaca in Peru, the Guayas Basin of Ecuador, the floodplain of the Rio San Jorge in Colombia, the Karinya Valley of Venezuela, and the coast of Surinam (e.g., Denevan 1970; J. Parsons and Denevan 1967). Further to the north, in Mesoamerica, they have been reported from numerous locales in the Maya Lowlands of Belize and the Yucatan Peninsula (e.g., B. Turner and Harrison 1981), several locations in the coastal state of Veracruz, Mexico (e.g., Siemens 1989), and a few of the lake basins in highland central Mexico (e.g., Crossley 1999; Rojas Rabiela 1983; Weigand 1993). Some of these raised fields, most notably those in the Llanos de Mojos of Bolivia (Denevan 1966), those in Pulltrouser Swamp (Harrison and Turner 1978) and on Albion Island (Pohl 1990) in Belize, those in the San Juan Basin of central Veracruz (Siemens 1983), and those on the southern margins of ancient Lake Texcoco (Avila López 1991) were subjected to rather detailed investigation, and, hence, have provided a wealth of information. Combined with a few studies of raised fields still in use in these regions (e.g., Wilken 1969) and other parts of the world today (e.g., Waddell 1972), a great deal is known about the mechanics and ecology of this agricultural strategy.

Raised fields in wetlands, be they modern or ancient, are, in effect, artificial islands. As in the case of natural islands, they vary considerably in length, width, height, spacing, number, shape, configuration, and cross-section. Some are exceptionally long, others are short. Some are wide, others quite narrow. Some are high, others are low; some have deep canals between them, others have shallow canals. Some seem too close together, others are separated by what appear to be disproportionately great spaces; they have wide canals. Some sites contain large numbers of raised fields, others are comprised of few fields. From the air, or in plan view, they can be rectangular, round, ovaloid, or linear, the latter being either straight or curved. They can also be aligned according to compass directions, arranged in any number of patterns, or laid out in seemingly random configurations. In cross-section, they can be flat-topped or convex. No two characteristics seem to be correlated. If there are any correlations, they are with their presence, the production of large amounts of food, and high labor requirements.

Every study conducted on raised fields to date has concluded that these are some of the most labor-intensive yet highly productive agricultural ecosystems. The excavation and subsequent piling of heavy wet earth is not easy. Furthermore, raised fields require a great deal of nearly constant maintenance. Rainfall and cultivation activities combine to result in erosion of the elevated field surfaces. Eroded materials then accumulate in the adjacent canals. These sediments, along with organic matter that also washes off the fields and that which is the product of aquatic life, need to be cleaned from the canals. They are typically placed on top of the fields—a process known as 'mucking'.

The combination of the high nutrient value of muck and the careful management of both soil and water resources contribute to an agricultural system that produces very high yields. Indeed, the yields from raised fields tend to be some of the highest per unit area of land. Given their high labor input requirements, it should not be surprising that raised fields are typically, although not always, associated with large and dense populations. They can feed several people, but only with much effort; the work they require can only be done by large groups.

Agricultural systems requiring drainage ditches and levees may not be quite as labor-intensive and as productive as those involving raised fields, but they are still on the high end of the input–yield spectrum. Accordingly, they, too, tend to be associated with sizable populations. In both South America and Mesoamerica, wetlands frequently contain numerous vestiges of ancient high cultures or civilizations—pyramids, temples, ballcourts, plazas, raised fields, drainage ditches, and levees. That many of these are in coastal settings and ring the Caribbean Sea and the Gulf of Mexico, has given rise to the notion that intensive wetland agriculture in North America prior to the arrival of Europeans diffused first into the Southeast whence it spread north and northwest (Keegan 1987*b*; T. Riley 1987: 301; Sears 1977: 7–9). Data used to support this notion are not terribly compelling (Knapp 1990). However, the data used to refute it are not very convincing either (Watson 1988: 45–6). Problems exist not so much with the arguments, but with the data themselves. Indeed, issues other than origins of wetland cultivation suffer understanding because of the paucity of data.

Documentary and Ethnographic Evidence

The evidence for drained and raised or ridged field agriculture in North America is sparse, to put it bluntly. There are no documentary sources and there have been no ethnographic studies that reveal insight into aboriginal dwellers of the continent practicing agriculture in locales that either retained standing water or had soils that remained saturated during the growing season. The reasons for this may be many. It may be the case that at the time Europeans appeared on the continent there were no native inhabitants cultivating wetlands. That Smith (1907: 93) reported seeing fields of onions in a marsh (see Chapter 11), and that Marquette (1903: 9–10) saw nondomesticated rice cultivated in lakes (see Chapter 2), however, suggests this conclusion is not entirely correct. An alternative explanation might be that native wetland farmers at the time were not using levees, drainage ditches, or raised fields that otherwise would have been seen by early explorers. It is also quite possible that such landscape modifications were being used, but that traveling chroniclers either did not see them or did see them and did not report their observations (T. Riley 1987: 301).

As for the lack of ethnographic evidence, a number of plausible explanations also exist. It is possible, although it seems most unlikely, that various ethnographers could have failed to see, or they saw and failed to report, landscape modifications due to the cultivation of wetlands by certain native North Americans in the 19th and 20th centuries. Alternatively, it may be that indigenous farmers were not excavating drainage ditches or raising fields or levees in historic times. If this was the case, it could have been due to the practices having fallen out of favor, for whatever reasons, by people who remained on their ancestral homelands. Also, the lack of ethnographic evidence for the agriculturally related modification of wetland landscapes could have been the result of people who formerly cultivated such environments emigrating territories they once occupied. In effect, once people moved (or were moved) away from wetlands, they would have no longer needed, and hence they would have abandoned, various labor-demanding technologies. Ethnographers, by virtue of studying what people do in the present, cannot be held culpable for not reporting what people did in the past.

Archaeological Evidence

As was the case with ridged fields used for ameliorating cool temperatures, strategies used by aboriginal people in North America for cultivating wetlands are best evidenced in the archaeological record. Although documentary accounts and ethnographic descriptions are lacking, relicts of various landscape modifications in the form of drainage ditches, levees, and ridged fields remain extant. Given the environmental parameters, it should not be surprising that vestiges of such features have been found in the Southeast, the Mississippi Valley, and in the Great Lakes area (Fig. 12.1).

One of the more important sites in regard to pre-European wetland agriculture in North America is the Fort Center site which is located in a savanna

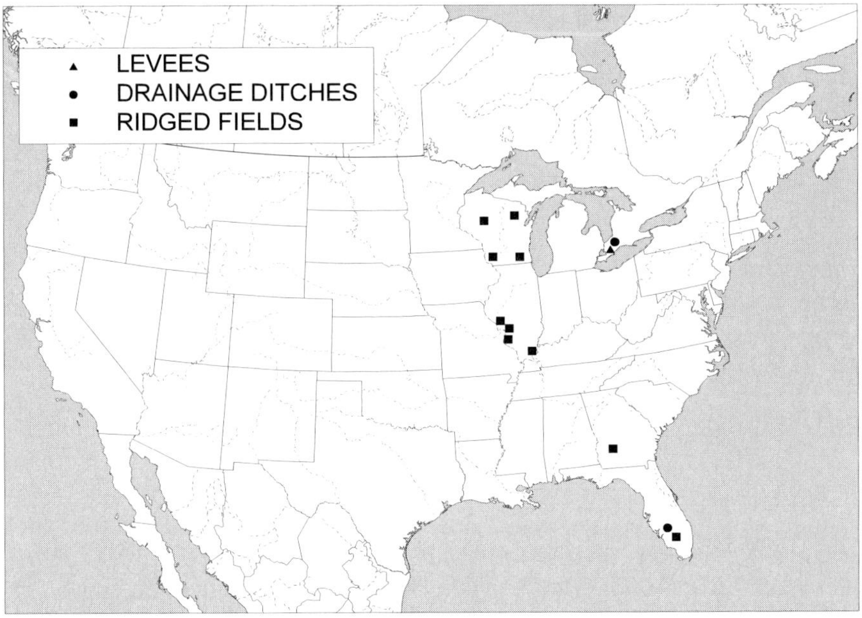

Fig. 12.1. Locations cited in the text for which there is archaeological evidence of wetland agriculture.

dotted with sloughs, and ponds, near Lake Okeechobee in south central Florida (Sears 1982: 3). According to archaeologist William H. Sears (1971: 325), it was occupied from at least five centuries BC into the 16th century AD. Sometime around 400 BC settlers excavated the first of a series of circular ditches with ends terminating in Fisheating Creek (Fig. 12.2), the smallest being more than 90 m and the largest being 360 m in diameter (Sears 1971: 327; 1982: 3–5). They averaged 7.5 to 9 m wide and 1.8 m deep (Sears 1982: 175). Because these ditches were cut into the underlying hardpan which inhibited percolation of soil moisture, it was proposed that they functioned to drain the enclosed land thereby making agriculture possible (Sears 1971: 326, 1982: 6, 178, 186). The discovery of maize pollen within the circle was seen as further evidence (Sears 1971: 328–9; Sears and Sears 1976). Also found at this site were a few small, single-house, mounds. Each had a 60 to 75 cm high flat-topped linear mound extending as much as 360 m length and being up to 90 m in width adjacent to it (Fig. 12.2). A ceremonial function for these embankments was ruled out and instead it was proffered that they were used as cultivation platforms, or raised fields (Sears 1971: 327, 329, 1977: 7, 9, 1982: 5). Pollen evidence was once again used to confirm the notion that these ridges were planted in maize (Sears 1982: 129).

The interpretations that these ditches and ridges served agricultural pur-

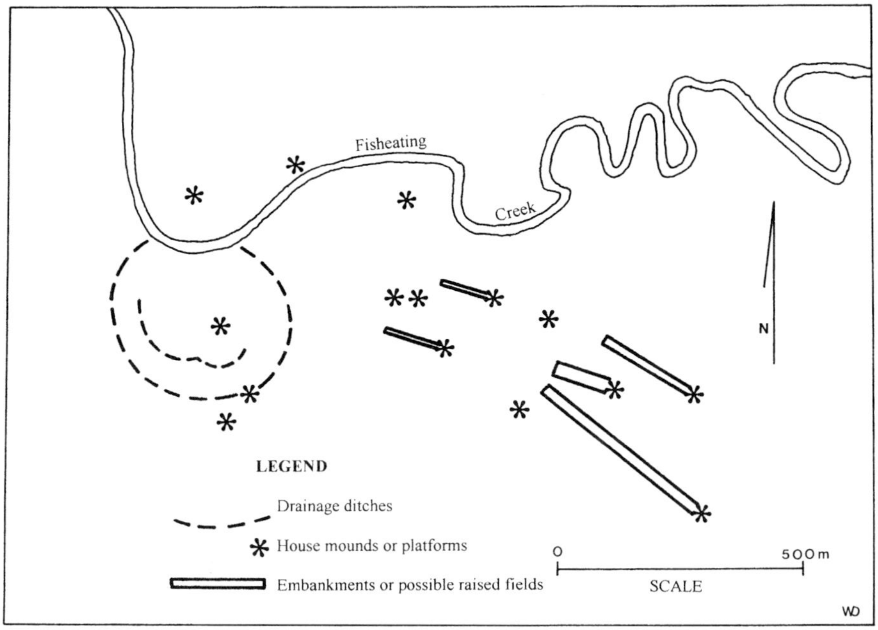

Fig. 12.2. Schematic map of the Fort Center site, Florida, showing ditches and embankments that have been interpreted as being related to ancient wetland cultivation (after Sears 1982: 4, fig. 1.1). (Redrawn by author.)

poses has not gone unchallenged. It has, for example, been argued on the basis of ethnographic accounts, that the ditches were designed to restrict the movement of spirits thereby protecting the enclosed area from unwanted supernatural influences (R. Hall 1976). The embankments are not as numerous, clustered, and involving as much area as would be expected of a site this important on the basis of comparisons with other raised field sites in South and Mesoamerica. Another study (Johnson and Collins 1992) concluded that the soils of this site are too acidic to support annual harvests of maize and that high levels of aluminum would have been toxic to crops. Regardless, others (e.g., Nabhan 1989: 147) do accept the agricultural interpretation.

Although the agricultural function of the Fort Center features is not beyond dispute, a similar interpretation for a site much further north, in far southern Ontario, is above reproach. A few kilometers east of the St. Clair River and south of Lake Huron, near the town of Corunna, one can find a site known as the Parker Earthwork. The distinguishing feature of this site is a large earthen ridge in the form of a fairly symmetrical oval along the banks of a small stream and near a former spring (Fig. 12.3). The ridge measures from crown to crown 100 m by 52 m. It varies in width from 3.5 to 5.3 m, but most of it is approximately 4.4 m wide (Lee 1958: 7). It is surrounded by a shallow

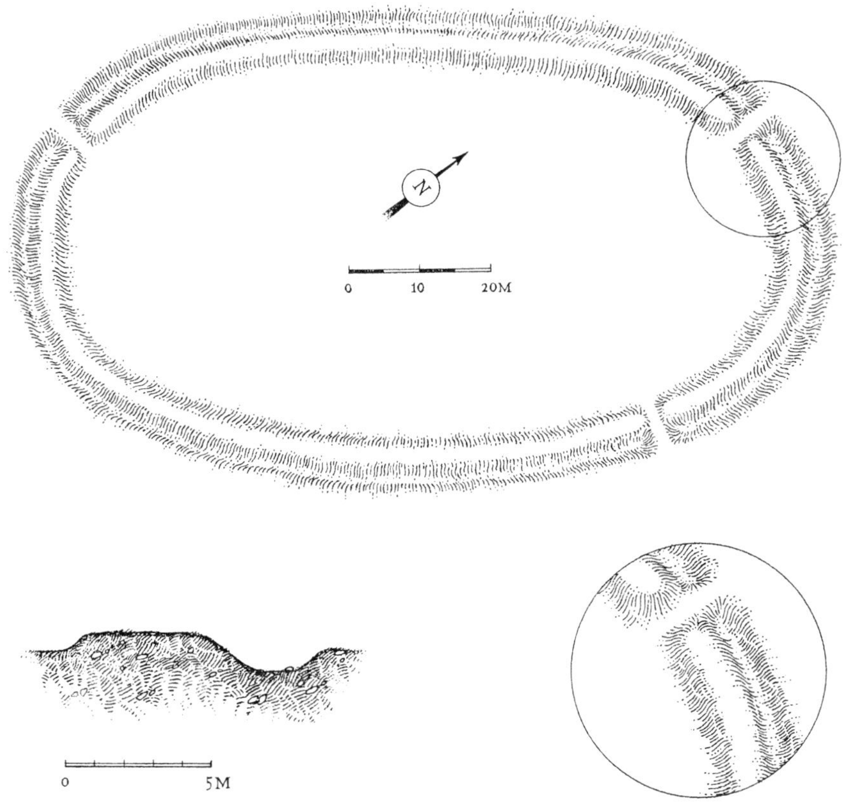

Fig. 12.3. Schematic diagram of the Parker Earthwork, Ontario, showing the oval-shaped levee, surrounding ditch, and breaks in the levee which presumably functioned to drain the enclosure. Not shown are the 1,100 corn hills within the confines of the feature (after a 1901 drawing reprinted in Lee 1958: 8, fig. 2). (Redrawn by, and published by permission of, Narda Lebo.)

ditch varying between 20 and 42 cm deep and from 1.8 to 3 m wide. The earthen ridge rises from 45 cm to 1 m above the bottom of the ditch (Lee 1958: 9). In other words, the volume of the ridge is approximately twice that of the ditch, meaning that fill was brought in from elsewhere, but presumably nearby.

In three places the ridge is broken by openings which clearly are not the result of erosion. They are therefore considered to be associated in function with the ditch. That the site is located in a low-lying area has given rise to the notion that the embankment, the openings, and the surrounding ditch were all intended to work in concert draining and keeping water out of the enclosure during the heavy rainfall month of August. That this enclosure had a drainage function related to agriculture is confirmed by the presence of nearly 1,100 corn hills within its confines (Lee 1958: 12–13). The presence of these features would seem to indicate a historic use, but other evidence indicates that the ridge and ditch were built 600 to 700 years ago (Lee 1958: 27). Presumably the site had a long-use life. Indeed, one hypothesis holds that an early group of people began building the ridge for defensive purposes, but never finished. A later group then moved in and adapted the vestiges into a wetland field complex (Lee 1958: 31). That people would have built a defensive structure in a wet lowland makes little sense. More work surely needs to be done on this site. What little is known, however, clearly points to a provocative agricultural system involving drainage.

The low-lying wetlands a little to the west of a line between the Fort Center and the Parker Earthwork sites, near the confluence of the Ohio, Mississippi, and Missouri rivers, have long been known as the center of a large population in prehistoric times. With many large sites, some of them containing substantial earthen pyramid mounds, and an abundance of highly fertile but frequently saturated floodplain soils, it has always seemed likely that native farmers in the area would have employed some type of field drainage or raising. Indeed, remnants of ridged fields have been associated with at least two sites in southern Illinois—the Texas, and Lundsford–Pulcher sites, and possibly a third site, Cahokia.

During the mid 1960s archaeological work conducted before the construction of Carlyle Reservoir in Clinton County involved the analysis of aerial photographs. Appearing on these photographs were some unusually light striations in an area characterized by dark organically rich soils, containing numerous fragments of prehistoric stone hoes, and adjacent to the Texas site (Morrell 1965*a*: 13). These striations were then investigated on the ground, mapped, and excavated archaeologically (Morrell 1965*b*). It was determined that the lighter discoloration resulted from the removal of topsoil and exposure of the subsoil, and that the darker color represented soil that remained intact (Morrell 1965*a*: 14). Accepted as being agricultural (Woods 1987: 281), the dark areas are vestiges of ridges and the lighter areas delineated the intervening spaces from which material was excavated in the construction of raised fields. The ridged fields were between 80 cm and 1 m in width and spaced 1.5 to 2 m apart, 2.5 to 3 m center to center (M. Fowler 1969: 371, 1992: 9). No data were obtained as to the height of these ridges as the area had been extensively disturbed by recent plowing, but given their wide spacing it seems probable that they were quite low.

Analysis of aerial photographs as part of archaeological investigations at the Lundsford–Pulcher site south of Cahokia (M. Fowler 1969: 366–7) also revealed light striations in areas with heavy organic soils. That these areas also contained several pieces of lithic hoes contributed to the interpretation that the lines were remnants of prehistoric ridged fields. Less is known about these possible ridged fields than those at the Texas site, largely because this site was subjected to limited investigation (see also M. Fowler 1992). One recent study (T. Riley and Said 1993), however, did find magnetite spherules—a product of the industrial era—in the overburden, and none in the soils of the ridges and furrows, thereby confirming the antiquity of the features.

The Cahokia site may well be the most interesting of valley bottom sites in the middle of the continent as it not only contains the largest prehistoric earthwork north of central Mexico—Monk's Mound—but it was clearly one of the largest ancient cities. It is also situated in what can only be described as a wetland and it has been the subject of a controversy concerning ridged fields. As part of a project undertaken prior to the construction of a new interpretive center for the site, archaeologists Melvin L. Fowler and Elizabeth Benchley discovered on some aerial photographs a series of lighter-colored parallel lines in a long-plowed low-lying field otherwise characterized by dark and frequently wet soil (M. Fowler and Benchley 1980: 22–4, figs 5 and 6). These lines were irregular in width, but averaged 4 m apart from center to center (M. Fowler and Benchley 1980: 27). That they were associated with a certain soil type and a low-lying area, and were oriented diagonally to the traditional direction of plowing, led to their being interpreted as remnants of aboriginal ridged fields intended to facilitate cultivation of a wetland (M. Fowler and Benchley 1980: 26, 28–9, 64).

The preliminary work on which this ridged field interpretation was made gave impetus to two separate follow-up studies, both of which included test excavations and soil analyses. One study, conducted in part by one of the initial investigators, concluded that the lighter-colored lines identified in the earlier study were the result of ancient people cutting trenches into the subsoil, and that these trenches helped to drain this low-lying area (Benchley and DePuydt 1982: 37–40). The second study questioned these findings. Its investigators did not find anything that could be positively identified as ridged fields. Indeed, it concluded that the lines identified in the original study may be the result of 19th-century deep plowing to enhance drainage (Nassaney *et al.* 1983: 30).

Geographer and archaeologist William I. Woods (1988, 1996 personal communication), who did the laboratory analysis of the soils for both projects, sees strengths and weaknesses in each. He concluded that there are some soil differences, but nothing conclusive. If ever there was a case that begged for more study to resolve a discrepancy over ridged fields, surely this is it. A site as large as Cahokia required much agricultural land to sustain it. The area in which it is situated is very fertile and productive, but standing water and saturated soils do pose problems. That remnants of ridged fields were identified at smaller sites in similar environments not too far away, suggests that ridged fields should have been used at Cahokia as well. The findings of the studies done thus far may be right for the wrong reasons, or they might be wrong for the right reasons.

The evidence of ridged fields built in ancient time for the purposes of cultivating wetlands in what is today Illinois is certainly less than impressive. Virtually all of the suspected ridged fields were identified by differences in soil color on aerial photographs. Ground truth verification substantiated these apparent differences. However, the more detailed the follow-up investigation, the less convincing the argument became. Much of the problem concerning positive identification of prehistoric ridged fields in this area involves the degree and extent to which features have been buried by subsequent slope wash and overbank deposits, or destroyed by cultivation in historic and recent times. Ever since the earliest suspicion of ridged fields in the 1960s, scholars commented that evidence would be next to impossible to obtain because of increased sedimentation and vast expanses of the area had been plowed repeatedly for decades (see e.g., M. Fowler 1969: 367). Although these arguments seem to have merit, remnants of ancient agricultural features have been easily identified in other areas much more heavily impacted than the American Midwest (Doolittle 1992*b*: 221). Also, remnants of prehistoric ridged fields remain visible in wetlands in other parts of the continent today.

The clearest evidence of wetland ridged fields is found in Wisconsin (Fig. 12.1). Although Lapham, the father of Wisconsin archaeology and a catalyst for the study of the state's remnant prehistoric earthworks, noted only ridged fields on well-drained upland prairies (see Chapter 6), several scholars who followed his lead identified morphologically similar fields in low-lying wetlands. The first among these was a minister, the Reverend S. D. Peet (1887). Indeed, to him goes credit for making the earliest observation that '[a]ncient cornfields are found by the sides of nearly all our lakes. Some of them are quite extensive, covering at times forty acres [16.2 ha] of land' (Peet 1887: 395). Although he did not provide specific details about any one set of ridged fields, Peet observed many during his travels and made some generalizations. He noted, for example that individual fields or 'beds' were generally 1.8 m wide, varied in length from 90 to 180 m, had 'sunken paths' between them 60 to 90 cm wide, and 'were placed on the bottom lands' (Peet 1887: 395–6).

A description similar to that of Peet's was made a few years later, but this one was site specific. Archaeologist George A. West, wrote of a ridged field site 3.2 km west of Racine. These fields were clearly built to cultivate wetlands. They were 'situated on a river bottom, only a few feet from Root River. The soil is river deposit and very rich' (West 1904: 23). The individual elevated planting surfaces averaged 1.2 m wide, with the paths between them being 38 cm wide and 15 cm deep. They were clustered in groups, one oriented east–west, with an adjacent one oriented north–south. On the basis of a fallen tree, the rings of which could be counted, it was determined that the site was cultivated *c.* AD 1500 (West 1904: 23).

A few years after West's report was published, Brown (1909) published his update of agricultural sites identified by members of the recently organized

Wisconsin Archaeological Society (see Chapter 6). In addition to those which were explicitly described as being on well-drained uplands, and more than likely used to offset frost hazards, some had less than complete descriptions of the biophysical environments in which they were located. That these were often located in proximity to rivers suggests they may have been used to facilitate cultivation of wetlands (see e.g., Brown 1909: 100). Of course, some of those on uplands were adjacent to bodies of water.

Brown, along with a colleague, did, however, find a most impressive set of ridged fields in what is indisputably a wetland setting (Brown and Becker 1917: 102–3). Along the western bank of Rice Lake, one of the lakes where non-domesticated rice was cultivated (see Chapter 2), they found two sets of fields near each other. One set contained 15 parallel ridged fields varying in length from 7 to 8 m. This set extended along the lake shore approximately 34 m. The second set began 15 m further north along the shore and extended for another 60 m. Ridged fields in this set varied between 4.5 and 11m long, and numbered 38. Individual elevated planting surfaces in both sets ranged in width between 90 cm and 1.1 m. The ditches between them were 30 to 35 cm wide. Distances between ditch bottoms and ridge tops varied between 10 and 15 cm (Brown and Becker 1917: 102). The ridged fields were oriented east–west, more or less perpendicular to the shoreline (Brown and Becker 1917; plate 5).

One of the most expansive sites containing ridged fields in Wisconsin has to be the Senator Lake site on the north shore of a lake of the same name. Located in the far northern part of the state, this site contained 228 individual elevated planting surfaces, and the water level appears to have been much higher at the time of occupation than it is today (Sperka 1964: 94). A map drawn of the site shows that the ridges are neither contiguous nor parallel to each other. They are clustered into at least twelve groups, and although many are parallel, others intersect at various angles; there is no standard orientation (Sperka 1964: 95). In addition to covering a substantial area, being located so far north, and clearly used in the cultivation of a wetland, the ridged fields at Senator Lake are also important as they were the first in Wisconsin to be excavated archaeologically. A test trench dug through a group of three ridges revealed some now not very startling facts. The widths of the individual ridges varied not at all; they were all 1.6 m wide. The heights, however, varied between 7.5 and 30 cm (Sperka 1964: 102). It was also found that they were much higher when in use than they were when studied in the 1960s, and that the ditches had filled in considerably (Sperka 1964: 101).

Whereas the Senator Lake site was spatially expansive and the first in Wisconsin to be subjected to archaeological excavation, it did not receive any follow-up attention. In contrast, the Sand Lake site north of La Crosse has a comparatively small number of remnant wetland ridged fields but has garnered much attention, principally by James P. Gallagher and his colleagues beginning in 1982. The Sand Lake site is located at the far downstream end of

Sand Lake Coulee, a small tributary valley of the Mississippi River. The mouth of this valley was blocked during the terminal Pleistocene by the rapid deposition of outwash sediments that also formed extensive terraces along this portion of the Mississippi. These sediments trapped the flow of water through the coulee, thereby resulting in the formation of a 247 ha wetland, albeit one without a perennially high water table (Knox 1985: 53). Known as Sand Lake, this wetland was drained in historic times for farming (Boszhardt, Bailey, and Gallagher 1985: 49–51; Gallagher *et al.* 1985: 607).

The ridged fields in this area are not visible on the surface (Fig. 12.4). They were found through excavations, and then only accidentally. The initial impetus for study was the several thousand artifacts found lying on the surface. Excavation began in light sandy alluvium. At a depth of 2.5 m the old wetland surface, represented by black silty loam was encountered. This surface was, however, not flat as would be expected under natural conditions; it was characterized by well-defined parallel ridges. The ridged wetland surface was found in no fewer than seven separate excavation units distributed over a 4.5 ha area. Because of their form, repetitive pattern, and uniformity, there was little doubt that these ridges were the product of human construction (Boszhardt 1983: 7).

The ridges (Fig. 12.5) varied in width from 23 cm to 1.2 m, in height from 5 to 26 cm, and the ditches between them from 47 to 90 cm wide (Gallagher *et*

Fig. 12.4. View of the former Sand Lake as it appeared in 1988. (Photograph by author.)

Fig. 12.5. Profile of ridged fields at the Sand Lake site. This specific excavation was not carried out with archaeological precision, but done only to expose features for follow-up inspection in 1988. The person in the photograph is archaeologist James P. Gallagher who discovered these ridged fields. (Photograph by author.)

al. 1985: 607). Many of the ridges appear to have been rebuilt in an attempt to counteract alluviation. Cross-sections reveal that some of the ditches between the ridges were filled-in with light sandy sediment. This material was then removed, mixed with black marsh soil, and used to cap a pair of standing ridges and intervening ditch, thereby creating one large flat-topped ridged field (Fig. 12.6). Palynological analysis of the black wetland ridge matrix resulted in the discovery of maize (*Zea mays*), bean family (*Leguminosae*), and squash (*Cucurbita* sp.) pollen (Boszhardt, Bailey, and Gallagher 1985: 56–63). Thus, there is little doubt that the ridges were constructed for agricultural purposes (Gallagher *et al.* 1985: 608). The ridged fields were also found to be covered with midden debris consisting of charcoal, ash, fish bones, and shell, suggesting that composting or fertilizing may have been employed. Elk scapula hoes were found and radiocarbon analysis indicated that the fields were used *c.* AD 1460 ± 70 years (Gallagher *et al.* 1985: 608).

Despite abundant evidence that the Sand Lake ridged fields were situated in a wetland, the conclusion that they were built in order to elevate planting surfaces above standing water has not gone unchallenged (see Overstreet 1987). Although there is always room for alternative explanations, the truly unfortunate thing about questions concerning the Sand Lake site is that rejection of the wetland hypothesis carried with it the notion that

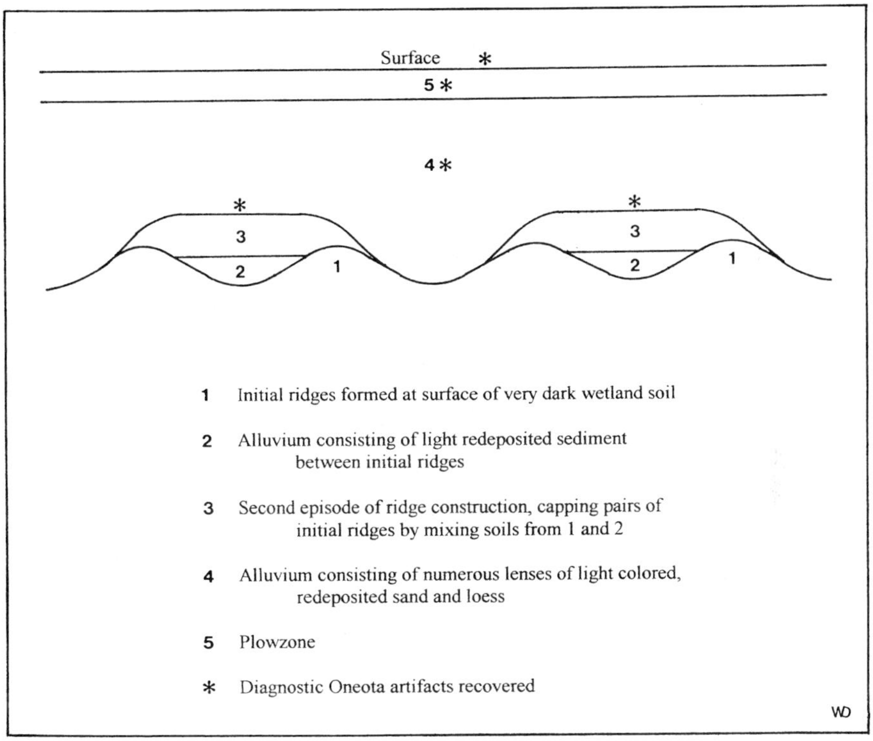

Fig. 12.6. Schematic profile of ridged fields at the Sand Lake site. No scale (after Boszhardt 1983: 8, fig. 2). (Redrawn by author.)

all ridged fields in Wisconsin served to facilitate cold air drainage (Overstreet 1987: 397). In a response to critics, Gallagher and his colleagues (1987) not only point out flaws in the arguments against their interpretation, but they also provide graphic evidence of how the Sand Lake Coulee can still fill with water periodically and present problems for present-day farming (Fig. 12.7). Perhaps equally as important as defending their findings, Gallagher *et al.* (1987: 402–3) also acknowledge that their initial interpretations are only partially correct. The Sand Lake ridged fields most certainly afforded moisture control. Surface drainage, however, was not a simple, straightforward function; previously unappreciated aspects of ridged fields in wetland environments exist.

To be sure, excess water is a problem that needs to be taken care of if agriculture is to succeed. However, there are other problems (in addition to cold air drainage as discussed in Chapter 6) that can be managed by means of ridged fields and these are not unrelated to inundation or soil saturation. Field seasons at the Sand Lake site in the mid 1980s allowed for not only more

Fig. 12.7. 'Inundated field and crops in Sand Lake floodplain following heavy rain in June 1984' (Gallagher *et al.* 1987: 401, fig. 2). (Provided courtesy of James P. Gallagher and reprinted by permission of the Society for American Archaeology.)

and closer examination of ridged fields, but they also allowed for more in-depth consideration of their ecology and function. In all areas investigated a full range of ridged fields, from low narrow ridges to broad flat-topped platforms, was encountered. The average width of the elevated surfaces was 1.25 m, with a range from 31 cm to 3.9 m. Ditches between the ridges varied in width in direct proportion to the sizes of the ridges themselves. The average height of the ridges was 18 cm, with a range of 4 to 40 cm (Gallagher 1992: 110, 115; Gallagher and Sasso 1987: 145–6).

All of the ridges were, as was found earlier, eventually covered with alluvial or colluvial materials, requiring the construction of new ridges. In some instances, the rebuilding of ridges with the admixture of light sediment obscured earlier stages of rebuilding. In others, however, evidence of the rebuilding in as many as ten stages was clear. It was also discovered that although individual ridged fields were parallel to each other, some groups changed orientation quite abruptly (Gallagher 1992: 117, fig. 4.17; Gallagher and Sasso 1987: 145–6). More than 600 m of ridges were found spread over an area of at least 5.3 ha. These appear to have been built and rebuilt in a relatively brief period of time, probably less than a century, and they may have encompassed at much as 81 ha (Gallagher 1992: 110; Gallagher and Arzigian 1994: 177–8).

Certainly one of the more interesting, and significant, of the more recent discoveries at the Sand Lake site is the evidence of corn hills or mounds on top of the raised fields (Fig. 12.8). 'Nodes' were found at 'fairly regular intervals' along the crests of elevated surfaces (Gallagher 1992: 115). In one 6m by 6m area that was excavated, no fewer than forty-nine such mounds on ridges were found (Gallagher 1992: 116, fig. 4.16). At this rate—one per 0.74 sq m—a total of 13,514 mounds were planted on each ha (Gallagher and Arzigian 1994: 178). This figure is sufficiently high that it indicates an intensive agricultural system, one that is not only high in labor inputs per unit area, but high in outputs or yields per unit area as well (Turner and Doolittle 1978). It also suggests that those who have argued that the use of ridged fields in the Upper Midwest in prehistoric times were as much a function of large, concentrated populations as they were extreme environmental conditions are not incorrect (see e.g., Hart 1990).

As for how these fields functioned, it was determined that elevating the planting surfaces in this wetland also served some ancillary tillage functions. First, grasses with dense root systems are the characteristic vegetation of

Fig. 12.8. 'Close-up view of ridges at the Sand Lake site' (Gallagher 1992: 114, fig. 4.15). This archaeologically excavated grid measures 6m by 6m. The small peaks along the ridge tops are interpreted as corn hills. (Provided courtesy of James P. Gallagher and reprinted by permission of the Illinois Historical Preservation Agency, Springfield.)

marshes, and the Sand Lake Coulee appears to have been no exception. Even if this wetland dried-out periodically during the growing season the natural vegetation and weeds would have been a major problem for agriculture. The construction of ridged fields eliminated this problem. Second, saturated soils have minimal pore space and are poorly aerated. Although some natural vegetation grows quite luxuriantly in such edaphic conditions, crops do not. They need open pore space and adequate aeration, factors that, in turn, affect root temperatures and decrease the susceptibility of crops to diseases caused by water- and soil-borne pathogens. Ridged fields remedied this problem. Third, wetland soils are fine-grained, heavy, and compact. Successful cultivation of most crops (rice is an obvious exception), and certainly all New World domesticates, requires moderate- to coarse-grained soil. The construction of ridged fields with admixtures of sand, presumably from the dunes atop the previously discussed terrace which blocks the mouth of the coulee, improved soil structure. Fourth, wetlands are often home to soil-dwelling insects which affect cultigens. The construction of ridged fields greatly affected the natural conditions of Sand Lake and thereby reduced this hazard (Gallagher 1989: 576–7; Gallagher and Sasso 1987: 147–8).

In addition to these benefits, the construction of ridged fields in wetlands can affect atmospheric temperatures in a manner not unlike, but not identical to, that discussed in Chapter 6. Extending the growing season may have been of special importance at Sand Lake and is related to both cold-air drainage and soil moisture control. The simple raising of planting surfaces, of course, results in the drainage of cold air into adjacent ditches, a phenomenon most important during conditions of radiation frost. In wetlands, however, this temperature inversion is aided by water standing in the ditches (Crossley 1999). Water heats up and cools down slower than air and land—anyone who has jumped into a swimming pool on an exceptionally warm spring day and felt the shock of the cold water, or has seen fog over a warm river on a cool fall day, has learned this lesson first hand. As a result, warm air rising off water in the ditches flowed onto the ridged field surfaces, thereby warming them, counteracting frost conditions, and extending the growing season further into the fall.

Were this not enough by itself, the elevation of growing surfaces in the Sand Lake wetland also resulted in drier soils which also had temperature-modifying properties. Dry soils warm quicker than wet soils. Accordingly seeds can be planted earlier in the year than under normal or wet conditions, and they will germinate early, and grow fast. The construction of ridged fields in wetlands can, therefore, extend the growing season considerably by simply creating a dry environment. When combined with construction for the purposes of promoting cold-air drainage, it is a technique that converts cool, wet lands otherwise impossible to cultivate into a highly productive agricultural system (Gallagher 1989: 577–80; Gallagher and Sasso 1987: 148–9).

Frost-free days and nights, unsaturated soil, and soil of the right temperature in the spring are things that will come to those who wait, but in the Upper Midwest, where the winters are long and ideal growing conditions limited, Native Americans apparently could not afford to wait. Instead, they constructed artificial earthen ridges which gave them all of the necessary benefits of tillage and enabled them to extend the growing season by controlling factors of air and soil temperature, and soil moisture (Gallagher and Sasso 1987: 150).

The ridged fields that were discovered quite inadvertently at the Sand Lake site have certainly yielded a wealth of information, and their detailed study by highly competent scholars has increased greatly our understanding and knowledge of ridged field agriculture, particularly as practiced prehistorically, and especially in North America. Revised interpretations demonstrate that single purpose explanations (e.g., wetland reclamation or temperature modification) are not always correct, although they are not always wrong either. In some cases, two or more functions can work together in concert. Recent findings also demonstrate that apparently simple explanations can be misleading. There can be many agronomic advantages to raising planting surfaces above the water level other than merely eliminating the problem of excess moisture. These findings suggest that perhaps it is time for someone to conduct a restudy of an important ridged field site that has long been known, but has received little systematic, in-depth investigation—the Ocmulgee site.

Located on the southern outskirts of Macon, Georgia, this site is sufficiently large, and contains such an impressive array of temple mounds, that it was designated as a national monument. The site lies on the edge of a bluff, more than 15 m above and overlooking the floodplain of the Ocmulgee River. Under the direction of A. R. Kelly, the site was subjected to extensive archaeological investigations as part of a Works Progress Administration (WPA) project beginning in 1933. Mound D, which is flat-topped and oval in shape, measuring 3 to 3.5 m high, 45 m long, and 37 m wide (Fig. 12.9), was initially described as 'not a particularly imposing mound' (Kelly 1935: 185), but it did reveal 'perhaps the most interesting developments' (Kelly 1935: 119).

Excavations revealed, quite unexpectedly, that the mound was constructed immediately over the site of an abandoned maize field (Kelly 1935: 134, fig. 2; 1938: plate 1b). The mound thus effectively sealed the cultivated plot of ground from the ravages of weather and erosion (Kelly 1935: 185). While removing a portion of the mound, workers found a series of distinctive ridges oriented in uniform, parallel lines (Fig. 12.10) and dating to *c.* AD 1000 (Kelly 1935: 185, 1965: 51). They also found what appeared to be the remains of a house in the middle of the plot, and clearly demarcated 'paths' oriented perpendicular to the ridges (Fig. 12.11) and dividing the prehistoric field into discrete 'small patches' (Kelly 1935: 185).

Precise measurements were not reported in early publications about this

Fig. 12.9. A view of what is today known as Cornfield Mound in the Ocmulgee National Monument near Macon, Georgia. Known originally as Mound D, the section to the left and rear is that part not excavated in the 1930s. The area in the fore and to the right, where geographers/archaeologists Michael D. Myers (left) and William Gustav Gartner (right) are walking, was removed during the excavations which revealed a plot of prehistoric ridged fields. Presumably, these fields were covered with earth after excavations were complete. Remnants of ridged fields should remain buried, especially under that portion of the mound not excavated. (Photograph by author.)

site, and, indeed, may not have been made. Thirty years later Kelly (1965: 51) recalled that the ridges 'were separated, from crest to crest, by a span of around 12–14 inches [30 to 35 cm] of sterile sand'. Another observer, Charles H. Fairbanks, provided different measurements, commenting only: '18 inches [45 cm], crest to crest, on the average' (see Kelly 1965: 51). On the basis of a compilation of some field notes, it was suggested that the distances between the peaks of the ridges ranged from 30 to 50 cm (see T. Riley 1994: 98). These dimensions seem to be too small, especially in the light of comparisons made with ridged fields elsewhere (M. Fowler 1969). From an inspection of photographs taken at the time Mound D was excavated, including a comparison of the ridges with wheelbarrows parked (Fig. 12.10) and workers standing (Fig. 12.11) nearby, one can only conclude that these ridged fields were larger than previously stated. Kelly did not report the heights of these ridges, but Fowler, who discovered some of the previously discussed midwestern ridged field sites, says they were 20 cm high (M. Fowler 1969: 372). Exactly how this

Fig. 12.10. 'Initial profiling and horizontal stripping to uncover prehistoric cultivated field beneath Mound D' (Kelly 1938: plate 2a).

Fig. 12.11. 'View from photographic tower of prehistoric cultivated field uncovered by the removal of about half of Mound D. The rectangular house and house platform in the foreground had also been built over the cultivated plot of ground and was covered by subsequent construction of Mound D. The photo was taken after a heavy rain' (Kelly 1938: plate 1a).

figure was derived is not known. It appears, however, to be equally as plausible as, if not more so than, the 13 cm claimed by others (see T. Riley 1994: 98).

Paralleling the controversy about the dimensions of the ridges under Mound D is a dispute of sorts over the presence of corn hills along their crests. Gallagher (1992: 115) claims that a statement made by Kelly suggests that the Ocmulgee ridges were topped with corn hills not unlike those found at the Sand Lake site. In his recollections, however, Kelly (1965: 50) noted only that there 'was a tendency for the "rows" to be constricted or "noded" at regular intervals in their continuity'. He goes on to explain that native farmers cultivated in elongated ridges, and that hoeing resulted in earth being scraped away from areas between plants, only to be redeposited, and therefore accumulating, at the base of plants (Kelly 1965: 50). This interpretation, although made by the discoverer of the field himself, cannot go unchallenged. The previously mentioned reexamination of field notes found that hills up to 13 cm high were found on the ridge tops (see T. Riley 1994). If this is so, one is led to wonder why corn hills that were as high again as the ridges on which they were made are not visible in photographs taken at the time of their alleged discovery (see Figs 12.10 and 12.11).

Much remains to be learned about the ridged fields at Ocmulgee. To begin with, some long unreported facts need to be obtained. Next, some inconsistencies in data have to be corrected. And, finally, some controversies or disputes of interpretation between scholars must be resolved. That portion of the field which was uncovered during excavations in the 1930s measured either 15 by 22 m (Kelly 1938: 10) or 16 m by 18 m (T. Riley 1994: 98) [yet another controversy], and is surely only a small portion of the area cultivated prehistorically. Only half of the mound was excavated; the other half remaining in place and presumably overlying yet more ridged fields (Kelly 1938: 10, plate 1b). A study of the ridges lying beneath the unexcavated portion of Mound D was called for as early as the middle of the 1960s (Kelly 1965: 51), *and still remains to be done!* Such a study, combined with a restudy of those ridges presumably backfilled after the initial investigations (see Fig. 12.9), would shed a great deal of light on the ecology of aboriginal ridged fields in North America.

The Ocmulgee fields are truly important. Unlike those in Wisconsin, the ridged field capital of the continent, these were not situated in a wetland. Similarly, given their southern location, they were not used to extend the growing season by means of cold air drainage. On the basis of their age—earlier than some ridged fields in the Midwest, but later than others—and their lack of correspondence to any major climatic shifts, the Ocmulgee fields were neither the result of the diffusion of technology nor were they a response to environmental change (see M. Fowler 1969; King 1993). Instead, it has been suggested that the planting of crops in rows, using ridges, rather than in hills or mounds, was a practice that was widespread and of great antiquity throughout the Eastern Woodlands (R. Ford, in S. Williams 1971: 16). Ridged fields in Georgia did differ from those in the Midwest in certain morphological traits, suggesting variations might have been due to different soil characteristics (Waselkov 1977: 516–17). Their function may well have had something to do with tillage and cultivation practices.

In addition to tillage functions identified in follow-up studies of wetland ridged fields at the Sand Lake site—for example, increased soil aeration, improved soil fertility through the mixing of decomposers, and restructured soil texture—there is one other function that is most worthy of future attention. Weed control was, of course, noted as a principal tillage function (see Gallagher 1989: 576; Gallagher and Sasso 1987: 147), but it was not explained in detail other than in regard to specific wetland conditions. There is, however, a scenario that has implications to a broader array of environments.

In addition to controlling frosts and extending the growing season, the experimental ridged fields built by Riley and Freimuth (1979) and discussed in Chapter 6 controlled the growth of weeds. This function was discovered quite by accident and not as part of any controlled experiment, although it has been noted from other parts of the world (Denevan and Turner 1974: 27; Miracle 1967: 162). As will be recalled, Riley and Freimuth (1979: 279) built their ridged fields in the spring of 1976, and carried out their experiments in the fall of 1976 and the fall of 1977. Quite to their surprise, the tops of their test plots remained weed free throughout the summers of both 1976 and 1977, while the spaces between the ridged fields, or paths, became 'choked with a typical Midwestern primary invader floral assemblage' (Riley and Freimuth 1979: 282–3).

A somewhat similar, although slightly different, situation was recorded even more recently in Northern Ireland. There, Michael D. Myers (1998) built some experimental ridged fields in the spring of 1994. By mid summer of that same year, both the elevated planting surfaces and the spaces in between, from which the fill used in field-raising was excavated, remained weed free. In contrast, an adjacent plot that was plowed at the same time the ridged fields were built was beginning to be overgrown with weeds, the furrows showing more weed invasion than the ridges (Fig. 12.12). Clearly, ridged fields serve tillage functions that can be unrelated to wetlands. In such environments, however, weed control is at least a secondary function and may be a primary one operating in conjunction with eliminating problems of excess water.

Overview

Some native North Americans not only cultivated wetlands prior to the arrival of Europeans, but they employed some interesting technology in the

Fig. 12.12. Experimental fields near Belfast, Northern Ireland. The plot on the left was plowed and shows weed encroachment, especially in the furrows. The plot on the right consists of raised fields and shows no weed invasion. (Provided courtesy of the fields' builder, Michael D. Myers 1998, and published by his permission.)

process. It is clear that drainage ditches, levees, and ridged fields were all used. However, none of these landscape modifications were commonplace or widespread. Also, the evidence for their use and existence is far short of impeccable. There exist neither documentary accounts nor ethnographic reports containing passages that can even loosely be interpreted as references to aboriginal agricultural landforms in wetlands. This is, of course, not to say that various strategies were not used by indigenous North Americans, either at the time of contact with explorers from across the Atlantic Ocean, or during the 1800s and 1900s as descendants of these latter people carried out in-depth studies of the former people. Yet again, absence of evidence should not be envisaged as evidence of absence.

The only solid evidence that exists for the use of drainage ditches, levees, and ridged fields in North America at and before contact is archaeological. Vestiges of each type of landform do exist. In some cases, however, these are single sites, and therefore suffer from the lack of corroborative evidence. In other cases there are multiple sites, but data on each site are incomplete, and often little more than speculation. Finally, the evidence for ridged fields has given rise to some controversy over the actual function, and this has clouded the issue as to how important the features were in offsetting problems with excess water as opposed to compensating for other ecological factors. Were this not problem enough, it is obvious that the ridged fields used in North American wetlands were not even close to resembling raised fields known elsewhere in the hemisphere. Whereas wetland raised fields elsewhere in the Americas are distinctive island-like features, those in the northern continent look more like the ridges and furrows of a plowed field.

The lack of morphological similarity between North American wetland ridged fields and their Mesoamerican and South American counterparts has not been given much consideration. Most of the scholars who have written on the North American fields seem to operate in a state of ignorant bliss, apparently maintaining that a field is a field. In fact, of course, there is a great variety in the sizes, shapes, ecologies, and functions. The failure to recognize differences in ridged fields has carried with it a high intellectual price. If one assumes that all fields are essentially alike, it is easy to look for, and find, a monocausal explanation. More often than not, this has resulted in diffusion being seen as the principal underlying factor.

There is, of course, nothing wrong with diffusion as a process—many innovations do today, and have in times past as well, spread from place to place. When it comes to explanatory value, however, diffusion is short on power. Just because things can and do diffuse, does not mean that they will. One is forced to ask why people adopt an innovation. The answer to this question is, in general, an explanation; diffusion is only a means to an end, and involves space and time. In the specific case of ancient North American ridged fields, the problem involves the locations where features were found, and the dates of their use.

In terms of regional distribution, ridged fields in wetlands were found in a few discrete pockets within the greater Eastern Woodlands. They are on record from Florida in the far Southeast, the state of Wisconsin, and in the vicinity of the confluences of the Mississippi, Missouri, and Ohio rivers (see Fig. 12.1). In terms of dates, they were apparently used earlier and longer in the Southeast than in the Midwest. Accordingly, scholars who have made the case for diffusion argue that the technology came out of either Mesoamerica or South America via the islands in the Caribbean, entering North America through Florida. The ridged fields argued to exist in Florida are claimed to be associated with the so-called 'mound-builder' cultures, a claim which carries with it the argument that some large and elaborate, but not agricultural mounds, were built as early as 4,500 years ago (Nassaney 1987: 139; Webb 1982). Clearly, more work is needed in order to clarify discrepancies in our understanding of technological developments and cultural affiliation of certain wetland agricultural practices.

Anthropological problems comprise only part of what needs to be done before a full and accurate comprehension of prehistoric North American wetland ridged fields is gained. The ecology of these enigmatic features is arguably more important, and certainly no less important. The long-term research on those fields at the Sand Lake site has resulted in interpretations and revisions. The latter did not negate the former, but lent qualification and insight. Initial interpretations were in the same vein as most of the research on this continent's ridged fields—simple. Revisions showed that what was originally thought to be straightforward turned out to be complex with several subtle nuances. Indeed, they point out the almost false dichotomy between water control and temperature amelioration that has characterized study of ancient ridged fields in Wisconsin. The ridged fields at Ocmulgee throw yet another proverbial monkey wrench in the works. These fields, by virtue of being on well-drained uplands, were clearly not for moisture control. Similarly, by being in the far Southeast, where the growing season is more than sufficiently long, they did not serve to modify temperatures and offset the effects of frost. What purpose did they serve? No one really knows, at least for the moment. They have garnered remarkably little professional attention, other than to be mentioned in various books and articles. The time is more than right for someone to conduct a thorough investigation into those that remain intact and preserved.

In summary, very little is really known about ancient North American ridged fields, especially those in wetlands. Much of what we think we know is actually flawed, and suffers shortcomings stemming from the paucity of data. While little is known of ridged fields, even less is known about levees and drainage. It is clear, however, that at least some industrious native farmers on the continent took measures to control these most difficult of environments.

Part VI

Conclusion

13

Regional Overview

The lands inhabited by native North Americans were certainly varied. In terms of biophysical characteristics, the continent's environment is a complex mosaic of soils, slopes, vegetation, temperature, and moisture conditions. Aboriginal farmers had a myriad of factors with which to deal—some favorable, some unfavorable—and they dealt with them in some rather ingenious ways. These people developed a plethora of techniques for increasing their food supply. And, as a result, they created a variety of cultivated landscapes.

New England

The evidence for native agriculture in New England is largely documentary. In no small way, this is because the region was visited and settled early on by people from northern Europe. There is scant ethnographic evidence of native agricultural landforms because most of the indigenous population was gone by the 19th century. There is not much in the way of archaeological evidence of fields either, and this condition can be attributed to the heavy landscape transformation that has gone on here for nearly five centuries. Relatively recent urbanization has added to the demise.

From the evidence that does exist, it is clear that aboriginal farmers were concentrated mainly in the south, along river valleys and the coast, and relied primarily on rainfed agriculture. Forests were cleared by slashing-and-burning, creating fields that were numerous and large. Stubble from previous harvests was burned and tilled. Between five and seven maize seeds were planted in holes made with digging sticks. Hoes were used to weed and mound the fields. Squashes and beans were intercropped, a practice that was ecologically sound by virtue of replacing soil nutrients depleted by maize. Contrary to long-held and popular thinking, fields were not rotated in a swidden fashion and fish were not buried as fertilizer.

The first farmers of this region did some gardening in addition to the fields they maintained. They also used seedbeds to germinate certain crops.

Fire was a major tool used in this region. Not only did it play an important role in the clearing for new fields, as well as for a variety of non-agricultural purposes, it was also used to promote and enhance the growth of some economically important wild plants. Fire, of course, needed to be controlled, and in so doing early New England cultivators also protected some stands of wild herbs and trees. Interestingly enough, they did nothing in the way of cranberry husbandry.

Upstate New York and French Canada

The evidence for native agriculture in northern and western New York state and the area along the St. Lawrence River from Montreal up to Lake Huron is much like that of New England. If there is a difference, it is that it is much more plentiful. In no small way, this is a function of the large number of early chroniclers involved in the so-called 'French and Indian Wars'. The Iroquois and the Huron were both quite numerous and resistant to European invasion. They persisted longer than their counterparts in New England.

Not only is the evidence from this area similar to that of the former, but the aboriginal practices themselves are not too different. Fields were large and numerous. Indeed, very large and very numerous. They were cleared by slashing-and-burning, and all evidence suggests they were permanently cultivated and not rotated as in true swidden agriculture. Fields were tilled and the stubble from the previous year's crop was burned at the beginning of the crop season. Eight to ten seeds of maize were planted in each hole. The larger number of seeds planted here as in comparison to New England appears to be related to cooler temperatures and a shorter growing season. Fields were weeded and mounded.

Gardens with seedbeds, and certain herbs and trees were protected, encouraged, and cultivated.

Delaware Valley

This rather small pocket in the landscape of northeastern North America is important in terms of native agriculture largely because of the writings of one early naturalist, Kalm. He, and a few others, reported a variety of husbandry practices for herbs and trees. Also reported were gardens with seedbeds.

As for fields, the evidence suggests rainfed agriculture. Burning stubble and tilling, as well as weeding and mounding were practiced. Numerous fields were recorded, but some were quite small. This suggests that true swidden might have been the principal form of cultivation. In fact, however, Kalm was rather late for an 'early' observer, and the native fields he saw were on the uplands, and not along valley bottoms. There is reason to believe that European settlers had displaced indigenous farmers, taking over their bottomland fields and relegating natives to the more marginal and distant uplands. The swidden that was practiced might well have been introduced by Europeans.

Virginia

The evidence of native cultivation from the Tidewater Area is principally documentary and the reasons why are exactly the same as those from New England. What little difference exists between the evidence from the two areas, involves abundance and detail; that of Virginia is greater and more specific than that from further north and east. Indeed, much of what can be generalized for all of the Eastern Woodlands is the result of detailed documentation from this area.

Much like the data, the aboriginal agriculture of Virginia itself, was also like that of New England. Large, numerous fields, dependent on direct rainfall, were tilled, weeded, and mounded. Intercropping was practiced and fertilization was not needed. Bottomland fields may have received a fresh layer of silt periodically as the result of flooding. Only three to five maize seeds were planted per hole. Quite in contrast to the number planted in New England and especially French Canada, this is probably the result of relatively warm temperatures and a longer growing season.

Gardens with seedbeds were not atypical near residences, and some herbs and woody plants were protected, encouraged, and cultivated. Foremost among these has to be grapes, a variety of which was actually domesticated in the area prior to the arrival of Europeans.

The Southeast

The area extending south from Virginia to Florida and then west to Mississippi, and including what is today Tennessee, is a large region, but one with a rather uniform landscape of native cultivation. Knowledge of aboriginal practices here is based on a variety of types of data. There is much documentary evidence, but there is also good ethnographic and archaeological evidence. The main factor accounting for the relative abundance of both latter types of evidence rests with the persistence of native peoples into the 19th century and the relative paucity of modern development, respectively.

As throughout the Eastern Woodlands agriculture was dependent entirely on direct rainfall. Fields were located along river floodplains, cleared of forests. They were large and numerous, tilled prior to planting with stubble often being burned. Weeding and mounding were typical. Periodic floods may have fertilized fields naturally, but intercropping played a larger role in this regard. Swidden was not practiced.

In at least one location, near Ocmulgee, Georgia, there is archaeological evidence of ridged fields. The exact purpose of this type of landscape modification is unclear. Although there is also some evidence that wetlands might have been drained and fields raised in the general region, ridging seems to have been for some other purpose, perhaps ease of cultivation.

Gardens were common near residences in addition to the fields which lay somewhat farther away. Similarly, native farmers manipulated a number of wild herbaceous and woody plants as well. The most impressive form of horticulture, however, involved youpon, a bush, the leaves of which were brewed.

Lower Mississippi Valley

Evidence of native agriculture is sparse from this region. It is also limited to being documentary and archaeological. In every regard, it is nothing appreciably different from the Southeast. If there is an exception, it may well be that the landscape did not undergo either the extensiveness or the intensiveness of modification. The aboriginal population was smaller and more scattered. Being restricted to a relatively small number of sites, farmers were active in certain locales.

Ohio–Mississippi Valley

Documentary evidence from this area is limited, undoubtedly as a result of its interior location. Ethnographic evidence is also anything but abundant, a function of few natives remaining after AD 1800. What is abundant here are archaeological data. And, indeed, these are some of the more impressive data from the continent.

Evidence from caves in Kentucky has been overwhelming in quantity, quality, and findings in recent years. It is now known that indigenous people here practiced husbandry on all types of plants for a very long time. They not only protected, encouraged, and cultivated several species of herbaceous and woody plants, they domesticated many. Foremost among the domesticates is one of North America's three major cultigens—squash.

In addition to the ethnobotanical evidence that continues to be found in archaeological contexts, material evidence of fields has been uncovered. Investigations along the floodplains of the Mississippi, Ohio, Missouri, and Illinois rivers have found ridges suggestive of wetland cultivation. Other evidence, however, indicates that rainfed agriculture was much more important and widespread, especially away from seasonally inundated areas. Although details are lacking, there is no reason to believe that this form of cultivation was any different from the principal form of rainfed agriculture discussed in other Eastern Woodland settings.

Upper Midwest and Great Lakes

Evidence of native agriculture is overwhelmingly archaeological, with some ethnographic, and little documentary data from this area. It also reveals a complexity not seen previously in cultivated landscapes. Several types of cultivation strategies, which involved different landscape modifications, were employed.

The principal form of cultivation practiced by aboriginal people in this region involved plain fields dependent on direct rainfall. Although the sizes, and number of fields are not known, the presence of corn hills indicates that mounding, and perhaps weeding and tillage were practiced. The location of many habitation sites near areas with suitable soils indicates that this form of cultivation was widespread. Also widespread, but perhaps not so much as plain fields, were ridged fields. Remnants of such fields have been found in two types of settings, suggesting that two types of cultivation were practiced.

Some ridged fields are located quite far north, and in well-drained upland settings. These were most likely used to facilitate cold-air drainage and allowed for crops to be grown in places where the temperatures would otherwise be too cold and the growing season too short. Other ridged fields were found in lowland settings where standing water would have been problematical. These fields have been interpreted as devices to compensate for saturated conditions. In a few places, archaeologists have found evidence that native farmers in this region built levees and actually drained plots of standing water prior to planting.

There exists some evidence that indigenous farmers in the Upper Midwest had gardens and seedbeds. The evidence of husbandry of wild plants is limited, with one exception. Certain farmers cultivated 'wild' rice in lakes in northern Wisconsin.

Northern Plains

The farther one goes north, the less information is available. Two factors account for this. First, the short growing season limited the extent and the dependability of cultivation. Second, the far inland setting put the region well beyond the range of early chroniclers.

From the evidence that does exist, it is clear that rainfed agriculture was the most common. Ethnographers have seen fields along rivers, and archaeologists have inferred likewise on the presence of habitation sites. Given the sparseness of the riparian woodlands here, clearing was probably limited to grasses. Annual flooding undoubtedly helped maintain soil fertility, but intercropping might have as well. Gardens were located near houses and some contained beds from which seeds were transplanted. Some wild plants were cultivated.

Southern Plains

The differences between the Southern Plains and the Northern Plains are in some respects minimal and in other respects great. This holds true for both the types of agricultural evidence and the practices themselves. Rainfed agriculture, perhaps augmented by floods was most common. Floodplain fields were seen by early Europeans and ethnographers, and inferred by archaeologists using other corroborative evidence. Gardens are suspected to have been cultivated, but remain unconfirmed. Ethnobotanical and archeological evidence shows that a number of plants were subjected to all forms of husbandry, and that this region was a corridor for plants, and perhaps by extension, agricultural information being transferred between east and west.

Perhaps the most unusual thing about native agriculture in this region is the possibility that canal irrigation was practiced in a few locales. That particular technology, of course, was probably diffused into the area from the region immediately to the west.

The Southwest

There is more evidence of more types of native cultivation from the Southwest than from any other part of North America. The abundance of data stems from the early arrival and intense colonization effort of Spaniards, the persistence of indigenous people to the present day, and the preservation of archaeological data in this arid environment. The Southwest was densely populated in prehistoric times and in order to support large numbers of people, farmers had to use every type of land available. Europeans were attracted to this region in part because of the sizable native population which was ripe for religious conversion, and economic exploitation. Although the native population declined precipitously after contact, a sufficient number of people survived to lend their character to the region's landscape. At least eight types of cultivation have been relied upon by native southwestern people.

Being mainly a dry land, water harvesting, both of slope runoff and the tapping of ephemeral streams by means of canals, was and remains widespread. This often inconspicuous form of agriculture eluded early chroniclers but not ethnographers and archaeologists. Quite in contrast, and visible to everyone, was canal irrigation dependent on perennial rivers. Be they small ones used in historic times in northern New Mexico, or the huge ones used prehistorically in southern Arizona, irrigation canals have long constituted a major component of the native agricultural landscape of this region. Native southwestern farmers were clearly on equal technological footing to irrigators anywhere on earth. Furthermore, they are probably responsible for having developed this technology themselves, independent of irrigators elsewhere, particularly Mesoamerica.

Some locales in the Southwest are currently dry farmed, while others were in times past. These are typically areas that were not suitable for either canal irrigation or water harvesting, but had good soils that were productive but only if what little rain that fell on them could be retained. The Zuni and especially the Hopi have long planted in sandy areas where water is trapped just below the surface. In prehistoric times their ancestors near Sunset Crater cultivated fields covered with a veneer of volcanic ash. Recently, archaeologists have found that rock-covered plots in northern New Mexico are the remains of fields dry farmed by aboriginal people long ago.

In addition to being arid, the Southwest is also a rugged land with some high peaks and steep slopes. Higher elevations receive more rainfall than lower ones and, not surprisingly, these areas were cultivated without the use of any permanent agriculture landforms built by native peoples. On the slopes, terracing was not uncommon. In some places the terraces were small, with low and short risers on slopes. In other places, they involved long and high walls across drainages. Terracing, like canals, was quite diverse in size and scope.

As dry as it is, the Southwest also has some locales that are best classified as wetlands. These are the riparian areas near the downstream ends of exotic streams. In such locales annual floods provide not only nutrient-rich sediments, but water which soaks in to great depths. The area near and downstream of the confluence of the Gila and Colorado rivers has long been an area characterized by native flood recession agriculture.

Gardens, some with seedbeds, are recorded in the documentary, ethnographic, and archaeological literature. Similarly, there is abundant evidence of the husbandry of many wild plants. The number of herbaceous plants either protected, encouraged, or cultivated here is staggering. Most impressive, however, is the recent discovery that *agave* has a long history of being cultivated by natives and over a very wide range.

Native farmers of the Southwest, be they prehistoric, historic, or modern, were and are some of the best anywhere.

Northwestern Mexico

In many ways Northwest Mexico is merely a southern extension of the Southwest. It possesses low deserts, high mountains, and environments in between. The evidence of native cultivation is as diverse as the environment itself and parallels that from lands lying just to the north. The only real differences are the lack of large irrigation canals and a paucity of archaeological research in northwestern Mexico.

Canal irrigation is known to have been practiced by aboriginal people in this region. Here, it seems to have facilitated double cropping as well as intercropping. Irrigated lands might not have been very extensive, but they were very productive. Water harvesting was also practiced here, but to what extent really is not known. Bottomlands at the mouths of large rivers were locales in which flood recession farming is well documented. The headwaters of these rivers lie in the Sierra Madre Occidental, an area well-known in archaeological circles for its abundant and sizable terraces. Non-terraced lands in these mountains continue to be farmed by indigenous people who rely only on direct rainfall. In some locales, they even rotate fields among plots cleared from the forest in a manner not unlike true swidden cultivation.

There is some very good evidence of native gardening from northwestern Mexico, including the use of seedbeds. Trees are often cultivated in gardens as are certain wild herbs.

California

Aboriginal Californians are not typically thought of as farmers. To be sure, there is not much in the way of evidence of fields from this area. However, there is an abundance of evidence to illustrate that these native people were masters when it came to plant husbandry.

Although evidence of plant protection is sparse, nearly every group in the region has a history of encouraging the growth of certain plants, using an array of techniques, and cultivating many as well. This holds true for herbs and woody plants. In all likelihood, the environments of California were so rich and abundant, that indigenous people there had no need to practice what most people think of as agriculture. These people did more than simply gather, however. They manipulated their environments so that they could increase harvests appreciably.

The actual occurrences of traditional agricultural landscapes are in the far south. There, along the banks of the Colorado River, flood recession agriculture—which is in reality an extension from the Southwest—was practiced. Also, there is at least one case where canal irrigation, albeit on a very small scale, was used.

The Great Basin

California natives were master plant manipulators, but right behind them in terms of expertise were inhabitants of the Great Basin. These people attracted no attention from early chroniclers, and left little in the way of archaeological evidence. Their remoteness, however, made them a most appealing set of subjects for ethnographers who during the 20th century have recorded numerous instances of them having encouraged plants.

There is some suspicion that these people might have had gardens in places near permanent residences. Clearly, however, the most impressive activity is their use of canals to irrigate stands of wild grasses in the Owens Valley. This irrigation without agriculture was first reported in the 1920s and has been little-studied since. What has been done, suggests that this practice probably predates any European influence. It may well be an extension of, as well as a combination of, native practices from the Southwest.

Pacific Northwest

If the California and Great Basin natives are only infrequently thought of as farmers, then it is natural that the indigenous people of the far Northwest never should be. Calling these people 'agriculturists' is certainly stretching the term. Doing so, however, certainly seems to be in order. The evidence, ethnographic and archaeological, is definitely sparse. But, as scant as it is, it is tantalizing. Some plants were protected, others encouraged, and yet others cultivated. This was observed by some of the earliest ethnographers and then forgotten. Recently, this old work was rediscovered and the topic of agriculture here is now undergoing somewhat of a renaissance. The greatest lesson from the Northwest, at least for now, is that native cultivation continues to be a promising line of research with much to be learned.

Conclusion

Some of the evidence presented and discussed here is certainly subject for debate. Most, however, indeed the overwhelming preponderance, is irrefutable. Taken as a whole, it clearly shows that North America has long been a very productive agricultural land. The natural landscape was converted into a cultural one, with aboriginal people being responsible for the transformation generations ago. These people took a productive land and made it produce even more. They developed a wide variety of techniques to do so, and these were reported by early Europeans. They were made more understandable by later ethnographers. Archaeologists extended knowledge of native agriculture further back in time by finding remains of ancient fields on the landscape and interpreting them.

To one extent or another, almost every part of North America was cultivated by its aboriginal people. These were some of the best farmers ever; best in the sense of being productive, and best in terms of being stewards for future generations. The impacts and the contributions of these people were nothing if not great.

BIBLIOGRAPHY

Aase, J. K., W. D. Kemper, and R. E. Davidson. 1968. 'Response of Corn to White and Black Ground Curves', *Arizona Journal* **60**: 234–6.

Aberle, David F. 1983. 'Navaho Economic Development', in *Handbook of North America Indians*, Vol. 10, edited by Alfonso Ortiz, pp. 641–58. Smithsonian Institution, Washington, D. C.

Ackerly, Neal W. 1982. 'Irrigation, Water Allocation Strategies, and the Hohokam Collapse', *The Kiva* **47**: 91–105.

—— 1988. 'False Causality in the Hohokam Collapse', *The Kiva* **53**: 305–19.

—— 1991. 'Constructing Analog Models of Hohokam Irrigation Using Historic Documents', in *Prehistoric Irrigation in Arizona: Symposium 1988*, edited by Cory Dale Breternitz, pp. 41–60. Soils Systems Publications in Archaeology 17, Phoenix.

—— 1996. *A Review of the Historical Significance of and Management Recommendations for Preserving New Mexico's Acequia Systems*. Dos Rios Consultants, Silver City, New Mexico.

—— 1997. 'Mimbreno and Gileno Apache Irrigation Systems, 1853–1859', *Kiva* **62**: 349–363.

—— and T. Kathleen Henderson (editors). 1989. *Prehistoric Agricultural Activities on the Lehi-Mesa Terrace: Perspectives on Hohokam Irrigation Cycles*. Northland Research, Flagstaff, Arizona.

—— Jerry B. Howard, and Randall H. McGuire. 1987. *La Ciudad Canals: A Study of Hohokam Irrigation Systems at the Community Level*. Anthropological Field Studies 17. Arizona State University, Tempe.

—— Michelle Behr, Michael Casaus, and Tina L. Powell. 1993. *Acequia Systems of the Rio Mimbres: Floodwater Irrigation in Southern New Mexico*. Center for Archaeological Research, New Mexico State University, Las Cruces.

Adair, James. 1968 [1775]. *The History of the American Indians*. Johnson Reprint Corporation, New York.

Adair, Mary J. 1988. *Prehistoric Agriculture in the Central Plains*. Publications in Anthropology 16. University of Kansas, Lawrence.

Adams, E. Charles. 1979. 'Cold Air Drainage and Length of Growing Season in the Hopi Mesas Area', *The Kiva* **4**: 285–96.

Adams, Robert McC. 1965. *Land Behind Baghdad: A History of Settlement on the Diyala Plains*. University of Chicago Press, Chicago.

Aikens, C. Melvin and David B. Madsen. 1986. 'Prehistory of the Eastern Area', in *Handbook of North American Indians*, Vol. 11, edited by Warren L. D'Azevedo, pp. 149–60. Smithsonian Institution, Washington, D.C.

Alcorn, Janice B. 1984. *Huastec Mayan Ethnobotany*. University of Texas Press, Austin.

Alderfer, R. B. and F. G. Merkle. 1943. 'The Comparative Effects of Surface Application versus Incorporation of Various Mulching Materials on Structure, Permeability, Runoff, and Other Soil Properties', *Soil Science Society of America Proceedings* **8**: 79–86.

Alegre, Francisco Javier. 1841 [1604]. *Historia de la Compañía de Jesus en Nueva-España*. J. M. Lara, Mexico.

——1958 [1604]. *Historia de la Provincia de la Compañia de Jesus de Nueva España*, Tomo II. Institutum Historicum S. J., Roma.

Allen, William. 1849. *The History of Norridgewock*. Edward J. Peet, Norridgewock.

Aller, Wilma F. 1954. 'Aboriginal Food Utilization of Vegetation by the Indians of the Great Lakes Region as Recorded in the Jesuit Relations', *The Wisconsin Archaeologist* **35**: 59–73.

Allouez, Claude Jean. 1899 [1667]. 'Relations of What Occurred Most Remarkable in the Missions of the Fathers of the Society of Jesus in New France', in *The Jesuit Relations and Allied Documents: Travels and Explorations of the Jesuit Missionaries in New France, 1610–1791*, Vol. 51, edited by Reuben Gold Thwaites, pp. 19–117. The Burrows Brothers, Cleveland.

——1903 [1677]. 'Narrative of a Voyage Made to the Illinois', in *Discovery and Exploration of the Mississippi Valley*, edited by John Gilmary Shea, pp. 70–81. Joseph McDonough, Albany.

Alvarado, Hernando de and Juan de Padilla. 1940 [1540]. 'Account of What Hernando de Alvarado and Fray Juan de Padilla Discovered While in Search of the South Sea', in *Narratives of the Coronado Expedition, 1540–1542*, edited by George P. Hammond and Agapito Rey, pp. 182–4. University of New Mexico Press, Albuquerque.

Amsden, Charles. 1929. 'The First Farmers of America', *Masterkey* **3**.2: 5–12; **3**.3: 13–17.

Anderson, Edgar. 1952. *Plants, Man and Life*. University of California Press, Berkeley.

Anderson, Kat. 1991*a*. 'Wild Plant Management: Cross-Cultural Examples of the Small Farmers of Jaumave, Mexico, and the Southern Miwok of the Yosemite Region', *Arid Lands Newsletter* **31**: 18–23.

——1991*b*. 'California Indian Horticulture: Management and Use of Redbud by the Southern Sierra Miwok', *Journal of Ethnobiology* **11**: 145–57.

——1993. 'Native Californians as Ancient and Contemporary Cultivators', in *Before the Wildnerness: Environmental Management by Native Californians*, compiled and edited by Thomas C. Blackburn and Kat Anderson, pp. 151–74. Ballena Press Anthropological Papers 40, Menlo Park.

Andresen, John M. 1987. 'Archival Excavation: Frank Midvale and the Casa Grande Canal', *The Kiva* **52**: 229–35.

Anonymous. 1804 [?]. 'Extract from an Indian History', in *Collections of the Massachusetts Historical Society*, Vol. 9, pp. 99–102. Hall and Hiller Publishers, Boston.

——1844 [1610]. 'A True Declaration of the Colonie in Virginia', in *Tracts and Other Papers Relating Principally to the Origin, Settlement, and Progress of the Colonies in North America, From the Discovery of the Country to the Year 1776*, collected by Peter Force, pp. 3–27. William Q. Force, Washington, D. C.

——1922 [1541]. 'Traslado de las Nuevas', in *The Journey of Coronado 1540–1542*, translated and edited by George Parker Winship, pp. 186–96. Allerton Book Co., New York.

——1988. 'What is Coppicing?' *Sustainable Living in Drylands: Networking Newsletter for Permaculture in the Southwest* **4**: 3.

Anschuetz, Kurt F. 1984. 'Prehistoric Change in Tijeras Canyon, New Mexico'.

Unpublished master's thesis, Department of Anthropology, University of New Mexico.

——1995. 'Saving a Rainy Day: The Integration of Diverse Agricultural Technologies to Harvest and Conserve Water in the Lower Rio Chama Valley, New Mexico', in *Soil, Water, Biology, and Belief in Prehistoric and Traditional Southwestern Agriculture*, edited by H. Wolcott Toll, pp. 25–39. New Mexico Archeological Council, Special Publication 2, Albuquerque.

——n.d. [1995]. 'Two Sides of a Coin: Early Pueblo Indian Farming Practices in the Rio Arriba and the Rio Abajo of the Northern Rio Grande Region'. Unpublished paper presented at the Annual Meeting of the Society for American Archaeology, Minneapolis, Minnesota.

Antieau, John M. 1977. 'An Archaeological Survey of Selected Portions of the Roosevelt Water Conservation District Floodway Pinal County, Arizona'. Report to the Soil Conservation Service, Phoenix.

Anza, Juan Bautista de. 1966*a* [1774]. 'Anza's Complete Diary', in *Anza's California Expeditions*, Vol. 2, translated and edited by Herbert Eugene Bolton, pp. 1–130. Russell and Russell, New York.

——1966*b* [1774]. 'Anza's Diary from Tubac to San Gabriel', in *Anza's California Expeditions*, Vol. 2, translated and edited by Herbert Eugene Bolton, pp. 131–212. Russell and Russell, New York.

Arber, Edward. 1967 [1613]. *Travels and Works of Captain John Smith*, American Classics in History and Social Science 15. Burt Franklin, New York.

Archer, Gabriel. 1969 [1607]. 'A Relaytion of the Discovery of Our River, From James Fort into the Maine', in *The Jamestown Voyages Under the First Charter 1606–1609*, edited by Philip L. Barbour, Vol. 1, pp. 80–98. The University Press, Cambridge.

Arnon, Nancy S. and W. W. Hill. 1979. 'Santa Clara Pueblo', in *Handbook of North American Indians*, Vol. 9, edited by Alfonso Ortiz, pp. 296–307. Smithsonian Institution, Washington, D. C.

Arzigian, Constance. 1987. 'The Emergence of Horticultural Economies in Southwestern Wisconsin', in *Emergent Horticultural Economies of the Eastern Woodlands*, edited by William F. Keegan, pp. 217–42. Occasional Paper 7. Center for Archaeological Investigations, Southern Illinois University, Carbondale.

Asch, David. 1994. 'Aboriginal Specialty—Plant cultivation in Eastern North America: Illinios Prehistory and a Post-context Perspective', in *Agricultural Origins and Development in the Midcontinent*, edited by William Green, pp. 25–86. Report 19. Office of the State Archeologist, the University of Iowa, Iowa City.

——and Nancy B. Asch. 1977. 'Chenopod as Cultigen: A Re-Evaluation of Some Prehistoric Collections From Eastern North America', *Mid-Continental Journal of Archaeology* **2**: 3–45.

————1978. 'The Economic Potential of *Iva Annua* and its Prehistoric Importance in the Lower Illinois Valley', in *The Nature and Status of Ethnobotany*, edited by Richard I. Ford, pp. 300–41. Anthropological Papers 67. University of Michigan, Ann Arbor.

————1985. 'Prehistoric Plant Cultivation in West-Central Illinois', in *Prehistoric Food Production in North America*, edited by Richard I. Ford, pp. 149–204. Anthropological Papers 75. Museum of Anthropology, University of Michigan, Ann Arbor.

Asch, David and William Green. 1992. *Crops of Ancient Iowa: Native Plant Use and Farming Systems*. Leopold Center for Sustainable Agriculture, Iowa State University, Ames.

Avila López, Raúl. 1991. *Chinampas de Iztapalapa, D.F.* Instituto Nacional de Antropología e Historia, Mexico City.

Axtell, James. 1981. *The European and the Indian: Essays in the Ethnohistory of Colonial North America*. Oxford University Press, New York and Oxford.

——1992. *Beyond 1492: Encounters in Colonial North America*. Oxford University Press, New York.

Ayres, James E. 1967. 'A Prehistoric Farm Site Near Cave Creek, Arizona', *The Kiva* **32**: 106–11.

Bahr, Donald M. 1983. 'Pima and Papago Social Organization', in *Handbook of North American Indians*, Vol. 10, edited by Alfonso Ortiz, pp. 178–92. Smithsonian Institution, Washington, D. C.

Baker, T. Lindsay, Steven R. Rae, Joseph E. Minor, and Seymour V. Conner. 1973. *Water for the Southwest: Historical Survey and Guide to Historic Sites*. ASCE Historical Publication 3. American Society of Civil Engineers, New York.

Baldanzi, Giampiero. 1960. 'Burning and Soil Fertility', *Transactions of the 7th International Congress of Soil Science*, 523–30.

Baldwin, Gordon C. 1965. *The Warrior Apaches*. Dale Stuart King Publishers, Tucson.

Bale, A. N. and T. C. Sheng. 1988. *Watershed Management Field Manual: Slope Treatment Measures and Practices*. FAO Conservation Guide 13/3. Food and Agriculture Organization of the United Nations, Rome.

Balmer, Frank E. 1926. 'The Farmer and Minnesota History', *Minnesota History* **7**: 199–217.

Bandelier, Adolph F. 1890. Final Report of Investigations Among the Indians of the Southwestern United States, Carried on Mainly in the Years from 1880 to 1885, Part I. *Papers of the Archaeological Institute of America, American Series IV.* University Press, Cambridge, Massachusetts.

——1892. Final Report of Investigations Among the Indians of the Southwestern United States, Carried on Mainly in the Years from 1880 to 1885, Part II. *Papers of the Archaeological Institute of America, American Series IV.* University Press, Cambridge, Massachusetts.

——1937. 'Documentary History of the Rio Grande Pueblos', in *Indians of the Rio Grande Valley*, edited by Adolph F. Bandelier and Edgar L. Hewett, pp. 115–246, University of New Mexico Press, Albuquerque.

——and Edgar L. Hewett. 1937. *Indians of the Rio Grande Valley*. University of New Mexico Press, Albuquerque.

Bandera (Vandera) Juan de la. 1990 [1566]. 'The "Short" Bandera Relation', in *The Juan Pardo Expeditions: Exploration of the Carolinas and Tennessee, 1566–1568*, edited by Charles Hudson, pp. 297–304. Smithsonian Institution, Washington, D. C.

Bannon, John F. 1955. *The Mission Frontier in Sonora, 1620–1687*. United States Catholic Historical Society Monograph Series 26, New York.

Barlow, M. Arthur. 1906 [1584]. 'Captain Arthur Barlow's Narrative of the First Voyage to the Coasts of America', in *Early English and French Voyages, Chiefly from Hakluyt*, edited by Henry S. Burrage, pp. 227–41. Barnes and Noble, New York.

Bartlett, John R. 1854. *Personal Narrative of Explorations and Incidents in Texas, New*

Mexico, California, Sonora, and Chihuahua Connected with the United States and Mexican Boundary Commission, 2 volumes. D. Appleton, New York.

Bartram, William. 1940 [1773]. *The Travels of William Bartram*, edited by Mark Van Doren. Facsimile Library, New York.

Basso, Keith H. 1983. 'Western Apache', in *Handbook of North American Indians*, Vol. 10, edited by Alfonso Ortiz, pp. 462–88. Smithsonian Institution, Washington, D. C.

Baumhoff, Martin A. 1978. 'Environmental Background', in *Handbook of North American Indians*, Vol. 8, edited by Robert F. Heizer, pp. 16–24. Smithsonian Institution, Washington, D. C.

Baxter, Sylvester. 1882. 'The Father of the Pueblos', *Harpers Magazine* **65**.385 (June): 81.

Beaglehole, Ernest. 1937. *Notes on Hopi Economic Life*. Yale University Publications in Anthropology 15. Yale University Press, New Haven.

Beals, Ralph L. 1945. *The Contemporary Culture of the Cáhita Indians*. Smithsonian Institution, Bureau of American Ethnology, Bulletin 142, Washington, D. C.

Bean, Lowell John. 1978. 'Cahuilla', in *Handbook of North American Indians*, Vol. 8, edited by Robert F. Heizer, pp. 575–86. Smithsonian Institution, Washington, D. C.

—— and Harry W. Lawton. 1973. 'Some Explanations for the Rise of Cultural Complexity in Native California with Comments on Proto-Agriculture and Agriculture', in *Patterns of Indian Burning in California: Ecology and Ethnohistory*, by Henry T. Lewis, pp. xi–xlvii. Ballena Press Anthropological Papers 1, Menlo Park, California.

—— and Katherine S. Saubel. 1972. *Temalpakh: Cahuilla Indian Knowledge and Usage of Plants*. Malki Museum Press, Morongo Indian Reservation, California.

—— and Florence C. Shipek. 1978. 'Luiseño', in *Handbook of North American Indians*, Vol. 8, edited by Robert F. Heizer, pp. 550–63. Smithsonian Institution, Washington, D. C.

—— and Charles R. Smith. 1978. 'Serrano', in *Handbook of North American Indians*, Vol. 8, edited by Robert F. Heizer, pp. 570–4. Smithsonian Institution, Washington, D. C.

Beaudry, Mary C. (editor). 1988. *Documentary Archaeology in the New World*. The University Press, Cambridge.

Beckwith, Stephen L. 1954. 'Ecological Succession on Abandoned Farm Lands and Its Relationship to Wildlife Management', *Ecological Monographs* **24**: 349–76.

Bee, Robert L. 1983. 'Quechan', in *Handbook of North American Indians*, Vol. 10, edited by Alfonso Ortiz, pp. 86–98. Smithsonian Institution, Washington, D. C.

Bell, Jonathan. 1984. 'A Contribution to the Study of Cultivation Ridges in Ireland', *Journal of the Royal Society of Antiquaries of Ireland* **114**: 80–97.

Benavides, Alonso de. 1954 [1630]. *Benavides Memorial of 1630*, translated by Peter P. Forrestal. Academy of American Franciscan History, Washington, D. C.

—— 1965 [1630]. *The Memorial of Fray Alonso de Benavides 1630*, translated by Mrs Edward E. Ayer. Horn and Wallace, Albuquerque.

Benchley, Elizabeth D. and Peter J. DePuydt. 1982. *Final Report of 1980 Test Excavations at the Interpretive Center Tract Cahokia Mounds Historic Site*. Report of Investigations 61. Archaeological Research Laboratory, University of Wisconsin–Milwaukee.

Bennett, Charles E. (compiler). 1968. *Settlement of Florida.* University of Florida Press, Gainesville.

Bennett, M. K. 1955. 'The Food Economy of the New England Indians, 1605–75', *The Journal of Political Economy* **63**: 369–99.

Bennett, Wendell C. and Robert M. Zingg. 1935. *The Tarahumara: An Indian Tribe of Northern Mexico*. University of Chicago Press, Chicago.

Bentley, Jeffery W. 1987. 'Water Harvesting on the Papago Reservation: Experimental Agricultural Technology in the Guise of Development', *Human Organization* **46**: 141–6.

Berlin, G. Lennis, J. Richard Ambler, Richard H. Hevly, and Gerald G. Schraber. 1977. 'Identification of a Sinagua Agricultural Field by Aerial Thermography, Soil Chemistry, Pollen/Plant Analysis, and Archaeology', *American Antiquity* **42**: 588–600.

—— David E. Salas, and Phil R. Geib. 1990. 'A Prehistoric Sinagua Agricultural Site in the Ashfall Zone of Sunset Crater, Arizona', *Journal of Field Archaeology* **17**: 1–16.

Bernard-Shaw, Mary. 1988. 'Hohokam Canal Systems and Late Archaic Wells: The Evidence from the Los Morteros Site', in *Recent Research on Tucson Basin Prehistory: Proceedings of the Second Tucson Basin Conference*, edited by William H. Doelle and Paul R. Fish, pp. 153–74. Anthropological Papers 10. Institute for American Research, Tucson.

—— and William H. Doelle. 1991. 'Tucson Basin Irrigation Systems', in *Prehistoric Irrigation in Arizona: Symposium 1988*, edited by Cory Dale Breternitz, pp. 25–40. Soils Systems Publications in Archaeology 17, Phoenix.

Bernstein, David J. 1990. 'Prehistoric Seasonality by Studies in Coastal Southern New England', *American Anthropologist* **92**: 96–115.

Beverly, Robert. 1947 [1705]. *The History and Present State of Virginia*, edited by Louis B. Wright. University of North Carolina Press, Chapel Hill.

Biddle, Nicholas. 1978 [1810]. 'The Nicholas Biddle Notes', in *Letters of the Lewis and Clark Expedition with Related Documents: 1783–1854*, Vol. 2, edited by Donald Jackson, pp. 497–545. University of Illinois Press, Urbana.

Bidwell, Percy Wells and John I. Falconer. 1925. *History of Agriculture in the Northern United States: 1620–1860*. Carnegie Institution of Washington, Washington, D. C.

Biedma, Luys Hernandez de. 1905 [1544]. 'Relation of the Conquest of Florida', in *Narratives of the Career of Hernando de Soto*, Vol. II, translated by Buckingham Smith and edited by Edward Gaylord Bourne, pp. 1–40. Williams-Barker Co., New York.

—— 1993 [1544]. 'Relation to the Island of Florida', translated and edited by John E. Worth, in *The De Soto Chronicles: The Expedition of Hernando De Soto to North America in 1539–1543,* Vol. 1, edited by Lawrence A. Clayton, Vernon James Knight, Jr., and Edward C. Moore, pp. 221–46. University of Alabama Press, Tuscaloosa.

Biggar, H. Howard. 1918. 'The Old and the New in Corn Culture', in *Yearbook of the Department of Agriculture*, pp. 123–36. United States Government Printing Office, Washington, D. C.

Billeck, William T. 1986. 'Garden Beds in Black Hawk County', *Iowa Archaeological Society Newsletter* **36**: 5–6.

Billy, Bahe. 1981. 'Water Harvesting for Dryland and Floodwater Farming on the Navajo Indian Reservation', in *Rainfall Collection for Agriculture in Arid and Semiarid Regions: Proceedings of a Workshop hosted by the University of Arizona,*

USA and The Chapingo Postgraduate College, Mexico, edited by G. R. Dutt, C. F. Hutchinson, and M. Anaya Garduno, pp. 3–7. Commonwealth Agricultural Bureaux, Slough, UK.
Bingham, Sam and Janet Bingham. 1979. *Navajo Farming*. Rock Point Community School, Chinle, Arizona.
Black, M. Jean. 1978. 'Plant Dispersal by Native North Americans in the Canadian Subarctic', in *The Nature and Status of Ethnobotany*, edited by Richard I. Ford, pp. 255–62. Anthropological Papers 67. University of Michigan, Ann Arbor.
Blackburn, Glen A. 1929. 'The White Man Took More than the Land', *Wallace's Farmer* **54**.3: 7.
Blackburn, Thomas and Kat Anderson. 1993. 'Introduction: Managing the Domesticated Environment', in *Before the Wilderness: Environmental Management by Native Californians*, compiled and edited by Thomas C. Blackburn and Kat Anderson, pp. 15–25. Ballena Press Anthropological Papers 40, Menlo Park, California.
Black Hawk. 1964 [1833]. *Black Hawk: An Autobiography*, edited by Donald Jackson. University of Illinois Press, Urbana.
Blakeman, Crawford Harris, Jr. 1974. 'The Late Prehistoric Paleoethnobotany of the Black Bottom, Pope and Massac Counties, Illinois'. Unpublished doctoral dissertation, Department of Anthropology, Southern Illinois University, Carbondale.
Blair, Emma Helen. 1912. *The Indian Tribes of the Upper Mississippi Valley and Region of the Great Lakes*, Vol. 2. Arthur H. Clark, Cleveland, Ohio.
Blaut, J. M. 1993. *The Colonizer's Model of the World: Geographical Diffusionism and Eurocentric History*. The Guilford Press, New York.
Boas, Franz. 1921. 'Ethnology of the Kwakiutl: Based on Data Collected by George Hunt', in *Thirty-fifth Annual Report of the Bureau of American Ethnology, 1913–1914*, pp. 57–780. Smithsonian Institution, Washington, D. C.
——1966. *Kwakiutl Ethnography*, edited by Helen Codere. University of Chicago Press, Chicago.
Bohrer, Vorsila L. 1960. 'Zuni Agriculture', *El Palacio* **67**: 181–202.
——1983. 'New Life from Ashes: The Tale of the Burnt Bush (*Rhus trilobata*)', *Desert Plants* **5**: 122–4.
——1991. 'Recently Recognized Cultivated and Encouraged Plants Among the Hohokam', *Kiva* **56**: 227–35.
——1992. 'New Life from Ashes II: A Tale of Burnt Bush', *Desert Plants* **10**: 122–5.
Bolton, Herbert Eugene (editor). 1908. *Spanish Exploration in the Southwest 1542–1706*. Barnes and Noble, New York.
——1927. *Fray Juan Crespi: Missionary Explorer on the Pacific Coast, 1769–1774*. University of California Press, Berkeley.
Bond, Stanley C., Jr. 1985. 'The Relationship Between Soils and Settlement Patterns in the Mohawk Valley', in *The Mohawk Valley Project: 1982 Field Season Report*, edited by Dean R. Snow, pp. 17–40. State University of New York, Albany.
Boserup, Ester. 1965. *The Conditions of Agricultural Growth: The Economics of Agrarian Change under Population Pressure*. Aldine, Chicago.
——1981. *Population and Technological Change: A Study of Long-Term Trends*. University of Chicago Press, Chicago.
Boszhardt, Robert F. 1983. 'Ridged Fields at the Sand Lake Site Near La Crosse, Wisconsin', *Minnesota Archaeological Newsletter* **30**: 6–8.
——Thomas W. Bailey, and James P. Gallagher. 1985. 'Oneota Ridged Fields at the

Sand Lake Site (47LC44): La Crosse County, Wisconsin', *Wisconsin Archaeologist* **66**: 47–67.

Bourbon, L. A. de. 1900 [1716]. 'Decrees of the Council of Marine respecting the Christian Savages in Canada', April 1, 1716, in *Jesuit Relations and Allied Documents: Travels and Explorations of the Jesuit Missionaries in New France, 1610–1791*, Vol. 67, edited by Reuben Gold Thwaites, pp. 24–37. The Burrows Brothers Co., Cleveland.

Bourke, John G. 1958. *An Apache Campaign in the Sierra Madre*. Charles Scribner's Sons, New York.

Botsford, Manton and Gay Kinkade. n.d. [1993]. 'Cultural Resources Assessment of the BLM Safford District Office Project'. Unpublished report on file. U. S. Department of Interior, Bureau of Land Management, Safford District, Arizona.

Bowen, Thomas. n.d. [1976]. 'A Survey and Re-evaluation of the Trincheras Culture, Sonora, Mexico'. Unpublished report on file. Arizona State Museum, University of Arizona, Tucson.

Bowers, Alfred W. 1950. *Mandan Social and Ceremonial Organization*. University of Chicago Press, Chicago.

Bradfield, Maitland. 1971. *The Changing Pattern of Hopi Agriculture*. Occasional Paper 30. Royal Anthropological Institute of Great Britain and Ireland, London.

Bradford, William and Edward Winslow. 1844 [1602–1625]. 'Bradford's and Winslow's Journal', in *Chronicles of the Pilgrim Fathers of the Colony of Plymouth from 1602–1625*, edited by Alexander Young, pp. 109–268. Charles C. Little and James Brown, Boston.

Brand, Donald D. 1939. 'The Origin and Early Distribution of New World Cultivated Plants', *Agricultural History* **13**: 109–17.

—— 1943. 'The Chihuahua Culture Area', *New Mexican Anthropologist* **6–7**: 115–58.

Brandt, Carol L. 1995. 'Traditional Agriculture on the Zuni Indian Reservation in the Recent Historic Period', in *Soil, Water, Biology, and Belief in Prehistoric and Traditional Southwestern Agriculture*, edited by H. Wolcott Toll, pp. 291–301. New Mexico Archaeological Council, Special Publication 2, Albuquerque.

Brasser, T. J. C. 1971. 'The Coastal Algonkians: People of the First Frontiers', in *North American Indian in Historical Perspective*, edited by Eleanor Burke Leacock and Nancy Oestreich Lurie, pp. 64–85. Random House, New York.

—— 1978. 'Mahican'. in *Handbook of North American Indians*, Vol. 15, edited by Bruce G. Trigger, pp. 198–212. Smithsonian Institution, Washington, D. C.

Bray, Francesca. 1986. *The Rice Economies: Technology and Development in Asian Societies*. Basil Blackwell, Oxford.

Bray, Warwick (editor). 1993. *The Meeting of Two Worlds: Europe and the Americas 1492–1650*. The British Academy, London.

Breed, Jack. 1948. 'Land of the Havasupai', *National Geographic* **93**.5: 655–74.

Breternitz, David A. 1957. 'Excavations Near Flagstaff', *Plateau* **30**: 22–30, 43–54.

—— 1967. 'The Eruption(s) of Sunset Crater: Dating and Effects', *Plateau* **40**: 72–6.

Bretting, Peter K. 1986. 'Changes in Fruit Shape in *Proboscidea parviflora* spp. *parviflora* (Martyniaceae) with Domestication', *Economic Botany* **40**: 170–6.

—— and Gary Paul Nabhan. 1981. 'Ethnobotany of Devil's Claw (*Proboscidea parviflora* spp. *parviflora*): Martyniaceae in the Greater Southwest', *Journal of California and Great Basin Anthropology* **8**: 226–37.

Brew, J. O. 1979. 'Hopi Prehistory and History to 1850', in *Handbook of North*

American Indians, Vol. 9, edited by Alfonso Ortiz, pp. 514–23. Smithsonian Institution, Washington, D. C.

Brierley, Kenneth. 1991. 'The Goodliest People', *Cape Cod Life* October–November: 54–9.

Bright, William. 1978. 'Karok', in *Handbook of North American Indians*, Vol. 8, edited by Robert F. Heizer, pp. 180–9. Smithsonian Institution, Washington, D. C.

Bromley, S. W. 1945. 'An Indian Relict Area', *The Scientific Monthly* **60**: 153–4.

Brooks, Danny. 1974. 'Prehistoric Soil and Water Control in the American Southwest: A Case Study'. Unpublished master's thesis, Department of Anthropology, Northern Arizona University, Flagstaff.

——and R. Gwinn Vivian. 1978. *Archaeological Investigations of the Queen Creek Floodway Project*. Arizona State Museum Archaeological Series 66, Tucson.

Brown, Charles E. 1909. 'Wisconsin Garden Beds', *Wisconsin Archaeologist* **8**: 97–105.

——and Robert H. Becker. 1917. 'The Chetek and Rice Lakes', *Wisconsin Archaeologist* **16**: 84–113.

Brown, James A. 1984. *Prehistoric Southern Ozark Marginality: A Myth Exposed*. Special Publications 6. Missouri Archaeological Society, Columbia.

Brown, Ralph H. 1948. *Historical Geography of the United States*. Harcourt Brace Jovanovich, New York.

Brugge, David M. 1983. 'Navaho Prehistory and History to 1850', in *Handbook of North American Indians*, Vol. 10, edited by Alfonso Ortiz, pp. 489–501. Smithsonian Institution, Washington, D. C.

Bryan, Kirk. 1929. 'Flood-Water Farming', *Geographical Review* **19**: 444–56.

——1941. 'Pre-Columbian Agriculture in the Southwest, as Conditioned by Periods of Alluviation', *Annals of the Association of American Geographers* **31**: 219–42.

Bugé, David E. 1978. 'Preliminary Report: 1978 Excavations at NM01-1407, Ojo Caliente, New Mexico'. Department of Sociology/Anthropology, Occidental College, Los Angeles, California.

——1984. 'Prehistoric Subsistence Strategies in the Ojo Caliente Valley, New Mexico', in *Prehistoric Agricultural Strategies in the Southwest*, edited by Suzanne K. Fish and Paul R. Fish, pp. 27–34. Anthropological Research Papers 33. Arizona State University, Tempe.

Buikstra, Jane E., Lyle W. Konigsberg, and Jill Bullington. 1986. 'Fertility and the Development of Agriculture in the Prehistoric Midwest', *American Antiquity* **51**: 528–46.

Bullock, Harold R. 1942. 'Lasley Point Mound Cruising', *Wisconsin Archaeologist* **23.2**: 32–6.

Buol, S. W., F. D. Hole, and R. J. McCracken. 1980. *Soil Genesis and Classification*. Iowa State University Press, Ames.

Burrus, Ernest J. 1971. *Kino and Manje: Explorers of Sonora and Arizona, Their Vision of the Future, A Study of Their Expeditions and Plans*. Jesuit Historical Institute, Rome.

Busch, C. D., L. Mark Raab, and R. C. Busch. 1976. '"Q = A*V" Prehistoric Water Canals in Southern Arizona', *American Antiquity* **41**: 531–4.

Buskirk, Winfred. 1949. 'Western Apache Subsistence Economy'. Unpublished doctoral dissertation, Department of Anthropology, University of New Mexico, Albuquerque.

Buskirk, Winfred 1986. *The Western Apache: Living with the Land Before 1950*. University of Oklahoma Press, Norman.

Bustamante, Pedro de. 1966 [1582]. 'Testimony of Pedro de Bustamante, May 16, 1582', in *The Rediscovery of New Mexico 1580–1594: The Explorations of Chamuscado, Espejo, Castaño de Sosa, Morlete, and Leyva de Bonilla and Humaña*, translated and edited by George P. Hammond and Agapito Rey, pp. 127–32. University of New Mexico Press, Albuquerque.

Butler, Eva L. 1948. 'Algonkian Culture and Use of Maize in Southern New England', *Bulletin of the Archaeological Society of Connecticut* **22**: 1–39.

Butzer, Karl W. 1976. *Early Hydraulic Civilization in Egypt: A Study in Cultural Ecology*. University of Chicago Press, Chicago.

——1990*a*. 'The Indian Legacy in the American Landscape', in *The Making of the American Landscape*, edited by Michael P. Conzen, pp. 27–50, 379–81, Unwin Hyman, Boston.

——1990*b*. 'A Human Ecosystem Framework for Archaeology', in *The Ecosystem Approach in Anthropology: From Concept to Practice*, edited by Emilio F. Moran, pp. 91–130. University of Michigan Press, Ann Arbor.

Bye, Robert A., Jr. 1979. 'Incipient Domestication of Mustards in Northwest Mexico', *The Kiva* **44**: 237–56.

——1981. 'Quelites: Ethnoecology of Edible Greens—Past, Present, and Future', *Journal of Ethnobiology* **1**: 109–23.

——1985. Botanical Perspectives of Ethnobotany of the Greater Southwest. *Economic Botany* **39**: 375–86.

——and Rita Shuster. 1984. 'Developing an Integrated Model for Contemporary and Archaeological Agricultural Subsistence Systems', in *Prehistoric Agricultural Strategies in the Southwest*, edited by Suzanne K. Fish and Paul R. Fish, pp. 125–46. Anthropological Research Papers 33. Arizona State University, Tempe.

Cabeza de Vaca, Alvar Nuñez. 1946 [1528]. 'The Narrative of Alvar Nuñez Cabeza de Vaca', in *Spanish Explorers in the Southern United States 1528–1543*, edited by Frederick W. Hodge, pp. 1–126. Barnes and Noble, New York.

Cable, John S. and David E. Doyel. 1985. 'Hohokam Land-Use Patterns along the Terraces of the Lower Salt River Valley: The Central Phoenix Project', in *Proceedings of the 1983 Hohokam Symposium*. Part I, edited by Alfred E. Dittert, Jr., and Donald E. Dove, pp. 263–310. Occasional Paper 2. Phoenix Chapter of the Arizona Archaeological Society, Phoenix.

Caddell, Gloria May. 1982. 'Plant Resources, Archaeological Plant Remains, and Prehistoric Plant-Use Patterns in the Central Tombigbee River Valley', *Bulletin Alabama Museum of National History* **7**. University of Alabama, Tuscaloosa.

Cadillac, Antoine de la Mothe. 1947 [1721]. 'The Memoir of Lamothe Cadillac', in *The Western Country in the 17th Century: The Memoirs of Lamothe Cadillac and Pierre Liette*, edited by Milo Milton Quaife, pp. 3–83. R. R. Donnelley and Sons, Chicago.

Callaway, Donald, Joel Janetski, and Omer C. Stewart. 1986. 'Ute', in *Handbook of North American Indians*, Vol. 11, edited by Warren L. D'Azevedo, pp. 336–67. Smithsonian Institution, Washington, D. C.

Callender, Charles. 1978*a*. 'Shawnee', in *Handbook of North American Indians*, Vol. 15, edited by William C. Sturtevant, pp. 622–29. Smithsonian Institution, Washington, D. C.

——1978*b*. 'Illinois', in *Handbook of North American Indians*, Vol. 15, edited by William C. Sturtevant, pp. 673–79. Smithsonian Institution, Washington, D. C.

Camilli, Eileen and Joseph C. Winter. 1975. 'Plant Distribution, Environmental Constraint and Farm Site Location at Hovenweep', in *Hovenweep 1974*, Joseph C. Winter, pp. 153–60. Archeological Report 1. Department of Anthropology, San Jose State University.

Cammisa, Alfred. 1984. 'A Comparison of Settlement Patterns and General Land Use Between the Native Americans and the English Settlers in Southern New England in the Seventeenth Century', *Bulletin of the Massachusetts Archaeological Society* **45**: 66–77.

Campbell, T. N. 1959. 'Choctaw Subsistence: Ethnographic Notes from the Lincecum Manuscript', *The Florida Anthropologist* **12**: 9–23.

Canby, Thomas Y. 1982. 'The Anasazi: Riddles in the Ruins', *National Geographic* **162**: 554–92.

Cannell, G. H. and L. V. Weeks. 1979. 'Erosion and its control in Semi-Arid Regions', in *Agriculture in Semi-Arid Environments*, edited by A. E. Hall, G. H. Cannell, and H. W. Lawton, pp. 238–56. Springer-Verlag, Berlin.

Carlson, Paul H. 1992. 'Indian Agriculture, Changing Subsistence Patterns, and the Environment on the Southern Great Plains', *Agricultural History* **66**: 52–60.

Carr, Lucien. 1893. 'The Mounds of the Mississippi Valley, Historically Considered', *Annual Report of the Board of Regents of the Smithsonian Institution*, pp. 503–33, Government Printing Office, Washington, D. C.

——1895. 'The Food of Certain American Indians and Their Methods of Preparing It', *Proceedings of the American Antiquarian Society* **10**: 155–90.

Carrera, Juan de la. 1953 [1600]. 'Relación de Juan de la Carrera, Marzo 3, 1600', in *The Spanish Jesuit Mission in Virginia 1570–1572*, edited by Clifford M. Lewis and Albert J. Loomie, pp. 123–42. Virginia Historical Society and University of North Carolina Press, Chapel Hill.

Carrier, Lyman. 1923. *The Beginnings of Agriculture in America*. McGraw-Hill, London.

Carse, Mary Howell. 1949. 'The Mohawk Iroquois', *Archaeological Society of Connecticut Bulletin* **23**: 5–53.

Carter, George F. 1945. *Plant Geography and Culture History in the American Southwest*. Viking Fund Publications in Anthropology 5. Johnson Reprint Corporation, New York.

Cartier, Jacques. 1906*a* [1534]. 'The First Relation to Jaques Carthier of S. Malo', in *Early English and French Voyages Chiefly from Hakluyt, 1534–1608*, edited by Henry S. Burrage, pp. 4–31. Barnes and Noble, New York.

——1906*b* [1536]. 'A Shorte and Brief Narration (Cartier's Second Voyage)', in *Early English and French Voyages Chiefly from Hakluyt, 1534–1608*, edited by Henry S. Burrage, pp. 37–88. Barnes and Noble, New York.

——1906*c* [1541]. 'The Third Voyage of Discovery Made by Captaine Jaques Cartier', in *Early English and French Voyages Chiefly from Hakluyt, 1534–1608*, edited by Henry S. Burrage, pp. 93–102. Barnes and Noble, New York.

Casañas de Jesús María, Francisco. 1927 [1691]. 'Fray Francisco Casañas de Jesús María to the Viceroy of Mexico', in *Descriptions of the Tejas or Asinai: Indians, 1691–1722*, I, translated by Mattie Austin Hatcher, *Southwestern Historical Quarterly* **30**: 206–18, 283–304.

Castañeda, Pedro de. 1907 [1642]. 'The Narrative of the Expedition of Coronado, by Pedro de Castañeda', in *Spanish Explorers in the Southern United States 1528–1543*, edited by Frederick W. Hodge, pp. 273–387. Barnes and Noble, New York.

Castaño de Sosa, Gaspar. 1966 [1590]. 'Castaño de Sosa's Memoria', in *The Rediscovery of New Mexico 1580–1594: The Explorations of Chamuscado, Espejo, Castaño de Sosa, Morlete, and Leyva de Bonilla and Humaña*, translated and edited by George P. Hammond and Agapito Rey, pp. 245–311. University of New Mexico Press, Albuquerque.

Castetter, Edward F. 1943. 'Early Tobacco Utilization and Cultivation in the American Southwest', *American Anthropologist* **45**: 320–25.

——and Willis H. Bell. 1942. *Pima and Papago Indian Agriculture*. University of New Mexico Press, Albuquerque.

————1951. *Yuman Indian Agriculture*. University of New Mexico Press, Albuquerque.

————and A. R. Grove. 1938. 'The Early Utilization and Distribution of *Agave* in the American Southwest', *University of New Mexico Bulletin* **6**.

——and Ruth M. Underhill. 1935. 'The Ethnobiology of the Papago Indians', Ethnobiological Studies in the American Southwest II. *Bulletin, University of New Mexico Biological Series* 4. Albuquerque.

Catlin, George. 1841. *Letters and Notes of the Manners, Customs, and Condition of the North American Indians*. Wiley and Putnam, New York.

——1989 [1832–1839]. *North American Indians*. Viking, New York.

Caywood, Louis R. 1933. 'The Archaeology of the Sulphur Spring Valley, Arizona'. Unpublished master's thesis, College of Letters, Arts, and Sciences, University of Arizona, Tucson.

Ceci, Lynn. 1975*a*. 'Fish Fertilizer: A Native North American Practice', *Science* **189**: 26–30.

——1975*b*. Letter, *Science* **189**: 946–9.

Chacón, J. C. and S. R. Gliessman. 1982. 'Use of the "Non-Weed" Concept in Traditional Tropical Agroecosystems of South-Eastern Mexico', *Agro-Ecosystems* **8**: 1–11.

Champlain, Samuel de. 1907 [1604–1618]. *Voyages of Samuel de Champlain*, edited by W. L. Grant. Charles Scribner's Sons, New York.

Chapman, Carl H. and Eleanor F. Chapman. 1983. *Indians and Archaeology of Missouri*. University of Missouri Press, Columbia.

Chapman, Jefferson, Paul A. Delcourt, Patricia A. Cridlebaugh, Andrea B. Shea, and Hazel R. Delcourt. 1982. 'Man-Land Interaction: 10,000 Years of American Indian Impact on Native Ecosystems in the Lower Little Tennessee River Valley, Eastern Tennessee', *Southeastern Archaeology* **1**: 115–21.

Charlevoix, Pierre Francois Xavier de. 1923 [1761]. *Journal of a Voyage to North America*, translated by Louise Phelps Kellogg. The Caxton Club, Chicago.

Chase, J. S. 1919. *California Desert Trails*. Houghton Mifflin, New York.

Chiarella, Joseph V. and William H. Beck. 1975. 'Water Harvesting Catchments on Indian Lands in the Southwest', *Proceedings of the Water Harvesting Symposium, Phoenix, Arizona, 26–28, 1974*, edited by Gary W. Frasier, pp. 104–14. US Department of Agriculture, Western Region, Phoenix.

Chisholm, Michael. 1968. *Rural Settlement and Land Use: An Essay in Location*. Hutchinson University Library, London.

Chomko, Stephen A. 1978. 'Phillips Spring, 23HI216: A Multicomponent Site in the Western Missouri Ozarks', *Plains Anthropologist* **23**: 235–55.

——and Gary W. Crawford. 1978. 'Plant Husbandry in Prehistoric Eastern North America: New Evidence for Its Development', *American Antiquity* **43**: 405–8.

Choriki, R. T., J. C. Hide, S. L. Kroll, and B. L. Brown. 1964. 'Rock and gravel Mulch Aid in Moisture Storage', *Crops and Soil* **16**: 24.

Ciolek-Torrello, Richard, Stephanie Whittlesey, and John R. Welch. 1994. 'A Synthetic Model of Prehistoric Land Use', in *The Roosevelt Rural Sites Study, Changing Land Use in the Tonto Basin*, Vol. 3, edited by Richard Ciolek-Torrello and John R. Welch, pp. 437–72. Technical Series 28. Statistical Research, Tucson, Arizona.

Clark, Ira G. 1987. *Water in New Mexico: A History of Its Management and Use*. University of New Mexico Press, Albuquerque.

Clark, J. W. 1978. 'The Upper Rio Grande', *Natural Resources Journal* **18**: 69–76.

Clark, S. P. 1928. *Lessons from Southwestern Indian Agriculture*. University of Arizona Press, Tucson.

Clayton, Sarah. 1973. 'The Potomac (Patawomeke) Indians', *Quarterly Bulletin of the Archaeological Society of Virginia* **27**: 180–9.

Clotts, H. V. 1915. *Nomadic Papago Surveys*. Report to the United States Indian Service.

——1917. *History of the Indians and Irrigation on Indian Reservations*. Report to the United States Indian Service.

Cocannouer, Joseph A. 1964. *Weeds: Guardians of the Soil*. The Devin–Adair Co., New York.

Coe, Joffre L. 1979. 'The Indian in North Carolina', *The North Carolina Historical Review* **56**: 158–61.

Coe, Michael D. 1964. 'The Chinampas of Mexico', *Scientific American* **211**: 90–8.

Coleman, Sister Bernard. 1953. 'The Ojibwa and the Wild Rice Problem', *Anthropological Quarterly* **26**: 79–88.

Colton, Harold S. 1932. 'Sunset Crater: The Effect of a Volcanic Eruption on an Ancient Pueblo People', *Geographical Review* **22**: 582–90.

——1945. 'Sunset Crater', *Plateau* **18**: 7–14.

——1949. 'The Prehistoric Population of the Flagstaff Area', *Plateau* **22**: 21–5.

——1960. *Black Sand: Prehistory in Northern Arizona*. University of New Mexico Press, Albuquerque.

——1965. 'Experiments in Raising Corn in the Sunset Crater Ashfall Area East of Flagstaff, Arizona', *Plateau* **37**: 77–9.

Condie, Carol J. 1989. 'Slivers of the Past: New Mexico's Last *Canovas*', in *From Chaco to Chaco: Papers in Honor of Robert H. Lister and Florence C. Lister*, edited by Meliha S. Duran and David T. Kirkpatrick, pp. 103–16. The Archaeological Society of New Mexico 15, Albuquerque.

Conklin, H. C. 1957. 'Hanunoo Agriculture', *Food and Agriculture Organization, Forestry Development Paper* 12. United Nations, Rome.

Connelly, John C. 1979. 'Hopi Social Organizations', in *Handbook of North American Indians*, Vol. 9, edited by Alfonso Ortiz, pp. 539–53. Smithsonian Institution, Washington, D. C.

Connole, Dennis A. 1976. 'Land Occupied by the Nipmuck Indians of Central New England 1600–1700', *Bulletin of the Massachusetts Archaeological Society* **38**: 14–20.

Cook, Frederick. 1887. *Journals of the Military Expedition of Major General John Sullivan Against the Six Nations of Indians in 1779*. Knapp, Peck and Thomson, Auburn, New York.

Cook, Sherburne F. 1960. 'Colonial Expeditions to the Interior of California: Central Valley, 1800–1820', *University of California Anthropological Records* **16**: 239–92. Berkeley.

Cordell, Linda S. 1975. 'Predicting Site Abandonment at Wetherill Mesa', *The Kiva* **40**: 189–202.

——1979. 'Prehistory: Eastern Anasazi', in *Handbook of North American Indians*, Vol. 9, edited by Alfonso Ortiz, pp. 131–51. Smithsonian Institution, Washington, D. C.

——1984. *Prehistory of the Southwest*. Academic Press, Orlando.

——Amy C. Earls, and Martha R. Binford. 1984. 'Subsistance Systems in the Mountanous Settings of the Rio Grande Valley', in *Prehistoric Agricultural Strategies in the Southwest*, edited by Suzanne K. Fish and Paul R. Fish, pp. 222–41. Anthropological Research Papers 33. Arizona State University, Tempe.

Córdova F. de A., Carlos and Alberto Vázquez C. 1991. 'Tipología para terrazza de Cultivo en el Contexto de los sistemas agrícolas tradicionales de Mesoamérica'. *Topicas de Investigacion y Postgrado* **2**: 3–14.

Core, Earl L. 1967. 'Ethnobotany of the Southern Appalachian Aborigines', *Economic Botany* **21**: 199–214.

Corey, A. T. and W. D. Kemper. 1968. 'Conservation of Soil Water by Gravel Mulches', *Colorado State University Hydrology Papers* 30.

Cortés, José. 1989 [1799]. *Views from the Apache Frontier: Report on the Northern Provinces of New Spain*, translated by John Wheat and edited by Elizabeth A. H. John. University of Oklahoma Press, Norman.

Covington, James W. 1978. 'Agriculture as Practiced by the Eastern Timucuans of Florida: 1564–1590', *Associates of the NAL Today* **3**: 9–12.

Cowan, C. Wesley. 1978. 'The Prehistoric Use and Distribution of Maygrass in Eastern North America: Cultural and Phytogeographical Implications', in *The Nature and Status of Ethnobotany*, edited by Richard I. Ford, pp. 263–88. Anthropological Papers 67. University of Michigan, Ann Arbor.

——1985. 'Understanding the Evolution of Plant Husbandry in Eastern North America: Lessons from Botany, Ethnography, and Archaeology', in *Prehistoric Food Production in North America*, edited by Richard I. Ford, pp. 205–43. Anthropological Papers 85. University of Michigan, Ann Arbor.

——1997. 'Evolutionary Changes Associated with the Domestication of *Cucurbita pepo*: Evidence from Eastern Kentucky', in *People, Plants, and Landscapes: Studies in Paleobotany*, edited by Kirsten J. Gremillion, pp. 63–85. University of Alabama Press, Tuscaloosa.

——and Patty Jo Watson (editors). 1992. *The Origins of Agriculture: An International Perspective*. Smithsonian Institution, Washington, D. C.

Cowardin, Lewis M., Virginia Carter, Francis C. Golet, and Edward T. LaRoe. 1979. *Classification of Wetlands and Deepwater Habitats of the United States*. FWS/OBS-79-31, United States Department of Interior, Washington, D. C.

Cox, George W. and Michael D. Atkins. 1979. *Agricultural Ecology: An Analysis of World Food Production Systems*. W. H. Freeman, San Francisco.

Creel, Darrell. 1990. 'Efforts to Date Water Control Features at Site NAN-20', in

Archaeology at the NAN Ruin (LA15049), 1989 Season Investigations, compiled by Harry J. Shafer, pp. 136–8. Special Report 10. Anthropology Laboratory, Department of Anthropology, Texas A&M University, College Station.

—— and Bradley Adams. 1986. 'Investigation of Water Control Features at NAN-20', in *Archaeology at NAN Ranch, 1985 Interim Report*, edited by Harry J. Shafer, pp. 50–66. Special Report 7. Anthropology Laboratory, Department of Anthropology. Texas A&M University, College Station.

Crisell, Rob. 1998. 'War or Peace in the Southwest?' *American Archaeology* **2**.2: 7–8.

Crites, Gary D. 1987. 'Human Plant Mutualism and Niche Expression in the Paleoethnobotanical Record: A Middle Woodland Example', *American Antiquity* **52**: 725–40.

—— 1993. 'Domesticated Sunflower in Fifth Millennium B.P. Temporal Context: New Evidence From Middle Tennessee', *American Antiquity* **58**: 146–8.

Cronon, William. 1983. *Changes in the Land: Indians, Colonists, and the Ecology of New England*. Hill and Wang, New York.

Crossley, Philip Lawrence. 1999. 'Sub-irrigation and Temperature Amelioration in Chinampa Agriculture'. Unpublished doctoral dissertation, Department of Geography, University of Texas.

Crosswhite, Frank. 1981. 'Desert Plants, Habitat, and Agriculture in Relation to the Major Pattern of Cultural Differentiation in the O'odham People of the Sonoran Desert', *Desert Plants* **3**: 47–76.

Crown, Patricia L. 1984*a*. 'Prehistoric Agricultural Technology in the Salt-Gila Basin', in *Hohokam Archaeology along the Salt-Gila Aqueduct Central Arizona Project*, Vol. VII, edited by Lynn S. Teague and Patricia L. Crown, pp. 207–60. Archaeological Series 150. Cultural Resource Management Division, Arizona State Museum, University of Arizona, Tucson.

—— 1984*b*. 'Adaptation through Diversity: An Examination of Population Pressure and Agricultural Technology in the Salt-Gila Basin', in *Prehistoric Agricultural Strategies in the Southwest*, edited by Suzanne K. Fish and Paul R. Fish, pp. 5–26, Anthropological Research Papers 33. Arizona State University, Tempe.

—— 1984*c*. 'Hohokam Subsistence and Settlement in the Salt-Gila Basin', in *Hohokam Archaeology along the Salt-Gila Aqueduct Central Arizona Project*, Vol. IX, edited by Lynn S. Teague and Patricia L. Crown, pp. 87–113. Archaeological Series 150. Cultural Resource Management Division, Arizona State Museum, University of Arizona, Tucson.

—— 1987. 'Water Storage in the Prehistoric Southwest', *The Kiva* **52**: 209–28.

Crumrine, N. Ross. 1983. 'Mayo', in *Handbook of North American Indians*, Vol. 10, edited by Alfonso Ortiz, pp. 264–75. Smithsonian Institution, Washington, D. C.

Culbertson, Thaddeus A. 1952. *Journal of an Expedition to the Mauvaises Terres and the Upper Missouri in 1850*, edited by John Francis McDermott. Smithsonian Institution, Bureau of American Ethnology, Bulletin 147, Washington, D. C.

Cully, Anne C., Marcia L. Donaldson, Mollie S. Toll, and Klara B. Kelley. 1982. 'Agriculture in the Bis Sa'ani Community', in *Bis Sa'ani: A Late Bonito Phase Community on Escavada Wash, Northwest New Mexico*, edited by Cory Dale Breternitz, David E. Doyel, and Michael P. Marshall, pp. 115–66. Navajo Nation Papers in Anthropology 14. Navajo Nation Cultural Resource Management Program, Window Rock, Arizona.

Cummings, Byron. 1926. 'Ancient Canals of the Casa Grande', *Progressive Arizona* **3**: 9–10.

Currie, Douglas R. 1994. 'Micromorphology of a Native American Cornfield', *Archaeology of Eastern North America* **22**: 63–72.

Cushing, Frank Hamilton. 1974 [1920]. 'Zuñi Breadstuff', *Indian Notes and Monographs*, Vol. 8. Museum of the American Indian, Heye Foundation, New York.

——1979. *Zuñi: Selected Writings of Frank Hamilton Cushing*, edited by Jesse Green. University of Nebraska Press, Lincoln.

Cutler, Hugh C. and Leonard W. Blake. 1985. 'Food Plant Remains from Eight Prehistoric Indian Sites in the Yazoo Delta Area of Mississippi', in *Anthology of Mississippi Archaeology: 1966–1979*, edited by Patricia Galloway, pp. 114–21. Mississippi Archaeological Association, Mississippi Department of Archives and History, Jackson.

——and Winton Meyer. 1965. 'Corn and Cucurbits from Wetherill Mesa', in *Contributions of the Wetherill Mesa Archeological Project*, assembled by Douglas Osborne and edited by Bernard S. Katz, pp. 136–52. Memoirs of the Society for American Archaeology 19.

Czaplicki, Jon. S. and Adrianne G. Rankin. 1984. *A Supplemental Class III Archaeological Survey of Phase A, Reach 3 Corridor, Tucson Aqueduct, Central Arizona Project: Late Sedentary and Early Classic Period Tucson Basin Hokoham Occupation in the Lower Santa Cruz River Basin, Marana to Rillito, Arizona*. Supplement to Archaeological Series 165. Cultural Resource Management Division, Arizona State Museum, University of Arizona, Tucson.

D'Ablon, Claude. 1899*a* [1672]. 'Relation of What Occurred Most Remarkable in the Missions of the Fathers of the Society of Jesus in New France, in the Years 1670 and 1671', in *The Jesuit Relations and Allied Documents: Travels and Explorations of the Jesuit Missionaries in New France, 1610–1791*, Vol 54, pp. 245–301; Vol. 55, pp. 19–225, edited by Reuben Gold Thwaites. The Burrows Brothers, Cleveland.

——1899*b* [1674]. 'Relation of the Discovery of Many Countries Situated to the South of New France, Made in 1673', in *The Jesuit Relations and Allied Documents: Travels and Explorations of the Jesuit Missionaries in New France, 1610–1791*, Vol. 58, pp. 91–109, edited by Reuben Gold Thwaites. The Burrows Brothers, Cleveland.

Daniel, Thomas E. 1962. *Vermont Indians*. Journal Press, Poultney, Vermont.

Danson, Edward Bridge.1957. 'An Archaeological Survey of West Central New Mexico and East Central Arizona', *Papers of the Peabody Museum of Archaeology and Ethnology* 44, Cambridge, Massachusetts.

Dart, Allen. 1983. 'Agricultural Features', in *Hohokam Archaeology along the Salt-Gila Aqueduct Central Arizona Project*, Vol. III, Part IV, edited by Lynn S. Teague and Patricia L. Crown. Archaeological Series 150. Cultural Resource Management Division, Arizona State Museum, University of Arizona, Tucson.

——1986. 'Sediment Accumulation along Hohokam Canals', *The Kiva* **51**: 63–84.

——1989. *Prehistoric Irrigation in Arizona: A Context for Canals and Related Cultural Resources*. Center for Desert Archaeology, Technical Report 89–7, Tucson.

Darwin, Charles. 1958 [1859]. *The Origin of Species: By Means of Natural Selection or the Preservation of Favoured Races in the Struggle for Life*. Mentor, New York.

Daubenmire, R. 1968. 'Ecology of Fire in Grasslands', *Advances in Ecological Research* **5**: 209–66.

Day, Clarence Albert. 1954. *A History of Maine Agriculture: 1604–1860*. University Press, Orono.

Day, Gordon M. 1953. 'The Indian as an Ecological Factor in the Northeastern Forest', *Ecology* **34**: 329–46.

——1965. 'The Indian Occupation of Vermont', *Vermont History* **33**: 365–74.

——1978*a*. 'Western Abenaki', in *Handbook of North American Indians*, Vol. 15, edited by Bruce G. Trigger, pp. 148–59. Smithsonian Institution, Washington, D.C.

——1978*b*. 'Algonquin,' in *Handbook of North American Indians*, Vol. 15, edited by Bruce G. Trigger, pp. 792–797. Smithsonian Institution, Washington, D. C.

Deagan, Kathleen A. 1985. 'Spanish–Indian Interaction in Sixteenth-Century Florida and Hispaniola', in *Cultures in Contact: The Impact of European Contacts on Native American Cultural Institutions, A.D. 1000–1800*, edited by William W. Fitzhugh, pp. 285–316. Smithsonian Institution, Washington, D. C.

Deale, Valentine B. 1958. 'The History of the Potawatomis Before 1722', *Ethnohistory* **5**: 305–60.

Dean, Glenna. 1995. 'In Search of the Rare Pollen Evidence of Prehistoric Agriculture', in *Soil, Water, Biology, and Belief in Prehistoric and Traditional Southwestern Agriculture*, edited by H. Wolcott Toll, pp. 353–9. New Mexico Archaeological Council, Special Publication 2, Albuquerque.

Debowski, Sharon S., Anique George, Richard Goddard, and Deborah Mullen. 1976. *An Archaeological Survey of the Buttes Reservoir*. Arizona State Museum Archaeological Series 93. University of Arizona, Tucson.

de Brébeuf, J[ean]. 1898 [1637]. 'Letter of Father Jean de Brébeuf to the Very Reverend Father Mutins Vitelleschi, General of the Society of Jesus, at Rome', in *The Jesuit Relations and Allied Documents: Travels and Explorations of the Jesuit Missionaries in New France, 1610–1791*, Vol. 11, edited by Reuben Gold Thwaites, pp. 6–11. The Burrows Brothers, Cleveland.

Decker-Walters, Deena A. 1993. 'New Methods for Studying the Origins of New World Domesticates: The Squash Example', in *Foraging and Farming in the Eastern Woodlands*, edited by C. Margaret Scarry, pp. 91–7. University Press of Florida, Gainesville.

De Forest, John W. 1964 [1851]. *History of the Indians of Connecticut: From the Earliest Known Period to 1850*. Shoe String Press, Hamden, Connecticut.

De Gannes (aka Pierre Deliette). 1934 [1721]. 'Memoir of De Gannes Concerning the Illinois Country', in *The French Foundations 1680–1693*, edited by Theodore Calvin Pease and Raymond C. Werner, pp. 302–95. Illinois State Historical Library, Springfield.

Delabarre, Edward B. and Harris H. Wilder. 1920. 'Indian Corn-Hills in Massachusetts', *American Anthropologist* **22**: 203–25.

De Laet, Johan. 1909 [1625, 1630, 1633, 1640]. 'The New World', in *Narratives of New Netherland, 1609–1664*, edited by J. Franklin Jameson, pp. 29–60. Charles Scribner's Sons, New York.

Delafield, Joseph. 1943. *The Unified Boundary: A Diary of the First Survey of the Canadian Boundary Line from St. Regis to the Lake of the Woods*, edited by Robert McElroy and Thomas Riggs. Privately printed, New York.

Delcourt, Paul A., Hazel R. Delcourt, Patricia A. Cridlebaugh, and Jefferson Chapman. 1986. 'Holocene Ethnobotanical and Paleoecological Record of Human Impact on Vegetation in the Little Tennessee River Valley, Tennessee', *Quaternary Research* **25**: 330–49.

Delcourt, Paul A., Hazel R. Delcourt, Cecil R. Ison, William E. Sharp, and Kristen J. Gremillion. 1998. 'Prehistoric Human Use of Fire, the Eastern Agricultural Complex, and Appalachian Oak–Chestnut Forests: Paleoecology of Cliff Palace Pond, Kentucky', *American Antiquity* **63**: 263–78.

de Leon, Alonso. 1968 [1689]. 'Carta de Alonso de León al Virrey, dándole cuenta de su jornada a la Bahía del Espíritu Santo, donde hallaron destruída la colonia de los franceses', in *Primeras Exploraciones Poblamiento de Texas (1686–1694)*, edited by Lino Gómez Canedo, pp. 113–16. Publicaciones del Instituto Tecnologico y de Estudios Superiores de Monterrey, Monterrey, Nuevo León, México.

Denevan, William M. 1966. *The Aboriginal Cultural Geography of the Llanos de Mojos of Bolivia*. Ibero-Americana 48. University of California Press, Berkeley.

——1970. 'Aboriginal Drained-Field Cultivation in the Americas', *Science* **169**: 647–54.

——1980. 'Tipología de configuraciones agrícolas prehispánicas', *América Indígena* **40**: 619–52.

——1989. 'The Geography of Fragile Lands in Latin America', in *Fragile Lands of Latin America: Strategies for Sustainable Development*, edited by John O. Browder, pp. 11–24. Westview Press, Boulder.

——1992*a*. *The Native Population of the Americas in 1492*. University of Wisconsin Press, Madison.

——1992*b*. 'Stone vs. Metal Axes: The Ambiguity of Shifting Cultivation in Prehistoric Amazonia', *Journal of the Steward Anthropological Society* **20**: 153–65.

——1998. 'Comments on Prehistoric Agriculture in Amazonia', *Culture and Agriculture* **20**: 54–9.

——and Christine Padoch (editors). 1988. 'Swidden-Fallow Agroforestry in the Peruvian Amazon', *Advances in Economic Botany* 5. New York Botanical Garden.

——and B. L. Turner II. 1974. 'Forms, Functions, and Associations of Raised Fields in the Old World Tropics', *The Journal of Tropical Geography* **39**: 24–33.

Denig, Edwin Thompson. 1930. 'Indian Tribes of the Upper Missouri', in *Forty-Sixth Annual Report of the Bureau of American Ethnology, 1928–1929*, pp. 375–654. Smithsonian Institution, Washington, D. C.

——1961 [1857]. *Five Indian Tribes of the Upper Missouri*, edited by John C. Ewers. University of Oklahoma Press, Norman.

Denmead, O. T. and R. H. Shaw. 1960. 'The Effects of Soil Moisture Stress at Different Stages of Growth or the Development of Yield of Corn', *Agronomy Journal* **52**: 272–4.

Dennis, Howard Willis. 1967. 'The Effect of Trincheras upon Agriculture in the Pompa Basin, Sierra Madre Occidental, Mexico'. Unpublished master's thesis, Department of Geography, University of Denver.

Denny, Sidney G. and William I. Woods. 1981. *Emergency Data Recovery at the Sumac Site (11-Fk-62): A Crab Orchard Hamlet in Franklin County, Illinois*. Report Submitted to the US Army Corps of Engineers, St. Louis District.

Denonville, M. de. 1850 [1687]. 'M. de Denonville to the Minister', in *The Documen-*

tary History of the State of New York, edited by E. B. O'Callaghan, pp. 146–8. Weed and Parsons, Albany.

Densmore, Frances. 1928. 'Uses of Plants by the Chippewa Indians', *Forty-Fourth Annual Report of the Bureau of American Ethnology, 1926–1927*, pp. 275–541. Smithsonian Institution, Washington, D. C.

Denys, Nicolas. 1908 [1632]. *The Description and Natural History of the Coasts of North America (Acadia)*, translated and edited by William F. Ganong. Champlain Society Publications II, Toronto.

DePratter, Chester B., Charles M. Hudson, and Marvin T. Smith. 1983. 'The Route of Juan Pardo's Explorations in the Interior Southeast, 1566–1568', *Florida Historical Quarterly* **62**: 125–58.

Dermer, Thomas. 1905 [1619]. 'To his Worshipfull Friend M. Samuel Pvrchas, Preacher of the Word, at the Church a Little within Ludgate, London', in *Sailors' Narratives of Voyages along the New England Coast: 1524–1564*, edited by George Parker Winship, pp. 249–258. Burt Franklin, New York.

de Schlippe, Pierre. 1956. *Shifting Cultivation in Africa: The Zande System of Agriculture*. Routledge and Kegan Paul, London.

Deur, Douglas. 1999. 'Salmon, Sedentism, and Cultivation: Toward an Environmental Prehistory of the Northwest Coast', in *Northwest Coast, Northwest Peoples: Readings in Environmental History*, edited by Dale D. Goble and Paul W. Hirt, pp. 129–55. University of Washington Press, Seattle.

———— forthcoming. 'Was the Northwest Coast Agricultural?: Ecological, Archaeological, and Ethnographic Evidence', in *Foragers or Cultivators?: Reassessments of Indigenous Resource Management on the Northwest Coast of North America*, edited by Douglas E. Deur and Nancy J. Turner. University of Washington Press, Seattle.

De Vries, David Pietersz. 1909 [1655]. 'The Korte Historiael Ende Journaels Aenteyckeninge', in *Narratives of New Netherland, 1609–1664*, edited by J. Franklin Jameson, pp. 181–234. Charles Scribner's Sons, New York.

de Wet, J. M. J. and J. R. Harlan. 1975. 'Weeds and Domesticates: Evolution in the Man-Made Habitat', *Economic Botany* **29**: 99–107.

Díaz, Juan. 1966 [1774]. 'Díaz's Diary from Tubac to San Gabriel', in *Anza's California Expeditions*, Vol. 2, translated and edited by Herbert Eugene Bolton, pp. 245–90. Russell and Russell, New York.

Díaz Vara Calderón, Gabriel. 1936 [1675]. 'A 17th Century Letter of Gabriel Díaz Vara Calderón, Bishop of Cuba, Describing the Indians and Indian Missions of Florida', in *Smithsonian Miscellaneous Collections*. Vol. 95, No. 16, transcribed and translated by Lucy L. Wenhold. Smithsonian Institution, Washington, D. C.

Dick, Herbert W. 1976. *Archaeological Excavations in the Llaves Area, Santa Fe National Forest, New Mexico, 1972–1974*. Part I. Archeological Report 13. USDA Forest Service, Southwestern Region, Albuquerque.

Dinkins, Stephanie. 1969. 'Lanzarote: The Strangest Canary', *National Geographic Magazine* **135**.1: 117–39.

DiPeso, Charles C. 1956. *The Upper Pima of San Cayetano Del Tumacacori: An Archaeohistorical Reconstruction of the Ootam of Pimeria Alta*. The Amerind Foundation, Dragoon, Arizona.

—— 1974. *Casas Grandes: A Fallen Trading Center of the Gran Chichimeca*, Vol. 2. Northland Press for the Amerind Foundation, Flagstaff and Dragoon, Arizona.

—— 1984. 'The Structure of the 11th Century Casas Grandes Agricultural System', in

Prehistoric Agricultural Strategies in the Southwest, edited by Suzanne K. Fish and Paul R. Fish, pp. 261–70. Anthropological Research Papers 33. Arizona State University, Tempe.

——John B. Renaldo, and Gloria J. Fenner. 1974. *Casas Grandes: A Fallen Trading Center of the Gran Chichimeca*, Vol. 5. Northland Press for the Amerind Foundation, Flagstaff and Dragoon, Arizona.

Dittert, Alfred E., Paul R. Fish, and Don E. Simonis. 1969. *A Cultural Inventory of the Proposed Granite Reef and Salt-Gila Aqueducts: Agua Fria River to Gila River, Arizona*. Anthropological Research Papers 1. Arizona State University, Tempe.

Dobyns, Henry F. 1951. 'Blunders with Bolsas: A Case Study of Diffusion of Closed-Basin Agriculture', *Human Organization* **10**: 25–32.

——1974. 'The Kohatk: Oasis and Akchin Horticulturalist', *Ethnohistory* **21**: 317–27.

——1988. 'Piman Indian Historic Agave Cultivation', *Desert Plants* **9**: 49–53.

——and Robert C. Euler. 1976. *The Walapai People*. Indian Tribal Services, Phoenix, Arizona.

Doddridge, Joseph. 1925 [1782]. 'Notes of the Settlement and Indian Wars of the Western Parts of Virginia and Pennsylvania From the year 1763 until the year 1783 inclusive', in *A History of the Valley of Virginia*, edited by Samuel Kercheval, pp. 183–301. Shenandoah Publishing House, Strasburg, Virginia.

Dodge, Ernest S. 1957. 'Ethnology of Northern New England and the Maritime Provinces', *Bulletin of the Massachusetts Archaeological Society* **18**: 68–71.

Dodge, J. P. 1871. 'Food Products of the North American Indians', in *Report of the Commissioner of Agriculture for the Year 1870*. United States Department of Agriculture, Washington, D. C.

Doelle, William H. 1981. 'The Gila Pima in the Late Seventeenth Century', in *The Protohistoric Period in the Northern American Southwest, A.D. 1450–1700*, edited by David R. Wilcox and W. Bruce Masse, pp. 57–70. Anthropological Research Papers 26. Arizona State University, Tempe.

Doggett, Rachel, Monique Hulvey, and John Ainsworth. 1992. *New World of Wonders: European Images of the Americas 1492–1700*. The Folger Shakespeare Library, Washington, D. C.

Dominguez, Francisco Atanasio. 1956 [1776]. *The Missions of New Mexico, 1776*, translated and edited by Eleanor B. Adams and Angelico Chavez. University of New Mexico Press, Albuquerque.

Donkin, R. A. 1970. 'Pre-Columbian Field Impliments and Their Distribution in the Highlands of Middle and South America', *Anthropos* **65**: 505–29.

——1979. *Agricultural Terracing in the Aboriginal New World*. Viking Fund Publications in Anthropology 56. Wenner–Gren Foundation, University of Arizona Press, Tucson.

Doolittle, William E. 1984*a*. 'Agricultural Change as an Incremental Process', *Annals of the Association of American Geographers* **74**: 124–37.

——1984*b*. 'Cabeza de Vaca's Land of Maize: An Assessment of its Agriculture', *Journal of Historical Geography* **10**: 246–62.

——1985. 'The Use of Check Dams for Protecting Downstream Agricultural Land in the Prehistoric Southwest: A Contextual Analysis', *Journal of Anthropological Research* **41**: 279–305.

——1988. *Pre-Hispanic Occupance in the Valley of Sonora, Mexico: Archaeological Confirmation of Early Spanish Reports*. Anthropological Papers of the University of Arizona 48. University of Arizona Press, Tucson.

——1990*a*. 'Terrace Origins: Hypotheses and Research Strategies', *Yearbook, Conference of Latin American Geographers* **12**: 99–108.

——1990*b*. *Canal Irrigation in Prehistoric Mexico: The Sequence of Technological Change*. University of Texas Press, Austin.

——1990*c*. 'Regional Ecology and Middle America: Teaching Geography in a Major Latin American Program', *Benchmark 1990, Conference of Latin Americanist Geographers* **17/18**: 309–12.

——1991. 'A Finger on the Hohokam Pulse', in *Prehistoric Irrigation in Arizona: Symposium 1988*, edited by Cory Dale Breternitz, pp. 139–51. Soil Systems Publications in Archaeology 17. Soil Systems, Inc., Phoenix.

——1992*a*. 'House-Lot Gardens in the Gran Chichimeca: Ethnographic Cause for Archaeological Concern', in *Gardens of Prehistory: The Archaeology of Settlement Agriculture in Greater Mesoamerica*, edited by Thomas W. Killian, pp. 69–91. University of Alabama Press, Tuscaloosa.

——1992*b*. 'Before There Was a Corn Belt: A Prospectus on Research', in *Late Prehistoric Agriculture: Observations from the Midwest*, edited by William I. Woods, pp. 217–40. Studies in Illinois Archaeology 8. Illinois Historic Preservation Agency, Springfield.

——1992*c*. 'Agriculture in North America on the Eve of Contact: A Reassessment', 'The Americas before and after 1492: Current Geographical Research', edited by Karl W. Butzer, *Annals of the Association of American Geographers* **82**: 386–401.

——1993. 'Canal Irrigation at Casas Grandes: A Technological and Developmental Assessment of Its Origins', in *Culture and Contact: Charles C. DiPeso's Gran Chichimeca*, edited by Anne I. Woosley and John C. Ravesloot, pp. 133–51. The Amerind Foundation and the University of New Mexico Press, Dragoon and Albuquerque.

——1995. 'Indigenous Development of Mesoamerican Irrigation', *Geographical Review* **85**: 301–23.

——1998. 'Innovation and Diffusion of Sand- and Gravel-Mulch Agriculture in the American Southwest: A Product of the Eruption of Sunset Crater', *Quaternaire* **9**: 61–9.

——and Charles D. Frederick. 1991. 'Phytoliths as Indicators of Prehistoric Maize (*Zea mays* subsp. *mays*, *Poaceae*) cultivation', *Plant Systematics and Evolution* **177**: 175–84.

——James A. Neely, and Michael D. Pool. 1993. 'A Method for Distinguishing Between Prehistoric and Recent Water and Soil Control Features', *Kiva* **59**: 7–25.

Dorsey, George A. and James R. Murie. 1940. 'Notes on Skidi Pawnee Society', *Field Museum of Natural History, Anthropological Series* 27.2.

Dosh, Steven G. 1988. 'Agave Study of Fools Hollow Ridge Ruin', in *Subsistence and Settlement Along the Mogollon Rim A.D. 1000–1150*, edited by Steven G. Dosh, pp. 581–84. The Museum of Northern Arizona University Press, Flagstaff.

Douay, Anastasius. 1903 [1697]. 'Narrative of La Salle's Attempt to Ascend the Mississippi', in *Discovery and Exploration of the Mississippi Valley*, edited by John Gilmary Shea, pp. 201–11. Joseph McDonough, Albany.

Douglas, James Sholto. 1973. 'Forest-Farming: An Ecological Approach to Increase Nature's Food Productivity', *Impact of Science on Society* **23**: 117–32.

Douglas, William Brown. 1917. 'Land of the Small House People', *El Palacio* **4**: 2–23.

Dove, Donald E. 1970. 'A Site Survey along the Lower Agua Fria River, Arizona', *The Arizona Archaeologist* **5**: 1–30.

——1985. 'Subsistence Issues and Population Stability in the Northern Hohokam Periphery', in *Proceedings of the 1983 Hohokam Symposium*, Part I, edited by Alfred E. Dittert, Jr. and Donald E. Dove, pp. 65–74. Occasional Paper 2. Phoenix Chapter of the Arizona Archaeological Society, Phoenix.

Downing, Theodore E. and McGuire Gibson (editors). 1974. *Irrigation's Impact on Society*. Anthropological Papers of the University of Arizona 25. University of Arizona Press, Tucson.

Downs, James F. 1966. 'The Significance of Environmental Manipulation', in *The Current Status of Anthropological Research in the Great Basin: 1964*, edited by Warren L. d'Azevedo, Wilber A. Davis, Don D. Fowler, and Wayne Suttles, pp. 39–56. Social Sciences and Humanities Publications No. 1, Technical Report Series S-H. Desert Institute, Reno, Nevada.

Downum, Christian E. 1993. *Between Desert and River: Hohokam Settlement and Land Use in the Los Robles Community*. Anthropological Papers of the University of Arizona 57. University of Arizona Press, Tucson.

——John E. Douglas, and Douglas B. Craig. 1985. 'Community Structure and Agricultural Strategies at Cerro Prieto (AZ AA:7:11)', in *Proceedings of the 1983 Hohokam Symposium*, edited by Alfred E. Dittert, Jr. and Donald E. Dove, pp. 545–56. Arizona Archaeological Society Occasional Paper 2, Phoenix.

——Paul R. Fish, and Suzanne K. Fish. 1994. 'Refining the Role of *Cerros Trincheras* in Southern Arizona Settlement', *Kiva* **59**: 271–96.

——Christopher L. Harper, and Richard L. Boston. 1995. *Walnut Canyon and Water: Capture, Storage, and Use of Water During the Sinagua Period of Occupation, ca. A.D. 1100–1300*. Northern Arizona University Archaeological Report 1125, Flagstaff.

Doyel, David E. 1977. 'Rillito and Rincon Period Settlement Systems in the Middle Santa Cruz River Valley: Alternative Models', *The Kiva* **43**: 93–110.

——1979. 'The Prehistoric Hohokam of the Arizona Desert', *American Scientist* **67**: 544–54.

——1984. 'Sedentary Period Hohokam Paleo-Economy in the New River Drainage, Central Arizona', in *Prehistoric Agricultural Strategies in the Southwest*, edited by Suzanne K. Fish and Paul R. Fish, pp. 35–52. Anthropological Research Papers 33. Arizona State University, Tempe.

——1985*a*. 'Prehistoric Agriculture and Water Control at the Antelope Glyph Site', in *Hohokam Settlement and Economic Systems in the Central New River Drainage, Arizona*, Vol. II, edited by David E. Doyel and Mark D. Elson, pp. 297–316. Soil Systems Publications in Archaeology 4. Soil Systems Inc., Phoenix.

——1985*b*. 'The New River-Palo Verde Community System', in *Hohokam Settlement and Economic Systems in the Central New River Drainage, Arizona*, Vol. II, edited by David E. Doyel and Mark D. Elson, pp. 681–99. Soil Systems Publications in Archaeology 4. Soil Systems Inc., Phoenix.

——1986. *Hohokam Prehistory in the Central New River Drainage, Arizona*. Pueblo Grande Museum, Phoenix.

——1987. 'Pueblo Grande National Landmark Celebrates a Centennial', *Masterkey* **60**: 3–11.

——1993. *Prehistoric Non-Irrigated Agriculture in Arizona*. Estrella Cultural Research Paper 3, Phoenix.
——1998. 'Vacant Ceremonial Centers Revisited', *Kiva* **64**: 225–44.
——and Mark D. Elson. 1985. 'Hohokam Settlement and Subsistence Systems in the Northern Periphery', in *Hohokam Settlement and Economic Systems in the Central New River Drainage, Arizona*, Vol. 2, edited by David E. Doyel and Mark D. Elson, pp. 701–14. Soil Systems Publications in Archaeology 4. Soil Systems Inc., Phoenix.
————and Teresa L. Hoffman. 1985. 'Summary of Excavations', *in Hohokam Settlement and Economic Systems in the Central New River Drainage, Arizona*, Vol. I, edited by David E. Doyel and Mark D. Elson, pp. 45–248. Soil Systems Publications in Archaeology 4. Soil Systems Inc, Phoenix.
Dozier, Edward P. 1954. *The Hopi–Tewa of Arizona*, University of California Publications in American Archaeology and Ethnology 44, pp. 259–376. Berkeley.
Driver, Harold E. and William C. Massey. 1957. 'Comparative Studies of North American Indians', *Transactions of the American Philosophical Society*, **47**.2.
Drucker, Phillip. 1937. 'Cultural Element Distributions: V. Southern California', *Anthropological Records, University of California* **1**: 1–52, Berkeley.
——1941. 'Cultural Element Distributions, XVIII: Yuman-Piman', *Anthropological Records, University of California* 6, Berkeley.
Dunbar, John B. 1880. 'The Pawnee Indians: Their History and Ethnology', *Magazine of American History* **4**: 221–81.
Dunbier, Roger. 1968. *The Sonoran Desert: Its Geography, Economy, and People*. University of Arizona Press, Tucson.
Dunn, M. E. 1983. 'Phytolith Analysis in Archaeology', *Midcontinent Journal of Archaeology* **8**: 287–97.
Dunnigan, Timothy. 1983. 'Lower Pima', in *Handbook of North American Indians*, Vol. 10, edited by Alfonso Ortiz, pp. 217–29. Smithsonian Institution, Washington, D. C.
du Peron, François. 1898 [1639]. 'Letter of Father Francois du Peron of the Society of Jesus, to Father Joseph Imbert du Peron, his Brother, Religious of the same Society', in *The Jesuit Relations and Allied Documents: Travels and Explorations of the Jesuit Missionaries in New France, 1610–1791*, Vol. 15, edited by Reuben Gold Thwaites, pp. 149–89. The Burrows Brothers, Cleveland.

Earls, J. 1986. 'Experimentación agrícola en el Perú precolombino y su factibilidad de reemple', in *Andenes y camellones en el Perú ardino*, edited by C. de la Torre and M. Burga, pp. 301–30. CONCYTEC, Lima.
Ebert, James I. and Thomas R. Lyons. 1980. 'Prehistoric Irrigation Canals Identified from Skylab III and Landsat Imagery in Phoenix, Arizona', in *Cultural Resources Remote Sensing*, edited by Thomas R. Lyons and Frances J. Mathien, pp. 209–28. National Park Service, Washington, D. C.
Eckstrom, Fannie H. 1919. 'The Indians of Maine', in *Maine: A History*, Vol. 1, edited by Louis C. Hatch, pp. 43–64. The American Historical Society, New York.
Edwards, Everett E., and Wayne D. Rasmussen. 1942. *A Bibliography on the Agriculture of the American Indian*. United States Department of Agriculture, Miscellaneous Publications 447, Washington, D. C.
Eggan, Fred. 1955 [1937]. 'The Cheyenne and Arapaho Kinship System', in *Social*

Anthropology of North American Tribes, edited by Fred Eggan, pp. 35–95. University of Chicago Press, Chicago.

Eidt, Robert C. 1984. *Advances in Abandoned Settlement Analysis: Application to Prehistoric Anthrosols in Colombia, South America*. The Center for Latin America University of Wisconsin–Milwaukee.

Eighmy, Jeffrey L. and Jerry B. Howard. 1991. 'Direct Dating of Prehistoric Canal Sediments Using Archaeomagnetism', *American Antiquity* **56**: 88–102.

Eixarch, Thomas. 1966 [1776]. 'Diary Written on the Colorado River', in *Anza's California Expeditions*, Vol. 3, translated and edited by Herbert Eugene Bolton, pp. 309–82. Russell and Russell, New York.

Elkin, Henry. 1940. 'The Northern Arapaho of Wyoming', in *Acculturation in Seven American Indian Tribes*, edited by Ralph Linton, pp. 207–55. D. Appleton-Century, New York.

Elliott, Jack D., Jr. 1997. 'Of Roads and Reifications: The Interpretation of Historical Roads and the Soto Entrada', in *The Hernando de Soto Expedition: History, Historiography, and 'Discovery' in the Southeast*, edited by Patricia Galloway, pp. 246–58. University of Nebraska Press, Lincoln.

Ellis, Florence Hawley. n.d.*a* [1970]. 'Irrigation and Water Works in the Rio Grande'. Unpublished paper presented at the 1970 Pecos Conference, Symposium on Prehistoric Southwestern Water Control Systems, Pecos National Monument, New Mexico, on file at the Arizona State Museum, University of Arizona, Tucson.

—— n.d.*b* [1979]. 'Summaries of the History of Water Use and the Tewa Culture of the Pojoaque Valley Pueblos'. Unpublished manuscript on file, Laboratory of Anthropology, Museum of New Mexico, Santa Fe.

—— 1975. 'Highways to the Past', *New Mexico Magazine* **53**: 18–25.

—— 1978. 'Small Structures Used by Historic Pueblo Peoples and their Immediate Ancestors,' in *Limited Activity and Occupation Sites: A Collection of Conference Papers*, compiled and edited by Albert E. Ward, pp. 59–68. Center for Anthropological Studies, Albuquerque.

—— 1979*a*. 'Isleta Pueblo', in *Handbook of North American Indians*, Vol. 9, edited by Alfonso Ortiz, pp. 351–65. Smithsonian Institution, Washington, D. C.

—— 1979*b*. 'Laguna Pueblo', in *Handbook of North American Indians*, Vol. 9, edited by Alfonso Ortiz, pp. 438–49. Smithsonian Institution, Washington, D. C.

—— and Andrea Ellis Dodge. 1989. 'Water and Solar Engineering in the Rio Grande Drainage: Fourteenth-Nineteenth Centuries', in *From Chaco to Chaco: Papers in Honor of Robert H. Lister and Florence C. Lister*, edited by Meliha S. Duran and David T. Kirkpatrick, pp. 35–44. The Archaeological Society of New Mexico, Albuquerque.

Elson, Mark D., David D. Doyel, and Teresa L. Hoffman. 1985. 'Hohokam Settlement and Economic Systems in the Northern Periphery: A Comparative Analysis', in *Proceedings of the 1983 Hohokam Symposium*, Part I, edited by Alfred E. Dittert, Jr., and Donald E. Dove, pp. 45–64. Occasional Paper 2. Phoenix Chapter of the Arizona Archaeological Society, Phoenix.

Elston, Robert G. 1986. 'Prehistory of the Western Area', in *Handbook of North American Indians*, Vol. 11, edited by Warren L. D'Azevedo, pp. 135–48. Smithsonian Institution, Washington, D. C.

Elvas, The Gentleman of. 1907 [1538–1543]. 'The Narrative of the Expedition of

Hernando de Soto', in *Spanish Explorers in the Southern United States 1528–1543*, edited by Theodore H. Lewis, pp. 127–272. Barnes and Noble, New York.

——1993 [1538–1543]. 'The Account by a Gentleman from Elvas', translated and edited by James Alexander Robertson, in *The De Soto Chronicles: The Expedition of Hernando De Soto to North America in 1539–1543*, Vol. 1, edited by Lawrence A. Clayton, Vernon James Knight, Jr., and Edward C. Moore, pp. 19–219. University of Alabama Press, Tuscaloosa.

Enloe, James G., Andrew T. Smith, and Stewart L. Peckham. 1974. *Archaeological Survey of the San Juan-To-Ojo 345 KV Transmission Line, Northwestern New Mexico*, Note 5. Laboratory of Anthropology, Santa Fe.

Erdman, James A., Charles L. Douglas, and John W. Marr. 1969. *Environment: Wetherill Mesa Studies of Mesa Verde, Colorado*. Archeological Research Series 7-B. National Park Service, Washington, D. C.

Ermigiotti, Paul. 1989. 'Mesa Verde National Park 1989 Experimental Garden,' *The Seedhead News* **26**: 6.

Erwin, A. T. and E. P. Lana. 1956. 'The Seminole Pumpkin', *Economic Botany* **10**: 33–7.

Escalante, Felipe de and Hernando Barrando. 1966 [1583]. 'Brief and True Account of the Discovery of New Mexico by Nine Men Who Set Out From Santa Bárbara in the Company of Three Franciscan Friars', in *The Rediscovery of New Mexico 1580–1594: The Explorations of Chamuscado, Espejo, Castaño de Sosa, Morlete, and Leyva de Bonilla and Humaña*, translated and edited by George P. Hammond and Agapito Rey, pp. 141–43. University of New Mexico Press, Albuquerque.

Espejo, Antonio de. 1966 [1582]. 'Report of Antonio de Espejo', in *The Rediscovery of New Mexico 1580–1594: The Explorations of Chamuscado, Espejo, Castaño de Sosa, Morlete, and Leyva de Bonilla and Humaña*, translated and edited by George P. Hammond and Agapito Rey, pp. 213–31. University of New Mexico Press, Albuquerque.

Espinosa, Isidro de. 1930 [1709]. 'The Espinosa–Olivares–Aguirre Expedition of 1709: Espinosa's Diary', translated by Gabriel Tous. *Preliminary Studies of the Texas Catholic Historical Society*, **1**: 5–14.

Espinosa, Isidro Felis de. 1927 [1722]. 'Fray Isidro Felis de Espinosa on the Asinai and Their Allies', in 'Descriptions of the Tejas or Asinai Indians, 1691–1722', translated by Mattie Austin Hatcher, *Southwestern Historical Quarterly* **31**: 150–80.

Euler, Robert C. 1954. 'Environmental Adaptation at Sia Pueblo', *Human Organization* **12**: 27–30.

Evanari, Michael, Leslie Shanan, and Naphtali Tadmore. 1971. *The Negev: The Challenge of a Desert*. Harvard University Press, Cambridge Massachusetts.

Ewers, John C. 1968. *Indian Life on the Upper Missouri*. University of Oklahoma Press, Norman.

Ezell, Paul H. 1961. 'The Hispanic Acculturation of the Gila River Pima', *American Anthropological Association Memoirs*, 90.

——1985. 'Ballcourt or Reservoir?', in *Southwestern Culture History: Collected Papers in Honor of Albert H. Schroeder*, pp. 59–68. Papers of the Archaeological Society of New Mexico 10. Ancient City Press, Santa Fe.

Fairbanks, Charles H. 1979. 'The Function of Black Drink Among the Creeks', in *Black Drink: A Native American Tea*, edited by Charles M. Hudson, pp. 120–49. University of Georgia Press, Athens.

Fairbourn, Merle. L. 1973. 'Effect of Gravel Mulch on Crop Yields', *Colorado Ranches and Farms* **24**: 925–8.

—— 1975. 'Field Evaluation of Microwatershed and Vertical Mulch Systems', in *Proceedings of the Water Harvesting Symposium, Phoenix, Arizona, March 26–28, 1974*, edited by Gary W. Frasier, pp. 233–43. US Department of Agriculture, Western Region, Phoenix.

Falcón Mariano, Joseph Agustín. 1979 [1777]. 'A Description of the Missions of Our Lord of Pópulo of Guaguachique and its Annexes: By Order of the King Our Lord Charles III; in God Our Lord May He Prosper', in *Rarámuri: A Tarahumara Colonial Chronicle, 1607–1791*, edited by Thomas E. Sheridan and Thomas H. Naylor, pp. 104–17. Northland Press, Flagstaff.

Farrington, Ian S. and James Urry. 1985. 'Food and the Early History of Cultivation', *Journal of Ethnobiology* **5**: 143–57.

Fausz, J. Frederick. 1985. 'Patterns of Anglo-Indian Aggression and Accommodation along the Mid-Atlantic Coast, 1584–1634', in *Cultures in Contact: The Impact of European Contacts on Native American Cultural Institutions,* AD *1000–1800*, edited by William W. Fitzhugh, pp. 225–61. Smithsonian Institution, Washington, D. C.

Feest, Christian F. 1978. 'Virginia Algonquians', in *Handbook of North American Indians*, Vol. 15, edited by William C. Sturtevant, pp. 253–70. Smithsonian Institution, Washington, D. C.

Felger, Richard Stephen and Mary Beck Moser. 1985. *People of the Desert and Sea: Ethnobotany of the Seri Indians*. University of Arizona Press, Tucson.

Fenton, William N. 1978. 'Northern Iroquoian Culture Patterns', in *Handbook of North American Indians*, Vol. 15, edited by William C. Sturtevant, pp. 296–302. Smithsonian Institution, Washington, D. C.

Ferguson, T. J. n.d. [1984]. 'A Century of Change in Zuni Traditional Farming and Water Control', Unpublished paper presented at the Annual Conference of the American Society for Ethnohistory, New Orleans.

—— 1985. *Patterns of Land Use and Environmental Change on the Zuni Indian Reservation, 1846–1985: Ethnohistorical and Archaeological Evidence*. Expert testimony submitted to the United States Claims Court as evidence in the case of *Zuni Indian Tribe* v. *United States*, Docket 327–81L.

—— and E. Richard Hart. 1985. *A Zuni Atlas*. University of Oklahoma Press, Norman.

Fernández, Santiago. 1967 [1788]. 'Diary of Santiago Fernández from Santa Fe to the Taovayas and Return to Santa Fe', in *Pedro Vial and the Roads to Santa Fe*, edited by Noel M. Loomis and Abraham P. Nassatip, pp. 316–68. University of Oklahoma Press, Norman.

Fewkes, Jesse Walter. 1904. 'Two Summers' Work in Pueblo Ruins', in *Twenty-Second Annual Report of the Bureau of American Ethnology, 1900–1901*, pp. 3–195. Smithsonian Institution, Washington, D. C.

—— 1917. 'A Prehistoric Mesa Verde Pueblo and its People', in *Thirty-eighth Annual Report of the Smithsonian Institution, 1916*, pp. 461–88. Government Printing Office, Washington, D. C.

Field, Chris. 1966. 'A Reconnaissance of Southern Andean Agricultural Terracing'. Unpublished doctoral dissertation, Department of Geography, University of California at Los Angeles.

Fish, Paul R. 1974. 'Prehistoric Land Use in the Perkinsville Valley', *The Arizona Archaeologist* **8**: 1–36.

——and Suzanne K. Fish. 1984. 'Agricultural Maximization in the Sacred Mountain Basin', in *Prehistoric Agricultural Strategies in the Southwest*, edited by Suzanne K. Fish and Paul R. Fish, pp. 147–60. Anthropological Research Papers 33. Arizona State University, Tempe.

————and Christian E. Downum. 1992. 'Nuevas observaciones sobre el fenómeno Trincheras'. *Noroeste de México* **11**: 61–74.

————and John H. Madsen. 1984. 'Northern Tucson Basin Survey: Research Summary and Recommendations for the Marana Complex', in *A Supplemental Class III Archealogical Survey of Phase A, Reach 3 Corridor, Tucson Aqueduct, Central Arizona Project: Late Sedentary and Early Classic Period Tucson Basin Hohokam Occupation in the Lower Santa Cruz River Basin, Marana to Rillito, Arizona*, edited by Jon S. Czaplicki and Adrianne G. Rankin, pp. 83–9. Supplement to Archaelogical Series 165. Cultural Resource Management Division, Arizona State Museum, University of Arizona, Tucson.

————Austin Long, and Charles Miksicek. 1986. 'Early Corn Remains from Tumamoc Hill, Southern Arizona', *American Antiquity* **51**: 563–71.

Fish, Suzanne K. 1984. 'Agricultural and Subsistence Implications of the Salt-Gila Aqueduct Project Pollen Analysis', in *Hohokam Along the Salt-Gila Aqueduct Central Arizona Project*, Vol. 7, edited by Lynn S. Teague and Patricia L. Crown, pp. 111–38. Archaeological Series 150. Cultural Resource Management Division, Arizona State Museum, Tucson.

——1985. 'Prehistoric Disturbance Floras of the Lower Sonoran Desert and their Implications', *AASP Contribution Series* **16**: 77–88.

——1994. 'Archeological Palynology of Gardens and Fields', in *The Archeology of Garden and Field*, edited by Naomi F. Miller and Kathryn L. Gleason, pp. 44–69. University of Pennsylvania Press, Philadelphia.

——1995. 'Mixed Agricultural Technologies in Southern Arizona and their Implications', in *Soil, Water, Biology, and Belief in Prehistoric and Traditional Southwestern Agriculture*, edited by H. Wolcott Toll, pp. 101–16. New Mexico Archaeological Council, Special Publication 2, Albuquerque.

——and Paul R. Fish. 1994. 'Prehistoric Desert Farmers of the Southwest', *Annual Review of Anthropology* **23**: 83–108.

——and Gary P. Nabhan, 1991. 'Desert as Context: The Hohokam Environment', in *Exploring the Hohokam: Prehistoric Desert People of the American Southwest*, edited by George Gumerman, pp. 29–60. University of New Mexico Press, Albuquerque.

——Paul R. Fish, and Christian Downum. 1984. 'Hohokam Terraces and Agricultural Production in the Tucson Basin', in *Prehistoric Agricultural Strategies in the Southwest*, edited by Suzanne K. Fish and Paul R. Fish, pp. 55–72. Anthropological Research Papers 33. Arizona State University, Tempe.

————and John H. Madsen. 1985. 'A Preliminary Analysis of Hohokam Settlement and Agriculture in the Northern Tucson Basin', in *Proceedings of the 1983 Hohokam Symposium*, Part I, edited by Alfred E. Dittert, Jr., and Donald E. Dove, pp. 75–100. Occasional Paper 2. Phoenix Chapter of the Arizona Archaeological Society, Phoenix.

——————1989. 'Differentiation and Integration in a Tucson Basin Classic Period

Hohokam Community', in *The Sociopolitical Structure of Prehistoric Southwestern Societies*, edited by Steadman Upham, Kent G. Lightfoot, and Robert A. Jewett, pp. 237–67. Westview Press, Boulder.

Fish, Suzanne K., Paul R. Fish, and John H. Madsen. 1990. 'Analyzing Regional Agriculture: A Hohokam Example', in *The Archaeology of Regions: A Case for Full-Coverage Survey*, edited by Suzanne K. Fish and Stephen A. Kowalewski, pp. 189–218. Smithsonian Institution Press, Washington, D. C.

————1992*a*. 'Parameters of Agricultural Production in the Northern Tucson Basin', in *The Marana Community in the Hohokam World*, edited by Suzanne K. Fish, Paul R. Fish, and John H. Madsen, pp. 41–52. Anthropological Papers of the University of Arizona 56. University of Arizona Press, Tucson.

————1992*b*. 'Evidence for Large-scale Agave Cultivation in the Marana Community', in *The Marana Community in the Hohokam World*, edited by Suzanne K. Fish, Paul R. Fish, and John H. Madsen, pp. 73–87. Anthropological Papers of the University of Arizona 56. University of Arizona Press, Tucson.

——Charles Miksicek, and John H. Madsen. 1985. 'Prehistoric Agave Cultivation in Arizona', *Desert Plants* **7**: 107–12, 100.

Fitzhugh, Ammadee B. 1926. 'North American Indian Agriculture and Plant Domestication'. Unpublished master's thesis, University of Texas, Austin.

Flack, J. R. 1970. 'The Spread and Domestication of the Pecan (Carya Illinoensis) in the United States'. Unpublished doctoral dissertation, Department of Geography, University of Wisconsin, Madison.

Fletcher, Roland. 1985. 'Intensification and Residential Stress: A Material Behavior Analysis', in *Prehistoric Intensive Agriculture in the Tropics*, edited by I. S. Farrington, pp. 656–82. BAR International Services 232, Oxford.

Fletcher, Stevenson W. 1950. *Pennsylvania Agriculture and Country Life: 1640–1840*. Pennsylvania Historical and Museum Commission, Harrisburg.

Fliedner, Dietrich. 1975. 'Pre-Spanish Pueblos in New Mexico', *Annals of the Association of American Geographers* **65**: 363–77.

Flores, Dan L. (editor). 1984. *Jefferson and Southwestern Exploration: The Freeman and Custis Accounts of the Red River Expedition of 1806*. University of Oklahoma Press, Norman.

Fogelson, Raymond D. and Paul Kutsche. 1961. 'Cherokee Economic Cooperatives: The Gadugi', in *Symposium on Cherokee and Iroquois Culture*, edited by William N. Fenton and John Gulick, pp. 83–121. Smithsonian Institution, Bureau of American Ethnology, Bulletin 180. Washington, D. C.

Font, Pedro. 1966 [1776]. 'Diary of an Expedition to Monterey by Way of the Colorado River', in *Anza's California Expeditions*, Vol. 4, translated and edited by Herbert Eugene Bolton. Russell and Russell, New York.

Fontana, Bernard L. 1983. 'Pima and Papago: Introduction', in *Handbook of North American Indians*, Vol. 10, edited by Alfonso Ortiz, pp. 125–36. Smithsonian Institution, Washington, D. C.

Ford, Richard I. 1968. 'An Ecological Analysis Involving the Population of San Juan Pueblo, New Mexico'. Unpublished doctoral dissertation, Department of Anthropology, University of Michigan.

——1974. 'Northeastern Archeology: Past and Future Directions', *Annual Review of Anthropology* **3**: 385–413.

——1977. 'The Technology of Irrigation in a New Mexico Pueblo', in *Material Cul-*

ture: Styles, Organization, and Dynamics of Technology, edited by Heather Lechtman and Robert S. Merrill, pp. 139–54. West Publishing, St. Paul, Minnesota.
——1979. 'Gathering and Gardening: Trends and Consequences of Hopewell Subsistence Strategies', in *Hopewell Archaeology: The Chillicothe Conference*, edited by David S. Brose and Naomi Greber, pp. 234–8. Kent State University Press, Kent, Ohio.
——1981. 'Gardening and Farming before AD 1000: Patterns of Prehistoric Cultivation North of Mexico', *Journal of Ethnobiology* **1**: 6–27.
——(editor). 1985*a*. *Prehistoric Food Production in North America*. Anthropological Papers 75. Museum of Anthropology, University of Michigan, Ann Arbor.
——1985*b*. 'The Processes of Plant Food Production in Prehistoric North America', in *Prehistoric Food Production in North America*, edited by Richard I. Ford, pp. 1–18. Anthropological Papers 75. Museum of Anthropology, University of Michigan, Ann Arbor.
——1985*c*. 'Patterns of Prehistoric Food Production in North America', in *Prehistoric Food Production in North America*, edited by Richard I. Ford, pp. 341–64. Anthropological Papers 75. Museum of Anthropology, University of Michigan, Ann Arbor.
——1985*d*. *Zuni Land Use and Damage to Trust Land*. Expert testimony submitted to the United States Claims Court as evidence in the case of *Zuni Indian Tribe* v. *United States*, Docket 327–81L.
——and David S. Brose. 1975. 'Prehistoric Wild Rice from the Dunn Farm Site, Leelanau County, Michigan', *Wisconsin Archaeologist* **56**: 9–15.
Forde, C. Daryll. 1931*a*. 'Ethnography of the Yuma Indians', *University of California Publications in American Archaeology and Ethnology* **28**: 83–278.
——1931*b*. 'Hopi Agriculture and Land Ownership', *The Journal of the Royal Anthropological Institute of Great Britain and Ireland* **61**: 357–405.
——1934. *Habitat, Economy and Society: A Geographical Introduction to Ethnology*. Methuen, London.
Foster, William C. 1995. *Spanish Expeditions into Texas, 1689–1768*. University of Texas Press, Austin.
Fowler, Catherine S. 1986. 'Subsistence', in *Handbook of North American Indians*, Vol. 11, edited by Warren L. D'Azevedo, pp. 64–97. Smithsonian Institution, Washington, D. C.
Fowler, Don D. and David B. Madsen. 1986. 'Prehistory of the Southeastern Area', in *Handbook of North American Indians*, Vol. 11, edited by Warren L. D'Azevedo, pp. 173–82. Smithsonian Institution, Washington, D. C.
Fowler, Melvin L. 1969. 'Middle Mississippian Agricultural Fields', *American Antiquity* **34**: 365–75.
——1971. 'Agriculture and Village Settlement in the North American East: The Central Mississippi Valley Area, a Case History', in *Prehistoric Agriculture*, edited by Stuart Struever, pp. 361–82. American Museum Sourcebooks in Anthropology. The American Museum of Natural History, The Natural History Press, Garden City, New York.
——1992. 'The Eastern Horticultural Complex and Mississippian Agricultural Fields: Studies and Hypotheses', in *Late Prehistoric Agriculture: Observations from the Midwest*, edited by William I. Woods, pp. 1–18. Illinois Historic Preservation Agency, Studies in Illinois Archaeology 8, Springfield.

Fowler, Melvin L. and Elizabeth Benchley. 1980. *Final Report of 1979 Archaeological Investigations at the Interpretive Center Tract Cahokia Mounds Historic Site*. Report of Investigations 40. Archaeological Research Laboratory, University of Wisconsin–Milwaukee.

Fowler, William S. 1954. 'Agricultural Tools and Techniques of the Northeast', *Bulletin of the Massachusetts Archaeological Society* **15**: 41–51.

——1960. 'Did Lafitau Draw What He Saw?', *Bulletin of the Massachusetts Archaeological Society* **21**: 38–43.

——1963. 'Classification of Stone Implements of the Northeast', *Bulletin of the Massachusetts Archaeological Society* **25**.

——1968. 'Aboriginal Grinding Equipment', *Bulletin of the Massachusetts Achaeological Society* **29**: 19–25.

——1970. 'Discovery of Fertilizer in Maize Cultivation', *Bulletin of the Massachusetts Archaeological Society* **31**: 23–6.

——1974. 'Comparative Study of Hoe and Spade Blades', *Bulletin of the Massachusetts Archaeological Society* **35**: 1–9.

——1976. 'Division of Labor: Archaeological Disclosures', *Bulletin of the Massachusetts Archaeological Society* **37**: 6–12.

Fox, George R. 1922. 'Stoneworks and Garden Beds in Winnebago County', *Wisconsin Archeologist* **1**: 48–55.

——1959. 'The Prehistoric Garden Beds of Wisconsin and Michigan and the Fox Indians', *Wisconsin Archeologist* **40**: 1–10.

Franke, Paul R. and Don Watson. 1936. 'An Experimental Corn Field in Mesa Verde National Park', in *Symposium on Prehistoric Agriculture*, University of New Mexico Bulletin, pp. 35–41. Anthropological Series, Vol. 1, No. 5, Albuquerque.

Franklin, Wayne, 1979. *Discoverers, Explorers, Settlers: The Diligent Writers of Early America*. University of Chicago Press, Chicago.

Frazier, Kendrick. 1986. *People of Chaco: A Canyon and Its Culture*. W. W. Norton, New York.

Fritz, Gayle J. 1984. 'Identification of Cultigen Amaranth and Chenopod from Rockshelters in Northwest Arkansas', *American Antiquity* **49**: 558–72.

——1997. 'A Three-Thousand-Year-Old Cache of Crop Seeds from Marble Bluff, Arkansas', in *People, Plants, and Landscapes: Studies in Paleoethnobotany*, edited by Kristen J. Gremillion, pp. 42–62. University of Alabama Press, Tuscaloosa.

Gabriel, Ralph Henry. 1926. *Toilers of Land and Sea*. Yale University Press, New Haven.

Gaillard, D. D. 1896. 'A Gigantic Earthwork in New Mexico', *American Anthropologist* **9**: 311–13.

Gallagher, James P. 1989. 'Agricultural Intensification and Ridged-Field Cultivation in the Prehistoric Upper Midwest of North America', in *Foraging and Farming: The Evolution of Plant Exploitation*, edited by David R. Harris and Gordon C. Hillman, pp. 572–84. Unwin Hyman, London.

——1992. 'Prehistoric Field Systems in the Upper Midwest', in *Late Prehistoric Agriculture: Observations from the Midwest*, edited by William I. Woods, pp. 95–135. Illinois Historic Preservation Agency, Studies in Illinois Archaeology 8, Springfield.

——and Constance M. Arzigian. 1994. 'A New Perspective on Late Prehistoric Agricultural Intensification in the Upper Mississippi River Valley', in *Agricultural*

Origins, and Development in the Midcontinent, edited by William Green, pp. 171–88. Office of the State Archaeologist of Iowa, Report 19, University of Iowa, Iowa City.

—— and Robert F. Sasso. 1987. 'Investigations into Oneota Ridged Field Agriculture on the Northern Margin of the Prairie Penninsula', *Plains Anthropologist* **32**: 141–52.

—— Robert F. Boszhardt, Robert F. Sasso, and Katherine Stevenson. 1985. 'Oneota Ridged Field Agriculture in Southwestern Wisconsin', *American Antiquity* **50**: 605–12.

—— —— —— —— 1987. 'Floodplain Agriculture in the Driftless Area: A Reply to Overstreet', *American Antiquity* **52**: 398–404.

Gallegos, Hernan. 1966 [1581]. 'Gallegos' Relation of the Chamuscado-Rodríguez Expedition', in *The Rediscovery of New Mexico 1580–1594: The Explorations of Chamuscado, Espejo, Castaño de Sosa, Morlete, and Leyva de Bonilla and Humaña*, translated and edited by George P. Hammond and Agapito Rey, pp. 67–114. University of New Mexico Press, Albuquerque.

Garcia-Mason, Velma. 1979. 'Acoma Pueblo', in *Handbook of North American Indians*, Vol. 9, edited by Alfonso Ortiz, pp. 450–66. Smithsonian Institution, Washington, D. C.

Gardiner, Ronald, W. Bruce Masse, and Carl D. Halbirt. 1987. 'Data Recovery at AZ U:9:71 (ASM)', in *Archaeological Investigations of Portions of the Las Acequias-Los Muertos Irrigation System: Testing and Partial Data Recovery within the Tempe Section of the Outer Loop Freeway System, Maricopa County, Arizona*, edited by W. Bruce Masse, pp. 25–56. Archaeological Series 176. University of Arizona, Tucson.

Garman, James C., Paul A. Russo, Stephen A. Mrozowski, and Michael A. Volmer. 1997. ' "This Great Wild Tract": Henry Thoreau, Native Americans, and the Archaeology of Estabrook Woods', *Historical Archaeology* **31. 4**: 59–80.

Gartner, William Gustav. 1991. 'A Field Report on the Hulbert Creek Ridged Fields, Wisconsin', in *Abstracts, The Association of American Geographers 1991 Annual Meeting*, p. 66. The Association of American Geographers, Washington, D. C.

—— 1993. 'Soils and Sediments as Artifacts in Prehistoric Wisconsin', in *Proceedings of the First International Conference on Pedo-Archaeology*, edited by J. E. Foss, M. E. Timpson, and M. W. Morris, pp. 113–16. Soil Science Society of America, Special Publication 93-03.

—— 1997*a*. 'Agricultural Potential as a Cultural Appraisal: The Northern Margins Reconsidered', in *1997 AAAS Annual Meeting and Science Innovation Exposition*, edited by Michael S. Strauss, p. A-44. American Association for the Advancement of Science, Washington, D. C.

—— 1997*b*. 'Four Worlds Without an Eden: Pre-Columbian Peoples and the Wisconsin Landscape', in *Wisconsin Land and Life*, edited by Robert C. Ostergren and Thomas R. Vale, pp. 331–50. University of Wisconsin Press, Madison.

—— 1998. 'An Atlas of Pre-Columbian Raised Fields in the Upper Midwest', *Abstracts, The Association of American Geographers 94th Annual Meeting*. The Association of American Geographers, Washington, D. C.

—— —— 1999. 'Late Woodland Landscapes of Wisconsin: Ridged Fields, Effigy Mounds and Territoriality.' *Antiquity* **73**: 671–83.

Gasser, Robert E. 1990. 'Ak-Chin Farming', in *Archaeology of the Ak-Chin Indian*

Community, West Side Farms Project, Vol. 2, compiled by Robert E. Gasser, Christine K. Robinson, and Cory Dale Breternitz, pp. 4.1–4.29. Soil Systems Publications in Archaeology 9, Phoenix, Arizona.

Gasser, Robert E. and Scott M. Kwiatkowski. 1991*a*. 'Food for Thought: Recognizing Patterns in Hohokam Subsistence', in *Exploring the Hohokam: Prehistoric Desert Peoples of the American Southwest*, edited by George J. Gumerman, pp. 417–59. University of New Mexico Press for the Amerind Foundation, Albuquerque.

—— —— 1991*b*. 'Regional Signatures of Hohokam Plant Use', *Kiva* **56**: 207–26.

Gaultier de Varennes, Pierre. 1980 [1738–39]. 'Explorations of 1738–39 to the Missouri River: The La Vérendrye "Journal"', in *The Explorations of the La Vérendryes in the Northern Plains, 1738–43*, edited by W. Raymond Wood, pp. 42–66, University of Nebraska Press, Lincoln.

Gebauer, Anne Birgitte and T. Douglas Price (editors). 1992. *Transitions to Agriculture in Prehistory*. Prehistory Press, Madison, Wisconsin.

Gentry, Howard Scott. 1942. *Rio Mayo Plants: A Study of the Flora and Vegetation of the Valley of the Rio Mayo, Sonora*. Carnegie Institution Publication 527, Washington, D. C.

—— 1963. *The Warihio Indians of Sonora-Chihuahua: An Ethnographic Survey*. Smithsonian Institution, Bureau of American Ethnology, Bulletin 186, pp. 61–144. Washington, D. C.

—— 1972. *The Agave Family in Sonora*. United States Department of Agriculture, Agricultural Handbook No. 399, Washington, D. C.

Gerhard, Peter. 1982. *The Northern Frontier of New Spain*. Princeton University Press, Princeton, New Jersey.

Geyik, M. P. 1986. *FAO Watershed Management Field Manual: Gully Control*. FAO Conservation Guide 12/2. Food and Agricultural Organization of the United Nations, Rome.

Gibson, Arrel M. 1971. 'Chickasaw Ethnography: An Ethnohistorical Reconstruction', *Ethnohistory* **18**: 99–117.

Gifford, Edward W. 1933. 'The Cocopa', *University of California Publications in American Archaeology and Ethnology* **31**: 257–334.

Giles, John. 1869 [1736]. *Memoirs of Odd Adventures, Strange Deliverances, etc. in the Captivity of John Giles, Esq.* William Dodge, Cincinnati.

Gilliland, Marion. 1975. *The Material Culture of Key Marco, Florida*. University of Florida Press, Gainesville.

Gilman, Patricia L. and Peter Sherman. 1975. *An Archaeological Survey of the Graham-Curtis Project: Phase II*. Arizona State Museum Archaeological Series 65. University of Arizona, Tucson.

Gilmore, Melvin R. 1915. 'A Glimpse at Nebraska Indian Geography', *The Journal of Geography* **13**: 178–85.

—— 1919. 'Uses of Plants by the Indians of the Missouri River Region', in *Thirty-third Annual Report of the Bureau of American Ethnology, 1911–1912*, pp. 43–154. Smithsonian Institution, Washington, D. C.

—— 1922. 'Aboriginal Tobaccos', *American Anthropologist* **22**: 480–91.

—— 1930. 'Dispersal by Indians a Factor in the Extension of Discontinuous Distribution of Certain Species of Native Plants', *Papers of the Michigan Academy of Science, Arts, and Letters* **13**: 89–94.

—— 1931. 'Vegetal Remains of the Ozark Bluff-Dweller Culture', *Papers of the Michigan Academy of Science, Arts, and Letters* **14**: 83–102.

——1933. 'Some Chippewa Uses of Plants', *Papers of the Michigan Academy of Science, Arts, and Letters* **17**: 119–43.

Ginsberg, Karen. 1987. 'Kitchen Gardens in the Petén Region of Guatemala: A Study of the Form, Crops, and Contributions of Gardens in Three Villages, 1978–1980'. Unpublished master's thesis, University of California, Davis.

Gladwin, Harold S., Emil W. Haury, Edwin B. Sayles, and Nora Gladwin. 1965 [1937]. *Excavations at Snaketown: Material Culture*. University of Arizona Press, Tucson.

Glass, Anthony. 1985 [1808]. *Journal of an Indian Trader: Anthony Glass and the Texas Trading Frontier, 1790–1810*, edited by Dan L. Flores. Texas A&M University Press, College Station.

Glassow, Michael A. 1980. *Prehistoric Agricultural Development in the Northern Southwest*. Anthropological Papers 16, Ballena Press, Socorro, New Mexico.

Gleason, Kathryn L. 1994. 'To Bound and to Cultivate: An Introduction to the Archeology of Gardens and Fields', in *The Archeology of Garden and Field*, edited by Naomi F. Miller and Katheryn L. Gleason, pp. 1–24. University of Pennsylvania Press, Philadelphia.

Gliessman, Stephen R. 1984. 'An Agroecological Approach to Sustainable Agriculture', in *Meeting the Expectations of the Land: Essays in Sustainable Agriculture and Stewardship*, edited by Wes Jackson, Wendell Berry, and Bruce Colman, pp. 160–71. North Point Press, San Francisco.

Golomb, Berl and Herbert M. Eder. 1964. 'Landforms Made by Man', *Landscape* **14**: 4–7.

González-Jacomé, Alba. 1985. 'Home Gardens in Central Mexico', in *Prehistoric Intensive Agriculture in the Tropics*, edited by I. S. Farrington, pp. 521–37. BAR, International Series 232, Oxford.

Goode, G. B. 1880. 'The Use of Agricultural Fertilizers by the American Indians and the Early English Colonists', *The American Naturalist* **14**: 473–8.

Goodwin, Grenville. 1942. *The Social Organization of the Western Apache*. University of Chicago Press, Chicago.

Gookin, Daniel. 1806 [1674]. 'Historical Collections of the Indians in New England', in *Collections of the Massachusetts Historical Society, For the Year 1792*. Munroe and Francis, Boston.

Gould, Richard A. 1978. 'Tolowa', in *Handbook of North American Indians*, Vol. 8, edited by Robert F. Heizer, pp. 128–36. Smithsonian Institution, Washington, D. C.

Graham, Martha. 1994. *Mobile Farmers: An Ethnoarchaeological Approach to Settlement Organization among the Rarámuri of Northwest Mexico*. International Monographs in Prehistory, Ann Arbor, Michigan.

Gray, Lewis Cecil. 1947 [1933]. *History of Agriculture in the Southern United States to 1860*, Vol. 1. Peter Smith, New York.

Graybill, Donald A., David A. Gregory, Fred L. Nials, Suzanne K. Fish, Robert E. Gasser, Charles H. Miksicek, and Christine R. Szuter. 1989. *The 1982–1984 Excavations at Las Colinas: Environment and Subsistence*. Archaeological Series 162, Vol. 5, Cultural Resource Management Division, Arizona State Museum, University of Arizona, Tucson.

Grebinger, Paul. 1976. 'Salado—Perspectives from the Middle Santa Cruz Valley', *The Kiva* **42**: 39–46.

Green, Roger C., Maryanne Atherton Danfelser, and Gwinn Vivian. 1958. 'Interpretation of Bg 91: A Specialized Largo-Gallina Surface Structure', *El Palacio* **65**: 41–60.

Green, William (editor). 1994. *Agricultural Origins and Development in the Midcontinent*. Report 19. Office of the State Archeologist, University of Iowa, Iowa City.

Greenblatt, Stephen (editor). 1993. *New World Encounters*. University of California Press, Los Angeles, Berkeley, and Oxford.

Greenleaf, J. Cameron and R. Gwinn Vivian. 1971. *Prehistoric Archaeological Tests on the Queen Creek Floodway Project*. Arizona State Museum Archaeological Series 6, Tucson.

Greenlee, Robert. 1933. *Archaeological Sites in the Chama Valley, and Report on Excavations at TSAMA, 1929–1933*. Laboratory of Anthropology, Santa Fe, New Mexico.

Greenman, E. F. 1957. 'The Garden Beds around Lake Michigan', *Michigan Archaeologist* **3**.2: 12–27.

Greenwald, David H. n.d. [1986]. 'Preliminary Results of Test Excavations and Data Recovery at the Dutch Canal Ruin, Phoenix, Arizona'. Unpublished paper presented at the 59th Pecos Conference at Shoofly Village Ruins, Payson, Arizona.

——and Richard Ciolek-Torrello. 1989. *Archaeological Investigations at the Dutch Canal Ruin, Phoenix, Arizona: Archaeology and History Along the Papago Freeway Corridor*. MNA Research Paper 38. Museum of Northern Arizona, Flagstaff.

Gregory, Clifford V. 1929. 'Farming Through the Ages: The First Farmers of America', *The Prairie Farmer* **77**: 5.

Gregory, David A. 1991. 'Form and Variation in Hohokam Settlement Patterns', in *Chaco and Hohokam: Prehistoric Regional Systems in the American Southwest*, edited by Patricia L. Crown and W. James Judge, pp. 159–93. School of American Research Press, Santa Fe.

——and Fred L. Nials. 1985. 'Observations Concerning the Distribution of Classic Period Hohokam Platform Mounds', in *Proceedings of the 1983 Hohokam Symposium*, Part I, edited by Alfred E. Dittert, Jr. and Donald E. Dove, pp. 373–88. Occasional Paper 2. Phoenix Chapter of the Arizona Archaeological Society, Phoenix.

Greiser, Sally T. and T. Weber Greiser. 1995. 'Prehistoric Irrigation in the Taos Valley', in *Soil, Water, Biology, and Belief in Prehistoric and Traditional Southwestern Agriculture*, edited by H. Wolcott Toll, pp. 221–37. New Mexico Archaeological Council, Special Publication 2, Albuquerque.

——and James L. Moore. 1995. 'The Case for Prehistoric Irrigation in the Northern Southwest', in *Soil, Water, Biology, and Belief in Prehistoric and Traditional Southwestern Agriculture*, edited by H. Wolcott Toll, pp. 189–95. New Mexico Archaeological Council, Special Publication 2, Albuquerque.

——T. Weber Greiser, and David Putnam. n.d. [1990]. 'Prehistoric Irrigation in the Taos Valley, New Mexico'. Unpublished paper delivered at the annual meeting of the Society for American Archaeology, Las Vegas, Nevada.

Gremillion, Kristen Johnson. 1989. 'The Development of a Mutualistic Relationship Between Humans and Maypops (*Passiflora incarnata* L.) in the Southeastern United States', *Journal of Ethnobiology* **9**: 135–55.

——1993. 'Crop and Weed in Prehistoric Eastern North America: The *Chenopodium* Example', *American Antiquity* **58**: 496–509.

——1994. 'Evidence of Plant Domestication from Kentucky Caves and Rockshelters', in *Agricultural Origins and Development in the Midcontinent*, edited by

William Green, pp. 87–104. Report 19. Office of the State Archeologist, University of Iowa, Iowa City.
——(editor). 1997*a*. *People, Plants, and Landscapes: Studies in Paleoethnobiology*. University of Alabama Press, Tuscaloosa.
——1997*b*. 'Introduction', in *People, Plants, and Landscapes: Studies in Paleoethnobiology*, edited by Kristen J. Gremillion, pp. 1–9. University of Alabama Press, Tuscaloosa.
——1997*c*. 'New Perspectives on the Paleoethnobiology of the Newt Kash Shelter', in *People, Plants, and Landscapes: Studies in Paleoethnobiology*, edited by Kristen J. Gremillion, pp. 23–41. University of Alabama Press, Tuscaloosa.
Grey, Zane. 1925. *Under the Tonto Rim*. Harper and Row, New York.
Griffen, William B. 1969. *Culture Change and Shifting Populations in Central Northern Mexico*. Anthropological Papers of the University of Arizona 13, University of Arizona Press, Tucson.
Griffin, Ernst. 1966. 'Some aspects of Alteration of Hydrology by the Construction of Trincheras in the Pompa Basin, Sierra Madre Occidental, Mexico'. Unpublished master's thesis, Department of Geography, University of Denver.
——and Howard W. Dennis. 1969. 'A Mexican Corporate Campaign in Conservation', *Professional Geographer* **21**: 358–9.
Griffin, James B. 1967. 'Eastern North American Archaeology: A Summary', *Science* **156**: 175–91.
Griffith, William Joyce. 1954. 'The Hasinai Indians of East Texas as Seen by Europeans, 1687–1772', in *Philological and Documentary Studies*, Vol. 2, No. 3, pp. 41–168. Middle American Research Institute, Tulane University, New Orleans.
Grigg, D. B. 1974. *The Agricultural Systems of the World: An Evolutionary Approach*. Cambridge University Press.
Grinnell, George Bird. 1923. *The Cheyenne Indians: Their History and Ways of Life*. Yale University Press, New Haven.
Gulick, Hubert. 1986. 'Landscape Ecological Aspects of Agro-Ecosystem', *Agriculture, Ecosystems and Environment* **16**: 79–86.
Gumerman, George J. and Emil W. Haury. 1979. 'Prehistory: Hohokam', in *Handbook of North American Indians*, Vol. 9, edited by Alfonso Ortiz, pp. 75–90. Smithsonian Institution, Washington, D. C.
——and R. Roy Johnson. 1971. 'Prehistoric Human Population Distribution in a Biological Transition Zone', in *The Distribution of Prehistoric Population Aggregates*, Proceedings of the Southwestern Anthropological Research Group, Anthropological Reports 1, edited by George J. Gumerman. Prescott College Press, Prescott, Arizona.
——and Thomas R. Lyons. 1971. 'Archeological Methodology and Remote Sensing', *Science* **172**: 126–32.

Haberman, Thomas J. 1984. 'Evidence for Aboriginal Tobaccos in Eastern North America', *American Antiquity* **49**: 269–78.
Hack, John T. 1942. *The Changing Physical Environment of the Hopi Indians of Arizona*, Papers of the Peabody Museum of American Archaeology and Ethnology, Vol. 35, No. 1. Harvard University, Cambridge, Massachusetts.
Hackbarth, Mark R. 1989. 'AZ T:4:38 (ASM)', in *Settlement, Subsistence, and Specialization in the Northern Periphery: The Waddell Project*, Vol. 1, edited by

Margerie Green, pp. 45–68. Cultural Resources Report 65. Archaeological Consulting Services, Tempe.

Hackenberg, Robert A. 1962. 'Economic Alternatives in Arid Lands: A Case Study of the Pima and Papago Indians', *Ethnology* **1**: 186–96.

——1983. 'Pima and Papago Ecological Adaptations', in *Handbook of North American Indians*, Vol. 10, edited by Alfonso Ortiz, pp. 161–77. Smithsonian Institution, Washington, D. C.

Hakimi, A. H. and R. P. Kachru. 1978. 'Silage Corn Responses to Different Mulch Tillage Treatments under Arid and Semiarid Climatic Conditions', *Journal of Agronomy and Crop Science* **147**: 15–23.

Hall, E. T., Jr. 1942. *Archaeological Survey of Walhalla Glades*. Museum of Northern Arizona Bulletin 20, Flagstaff.

Hall, Edward Hagaman. 1910. 'Henry Hudson and the Discovery of the Hudson River', in *Fifteenth Annual Report of the American Scenic and Historic Preservation Society*, pp. 227–346. J. B. Lyon Co., Albany.

Hall, Robert L. 1976. 'Ghosts, Water Barriers, Corn, and Sacred Enclosures in the Eastern Woodlands', *American Antiquity* **41**: 360–4.

Hall, Stephen A. 1985. 'Quaternary Pollen Analysis and Vegetational History of the Southwest', in *Pollen Records of Late-Quaternary North American Sediments*, edited by V. M. Bryant, Jr. and R. G. Holloway, pp. 95–123. American Association of Stratigraphic Palynologists Foundation, Dallas.

——1988. 'Prehistoric Vegetation and Environment at Chaco Canyon', *American Antiquity* **53**: 582–92.

Hallett, L. F. 1954. 'Cultural Traits of the Southern New England Indians', *Bulletin of the Massachusetts Archaeological Society* **15**: 59–64.

Hallowell, A. I. 1921. 'Indian Corn Hills', *American Anthropologist* **24**: 233.

Halseth, Odd S. 1932. 'Prehistoric Irrigation in Central Arizona', *The Masterkey* **5**: 164–75.

——1936. 'Prehistoric Irrigation in the Salt River Valley', in *Symposium on Prehistoric Agriculture*, University of New Mexico Bulletin, pp. 42–7. Anthropological Series, Vol. 1, No. 5, Albuquerque.

Hammett, Julia E. 1997. 'Interregional Pattern of Land Use and Plant Management in Native North America', in *People, Plants, and Landscapes: Studies in Paleoethnobotany*, edited by Kristen J. Gremillion, pp. 195–216. University of Alabama Press, Tuscaloosa.

Hammond, George P. and Agapito Rey. (translators and editors). 1929. *Expedition into New Mexico Made by Antonio de Espejo 1582–1583, as Revealed in the Journal of Diego Perez de Luxan, a Member of the Party*. The Quivira Society, Los Angeles.

————1940. *Narratives of the Coronado Expedition 1540–1542*. University of New Mexico Press, Albuquerque.

————1953. *Don Juan de Oñate: Colonizer of New Mexico*, Vols 1 and 2. University of New Mexico Press, Albuquerque.

Hanks, Lucien M. 1972. *Rice and Man: Agricultural Ecology in Southeast Asia*. Aldine-Atherton, Chicago and New York.

Hann, John H. 1988. *Apalachee: The Land between the Rivers*. University of Florida Press and the Florida State Museum, Gainesville.

Hard, Robert J. and John R. Roney. 1998. 'A Massive Terraced Village Complex in Chihuahua, Mexico, 3000 Years Before Present', *Science* **279**: 1661–4.

Hargrave, Michael L., Geral A. Oetelaar, Neal H. Lopinot, Brian M. Butler, and Deborah A. Billings. 1983. *The Bridges Site (11-Mr-11): A Late Prehistoric Settlement in the Central Kaskaskia Valley*. Center for Archaeological Investigations Research Paper 38. Southern Illinois University, Carbondale.

Harlan, Jack R. 1992. *Crops and Man*. American Society of Agronomy and Crop Science Society of America, Madison.

——J. M. J. de Wet, and Ann Stemler. 1976. 'Plant Domestication and Indigenous African Agriculture', in *Origins of African Plant Domestication*, edited by Jack R. Harlan, Jan M. J. de Wet, and Ann B. L. Stemler, pp. 3–19. Mouton, The Hague.

Harpe, Bérnard de la. 1958–1959 [1719]. 'Account of the Journey of Bérnard de la Harpe: Discovery Made by Him of Several Nations Situated in the West', translated and annotated by Ralph A. Smith, *Southwestern Historical Quarterly* **62**: 75–86, 246–59, 371–85, 525–41.

Harper, Kimball T. 1986. 'Historical Environments', in *Handbook of North American Indians*, Vol. 11, edited by Warren L. D'Azevedo, pp. 51–63. Smithsonian Institution, Washington, D. C.

Harrington, John P. 1932. *Karuk Indian Myths*. Smithsonian Institution, Bureau of American Ethnology, Bulletin 107, Washington, D. C.

Harris, David R. 1971. 'The Ecology of Swidden Cultivation in the Upper Orinoco Rain Forest, Venezuela', *Geographical Review* **61**: 475–95.

——1977. 'Alternative Pathways Toward Agriculture', in *World Anthropology*, edited by Sol Tax, pp. 179–243. Mouton, The Hague.

——and Gordon C. Hillman (editors). 1989. *Foraging and Farming: The Evolution of Plant Exploitation*. Unwin Hyman, London.

Harris, George H. 1812 [early 1600s]. 'The Discoveries and Settlements Made by the English in Different Parts of America', in *A General Collection of the Best and Most Interesting Voyages and Travels in All Parts of the World*, edited by John Pinkerton, pp. 156–489. Longman, Hurst, Rees, Orme, and Brown, London.

Harrison, Peter D. and B. L. Turner II (editors). 1978. *Pre-Hispanic Maya Agriculture*. University of New Mexico Press, Albuquerque.

Hart, John P. 1990. 'Modeling Oneota Agricultural Production: A Cross-cultural Evaluation', *Current Anthropology* **31**: 569–77.

——and Nancy Asch Sidell. 1997. 'Additional Evidence for Early Cucurbit Use in the Northern Eastern Woodlands of the Allegheny Front', *American Antiquity* **62**: 523–37.

Harwell, Henry O. and Marsha C. S. Kelly. 1983. 'Mohave', in *Handbook of North American Indians*, Vol. 10, edited by Alfonso Ortiz, pp. 71–85. Smithsonian Institution, Washington, D. C.

Harwood, Richard R. 1979. *Small Farm Development: Understanding and Improving Farming Systems in the Humid Tropics*. Westview Press, Colorado.

Hastings, James R. (editor). 1964. *Climatological Data for Sonora and Northern Sinaloa*. University of Arizona Institute of Atmospheric Physics Technical Reports on the Meteorology and Climatology of Arid Regions 15, Tucson.

——and Raymond M. Turner. 1965. *The Changing Mile: An Ecological Study of Vegetation Change with Time in the Lower Mile of an Arid and Semiarid Region*. University of Arizona Press, Tucson.

Haury, Emil. 1936. 'The Snaketown Canal', in *Symposium on Prehistoric Agriculture*,

The University of New Mexico Bulletin, pp. 48–50. Anthropological Series, Vol. 1, No. 5, Albuquerque.

Haury, Emil. 1937. 'The Snaketown Canal', in *Excavations at Snaketown: Material Culture*, edited by Harold S. Gladwin, Emil W. Haury, E. B. Sayles, and Nora Gladwin, pp. 50–8. Medallion Papers 25, Gila Pueblo, Globe, Arizona.

——1945*a*. 'Arizona's Ancient Irrigation Builders', *Natural History* **54**: 300–10, 335.

——1945*b*. *The Excavation of Los Muertos and Neighboring Ruins in the Salt River Valley, Southern Arizona*. Papers of the Peabody Museum of American Archaeology and Ethnobiology, Vol. 24, No. 1. Harvard University, Cambridge, Massachusetts.

——1976. *The Hohokam: Desert Farmers and Craftsmen, Excavations at Snaketown, 1964–1965*. University of Arizona Press, Tucson.

Havard, V. 1895. 'Food Plants of the North American Indians', *Bulletin of the Torrey Botanical Club* **22**: 98–123.

——1896. 'Drink Plants of the North American Indians', *Bulletin of the Torrey Botanical Club* **23**: 33–46.

Haviland, William A. and Majory W. Power. 1981. *The Original Vermonters: Native Inhabitants, Past and Present*. University Press of New England, Hanover New Hampshire.

Hawkins, Benjamin. 1848. 'A Sketch of the Creek Country, in the Years 1798 and 1799', *Collections of the Georgia Historical Society* III.1. Savanna.

Hayes, Alden C. 1964. *The Archeological Survey of Wetherill Mesa, Mesa Verde National Park—Colorado*. National Park Service, Archeological Research Series Number Seven-A. US Department of the Interior, Washington, D. C.

Hedrick, Ulysses Prentiss. 1933. *A History of Agriculture in the State of New York*. New York State Agricultural Society, Albany.

Hegmon, Michelle. 1989. 'Risk Reduction and Variation in Agricultural Economies: A Computer Simulation of Hopi Agriculture', *Research in Economic Anthropology* **11**: 89–121.

Heidenreich, Conrad. 1971. *Huronia: A History and Geography of the Huron Indians, 1600–1650*. McClelland and Steward, Toronto.

——1974. 'A Relict Indian Corn Field Near Creemore, Ontario', *Canadian Geographer* **18**: 379–94.

——1978. 'Huron', in *Handbook of North American Indians*, Vol. 15, edited by William C. Sturtevant, pp. 368–88. Smithsonian Institution, Washington, D. C.

Heiser, Charles B., Jr. 1951. 'The Sunflower among the North American Indians', *Proceedings of the American Philosophical Society* **95**: 432–48.

——1955. 'The Origin and Development of the Cultivated Sunflower', *The American Biology Teacher* **17**: 161–7.

——1985. 'Some Botanical Considerations of Early Domesticated Plants North of Mexico', in *Prehistoric Food Production in North America*, edited by Richard I. Ford, pp. 57–72. Anthropological Papers 85. University of Michigan, Ann Arbor.

——1989. 'Domestication of Cucurbitaceae: *Cucurbita and Lagenaría*', in *Foraging and Farming: The Evolution of Plant Exploitation*, edited by David R. Harris and Gordon C. Hillman, pp. 471–80. Unwin Hyman, London.

Heizer, Robert F. 1958. 'Prehistoric Central California: A Problem in Historical Development Classification', *University of California Archaeological Survey Report* **41**: 19–26.

——1978. 'Natural Forces and Native World View', in *Handbook of North American Indians*, Vol. 8, edited by Robert F. Heizer, pp. 649–53. Smithsonian Institution, Washington, D. C.

Henderson, T. Kathleen. 1979. 'Archaeological Survey at Chavez Pass Ruin, Coconino National Forest, Arizona: The 1978 Field Season'. Unpublished report on file USDA Forest Service, Flagstaff.

——(editor). 1989. *Prehistoric Agricultural Activities of the Lehi-Mesa Terrace: Excavations at La Cuerca del Sedimento*. Northland Research Inc., Flagstaff.

——and James B. Rodgers. 1979. *Archaeological Investigations in the Cave Creek Area, Maricopa County, South-central Arizona*. Anthropological Research Papers 17. Arizona State University, Tempe.

Hennepin, Louis. 1880 [1683]. *A Description of Louisiana*, translated by John Gilmary Shea. John G. Shea, New York.

——1903 [1683]. 'Narrative of the Voyage to the Upper Mississippi', in *Discovery and Exploration of the Mississippi Valley*, edited by John Gilmary Shea, pp. 111–50. Joseph McDonough, Albany.

Henry, Alexander and David Thompson. 1897 [1799–1814]. *The Manuscript Journals of Alexander Henry and David Thompson*, edited by Elliott Coues, 3 vols. Francis P. Harper, New York.

Herhahn, Cynthia L. 1995. '14th Century Dry Farming Features in the Northern Rio Grande Valley, New Mexico', in *Soil, Water, Biology, and Belief in Prehistoric and Traditional Southwestern Agriculture*, edited by H. Wolcott Toll, pp. 77–84. New Mexico Archaeological Council, Special Publication 2, Albuquerque.

Heriot, Thomas. 1812 [1587]. 'A Brief and true Report of the found new Land of Virginia, of the Commodities found there, and to be raised, as well merchantable as others', in *A General Collection of the Best and Most Interesting Travels in all Parts of the World*, edited by John Pinkerton, Vol. 12, pp. 590–609. Longman, Hurst, Rees, Orme, and Brown, Paternoster-Row; and Cadell and Davies, London.

Herlihy, Peter H. n.d. [1984]. 'The Maya Dooryard Orchard-Garden: An Ancient and Contemporary Agricultural System'. Unpublished paper presented at the Conference of Latin Americanist Geographers, Ottawa, Ontario, Canada.

Herndon, G. Melvin. 1967. 'Indian Agriculture in the Southern Colonies', *North Carolina Historical Review* **44**: 283–98.

Herold, Joyce. 1961. 'Prehistoric Settlement and Physical Environment in the Mesa Verde Area', *Anthropological Papers*, 53. University of Utah Press, Salt Lake City.

Herold, Laurance C. 1965*a*. 'A Progress Report on Trincheras Study in Chihuahua', *Rocky Mountain Social Science Journal* **2**: 80–4.

——1965*b*. *Trincheras and Physical Environment along the Rio Gavilan, Chihuahua, Mexico*, Publications in Geography 65–1. Department of Geography, University of Denver.

——1968. 'An Archaeological-Geographical Survey of the Rio Grande de Ranchos', in *Papers on Taos Archaeology*, edited by Laurance C. Herold and Ralph A. Luebben, pp. 9–42. Fort Burgwin Research Center 7, Taos, New Mexico.

——n.d. [1980]. 'A Terracing System Associated with a Medio Period Site in the Casas Grandes Area'. Paper presented at the Anthropology Program, Western Social Sciences Association, Albuquerque.

——and Reuben F. Miller. 1993. 'Soil Moisture Conditions in Agricultural Terraces Associated with a Highland Variant of the Casas Grandes Culture, Chihuahua,

Mexico', in *Proceedings of the First International Conference on Pedo-Archaeology*, edited by J. E. Foss, M. E. Timpson, and M. W. Morris, pp. 185–90. Soil Science Society of America Special Publications, 93-03.

Herold, Laurence C. and Reuben F. Miller. 1994. 'Soil Moisture Environments of Pre-Columbian Agricultural Terraces and Settlement, Rio Gavilan, Chihuahua, Mexico', in *Proceedings of the Second International Conference of Pedo-Archaeology*, edited by Albert C. Goodyear, John E. Foss, and Kenneth E. Sassaman, pp. 97–102. Anthropological Studies 10, Occasional Papers of the South Carolina Institute of Archaeology and Anthropology, University of South Carolina.

———— 1995. 'Water Availability for Plant Growth in Precolumbian Terrace Soils, Chihuahua, Mexico', in *Soil, Water, Biology, and Belief in Prehistoric and Traditional Southwestern Agriculture*, edited by H. Wolcott Toll, pp. 145–53. New Mexico Archeological Council, Special Publication 2, Albuquerque.

Herrington, Selma LaVerne Coldt. 1979. 'Settlement Patterns and Water Control Systems of the Mimbres Classic Phase, Grant County, New Mexico'. Unpublished doctoral dissertation, Department of Anthropology, University of Texas, Austin.

—— 1982. 'Water-Control Systems of the Mimbres Classic Phase', in *Mogollon Archaeology: Proceedings of the 1980 Megollon Conference*, edited by Patrick H. Beckett and Kira Silverbird, pp. 75–90. Acoma Books, Ramon, California.

—— and Darrell Creel. 1991. 'Treasure Hill: An Agricultural Center and Type Site Revisited', in *Mogollon V*, edited by Patrick H. Beckett, pp. 50–61. COAS Publishing and Research, Las Cruces.

Herskovitz, Robert M. 1981. 'Arizona U:9:46: A Dual Component Hohokam Site in Tempe, Arizona', *The Kiva* **47**: 1–90.

Hevly, Richard H., Roger E. Kelly, Glenn A. Anderson, and Stanley J. Olsen. 1979. 'Comparative Effects of Climate Change, Cultural Input, and Volcanos in the Paleoecology of Flagstaff, Arizona, A.D. 900–1300', in *Volcanic Activity and Human Ecology*, edited by Payson D. Sheets and Donald K. Grayson, pp. 487–523. Academic Press, New York.

Hewitt, Edgar L. 1906. *Antiquities of the Jemez Plateau, New Mexico*. Smithsonian Institution Bureau of American Ethnology, Bulletin 32, Washington, D. C.

Hibbard, Benjamin Horace. 1904. 'Indian Agriculture in Southern Wisconsin', *Proceedings of the State Historical Society of Wisconsin*, pp. 145–55.

Hibben, Frank C. 1937. *Excavation of the Riana Ruin and Chama Valley Survey*. Anthropological Series, 2. University of New Mexico Press, Albuquerque.

—— 1940. 'The Gallina Culture of North Central New Mexico'. Unpublished doctoral dissertation, Department of Anthropology, Harvard University, Cambridge, Massachusetts.

Higgeson, Francis H. 1929 [1630]. 'New-Englands Plantation, or A Short and Trve Description of the Commodities and Discommodities of that Country', *Massachusetts Historical Society*, pp. 305–18.

Hill, W. W. 1938. *The Agricultural and Hunting Methods of the Navajo Indians*. Yale University Publications in Anthropology 18. Yale University Press, New Haven.

Hinsdale, Wilbert B. 1928. 'Indian Corn Culture in Michigan', in *Papers of the Michigan Academy of Science, Arts and Letters*, edited by Eugene S. McCartney and Peter Okkelberg, pp. 31–49. Macmillan, New York.

Hinsley, Curtis and David R. Wilcox (editors). 1995. 'A Hemenway Portfolio: Voices and Views from the Hemenway Archaeological Expedition, 1886–1889', *Journal of the Southwest* **37**.4.

Hinton, Thomas B. 1983. 'Southern Periphery: West', in *Handbook of North American Indians* Vol. 10, edited by Alfonso Ortiz, pp. 315–28. Smithsonian Institution, Washington, D. C.

Hodge, F. W. 1893. 'Prehistoric Irrigation in Arizona', *The American Anthropologist* **6**: 323–30.

——1907. 'Irrigation', in *Handbook of American Indians North of Mexico*, Part 1, edited by Frederick Webb Hodge, pp. 620–1, Smithsonian Institution, Bureau of American Ethnology, Bulletin 30, Washington, D. C.

——1959. 'Women', in *Handbook of American Indians North of Mexico*, Part 2, edited by Frederick Webb Hodge, pp. 968–73. Pageant Books, New York.

Hodgson, Wendy, Gary Nabhan, and Liz Ecker. 1989. 'Conserving Rediscovered Agave Cultivars', *Agave: Quarterly Magazine of the Desert Botanical Garden* **3**: 9–11.

Hoebel, E. Adamson. 1979. 'Zia Pueblo', in *Handbook of North American Indians*, Vol. 9, edited by Alfonso Ortiz, pp. 407–17. Smithsonian Institution, Washington, D. C.

Hogan, Patrick F. 1987. 'Prehistoric Agricultural Strategies in West-Central New Mexico'. Unpublished doctoral dissertation, Department of Anthropology, Washington State University, Pullman.

Holden, W. C., C. C. Seltzer, R. A. Studhalter, C. J. Wagner, and W. G. McMillan. 1936. *Studies of the Yaqui Indians of Sonora, Mexico*. Texas Technological College Bulletin 7. Texas Technological College, Lubbock.

Holder, Preston. 1991. *The Hoe and the Horse on the Plains: A Study of Cultural Development among North American Indians*. University of Nebraska Press, Lincoln.

Holiday, William G. 1974. *Archeological Investigations in the Cave Creek Drainage, Tonto National Forest, Arizona*. Archaeological Report 1. USDA Forest Service, Southwestern Region, Albuquerque, New Mexico.

Holley, George Raymond. 1978. 'Mississippian Agriculture at Kincaid: An Economic Approach'. Unpublished master's thesis, Department of Anthropology, Southern Illinois University, Carbondale.

Holmes, G. K. 1912. 'Aboriginal Agriculture—The American Indians', in *Cyclopedia of American Agriculture*, edited by L. H. Bailey, pp. 24–39. The Macmillan Company, New York.

Homburg, Jeffrey A. 1997. 'Prehistoric Dryland Agricultural Fields of the Lower Verde', in *Vanishing River: Landscapes and Lives of the Lower Verde Valley, The Lower Verde Archaeological Project*, Vol. 2, edited by Jeffrey A. Homburg and Richard Ciolek-Torrello, pp. 103–26. SRI Press, Tucson, Arizona.

——and Richard Ciolek-Torrello. 1997. 'The Crash Landing Complex: Agricultural Sites and Field Houses', in *Vanishing River: Landscapes and Lives of the Lower Verde Valley, The Lower Verde Archaeological Project*, Vol. 2, edited by Jeffrey A. Homburg and Richard Ciolek-Torrello, pp. 57–84. SRI Press, Tucson, Arizona.

————and Steven D. Shelley. n.d. [1993]. 'Dry Farming in the Sonoran Uplands: A Methodological Study'. Unpublished paper presented at the 58th Annual meeting of the Society for American Archaeology, St. Louis, Missouri.

Horst, Oscar H. 1989. 'The Persistence of *Milpa* Agriculture in Highland Guatemala', *Journal of Cultural Geography* **9**: 13–29.

Horwitz, Elinor Lander. 1978. *Our Nation's Wetlands: An Interagency Task Force Report*. Council on Environmental Quality, Washington, D. C.

Hough, Walter. 1907. *Antiquities of the Upper Gila and Salt River Valleys in Arizona and New Mexico*. Smithsonian Institution, Bureau of American Ethnology, Bulletin 35, Washington, D. C.

Howard, Jerry B. 1993. 'A Paleohydraulic Approach to Examing Agricultural Intensification in Hohokam Irrigation Systems', in *Economic Aspects of Water Management in the Prehispanic New World*, edited by Vernon L. Scarborough and Barry L. Issac, pp. 263–324. Research in Economic Anthropology, Supplement 7. JAI Press, Greenwich, Connecticut.

—— n.d. [1995]. 'Agricultural Stategies in Hohokam Irrigation', *Abstracts of the 60th Annual Meeting*, p. 98. Society for American Archaeology, Washington, D. C.

—— and Gary Huckleberry. 1991. *The Operation and Evolution of an Irrigation System: The East Papago Canal Study*. Soil Systems Publications in Archaeology 18. Soil Systems Inc., Phoenix, Arizona.

Howard, William A. and Thomas M. Griffiths. 1966. *Trinchera Distribution in the Sierra Madre Occidental, Mexico*. Publications in Geography Technical Paper 66-1. Department of Geography, University of Denver.

Howarth, O. H. 1895. 'The Western Sierra Madre of Mexico', *Geographical Journal* **6**: 422–35.

Howe, Henry. 1898. *Ohio in Two Volumes: An Encyclopedia of the State*. The Laning Printing Co., Norwalk, Ohio.

Howes, W. J. 1946. 'Artifacts found on the Judd Tract in South Hadley, Mass.', *Massachusetts Archaeological Society Bulletin* **74**: 68–70.

Hubbard, Bela. 1878. 'Ancient Garden Beds of Michigan', *The American Antiquarian* **1**: 1–9.

Hubbard, Richard A. 1987. 'Prehistoric Agricultural Fields and Water Control Systems of South-Central New Mexico'. Unpublished master's thesis, Department of Anthropology, University of Texas at Austin.

—— 1992. *An Archaeological Reconnaissance of the Hudspeth County Conservation and Reclamation District #1 Proposed Regulations Reservoir #3, Hudspeth Co., Texas*. Texas Water Development Board, Austin.

Huckleberry, Gary. 1992. 'Soil Evidence of Hohokam Irrigation in the Salt River Valley, Arizona', *Kiva* **57**: 237–49.

Hudson, Charles. 1976. *The Southeastern Indians*. The University of Tennessee Press, Knoxville.

—— 1979. 'Introduction', in *Black Drink: A Native American Tea*, edited by Charles M. Hudson, pp. 1–9. University of Georgia Press, Athens.

—— 1990. *The Juan Pardo Expeditions: Exploration of the Carolinas and Tennessee, 1566–1568*. Smithsonian Institution, Washington, D. C.

—— Marvin Smith, David Hally, Richard Polhemus, and Chester DePratter. 1985. 'Coosa: A Chiefdom in the Sixteenth-Century Southeastern United States', *American Antiquity* **50**: 723–37.

Hughes, J. Donald. 1983. *American Indian Ecology*. Texas Western Press, El Paso.

Hunt, George T. 1940. *The Wars of the Iroquois: A Study in Intertribal Trade Relations*. University of Wisconsin Press, Madison.

Hunt, Robert C. 1988. 'Size and the Structure of Authority in Canal Irrigation Systems', *Journal of Anthropological Research* **44**: 335–55.

Hunter, Andrew F. 1899. 'Notes on Sites of Huron Villages in the Township of Tiny (Simcoe County)', *Archaeological Report, 1898, Being Part of the Appendix to the Report of the Minister of Education, Ontario*, pp. 5–42. Warwick Bro's. and Rutter, Toronto.

——1900. 'Notes on Sites of Huron Villages in the Township of Tay (Simcoe County)', *Archaeological Report, 1899, Being Part of the Appendix to the Report of the Minister of Education, Ontario*, pp. 51–82. Warwick Bro's. and Rutter, Toronto.

——1901. 'Notes on Sites of Huron Villages in the Township of Medonte (Simcoe County)', *Archaeological Report, 1900, Being Part of the Appendix to the Report of the Minister of Education, Ontario*, pp. 56–100. Warwick Bro's. and Rutter, Toronto.

Hunter-Anderson, Rosalind L. 1986. *Prehistoric Adaptation in the American Southwest*. Cambridge University Press.

Huntington, Ellworth. 1913. 'The Fluctuating Climate of North America'. *Annual Report of the Smithsonian Institution, 1912*, pp. 383–412. Washington, D. C.

——1914. *The Climatic Factor as Illustrated in Arid America*. Carnegie Institution, Washington, D. C.

Hurt, R. Douglas. 1985. 'The First Farmers in the Ohio Country', *Agriculture and Human Values*, **2.**3: 5–13.

——1987. *Indian Agriculture in America: Prehistory to the Present*. University Press of Kansas, Lawrence.

Hutchins, Wells A. 1928. 'The Community Acequia: Its Origin and Development', *Southwestern Historical Quarterly* **31**: 261–84.

Innis, Donald Q. 1980. 'The Future of Traditional Agriculture', *Focus* **30.**3: 1–8.

Irwin-Williams, C., Amy Dansie, Jonathan O. Davis, and Roger Jacobson. n.d. [1986]. 'Prehistoric Water Harvesting Techniques in the Northern Great Basin'. Unpublished paper presented at the joint meeting of the Association for Arid Land Studies and the Western Social Science Association, Reno, Nevada.

Iversen, Johannes. 1956. 'Forest Clearance in the Stone Age', *Scientific American* **194**: 36–41.

Iverson, Peter. 1992. 'Taking Care of the Earth and Sky', in *America in 1492: The World of the Indian Peoples Before the Arrival of Columbus*, edited by Alvin M. Josephy, Jr., pp. 85–117. Alfred A. Knopf, New York.

Ives, Ronald L. 1949. 'Climate of the Sonoran Desert Region', *Annals of the Association of American Geographers* **39**: 143–87.

Jacob, John Soren. 1992. 'The Agroecological Evolution of Cobweb Swamp, Belize'. Unpublished doctoral dissertation, Department of Soil Science, Texas A&M University, College Station.

James, Gregory and Elizabeth Benchley. 1980. *Final Report to the State Historical Society of Wisconsin for Archaeological Investigations in Milwaukee County, Wisconsin 1979–1980*. Report of Investigations 44. Archaeological Research Laboratory, University of Wisconsin–Milwaukee.

Janick, Jules, Robert W. Schery, Frank W. Woods, and Vernon W. Ruttan. 1974. *Plant Science: An Introduction to World Crops*. W. H. Freeman and Co., San Francisco.

Jaramillo, Juan. 1940 [1541]. 'Jaramillo's Narrative', in *Narratives of the Coronado Expedition 1540–1542*, translated and edited by George P. Hammond and Agapito Rey, pp. 295–307. University of New Mexico Press, Albuquerque.

Jeancon, J. A. 1923. *Excavations in the Chama Valley, New Mexico*, Smithsonian Institution, Bureau of American Ethnology, Bulletin 81, Washington, D. C.

——1929. *Archeological Investigations in the Taos Valley, New Mexico, During 1920*, Smithsonian Miscellaneous Collections, Vol. 81, No. 12. Smithsonian Institution, Washington, D. C.

Jenks, Albert Ernest. 1900. 'The Wild Rice Gatherers of the Upper Lakes: A Study in American Primitive Economics', in *Nineteenth Annual Report of the Bureau of American Ethnology*, Part 2, pp. 1013–79. Smithsonian Institution, Washington D. C.

Jenness, Diamond. 1931. 'Wild Rice', *Canadian Geographical Journal* **11**: 477–82.

——1955. *The Indians of Canada*. Anthropological Series 15. The National Museum of Canada, Bulletin 65, Ottawa.

Jennings, Francis. 1975. *The Invasion of America: Indians, Colonialism, and the Cant of Conquest*. University of North Carolina Press, Chapel Hill.

Jennings, Jesse D. 1966. *Glen Canyon: A Summary*. University of Utah, Anthropological Papers 81. Salt Lake City.

——1986. 'Prehistory: Introduction', in *Handbook of North American Indians*, Vol. 11, edited by Warren L. D'Azevedo, pp. 113–19. Smithsonian Institution, Washington, D. C.

Jenny, Hans. 1958. 'Role of the Plant Factor in the Pedogenic Futures', *Ecology* **39**: 5–16.

Johannessen, Sissel, and Christine A. Hasdorf (editors). 1994. *Corn and Culture in the Prehistoric New World*. Westview Press, Boulder, Colorado.

John, Elizabeth A. H. (editor), John Wheat (translator). 1991. 'Views from a Deck in Chihuahua: Manuel Merino's Report on Apaches and Neighboring Nations, ca. 1804', *Southwestern Historical Quarterly* **95**: 139–75.

Johnson, Edward. 1910 [1654]. *Johnson's Wonder-Working Providence: 1628–1651*, edited by J. Franklin Jameson. Charles Scribner's Sons, New York.

Johnson, Jean B. 1950. *The Opata: An Inland Tribe of Sonora*. University of New Mexico Pulblications in Anthropology 6. University of New Mexico Press, Albuquerque.

Johnson, William Gray and Mary E. Collins. 1992. 'Can Soil Evidence Prove that Prehistoric Maize was not the Basis for Complex Cultural Developments in Okeechobee Basin?', in *Proceedings of the First International Conference on Pedo-Archaeology*, edited by J. E. Foss, M. E. Timpson, and M. W. Morris, pp. 157–60. Soil Science Society of America Special Publications, 93-03.

Johnston, A. 1969. 'Man's Utilization of the Flora of the Northwest Plains,' in *Post-Pleistocene Man and His Environment on the Northern Plains*, edited by R. G. Forbis, L. B. Davis, O. A. Christensen, and G. Fedirchuk, pp. 109–76. The University of Calgary Archaeological Association, Calgary, Alberta.

Jones, Charles C. 1873. *Antiquities of the Southern Indians*. D. Appleton, New York.

Jones, Robert Leslie. 1983. *History of Agriculture in Ohio to 1880*. Kent State University Press, Kent, Ohio.

Jones, Volney H. 1930. Ethnobotany of Isleta. Unpublished master's thesis Department of Anthropology, University of New Mexico.

————1936*a*. 'A Summary of Data on Aboriginal Cotton in the Southwest', in *Symposium on Prehistoric Agriculture*. University of New Mexico Bulletin, pp. 51–64. Anthropological Series Vol. 1, No. 5, Albuquerque.

——1936*b*. 'The Vegetal Remains of Newt Kash Hollow Shelter', in *Rockshelters in Menifee County, Kentucky*, edited by William S. Webb and William D. Funkhouser, pp. 147–65. University of Kentucky Reports in Anthropology and Archaeology **3**.4, Lexington.

——and Elizabeth Ann Morris. 1960. 'A Seventh-Century Record of Tobacco Utilization in Arizona', *El Palacio* **67**: 115–17.

Joosten, J. H. L. 1962. *Wirtschaftliche und Agrarpolitische Aspekte Tropischer Landbausysteme*. Vortrag, Göttingen.

Jordan, Terry G. 1973. 'Pioneer Evaluation of Vegetation in Frontier Texas', *Southwestern Historical Quarterly* **76**: 233–54.

——and Matti Kaups. 1989. *The American Backwoods Frontier: An Ethnic and Ecological Interpretation*. Johns Hopkins University Press, Baltimore.

Jorgensen, Joseph G. 1983. 'Comparative Traditional Economics and Ecological Adaptations', in *Handbook of North American Indians*, Vol. 10, edited by Alfonso Ortiz, pp. 684–710. Smithsonian Institution, Washington, D. C.

Joseph, Alice, Rosamond B. Spicer, and Jane Chesky. 1949. *The Desert People: A Study of the Papago Indians*. University of Chicago Press, Chicago.

Josephy, Alvin M., Jr. (editor). 1993. *American in 1492: The World of the Indian Peoples Before the Arrival of Columbus*. Alfred A. Knopf, New York.

Josselyn, John. 1833 [1674]. 'An Account of Two Voyages to New England', *Collections of the Massachusetts Historical Society*, Vol. III, pp. 211–99. E. W. Metcalf, Cambridge, Massachusetts.

——1972 [1672]. *New-Englands Rarities Discovered*. A Massachusetts Historical Society Picture Book, Boston.

Joutel, Henri. 1906 [1687]. *Joutel's Journal of LaSalle's Last Voyage: 1684–1687*. Joseph McDonough, Albany, New York.

Judd, Neil M. 1930. 'Arizona Sacrifices Her Prehistoric Canals', in *Explorations and Field-Work of the Smithsonian Institution in 1929*, pp. 177–82. Washington, D. C.

——1931. 'Arizona's Prehistoric Canals, from the Air', in *Explorations and Field-Work of the Smithsonian Institution in 1930*, pp. 157–66. Washington, D. C.

Juet, Robert. 1909 [1610]. 'The Third Voyage of Master Henry Hudson', in *Narratives of New Netherland, 1609–1664*, edited by J. Franklin Jameson, pp. 11–28. Charles Scribner's Sons, New York.

Jurney, David H., Jr. 1981. 'The Source and Distribution of Specialized Stone Tools in the Ozarks', *Mid-Continental Journal of Archaeology* **6**: 17–31.

Kaldy, M. S., A. Johnston, and D. B. Wilson. 1980. 'Nutritive Value of Indian Breadroot, Squaw-root, and Jerusalem Artichoke', *Economic Botany* **34**: 352–7.

Kalm, Pehr. 1812 [1775]. 'Travels Into North America: Containing its Natural History, and a Circumstantial Account of its Plantations and Agriculture in General', in *A General Collection of the Best and Most Interesting Travels in all Parts of the World*, by John Pinkerton, Vol. 13, pp. 374–705. Longman, Hurst, Rees, Orme, and Brown, Paternoster-Row; and Cadell and Davies, London.

——1935 [1751–1752]. 'Pehr Kalm's Description of Maize, How it is Planted and Cultivated in North America, Together with the Many Uses of this Crop Plant', *Agricultural History* **9**: 98–117.

——1972. *Travels into North America*, translated by John Reinhold Forster. The Imprint Society, Barre, Massachusetts.

Kayser, David W. 1973. 'Castle Rock Project: Archaeological Salvage Along State Highway 32 in Agua Fria and Largo Creek Canyons, Catron County, New Mexico', *Laboratory of Anthropology Note* 71. Santa Fe, New Mexico.

——1974. 'A Prehistoric Water System', *El Palacio* **79.**3: 30–1.

Kearnes, Timothy M., Michael R. Polk, and Lynn S. Teague. 1973. 'Archaeological Resources, Orme Reservoir, Phase II, Verde Arm', in *An Archaeological Survey of the Orme Reservoir*, assembled by Veletta Canouts and Mark Grady, pp. 249–258. Archaeological Series 92. Arizona State Museum, University of Arizona, Tucson.

Keegan, William F. 1986. 'The Optimal Foraging Analysis of Horticultural Production', *American Anthropologist* **88**: 92–107.

——(editor). 1987*a*. *Emergent Horticultural Economies of the Eastern Woodlands*. Occasional Paper 7. Center for Archaeological Investigations, Southern Illinois University at Carbondale.

——1987*b*. 'Diffusion of Maize from South America: The Antilles Connection Reconstructed', in *Emergent Horticultural Economies of the Eastern Woodlands*, edited by William F. Keegan, pp. 329–40. Occasional Paper 7. Center for Archaeological Investigations, Southern Illinois University at Carbondale.

——and Brian M. Butler. 1987. 'The Microeconomic Logic of Horticultural Intensification in the Eastern Woodlands', in *Emergent Horticultural Economies of the Eastern Woodlands*, edited by William F. Keegan, pp. 109–28. Occasional Paper 7. Center for Archaeological Investigations, Southern Illinois University at Carbondale.

Kelley, J. Charles. 1952. 'Factors Involved in the Abandonment of Certain Peripheral Southwestern Settlements', *American Antiquity* **54**: 356–87.

——1956. 'Settlement Patterns in North-Central Mexico', in *Prehistoric Settlement Patterns in the New World*, edited by Gordon R. Willey, pp. 128–39. Viking Fund Publications in Anthropology 23. Wenner-Gren Foundation for Anthropological Research, New York.

Kelly, A. R. 1935. 'Exploring Prehistoric Georgia', *Scientific American* **152**: 117–20, 184–7, 244–6.

——1938. *A Preliminary Report on Archeological Explorations at Macon, Ga.* Smithsonian Institution, Bureau of American Ethnology, Bulletin 119, Anthropological Papers 1, Washington, D. C.

——1965. 'Notes on a Prehistoric Cultivated Field in Macon, Georgia', in *Proceedings of the Twenty-First Southeastern Archaeological Conference*, Bulletin 3, pp. 49–51. Southeastern Archaeological Conference, Cambridge, Massachusetts.

Kelly, Isabel T. and Catherine S. Fowler. 1986. 'Southern Paiute', in *Handbook of North American Indians*, Vol. 11, edited by Warren L. D'Azevedo, pp. 368–97. Smithsonian Institution, Washington, D. C.

Kelly, William H. 1977. *Cocopa Ethnography*. Anthropological Papers of the University of Arizona 29. University of Arizona Press, Tucson.

Kempton, James H. 1936. 'Maize as a Measure of Indian Skill', in *Symposium on Prehistoric Agriculture*, University of New Mexico Bulletin, pp. 19–28. Anthropological Series, Vol. 1, No. 5, Albuquerque.

Kemrer, Sandra, Sandra Schultz, and William Dodge. 1972. *An Archaeological Survey of the Granite-Reef Aqueduct*. Arizona State Museum, University of Arizona, Tucson.

Kennard, Edward A. 1979. 'Hopi Economy and Subsistence', in *Handbook of North*

American Indians, Vol. 9, edited by Alfonso Ortiz, pp. 554–63. Smithsonian Institution, Washington, D. C.

Keys, Eric Glenn. 1996. 'Three Yards: A Measure of Women's Agricultural Space in Highland Guatemala'. Unpublished master's thesis, Department of Geography, University of Texas, Austin.

Kidder, A. V. and S. J. Guernsey. 1919. *Archaeological Explorations in Northeastern Arizona*. Smithsonian Institution, Bureau of American Ethnology, Bulletin 65, Washington, D. C.

Killian, Thomas W. 1990. 'Cultivation Intensity and Residential Site Structure: An Ethnoarchaeological Examination of Peasant Agriculture in the Sierra de los Tuxtlas, Veracruz, Mexico', *Latin American Antiquity* **1**: 191–215.

Kimber, Clarissa T. 1973. 'Spatial Patterning in the Dooryard Gardens of Puerto Rico', *Geographical Review* **63**: 6–26.

—— n.d. [1989]. 'Dooryard Gardens, A Special Class of Agro-ecosystem: Some Preliminary Comments'. Unpublished paper on file with author.

Kinietz, W. Vernon. 1940. *The Indians of the Western Great Lakes: 1615–1760*. University of Michigan Press, Ann Arbor.

King, Frances B. 1993. 'Climate, Culture, and Oneota Subsistence in Central Illinois', in *Foraging and Farming in the Eastern Woodlands*, edited by C. Margaret Scarry, pp. 232–54. University Press of Florida, Gainesville.

Kinkade, Gay M. and Gordon L. Fritz. 1975. *The Tucson Sewage Project: Studies at Two Archaeological Sites in the Tucson Basin*. Archaeological Series 64. Cultural Resource Management Section, Arizona State Museum, University of Arizona, Tucson.

Kintigh, Keith W. 1984. 'Late Prehistoric Agricultural Strategies and Settlement Patterns in the Zuni Area', in *Prehistoric Agricultural Strategies in the Southwest*, edited by Suzanne K. Fish and Paul R. Fish, pp. 215–52. Anthropological Research Papers 33. Arizona State University, Tempe.

—— 1985. *Settlement, Subsistence, and Society in Late Zuni Prehistory*. Anthropological Papers of the University of Arizona 44. University of Arizona Press, Tucson.

Kino, Eusebio Francisco. 1948 [1683–1710]. *Kino's Historical Memoir of Pimeria Alta*, translated and edited by Herbert Eugene Bolton. University of California Press, Berkeley and Los Angeles.

Kirk, Ruth. 1973. *Desert: The American Southwest*. Houghton Mifflin, Boston.

Kluckholm, Clyde and Dorothea Leighton. 1974. *The Navaho*. Harvard University Press, Cambridge.

Knapp, Gregory. 1985. Review of '*Advances in Abandoned Settlement Analysis*', *Geographical Review* **75**: 370–1.

—— 1990. 'Comment', *Current Anthropology* **31**: 534.

Kniffen, Fred B. 1931. 'The Primitive Cultural Landscape of the Colorado Delta', *University of California Publications in Geography* **5**.2: 43–66.

Knox, James C. 1985. 'Holocene Geomorphic History of the Sand Lake Site', in *Prehistoric Ridged Field Agriculture in the Upper Mississippi Valley*, edited by James P. Gallagher, pp. 24–55. Reports of Investigations 38. Mississippi Valley Archaeology Center, University of Wisconsin–La Crosse.

Kohler, Timothy A. and Meredith H. Matthews. 1988. 'Long-Term Anasazi Land Use and Forest Reduction: A Case Study from Southwest Colorado', *American Antiquity* **53**: 537–64.

Kosok, Paul. 1965. *Land, Life, and Water in Ancient Peru*. Long Island University Press, New York.

Krantz, William B., Kevin J. Gleason, and Nelson Caine. 1988. 'Patterned Ground', *Scientific American*, **259**: 68–76.

Krech, Shepard, III. 1994. 'Ecology and the American Indians', *Ideas from the National Humanities Center* **3**: 4–22.

Kroeber, A. L. 1925. *Handbook of the Indians of California*. Smithsonian Institution, Bureau of American Ethnology, Bulletin 78, Washington, D. C.

——1941. 'Culture Element Distributions: Salt, Dogs, and Tobacco XV', *Anthropological Records* **6**.1.

——1953. *Cultural and Natural Areas of Native North America*. University of California Press, Berkeley and Los Angeles.

Kunstadter, Peter. 1978. 'Subsistence Agricultural Economics of "Lua" and Karen Hill Farmer, Mae Sariang District, Northwestern Thailand', in *Farmers in the Forest: Economic Development and Marginal Agriculture in Northern Thailand*, edited by Peter Kunstadter, E. C. Chapman, and Sanga Sabhesri, pp. 74–133. University Press of Hawaii, Honolulu.

Kupperman, Karen Ordahl. 1979. 'Nature's "Rude Garden": English and Indians as Producers and Consumers of Food in Early New England', *Comparative Civilizations Review* **1**: 64–77.

Ladd, Edmund J. 1979. 'Zuni Economy', in *Handbook of North American Indians*, Vol. 9, edited by Alfonso Ortiz, pp. 492–8. Smithsonian Institution, Washington, D. C.

Laferriére, Joseph E., and Willard Van Asdall. 1992. 'Contemporary Agricultural Patterns in a Mountain Pima Village', *Kiva* **58**: 155–75.

Lafitau, Joseph François. 1974, 1977 [1724]. *Customs of the American Indians Compared with the Customs of Primitive Times*, 2 vols, edited by William N. Fenton and Elizabeth L. Moore. The Champlain Society, Toronto.

Lafora, Nicolas de. 1958 [1767]. *The Frontiers of New Spain: Nicolas de Lafora's Description, 1766–1768*, translated and edited by Lawrence Kinnard. The Quivira Society, Berkeley.

Lagasse, Peter F., William B. Gillespie, and Kenneth G. Eggert. 1984. 'Hydraulic Engineering Analysis of Prehistoric Water-Controlled Systems at Chaco Canyon', in *Recent Research on Chaco Prehistory*, edited by W. James Judge and John D. Schelberg, pp. 187–211. Reports of the Chaco Center 8, Albuquerque.

Laidlaw, George E. 1917. 'Indian Village Sites, etc. in Victoria County and Vicinity Recorded Since 1912', *Twenty-ninth Annual Archaeological Report, 1917, by Dr. R. B. Orr, Being Part of the Appendix to the Report of the Minister of Education, Ontario*, pp. 91–105. A. T. Wilgress, Toronto.

Lamb, J., Jr. and J. E. Chapman. 1943. 'Effect of Surface Stress or Erosion Evaporation, Soil Temperature, and Soil Moisture', *Journal of the American Society of Agronomy* **35**: 567–78.

Lang, Richard W. 1980. *Archaeological Investigations at a Pueblo Agricultural Site, and Archaic and Puebloan Encampments on the Rio Ojo Caliente, Rio Arriba County, New Mexico*. School of American Research, Contract Achaeology Division Report 007.

——1995. 'The Fields of San Marcos: Agriculture of a Great Town of the Galisteo

Basin, Northern New Mexico', in *Soil, Water, Biology, and Belief in Prehistoric and Traditional Southwestern Agriculture*, edited by H. Wolcott Toll, pp. 41–76. New Mexico Archeological Council, Special Publication 2, Albuquerque.
Lange, Charles H. 1959. *Cochiti: A New Mexico Pueblo, Past and Present*. University of Texas Press, Austin.
——1979*a*. 'Cochiti Pueblo', in *Handbook of North American Indians*, Vol. 9, edited by Alfonso Ortiz, pp. 366–78. Smithsonian Institution, Washington, D. C.
——1979*b*. 'Santo Domingo Pueblo', in *Handbook of North American Indians*, Vol. 9, edited by Alfonso Ortiz, pp. 379–89. Smithsonian Institution, Washington, D. C.
Lantis, L. O. 1929. 'The First American Farmers: Authentic Accounts of the Red Men's Agriculture', *The Ohio Farmer* **12**: 60–1.
Lapham, I. A. 1855. *The Antiquities of Wisconsin as Surveyed and Described*. Smithsonian Institution, Washington, D. C.
Larson, A. 1926. 'Map of Prehistoric Canal System Gila Valley'. Unpublished map on file, Arizona State Museum, University of Arizona, Tucson.
Larson, Esther Louise. 1935. 'Pehr Kalm's Description of Maize, How It Is Planted and Cultivated in North America, Together with the Many Uses of this Crop Plant', *Agricultural History* **9**: 98–117.
Larson, Lewis H., Jr. 1980. *Aboriginal Subsistence Technology on the Southeastern Coastal Plain during the Late Prehistoric Period*. University of Florida Press, Gainesville.
Laudonniére, René. 1975 [1562–1565]. *Three Voyages*. University of Florida Press, Gainesville.
Lawson, John. 1937 [1714]. *Lawson's History of North Carolina*. Garrett and Massie, Richmond.
Lawson, Publins V. 1903. 'Summary of the Archaeology of Winnebago County, Wisconsin', *Wisconsin Archaeologist* **2**.2 and **2**.3: 40–85.
Lawton, Harry W. and Lowell J. Bean. 1968. 'A Preliminary Reconstruction of Aboriginal Agricultural Technology among the Cahuilla', *The Indian Historian* **1**: 18–24.
——and P. J. Wilke. 1979. 'Ancient Agricultural Systems in Dry Regions', in *Agriculture in Semi-Arid Environments*, edited by A. E. Hill, G. H. Cannell, and H. W. Lawton, pp. 1–44. Springer-Verlag, Berlin.
————Mary DeDecker, and William M. Mason. 1976. 'Agriculture Among the Paiute of Owens Valley', *The Journal of California Anthropology* **3**: 13–50.
Leach, Edmund R. 1959. 'Hydraulic Society in Ceylon', *Past and Present* **15**:2–26.
LeBlanc, Steven A. 1977. 'The 1976 Field Season of the Mimbres Foundation in Southwestern New Mexico', *Journal of New World Archaeology* **II**. 2. The Institute of Archaeology, University of California, Los Angeles.
——1983. *The Mimbres People: Ancient Pueblo Painters of the American Southwest*. Thames and Hudson, New York.
Lederer, John. 1958 [1672]. *The Discoveries of John Lederer With Unpublished Letters by and about Lederer to Governor John Winthrop, Jr.*, edited by William P. Cumming. University of Virginia Press, Charlottesville.
Lee, Thomas E. 1958. 'The Parker Earthwork, Corunna, Ontario,' *Pennsylvania Archaeologist* **28**: 3–30.
le Jeune, Paul. 1638 [1898]. 'Relation of What Occurred in the Mission of the Society of Jesus in the Country of the Hurons, in the Year 1637 and 38', in *The Jesuit*

Relations and Allied Documents: Travels and Explorations of the Jesuit Missionaries in New France, 1610–1791, Vol. 15, edited by Reuben Gold Thwaites, pp. 12–14. The Burrows Brothers, Cleveland.

Le Moyne, Jacques. 1968 [1564–1565]. 'The Engravings and the Narrative of le Moyne,' in *Settlement of Florida*, compiled by Charles E. Bennett, pp. 1–122. University of Florida Press, Gainesville.

Lensink, Stephen, Fred A. Finney, and Richard Fishel. n.d. [1995]. 'Rethinking Mill Creek Settlement and Agricultural Systems: New Data from the Phipps, Double Ditch, and Litka Sites', in *Abstracts of the 60th Annual Meeting*, pp. 120–1. Society for American Archaeology, Washington, D. C.

Leopold, Aldo. 1937. 'Conservationist in Mexico', *American Forests* **43**: 118–20, 146.

—— 1966. *A Sand County Almanac and Sketches Here and There*. Bantam, New York.

Leopold, A. Starker. 1949. 'Adíos, Gavilán', *Pacific Discovery* **2**: 4–13.

Le Page du Pratz, Antoine Simon. 1763. *The History of Louisiana, or of the Western Parts of Virginia and Carolina: Containing a Description of the Countries that lye on both sides of the River Mississippi with an Account of the Settlements, Inhabitants, Soil, Climate, and Products*, translated by M. Le Page du Pratz. T. Becket and P. A. De Hondt. London.

Le Petit, Father. 1900 [1730]. 'Letter from Father le Petit, Missionary, to Father d'Avangour, Procurator of the Missions in North America', in *The Jesuit Relations and Allied Documents: Travels and Explorations of the Jesuit Missionaries in New France, 1610–1791*, Vol. 68, edited by Reuben Gold Thwaites, pp. 121–223. The Burrows Brothers, Cleveland.

Lescarbot, Marc. 1977 [1609]. *Nova Francia*. Theatrum Orbis Terrarum, Amsterdam.

Levett, Christopher. 1905 [1628]. 'My Discouery of diverse Riuers and Harbours, with their names, and which are fit for Plantations, and which not', in *Sailors Narratives of Voyages Along the New England Coast: 1524–1624*, edited by George Parker Winship, pp. 259–92. Burt Franklin, New York.

Levy, Richard. 1978. 'Eastern Miwok', in *Handbook of North American Indians*, Vol. 8, edited by Robert F. Heizer, pp. 398–413. Smithsonian Institution, Washington, D. C.

Lewandowski, Stephen. 1987. 'Farming Like the Senecas', *The Seedhead News* **20**: 8.

Lewton, Frederick L. 1912. 'The Cotton of the Hopi Indians: A New Species of Gossypium', *Smithsonian Miscellaneous Collections* **6**.6: 1–17. Washington, D. C.

Liette, Pierre de. 1947 [1721]. 'The Memoir of Pierre Liette', in *The Western Country in the 17th Century: The Memoirs of Lamothe Cadillac and Pierre Liette*, edited by Milo Milton Quaife, pp. 87–171. R. R. Donnelley and Sons, Chicago.

Lightfoot, Dale. 1990. 'The Prehistoric Pebble-Mulched Fields of the Galisteo Anasazi: Agricultural Innovation and Adaptation to Environment.' Unpublished doctoral dissertation, Department of Geography, University of Colorado, Boulder.

—— 1993*a*. 'The Cultural Ecology of Puebloan Pebble-Mulch Gardens', *Human Ecology* **21**: 115–43.

—— 1993*b*. 'The Landscape Context of Anasazi Pebble-Mulched Fields in the Galisteo Basin, Northern New Mexico,' *Geoarchaeology* **8**: 349–70.

—— 1994. 'Morphology and Ecology of Lithic-Mulch Agriculture,' *Geographical Review* **84**: 172–85.

—— and Frank W. Eddy. 1994. 'The Agricultural Utility of Lithic-Mulch Gardens: Past and Present', *GeoJournal* **34**: 425–37.

——— 1995. 'The Construction and Configuration of Anasazi Pebble-Mulch Gardens in the Northern Rio Grande', *American Antiquity* **60**: 459–70.

Lightfoot, Kent G. 1981. 'Prehistoric Political Development in the Little Colorado Region, East-Central Arizona'. Unpublished doctoral dissertation, Department of Anthropology, Arizona State University, Tempe.

—— 1995. 'Culture Contact Studies: Redefining the Relationship Between Prehistoric and Historic Archaeology', *American Antiquity* **60**: 199–217.

—— and Fred Plog. 1984. 'Intensification along the North Side of the Mogollon Rim,' in *Prehistoric Agricultural Strategies in the Southwest*, edited by Suzanne K. Fish and Paul R. Fish, pp. 179–96. Anthropological Research Papers 33. Arizona State University, Tempe.

Liljeblad, Sven and Catherine S. Fowler. 1986. 'Owens Valley Paiute,' in *Handbook of North American Indians*, Vol. 11, edited by Warren L. D'Azevedo, pp. 412–34. Smithsonian Institution, Washington, D. C.

Lindauer, Owen. 1984. 'Sherd Areas as Loci of Subsistence in the Middle Gila,' in *Prehistoric Agricultural Strategies in the Southwest*, edited by Suzanne K. Fish and Paul R. Fish, pp. 73–84. Anthropological Research Papers 33. Arizona State University, Tempe.

Lindestrom, Peter. 1925 [1654–1656]. *Geographia Americae with An Account of the Delaware Indians*, translated by Amandus Johnson. Swedish Colonial Society, Philadelphia.

Lindsay, Alexander J., Jr. 1961. 'The Beaver Creek Agricultural Community on the San Juan River, Utah', *American Antiquity* **27**: 174–87.

—— n.d. [1970]. 'A Descriptive Summary of Soil and Water Conservation Devices and Settlement Patterns in the Kayenta Anasazi Region'. Unpublished paper presented at the 1970 Pecos Conference, symposium on Prehistoric Water Control Systems in the American Southwest, Pecos, National Monument, New Mexico, on file at the Arizona State Museum, University of Arizona, Tucson.

—— J. Richard Ambler, Mary Anne Stein, and P. M. Hobler. 1969. *Survey and Excavations North and East of Navajo Mountain, Utah, 1959–62.* Museum of Northern Arizona Bulletin 20, Flagstaff.

Linton, Ralph. 1924. 'The Significance of Certain Traits in North American Maize Culture', *American Anthropologist* **26**: 345–9.

Lister, Robert H. 1958. *Archaeological Excavations in the Northern Sierra Madre Occidental, Chihuahua and Sonora, Mexico*. University of Colorado Studies, Series in Anthropology 7, Boulder.

Litzinger, William J. 1976. 'The Experimental Garden Project, 1975', in *Hovenweep 1975*, Joseph C. Winters, pp. 177–90. Archeological Report 2. Department of Anthropology, San Jose State University.

Loose, Richard W. and Thomas R. Lyons. 1977. 'The Chetro Ketl Field: A Planned Water Control System in Chaco Canyon', in *Remote Sensing Experiments in Cultural Resource Studies: Non-Destructive Methods of Archeological Exploration, Survey, and Analysis*, assembled by Thomas R. Lyons, pp. 133–56. Reports of the Chaco Center 1. National Park Service and University of New Mexico, Albuquerque.

Lopinot, Neal H. 1992. 'Spatial and Temporal Variability in Mississippian Subsistence: The Archaeobotanical Record', in *Late Prehistoric Agriculture: Observations from the Midwest*, edited by William I. Woods, pp. 44–94. Illinois Historic Preservation Agency, Studies in Illinois Archaeology 8, Springfield.

Lopinot, Neal H. and William I. Woods. 1993. 'Wood Overexploitation and the Collapse of Cahokia', in *Foraging and Farming in the Eastern Woodlands*, edited by C. Margaret Scarry, pp. 206–31. University of Florida Press, Gainesville.

Lorant, Stefan (editor and annotator). 1946. *The New World: The First Pictures of America Made by John White and Jacques Le Moyne and Engraved by Theodore De Bry with Contemporary Narratives of the Huguenot Settlement in Florida, 1562–1565, and the Virginia Colony, 1585–1590*. Duel, Sloan, and Pearce, New York.

Loskiel, George H. 1794. *History of the Mission of the United Brethren among the Indians in North America*. London.

Lowie, Robert H. 1954. *Indians of the Plains*. Anthropological Handbook No. One. The American Museum of Natural History. McGraw-Hill, New York.

Luebben, Ralph A., Jonathan G. Andelson, and Laurance C. Herold. 1986. 'Elvino Whetten Pueblo and its Relationship to Terraces and Nearby Small Structures, Chihuahua, Mexico', *The Kiva* **51**: 165–87.

Lumholtz, Carl. 1902. *Unknown Mexico*, Vol. 1. Charles Scribner's Sons, New York.

—— 1912. *New Trails in Mexico*. Charles Scribner's Sons, New York.

Luomala, Katharine. 1978. 'Tipai-Ipai', in *Handbook of North American Indians*, Vol. 8, edited by Robert F. Heizer, pp. 592–609. Smithsonian Institution, Washington, D.C.

Lynott, Mark J., Thomas W. Boutton, James E. Price, and Dwight E. Nelson. 1986. 'Stable Carbon Isotopic Evidence for Maize Agriculture in Southeast Missouri and Northeast Arkansas', *American Antiquity* **51**: 51–65.

Lyons, Thomas R., Robert K. Hitchcock, and Basil G. Pouls. 1977. 'The Kin Bineola Irrigation Study: An Experiment in the Use of Aerial Remote Sensing Techniques in Archeology', in *Remote Sensing Experiments in Cultural Resource Studies: Nondestructive Methods of Archeological Exploration, Survey, and Analysis*, assembled by Thomas R. Lyons, pp. 115–31. Reports of the Chaco Center 1. National Park Service and University of New Mexico, Albuquerque.

Mabry, Jonathan B. 1997. 'The Canal Site, AZ BB:13:468 (ASM)', in *Archaeological Investigations of Early Village Sites in the Middle Santa Cruz Valley: Descriptions of the Santa Cruz Bend, Square Hearth, Stone Pipe and Canal Sites*, edited by Jonathan B. Mabry, pp. 419–28. Anthropological Papers 18. Center for Desert Archaeology, Tucson.

—— and James P. Holmlund. 1998. 'Canals', in *Archaeological Investigations of Early Village Sites in the Middle Santa Cruz Valley: Analyses and Synthesis*, Part 1, edited by Jonathan B. Mabry, pp. 283–98. Anthropological Papers 19. Center for Desert Archaeology, Tucson.

—— and J. Homer Thiel. 1995. 'A Thousand Years of Irrigation in Tucson', *Archaeology in Tucson* **9**.4: 1–6.

Madsen, David and Lamar Lindsay. 1977. *Backhoe Village*. Utah State Historical Society Publications 4, Salt Lake City.

Mahr, August C. 1955. 'Eighteenth Century Terminology of Delaware Indian Cultivation and Use of Maize: A Semantic Analysis', *Ethnohistory* **2**: 209–40.

Malone, James H. 1922. *The Chickasaw Nation: A Short Sketch of a Noble People*. Morton and Company, Louisville.

Manglesdorf, Paul C. 1974. *Corn: Its Origin, Evolution, and Improvement*. The Belknap Press of Harvard University, Cambridge, Massachusetts.

Manje, Juan Mateo. 1954 [1694–1701]. *Luz de Tierra Incógnita: Unknown Arizona and Sonora, 1693–1721*, translated and edited by Harry J. Karns. Arizona Silhouettes, Tucson.

——1971 [1694–1701]. 'Manje's Journals, and Appendix of Documents', in *Kino and Manje: Explorers of Sonora and Arizona, Their Vision of the Future, A Study of Their Expeditions and Plans*, translated and edited by Ernest J. Burrus, pp. 164–514. Jesuit Historical Institute, Rome.

Mannering, J. V. and L. D. Meyer. 1963. 'The Effects of Various Rates of Surface Mulch on Infiltration and Environment', *Soil Science Society of America Proceedings* **27**: 84–6.

Manshard, Walther. 1974. *Tropical Agriculture: A Geographical Introduction and Appraisal*. Longman, London.

Manypenny, George W. 1972. *Our Indian Wards*. Da Capo Press, New York.

Marozas, Bryan A. 1983. 'The Use of Aerial Photography from Wupatki National Monument, Arizona for the Discovery and Analysis of Prehistoric Agricultural Fields and Related Features'. Unpublished master's thesis, Department of Anthropology, Northern Arizona University, Flagstaff.

Marquette, James. 1903 [1673]. 'The Voyages and Discoveries in the Valley of the Mississippi', in *Discovery and Exploration of the Mississippi Valley*, edited by John Gilmary Shea, pp. 3–69. Joseph McDonough, Albany.

Marston, Richard A. 1982. 'The Geomorphic Significance of Log Steps in Forest Streams', *Annuals of the Association of American Geographers* **72**: 99–108.

Martin, Calvin. 1973. 'Fire and Forest Structure in the Aboriginal Eastern Forest', *The Indian Historian* **6**.4: 38–42.

——1981. 'The American Indian as Miscast Ecologist', *The History Teacher* **14**: 243–52.

Martin, Paul, S. and William Byers. 1965. 'Pollen and Archaeology at Wetherill Mesa', in *Contributions of the Wetherill Mesa Archeological Project*, assembled by Douglas Osborne and edited by Bernard S. Katz, pp. 122–35. Memoirs of the Society for Archaeology 19. Salt Lake City.

——and Fred Plog. 1973. *The Archaeology of Arizona: A Study of the Southwest Region*. Doubleday/Natural History Press, Garden City, New York.

——and James Schoenwetter. 1960. 'Arizona's Oldest Cornfield', *Science* **132**: 33–4.

——George I. Quimby, and Donald Collier. 1947. *Indians Before Columbus: Twenty Thousand Years of North American History Revealed by Archeology*. University of Chicago Press, Chicago.

Martinez del Rio, P. 1954. 'La comarca lagunera a fines del siglo 16 y principios del 17 segun las fuentes escritas', *Publicaciones del Instituto de Historia* **30**: 63–98.

Mártir de Anglería, Pedro. 1944 [1530]. *Décades de Nuevo Mundo: Colección de Fuentes Para la Historia de América*. Editorial Bajel, Buenos Aires.

Marwitt, John P. 1986. 'Fremont Cultures', in *Handbook of North American Indians*, Vol. 11, edited by Warren L. D'Azevedo, pp. 161–72. Smithsonian Institution, Washington, D. C.

Masse, W. Bruce. n.d. [1974]. 'Prehistoric Southwestern Agricultural and Water Control Features: Terminology and Description'. Unpublished paper on file, Arizona State Musuem, University of Arizona, Tucson.

——1976. *The Hohokam Expressway Project: A Study of Prehistoric Irrigation in the*

Salt River Valley, Arizona. Archaeological Salvage Project, Arizona State Museum, University of Arizona, Tucson.

Masse, W. Bruce. 1979. 'An Intensive Survey of Prehistoric Dry Farming Systems Near Tumamoc Hill in Tucson, Arizona', *The Kiva* **45**: 141–86.

——1980. *Excavations at Gu Achi: A Reappraisal of Hohokam Settlement and Subsistence in the Arizona Papagueria*. Western Archaeological and Conservation Center Publications in Anthropology 12. National Park Service, Washington, D. C.

——1981. 'Prehistoric Irrigation Systems in the Salt River Valley, Arizona', *Science* **214**: 408–15.

——1991. 'The Quest for Subsistence Sufficiency and Civilization in the Sonoran Desert', in *Chaco and Hohokam: Prehistoric Regional Systems in the American Southwest*, edited by Patricia L. Crown and W. James Judge, pp. 195–223. School of American Research Press, Santa Fe, New Mexico.

Mathes, W. Michael. 1997. 'Jesuit Chroniclers and Chronicles of Northwestern New Spain', in *Jesuit Encounters in the New World: Jesuit Chroniclers, Geographers, Educators and Missionaries in the Americas, 1549–1767*, edited by Joseph A. Gagliano and Charles E. Ronan, pp. 37–80. Institutum Historicum S. I., Roma.

Mathews, John Joseph. 1961. *The Osages: Children of the Middle Waters*. University of Oklahoma Press, Norman.

Mathews, Washington. 1877. *Ethnography and Philology of the Hidatsu Indians*. United States Geological and Geographical Survey, Washington, D. C.

Matson, R. G. 1991. *The Origins of Southwestern Agriculture*. University of Arizona Press, Tucson.

——William D. Lipe, and William R. Haase IV. 1988. 'Adaptational Continuities and Occupational Discontinuities: The Cedar Mesa Anasazi', *Journal of Field Archaeology* **15**: 245–64.

Maule, Stuart H. 1963. 'Corn Growing at Wupatki', *Plateau* **36**: 29–32.

Maximilian, Prince of Wied. 1906 [1832–1834]. *Travels in the Interior of North America*, translated by Hannibal Evans Lloyd and edited by Reuben Gold Thwaites, Vol. 2. The Arthur H. Clark Company, Cleveland.

Maxwell, Hu. 1910. 'The Use and Abuse of Forests by the Virginia Indians', *William and Mary College Quarterly Historical Magazine* **19**: 73–103.

Maxwell, Timothy D. 1995*a*. 'The Use of Comparative and Engineering Analyses in the Study of Prehistoric Agriculture', in *Evolutionary Archaeology: Methodological Issues*, edited by Patrice A. Teltser, pp. 113–128. University of Arizona Press, Tucson.

——1995*b*. 'A Comparative Study of Prehistoric Farming Strategies', in *Soil, Water, Biology, and Belief in Prehistoric and Traditional Southwestern Agriculture*, edited by H. Wolcott Toll, pp. 2–12. New Mexico Archeological Council, Special Publication 2, Albuquerque.

——and Kurt F. Anschuetz. 1992. 'The Southwestern Ethnographic Record and Prehistoric Agricultural Diversity', in *Gardens of Prehistory: The Archaeology of Settlement Agriculture in Greater Mesoamerica*, edited by Thomas W. Killion, pp. 35–68. University of Alabama Press, Tuscaloosa.

McBride, Kevin A. and Robert E. Dewar. 1987. 'Agriculture and Cultural Evolution: Causes and Effects in the Lower Connecticut River Valley', in *Emergent Horticultural Economies of the Eastern Woodlands*, edited by William F. Keegan, pp. 305–28.

Occasional Paper 7. Center for Archaeological Investigations, Southern Illinois University at Carbondale.

McCabe, Robert A. 1955. 'The Prehistoric Engineer-Farmers of Chihuahua', *Transactions of the Wisconsin Academy of Sciences, Arts and Letters* **44**: 75–90.

McCarthy, Helen. 1993. 'Managing Oaks and the Acorn Crop', in *Before the Wilderness: Environmental Management by Native Californias*, compiled and edited by Thomas C. Blackburn and Kat Anderson, pp. 213–28. Ballena Press Anthropological Papers 40, Menlo Park, California.

McCary, Ben C. 1957. *Indians in Seventeenth-Century Virginia*. University Press of Virginia, Charlottesville.

McCarry, Nancy Stanford. 1990. 'Variation Among Home Gardens in Guatemala: Reflections of Household Characteristics'. Unpublished master's thesis, Department of Geography, University of Texas at Austin.

McGuire, Randall H. and María Elisa Villalpando C. 1991. *Proyecto Reconocimiento Arqueológico en el Valle de Altar: Informe Final Sobre la Temporada de Campo 1988*. Instituto Nacional de Antropología e Historia, Hermosillo, Sonora, Mexico.

——— ——James P. Holmlund, and Maria O'Donovan. 1993. *Cerro de Trincheras Mapping Project*. Final Technical Report to the National Geographic Society, Washington, D. C.

McGuire, Thomas R. 1983. 'Walapai', in *Handbook of North American Indians*, Vol. 10, edited by Alfonso Ortiz, pp. 25–37. Smithsonian Institution, Washington, D. C.

McManis, Douglas R. 1972. *European Impressions of the New England Coast: 1497–1620*. Research Paper 139. Department of Geography, University of Chicago.

Meals for Millions Foundation. 1980. *O'odham I:waki: Wild Greens of the Desert People*. Meals for Millions Foundation, Southwest Program, Tucson.

Mecham, J. Lloyd. 1927. *Francisco de Ibarra and Nueva Vizcaya*. Duke University Press, Durham, North Carolina.

Megapolensis, Johannes, Jr. 1909 [1644]. 'A Short Account of the Mohawk Indians', in *Narratives of New Netherland, 1609–1664*, edited by J. Franklin Jameson, pp. 163–80. Charles Scribner's Sons, New York.

Membré, Zenobius. 1903*a* [1681]. 'Narrative of the Adventures of La Salle's Party at Crevecoeur', in *Discovery and Exploration of the Mississippi Valley*, edited by John Gilmary Shea, pp. 151–68. Joseph McDonough, Albany.

——1903*b* [1681]. 'Narrative of La Salle's Voyage Down the Mississippi', in *Discovery and Exploration of the Mississippi Valley*, edited by John Gilmary Shea, pp. 169–90. Joseph McDonough, Albany.

Merrell, James Hart. 1982. 'Natives in a New World: The Catawba Indians of Carolina, 1650–1800'. Unpublished doctoral dissertation, Department of Anthropology, The Johns Hopkins University, Baltimore, Maryland.

Merrill, William L. 1979. 'The Beloved Tree: *Ilex vomitoria* among the Indians of the Southeast and Adjacent Regions', in *Black Drink: A Native American Tea*, edited by Charles M. Hudson, pp. 40–82. University of Georgia Press, Athens.

Merton, Robert K. 1968. *Social Theory and Social Structure*. The Free Press, New York.

Meyer, Roy W. 1977. *The Village Indians of the Upper Missouri: The Mandans, Hidatsas, and Arikaras*. University of Nebraska Press, Lincoln.

Midvale, Frank. 1946. 'Prehistoric Ruins and Irrigation Features in the Horseshoe

Reservoir Region of the Verde River, Central Arizona'. Unpublished map on file, Department of Anthropology, Arizona State University, Tempe.

Midvale, Frank. 1963. 'The Prehistoric Irrigation of the Casa Grande Ruins Area of the Gila River in Southern Arizona'. Unpublished map on file, Department of Anthropology, Arizona State University, Tempe.

——1965. 'Prehistoric Irrigation of the Casa Grande Ruins Area', *The Kiva* **30**: 82–6.

——1968. 'Prehistoric Irrigation in the Salt River Valley, Arizona', *The Kiva* **34**: 28–32.

——1969. 'Prehistoric Irrigation of the New River Area of Central Arizona'. Unpublished map on file, Department of Anthropology, Arizona State University, Tempe.

——n.d. [1970]. 'Prehistoric "Canal-Irrigation" in the Buckeye Valley and Gila Bend Areas in Western Maricopa County, Arizona'. Unpublished paper presented at the 1970 Pecos Conference, Symposium on Prehistoric Southwestern Water Control Systems, Pecos National Monument, New Mexico, on file at the Arizona State Museum, University of Arizona, Tucson.

——1974. 'Prehistoric Ruins and Irrigation in the Eastern Buckeye Valley', *Arizona Archaeologist* **8**: 37–9.

Miksicek, Charles H. 1984. 'Historic Desertification, Prehistoric Vegetation Change, and Hohokam Subsistence in the Salt-Gila Basin', in *Hohokam Archaeology Along the Salt-Gila Aqueduct, Central Arizona Project*, Vol. 7, edited by Lynn S. Teaque and Patricia L. Crown, pp. 53–80. Archaeological Series 150(7). Arizona State Museum, University of Arizona, Tucson.

Milanich, Jerald T. 1979. 'Origins and Prehistoric Distributions of Black Drink and the Ceremonial Shell Drinking Cup', in *Black Drink: A Native American Tea*, edited by Charles M. Hudson, pp. 83–119. University of Georgia Press, Athens.

Miller, Naomi F. and Kathryn L. Gleason (editors). 1994. *The Archeology of Garden and Field*. University of Pennsylvania Press, Philadelphia.

Mindeleff, Cosmos. 1896. 'Aboriginal Remains in Verde Valley, Arizona', in *Thirteenth Annual Report of the Bureau of Ethnology, 1891–'92*, pp. 185–261. Smithsonian Institution, Washington, D. C.

Minnis, Paul E. 1985. *Social Adaptation to Food Stress: A Prehistoric Southwestern Example*. University of Chicago Press, Chicago.

——and Stephen E. Plog. 1976. 'A Study of the Site Specific Distribution of *Agave parryi* in East Central Arizona', *The Kiva* **41**: 299–308.

Miracle, Marvin P. 1967. *Agriculture in the Congo Basin*. University of Wisconsin Press, Madison.

Mobley-Tanaka, Jeanette L. and Frank W. Eddy. 1995. 'Check Dams at Chimney Rock Mesa: Agricultural Strategies in a Marginal Environment', *Kiva* **60**: 429–47.

Moffat, Charles R. n.d. [1977]. 'Report of an Archaeological Survey of Prehistoric Ridged-Fields in Eastern and Central Wisconsin'. Unpublished report. University of Illinois, Urbana.

——1979. 'Some Observations on the Distribution and Significance of the Garden Beds of Wisconsin', *Wisconsin Archeologist* **60**: 222–48.

——Robert F. Boszhardt, and Constance M. Arzigian. 1995. *Archaeological Investigations in the Castle Rock and Petenwell Hydroelectric Projects: The 1994 Field Season*, Vol. 1. Reports of Investigations 196. Mississippi Valley Archaeological Center, University of Wisconsin–La Crosse.

Monkhouse, F. J. 1965. *A Dictionary of Geography*. Aldine, Chicago.

Mood, Fulmer. 1937. 'John Winthrop, Jr., On Indian Corn', *New England Quarterly* **10**: 121–33.
Moodie, D. W. and Barry Kaye. 1969. 'The Northern Limit of Indian Agriculture in North America', *Geographical Review* **59**: 513–29.
Moore, James L. 1981. 'Prehistoric Water and Soil Conservation in the Middle Puerco River Valley'. Unpublished master's thesis, Department of Anthropology, University of New Mexico.
——1995*a*. 'Anasazi Field Systems in the Taos District', in *Soil, Water, Biology, and Belief in Prehistoric and Traditional Southwestern Agriculture*, edited by H. Wolcott Toll, pp. 13–23. New Mexico Archaeological Council, Special Publication 2, Albuquerque.
——1995*b*. 'Prehistoric Irrigation at Pot Creek Pueblo', in *Soil, Water, Biology, and Belief in Prehistoric and Traditional Southwestern Agriculture*, edited by H. Wolcott Toll, pp. 209–20. New Mexico Archaeological Council, Special Publication 2, Albuquerque.
Moore, W. G. 1974. *A Dictionary of Geography: Definitions and Explanations of Terms Used in Physical Geography*. Penguin Books, Baltimore.
Moorhead, Max L. 1975. *The Presidio: Bastion of the Spanish Borderlands*. University of Oklahoma Press, Norman.
Morissonneau, Christian. 1978. 'Huron of Lorette', in *Handbook of North American Indians*, Vol. 15, edited by Bruce G. Trigger, pp. 389–93. Smithsonian Institution, Washington, D. C.
Morrell, L. Ross. 1965*a*. *The Texas Site: Carlyle Reservoir, Clinton County, Illinois*. Archaeological Salvage Report 23. Southern Illinois University Museum, Carbondale.
——1965*b*. 'A Suggested Method for Locating Aboriginal Garden Areas', in *Bulletin, Southeastern Archeological Conference* **3**: 38–41.
Morris, Earl H. 1939. *Archaeological Studies in the La Plata District Southwestern Colorado and Northwestern New Mexico*. Carnegie Institution of Washington, Washington, D. C.
Morrisey, Richard J. 1949. 'Early Agriculture in Pimeria Alta', *Mid-America* **31**: 101–8.
Morton, Thomas. 1947 [1632]. 'New English Canaan or New Canaan', in *Tracts and Other Papers, Relating Principally to the Origin, Settlement, and Progress of the Colonies in North America, From the Discovery of the Country to the Year 1776*, Vol. 2, collected by Peter Force, pp. 1–69. Peter Smith, New York.
Moseley, Edwin Lincoln. 1930. 'Some Plants That Were Probably Brought to Northern Ohio from the West by Indians', *Papers of the Michigan Academy of Sciences, Arts, and Letters* **13**: 169–72.
Mrozowski, Stephen A. 1994. 'The Discovery of a Native American Cornfield on Cape Cod', *Archaeology of Eastern North America* **22**: 47–62.
Muenchrath, Deborah A. and Ricardo J. Salvador. 1995. 'Maize Productivity and Agroecology: Effects of Environment and Agricultural Practices on the biology of Maize', in *Soil, Water, Biology, and Belief in Prehistoric and Traditional Southwestern Agriculture*, edited by H. Wolcott Toll, pp. 303–33. New Mexico Archaeological Council, Special Publication 2, Albuquerque.
Mulholland, Mitchell T. 1988. 'Territoriality and Horticulture: A Perspective for Prehistoric Southern New England', in *Holocene Human Ecology in Northeastern*

North America, edited by George P. Nicholas, pp. 137–66. Plenum Press, New York.

Muller, Jon. 1986. *Archaeology of the Lower Ohio River Valley*. Academic Press, Orlando.

——1987. 'Lower Ohio Valley Emergent Horticulture and Mississippian', in *Emergent Horticultural Economies in the Eastern Woodlands*, pp. 243–73, edited by William F. Keagan. Center for Archaeological Investigations, Occasional Paper 7. Southern Illinois University, Carbondale.

Muro, Mark. 1998*a*. 'New Finds Explode Old Views of the American Southwest', *Science* **279**: 653–4.

——1998*b*. 'Not Just Another Roadside Attraction', *American Archaeology* **2**.4: 10–16.

Murphy, Christopher and Charles Hudson. 1968. 'On the Problem of Intensive Agriculture in the Aboriginal Southeastern United States', *Working Papers in Sociology and Anthropology* **2**: 24–34.

Myers, Michael D. 1998. 'Cultivation Ridges in Theory and Practice: Cultural Ecological Insights from Ireland'. Unpublished doctoral dissertation, Department of Geography, University of Texas, Austin.

Nabhan, Gary Paul. 1979*a*. 'Amaranth Cultivation in the U.S. Southwest and Northwest Mexico', *1979 Amaranth Proceedings*, pp. 129–33. Rodale Press, Emmaus, Pennsylvania.

——1979*b*. 'The Ecology of Floodwater Farming in Arid Southwestern North America', *Agro-Ecosystems* **5**: 245–55.

——1979*c*. 'Southwestern Indian Sunflowers', *Desert Plants* **1**: 23–6.

——1981. 'Runoff Farming for Dry Lands', *The New Farm* **3**: 50–3.

——1983. 'Papago Fields: Arid Lands Ethnobotany and Agricultural Ecology'. Unpublished doctoral dissertation, Program in Arid Lands Resource Science, University of Arizona, Tucson.

——1984. 'Soil Fertility Renewal and Water Harvesting in Sonora Desert Agriculture', *Arid Lands Newsletter* **20**: 20–8.

——1985*a*. *Gathering the Desert*. University of Arizona Press, Tucson.

——1985*b*. 'Native Crop Diversity in Aridoamerica: Conservation of Regional Gene Pools', *Economic Botany* **39**: 387–99.

——1986*a*. 'Papago Indian Desert Agriculture and Water Control in the Sonoran Desert, 1697–1934', *Applied Geography* **6**: 43–59.

——1986*b*. '*Ak-ciñ* "Arroyo Mouth" and the Environmental Setting of the Papago Indian Fields in the Sonoran Desert', *Applied Geography* **6**: 61–75.

——1989. *Enduring Seeds: Native American Agriculture and Wild Plant Conservation*. North Point Press, San Francisco.

——1995. 'Finding the Hidden Gardens', *Journal of the Southwest* **37**: 401–15.

——and J. M. J. de Wet. 1984. '*Panicum sonorum* in Sonoran Desert Agriculture', *Economic Botany* **37**: 65–82.

——and Thomas Edward Sheridan. 1977. 'Living Fence Rows of the Rio San Miguel, Sonora, Mexico: Traditional Technology for Floodplain Management', *Human Ecology* **5**: 97–111.

——and Helga Teiwes. 1983. 'Tepary Beans, O'odham Farmers, and Desert Fields', *Desert Plants* **5**: 15–37.

——Alfred Whiting, Henry Dobyns, Richard Hevly, and Robert Euler. 1981. 'Devil's Claw Domestication: Evidence from Southwestern Indian Fields', *Journal of Ethnobiology* **1**: 135–64.

——Amadeo M. Rea, Karen L. Reichhardt, Eric Mellink, Charles F. Hutchinson. 1982. 'Papago Influences on Habitat and Biotic Diversity: Quitovac Oasis Ethnoecology', *Journal of Ethnobiology* **2**: 124–43.

Nabokov, Peter and Dean Snow. 1992. 'Farmers of the Woodlands', in *America in 1492: The World of the Indian Peoples Before the Arrival of Columbus*, edited by Alvin M. Josephy, Jr., pp. 119–45, Alfred A. Knopf, New York.

Nagata, Shuichi. 1970. *Modern Transformations of Moenkopi Pueblo*. Illinois Studies in Anthropology. 6. University of Illinois Press, Urbana.

Nassaney, Michael S. 1987. 'On the Causes and Consequences of Subsistence Intensification in the Mississippi Alluvial Valley', in *Emergent Horticultural Economies of the Eastern Woodlands*, edited by William F. Keegan, pp. 129–52. Occasional Paper 7. Center for Archaeological Investigations, Southern Illinois University at Carbondale.

——Neal H. Lopinot, Brian M. Butler, and Richard W. Jefferies. 1983. *The 1982 Excavations at the Cahokia Interpretive Center Tract, St. Clair County, Illinois*. Center for Archaeological Investigations, Research Paper 37. Southern Illinois University, Carbondale.

National Academy of Science. 1974. *More Water for Arid Lands: Promising Technologies and Research Opportunities*. National Academy of Science, Washington, D. C.

Neely, James A. 1993*a*. 'Stone Hoes from the Proposed Bason Land Exchange Project Area', in *La Cuchilla de Piedra: Cultural Resources of the Bason Land Exchange*, edited by John A. Peterson and Christian B. Nightengale, pp. 221–30. Archaeological Research Inc., El Paso.

——1993*b*. 'Prehispanic and Historic Agricultural Pursuits within the Bason Land Exchange Project Area', in *La Cuchilla de Piedra: Cultural Resources of the Bason Land Exchange*, edited by John A. Peterson and Christian B. Nightengale, pp. 231–51, Archaeological Research Inc., El Paso.

——1995*a*. 'Prehistoric Irrigation at Pot Creek Pueblo', in *Soil, Water, Biology, and Belief in Prehistoric and Traditional Southwestern Agriculture*, edited by H. Wolcott Toll, pp. 209–20. New Mexico Archaeological Council, Special Publication 2, Albuquerque.

——1995*b*. 'Mogollon/Western Pueblo Soil and Water Control Systems of the Reserve Phase: New Data from West-Central New Mexico', in, edited *Soil, Water, Biology, and Belief in Prehistoric and Traditional Southwestern Agriculture* by H. Wolcott Toll, pp. 239–62. New Mexico Archaeological Council, Special Publication 2, Albuquerque.

——n.d.[1995*c*]. 'A Progress Report on the Survey and Mapping of Six (6) Water Management Sites (AZ CC:1:10, 33–37 [ASM]) in the Vicinity of Pima, Arizona'. Unpublished report on file in the Arizona State Museum, University of Arizona, Tucson.

Nentvig, Juan. 1980 [1764]. *Rudo Ensayo: A Description of Sonora and Arizona in 1764*, translated, clarified, and annoted by Alberto Francisco Pradeau and Robert B. Rasmussen. University of Arizona Press, Tucson.

Neuman, Robert W. 1961. 'Domesticated Corn from a Fort Walton Mound Site in Houston County, Alabama', *The Florida Anthropologist* **14**: 75–80.

Newcomb, W.W., Jr. 1961. *The Indians of Texas: From Prehistoric to Modern Times*. University of Texas Press, Austin.

——1976. *The People Called Wichita*. Indian Tribal Series, Phoenix.

——and W. T. Field. 1967. 'An Ethnohistoric Investigation of the Wichita Indians in the Southern Plains', in *A Pilot Study of Wichita Indian Archeology and Ethnohistory*, assembled by Robert E. Bell, Edward B. Jelks, and W. W. Newcomb, pp. 240–398. Final Report for Grant GS-964, National Science Foundation, Washington, D. C.

Nials, Fred L. 1980. 'Geology of the Salmon Ruin Area', in *Investigations at the Salmon Site: The Structure of Chacoan Society in the Northern Southwest*, Vol. 4, edited by Cynthia Irwin-Williams and Phillip H. Shelley, pp. 3–25. Final Report to Funding Agencies.

Nicholas, Linda. 1981. 'Irrigation and Sociopolitical Development in the Salt River Valley, Arizona: An Examination of Three Prehistoric Canal Systems'. Unpublished master's thesis, Department of Anthropology, Arizona State University, Tempe.

——and Gary M. Feinman. 1989. 'A Regional Perspective on Hohokam Irrigation in the Lower Salt River Valley, Arizona', in *The Sociopolitical Structure of Prehistoric Southwestern Societies*, edited by Steadman Upham, Kent G. Lightfoot, and Robert A. Jewett, pp. 199–235. Westview Press, Boulder.

——and Jill Neitzel. 1984. 'Canal Irrigation and Sociopolitical Organization in the Lower Salt River Valley: A Diachronic Analysis', in *Prehistoric Agricultural Strategies in the Southwest*, edited by Suzanne K. Fish and Paul R. Fish, pp. 161–78. Anthropological Research Papers. 33. Arizona State University, Tempe.

Nichols, Robert F. and David G. Smith. 1965. 'Evidence of Prehistoric Cultivation of Douglas-fir Trees at Mesa Verde', *Contributions of the Wetherill Mesa Archeological Project*, assembled by Douglas Osborne and edited by Bernard S. Katz, pp. 57–64. Memoirs of the Society for American Archaeology 19, Salt Lake City.

Niñez, Vera K. 1984. *Household Gardens: Theoretical Considerations on an Old Survival Strategy*. International Potato Center, Lima Peru.

Niza, Marcos de. 1987*a* [1539]. 'Relación', in *The Journey of Fray Marcos de Niza*, Cleve Hallenbeck, pp.liv–lxix. Southern Methodist University Press, Dallas.

——1987*b* [1539]. 'Narrative of Fray Marcos', in *The Journey of Fray Marcos de Niza*, Cleve Hallenbeck, pp. 15–36. Southern Methodist University Press, Dallas.

Noisat, Bradley A. 1975. 'Water Control at Hovenweep', in *Hovenweep 1974*, Joseph C. Winter, pp. 147–52. Archeological Report No. 1, Department of Anthropology, San Jose State University.

——1976. 'Water Control Experimentation', in *Hovenweep 1975*, Joseph C. Winter, pp. 191–6. Archaelogical Report. 2. Department of Anthropology, San Jose State University.

Nordenskiöld, Gustav. 1893. *The Cliff Dwellers of the Mesa Verde*. Translated by D. Lloyd Morgan. P. A. Norstedt and Soner, Stockholm–Chicago.

Norton, Jay Byron. 1996. 'Soil, Geomorphic, and Ecological Factors in Zuni Runoff Agriculture'. Unpublished master's thesis, Department of Agronomy, Iowa State University.

Nurge, Ethel. 1970. *The Modern Sioux: Social Systems and Reservation Culture*. University of Nebraska Press, Lincoln.

Nye, P.H. and D.J. Greenland. 1960. *The Soil Under Shifting Cultivation*. Technical Communication 51. Commonwealth Bureau of Soils, Harpenden.

Obregón, Baltasar de. 1924 [1584]. *Historia de los Descubrimientos Antiguos y Modernos de la Nueva España*, descubierta por Mariano Cuevas. Departamento Editorial de la Secretaria de Educación Pública, México.

——1928 [1584]. *Obregón's History of 16th Century Explorations in Western America Entitled Chronicle, Commentary, or Relation of the Ancient and Modern Discoveries in New Spain and New Mexico, Mexico, 1584*, translated, edited and annotated by George P. Hammond and Agapito Rey. Wetzel, Los Angeles.

Office of Indian Affairs. 1921. 'Primitive Agriculture of the Indians', *Bulletin, Office of Indian Affairs* **1**: 1–5.

Officer, James E. 1993. 'Kino and Agriculture in the Pimeria Alta', *The Journal of Arizona History* **34**: 287–306.

Oliver, Symmes C. 1962. 'Ecology and Cultural Continuity as Contributing Factors in the Social Organization of the Plains Indians', *Publications in American Archaeology and Ethnology* **48**: 1–90. University of California Press, Berkeley and Los Angeles.

Olmsted, D.L. and Omer C. Stewart. 1978. 'Achumawi', in *Handbook of North American Indians*, Vol. 8, edited by Robert F. Heizer, pp. 225–35. Smithsonian Institution, Washington, D. C.

Oñate, Juan de. 1908 [1601]. 'True Account of the Expedition of Oñate Toward the East', in *Spanish Exploration in the Southwest, 1542–1706*, edited by Herbert Eugene Bolton, pp. 250–67. Barnes and Noble, New York.

——1953*a* [1598]. 'Record of the Marches by the Army, New Spain to New Mexico, 1596–98', in *Don Juan de Oñate: Colonizer of New Mexico, 1595–1628*, Vol. 1, translated and edited by George P. Hammond and Agapito Rey, pp. 309–28. University of New Mexico Press, Albuquerque.

——1953*b* [1601]. 'Faithful and True Report of the Events that Took Place on the Entrada from New Mexico toward the North', in *Don Juan de Oñate: Colonizer of New Mexico, 1595–1628*, Vol. 2, translated and edited by George P. Hammond and Agapito Rey, pp. 746–60. University of New Mexico Press, Albuquerque.

Onion, Daniel K. 1964. 'Corn in the Culture of the Mohawk Iroquois', *Economic Botany* **18**: 60–6.

Opler, Morris E. 1941. *An Apache Life-way: The Economic, Social, and Religious Institutions of the Chiricahua Indians*. University of Chicago Press, Chicago.

——1983*a*. 'The Apachean Culture Pattern and Its Origins', in *Handbook of North American Indians*, Vol. 10, edited by Alfonso Ortiz, pp. 368–92. Smithsonian Institution, Washington, D. C.

——1983*b*. 'Chiricahua Apache', in *Handbook of North American Indians*, Vol. 10, edited by Alfonso Ortiz, pp. 401–18. Smithsonian Institution, Washington, D. C.

Oram, Peter A. 1979. 'Crop Production Systems in the Arid and Semi-Arid Warm Temperate and Mediterranean Zones', in *Soil, Water and Crop Production*, edited by D. Wynne Thorne and Marlowe D. Thorne, pp.193–228. AVI Publishing, Westport, Connecticut.

Ortiz, Bev. 1993. 'Contemporary California Indian Basketweavers and the Environment', in *Before the Wilderness: Environmental Management by Native*

Californians, compiled and edited by Thomas C. Blackburn and Kat Anderson, pp. 195–211. Ballena Press Anthropological Papers 40, Menlo Park, California.

Osborne, Douglas. 1964. 'Solving the Riddles of Wetherill Mesa', *National Geographic Magazine* **125**.2: 155–73.

——1976. 'Slow Exodus from Mesa Verde', *Natural History* **85**: 38–45.

Othieno, C. O. and I. M. Ahn. 1980. 'Effects of Mulches on Soil Temperature and Growth of Teaplants Plants in Kenya', *Experimental Agriculture* **16**: 287–94.

Overstreet, David F. 1987. 'Comment on Oneota Ridged Field Agriculture in Southwestern Wisconsin', *American Antiquity* **52**: 396–98.

Page, Diana. 1986. Growing Hope in Santiago's Urban Organic Gardens. *Grassroots Development* **10**.2: 38–43.

Page, Gordon B. n.d. [1964]. 'Hopi Personality and Culture'. Unpublished paper on file, Arizona State Museum, University of Arizona, Tucson.

Palmer, Rose A. 1929. *The North American Indians: An Account of the American Indians North of Mexico, Compiled from the Original Sources*. Smithsonian Scientific Series 4. Smithsonian Institution, Washington, D. C.

Pargellis, Stanley. 1959. 'An Account of the Indians in Virginia', *The William and Mary Quarterly* **16**: 228–43.

Parker, Arthur C. 1910. *Iroquois Uses of Maize and Other Food Plants*. Education Department Bulletin 144, 482. New York State Museum, Albany.

Parkins, A. E. 1918. 'The Indians of the Great Lakes Region and their Environment', *Geographical Review* **6**: 504–12.

Parsons, David B. 1981. *Maíz*. Editorial Trillas, Mexico City.

Parsons, E. C. 1920. 'Notes on Ceremonialism at Laguna', *American Museum of Natural History Anthropological Papers* **19**: 85–93.

Parsons, James J. 1985. 'The Canary Islands Search for Stability', *Focus* **35**.2: 22–9.

——and William M. Denevan. 1967. 'Pre-Columbian Ridged Fields', *Scientific American* **217**: 92–100.

Patch, Richard W. 1951. 'Irrigation in East Central California', *American Antiquity* **17**: 50–2.

Patencio, Francisco. 1943. *Stories and Legends of the Palm Springs Indians*. Palm Springs Desert Museum, Palm Springs.

Patrick, H. R. 1903. *The Ancient Canal Systems and Pueblos of the Salt River Valley, Arizona*. Free Museum Bulletin 1, Phoenix.

Patrick, Larry L. 1980. 'Los origenes de las terrazas de cultivo', *América Indígena* **40**: 757–72.

Patterson, William A., III and Kenneth E. Sassaman. 1988. 'Indian Fires in the Prehistory of New England', in *Holocene Human Ecology in Northeastern North America*, edited by George P. Nicholas, pp. 107–35. Plenum Press, New York.

Pattison, Natalie Batchelder. 1968. 'Nogales Cliff House: A Largo-Gallina Site'. Unpublished master's thesis, Department of Anthropology, University of New Mexico.

Pauketat, Timothy R. and William I. Woods. 1986. 'Middle Mississippian Structure Analysis: The Lawrence Primas Site (11-Ms-895) in the American Bottom', *The Wisconsin Archaeologist* **67**: 104–27.

Pawluk, Roman R. 1995. 'Indigenous Knowledge of Soil and Agriculture at Zuni

Pueblo, New Mexico'. Unpublished master's thesis, Department of Agronomy, Iowa State University.

Peacock, Evan. 1998. 'Historical and Applied Perspectives on Prehistoric Land Use in Eastern North America', *Environment and History* **4**: 1–29.

Pedrosa, Martín de. 1966 [1582]. 'Pedrosa's List of Pueblos', in *The Rediscovery of New Mexico 1580–1594: The Explorations of Chamuscado, Espejo, Castaño de Sosa, Moulette, and Leyva de Bonilla and Humana*, translated and edited by George P. Hammond and Agapito Rey, pp. 115–20. The University of New Mexico Press, Albuquerque.

Peet, S. D. 1887. 'Prehistoric Farming', *Wisconsin State Agricultural Society* **2**: 390–408.

Pennington, Campbell W. 1963. *The Tarahumar of Mexico: Their Environment and Material Culture*. University of Utah Press, Salt Lake City.

——1969. *The Tepehuan of Chihuahua: Their Material Culture*. University of Utah Press, Salt Lake City.

——(editor). 1977. *The Pima Bajo of Central Sonora, Mexico: Vocabulario en la Lengua Nevome*. University of Utah Press, Salt Lake City.

——1980. *The Pima Bajo of Central Sonora, Mexico: The Material Culture*. University of Utah Press, Salt Lake City.

——(editor). 1981. *Arte y Vocabulario de la Lengua Doheme, Heve o Eudeve*. Universidad Nacional Autónomo de México, Mexico City.

——1983*a*. 'Tarahumara', in *Handbook of North American Indians*, Vol. 10, edited by Alfonso Ortiz, pp. 276–89. Smithsonian Institution, Washington, D. C.

——1983*b*. 'Northern Tepehuan', in *Handbook of North American Indians*, Vol. 10, edited by Alfonso Ortiz, pp. 306–14. Smithsonian Institution, Washington, D. C.

Percy, George. 1969 [1608]. 'Observations Gathered Out of a Discourse of the Plantation of the Southern Colonies in Virginia by the English', in *The Jamestown Voyages Under the First Charter 1606–1609*, edited by Philip L. Barbour, Vol. 1, pp. 129–46. The University Press, Cambridge.

Perea, Estavan de. 1937 [1633]. 'Case Brought by Fray Estevan de Perea Against Juan López', in *Historical Documents Relating to New Mexico, Nueva Vizcaya, and Approaches Thereto, to 1773*, Vol. III, edited by Charles Witson Hackett, pp. 129–31. Carnegie Institution, Washington, D. C.

Pérez de Luxán, Diego. 1966 [1582]. 'Diego Pérez de Luxán's Account of the Antonio de Espejo Expedition into New Mexico, 1582', in *The Rediscovery of New Mexico 1580–1594: The Explorations of Chamuscado, Espejo, Castaño de Sosa, Morlete, and Leyva de Bonilla and Humaña*, translated and edited by George P. Hammond and Agapito Rey, pp. 153–212. The University of New Mexico Press, Albuquerque.

Pérez de Muzquía, Pedro. 1938 [1718]. 'The Muzquía Diary of the Alarcón Expedition into Texas, 1718', *Southwestern Historical Quarterly* **41**: 312–23.

Pérez de Ribas, Andrés. 1944 [1645]. *Trivnfos de nvestra santa Fe entre gentes las mas barbaras y fieras del Nvevo Orbe Consequidos por los soldados de la Milicia de la Compania de JESUS en las Misiones de las Provincias de Nueva Espana*. Layac, Mexico.

——1968 [1645]. *My Life Among the Savage Nations of New Spain*, translated and condensed by Tomas Antonio Robertson. The Ward Ritchie Press, Los Angeles.

Peri, David W. and Scott M. Patterson. 1993. 'The Basket is in the Rocks, That's Where it Begins', in *Before the Wilderness: Environmental Management by Native*

Californians. Compiled and edited by Thomas C. Blackburn and Kat Anderson, pp. 175–93. Ballena Press Anthropological Papers 40, Menlo Park, California.

Perkl, Bradley E. 1998. '*Cucurbita pepo* from King Coulee, Southeastern Minnesota', *American Antiquity* **63**: 279–88.

Perramond, Eric Phillipe. 1994. 'An Historical Geography of Zuni Agriculture'. Unpublished master's thesis, Department of Geography and Anthropology, Louisiana State University.

Perrot, Nicholas. 1911 [1642]. 'Memoir on the Manners, Customs, and Religion of the Savages of North America', in *The Indian Tribes of the Upper Mississippi Valley and the Region of the Great Lakes*, Vol. 1, translated and edited by Emma Helen Blair, pp. 31–272. The Arthur H. Clark Co., Cleveland.

Perttula, Timothy K. 1992. *The Caddo Nation: Archaeological and Ethnohistoric Perspectives*. University of Texas Press, Austin.

Peske, G. Richard. 1966. 'Oneota Settlement Patterns and Agricultural Patterns in Winnebago County', *Wisconsin Archeologist* **47**: 188–95.

Peters, Bernard C. 1980. 'Comments on the Distribution of Garden Beds in Kalamazoo County', *Ecumene* **12**: 31–7.

Peterson, John A. 1988. 'Settlement and Subsistence Patterns in the Reserve Phase and Mountain Mogollon: A Test Case in Devils Park, New Mexico', *The Kiva* **53**: 113–27.

Pfefferkorn, Ignaz. 1989 [1794–95]. *Sonora: A Description of the Province*, translated and annotated by Theodore E. Treutlein. University of Arizona Press, Tucson.

Phinney, S. B. 1863. 'Cranberry Culture', *United States Department of Agriculture, Report*, pp. 131–9. United States Government Printing Office, Washington, D. C.

Pilles, Peter J., Jr. 1978. 'The Field House and Sinagua Demography', in *Limited Activity and Occupation Sites: A Collection of Conference Papers*, compiled and edited by Albert E. Ward, pp. 119–33. Center for Anthropological Studies, Albuquerque, New Mexico.

—— 1979. 'Sunset crater and the Sinagua: A New Interpretation', in *Volcanic Activity and Human Ecology*, edited by Peyson D. Sleets and Donald K. Grayson, pp. 459–85. Academic Press, New York.

Pitty, Alistair F. 1971. *Introduction to Geomophology*. Methuen, London.

Plog, Fred T. n.d. [1970]. 'Water Control and the upper Little Colorado'. Unpublished paper presented at the 1970 Pecos Conference symposium on Prehistoric Southwestern Water Control Systems, Pecos, National Monument, New Mexico, on file at the Arizona State Museum, University of Arizona, Tucson.

—— 1978. 'The Keresan Bridge: An Ecological and Archeological Account', in *Social Archeology: Beyond Subsistence and Dating*, edited by Charles L. Redman, Mary Jane Berman, Edward V. Curtin, William T. Langhorne, Jr., Nina M. Versaggi, and Jeffery C. Wanser, pp. 349–72. Academic Press, New York.

—— 1979. 'Prehistory: Western Anasazi', in *Handbook of North American Indians*, Vol. 9, edited by Alfonso Ortiz, pp. 108–30. Smithsonian Institution, Washington, D. C.

—— and Cheryl K. Garrett. 1972. 'Explaining Variability in Prehistoric Southwestern Water Control Systems', in *Contemporary Archaeology: A Guide to Theory and Contributions*, edited by Mark P. Leone, pp. 280–88. Southern Illinois University Press, Carbondale.

Pohl, Mary DeLand. 1990. *Ancient Maya Wetland Agriculture: Excavations on Albion Island, Northern Belize*. Westview Press, Boulder.

Polzer, Charles W. 1976. *Rules and Precepts of the Jesuit Missions of Northwestern New Spain*. University of Arizona Press, Tucson.

Pool, Michael David. 1985. 'The Western Apache Settlement System and Its Implications for the Prehistoric Early Mogollon Period'. Unpublished master's thesis, Department of Anthropology, University of Texas, Austin.

——1994. 'Implications of the Western Apache Settlement System for the Early Mogollon Period,' in *Mogollon VII: The Collected Papers of the 1992 Mogollon Conference Held in Las Cruces, New Mexico*, pp. 89–103. COAS Publishing and Research, Las Cruces, New Mexico.

Pourade, Richard F. 1960. *The History of San Diego*, Vol. 2. The Union-Tribune Publishing Co., San Diego.

Powell, E. P. 1916. 'The Indians as Gardeners', *Country Life in America* **29**: 130–31.

Price, T. Douglas and Anne Birgitte Gebauer (editors). 1995*a*. *Last Hunters—First Farmers: New Perspectives on the Prehistoric Transition to Agriculture*. School of American Research Press, Santa Fe.

——1995*b*. 'New Perspectives or the Transition to Agriculture', in *Last Hunters—First Farmers: New Perspectives on the Prehistoric Transition to Agriculture*, edited by T. Douglas Price and Anne Birgitte Gebauer, pp. 3–19. School of American Research Press, Santa Fe.

Pring, Martin. 1905 [1605]. 'A Voyage set out from the Citie of *Bristoll* at the charge of the chiefest Merchants and Inhabitants of the said Citie with a small Ship and a Barke for the Discoverie of the North part of *Virginia*,' in *Sailors' Narratives of Voyages Along the New England Coast: 1524–1564*, edited by George Parker Winship, pp. 51–63. Burt Franklin, New York.

Prudden, T. Mitchell. 1903. 'The Prehistoric Ruins of the San Juan Watershed in Utah, Arizona, Colorado, and New Mexico', *American Anthropologist* **5**: 224–88.

Prufer, Olaf H. and Orrin C. Shane. 1970. *Blain Village and the Fort Ancient Tradition in Ohio*. Kent State University Press, Kent, Ohio.

Puleston, Dennis E. 1968. '*Brosimum alicastrum* as a Subsistence Alternative for the Classic Maya of the Central Southern Lowlands'. Unpublished master's thesis, Department of Anthropology, University of Pennsylvania.

Pyne, Stephen J. 1982. *Fire in America: A Cultural History of Wildfire and Rural Fire*. Princeton University Press, Princeton.

Quimby, George Irving. 1960. *Indian Life in the Upper Great Lakes: 11,000 B.C. to A.D. 1800*. University of Chicago Press, Chicago.

Radding, Cynthia. 1979. 'Estructuras socioeconómicas de las misiones de la Pimería Alta, 1768–1850', *Noroeste de México* 3. Instituto Nacional de Antropologia e Historia, Mexico City.

——1995. *Entre el desierto y la sierra: Las naciones O'odham y teguima de Sonora, México*. Centro de Investigaciones y Estudios, Mexico City.

Rainey, Steven James. 1991. 'Cultivators in the Clouds: A Study of Erosion in Relation to Agricultural Practices in the Sierra Alta of Hidalgo, Mexico'. Unpublished master's thesis, Department of Geography, University of Texas, Austin.

Ranjel, Rodrigo. 1904 [1546]. 'A Narrative of de Soto's Expedition based on the Diary of Rodrigo Ranjel, his Private Secretary', translated from *Oviedo's Historia General y Natural de las Indias* by Edward Gaylord Bourne, in *Narratives of the Career of Hernando de Soto*, Vol. II, edited by Edward Gaylord Bourne, pp. 41–149. Williams-Barker, Co., New York.

Ranjel, Rodrigo. 1993 [1546]. 'Account of the Northern Conquest and Discovery of Hernando De Soto', translated and edited by John E. Worth, in *The De Soto Chronicles: The Expedition of Hernando De Soto to North America in 1539–1543,* Vol. 1, edited by Lawrence A. Clayton, Vernon James Knight, Jr., and Edward C. Moore, pp. 247–306. The University of Alabama Press, Tuscaloosa.

Rankin, Adrianne G. 1989*a*. 'AZ T:4:40 (ASM)', in *Settlement, Subsistence, and Specialization in the Northern Periphery: The Waddell Project,* Vol. 1, edited by Margerie Green, pp. 69–175. Cultural Resources Report 65. Archaeological Consulting Services, Tempe, Arizona.

—— 1989*b*. 'Agricultural Features', in *Settlement, Subsistence, and Specialization in the Northern Periphery: The Waddell Project*, Vol. 2, edited by Margerie Green, pp. 949–79. Cultural Resources Report 65. Archaeological Consulting Services, Tempe, Arizona.

—— and Keith Katzer. 1989. 'Agricultural Systems in the ACS Waddell Project Area', in *Settlement, Subsistence, and Specialization in the Northern Periphery: The Waddell Project,* Vol. 2, edited by Margerie Green, pp. 981–1020. Cultural Resources Report 65. Archaeological Consulting Services, Tempe, Arizona.

Rasles, Sebastien. 1847 [1723]. 'From Father Sebastien Rasles, Missionary of the Society of Jesus in New France, to Monsieur his Brother', in *The Early Jesuit Missions in North America*, compiled and translated by William Ingraham Kip, pp. 21–66. Wiley and Putnam, London.

Rawitz, E. 1973. 'Gravity Irrigation', in *Arid Zone Irrigation*, edited by B. Yaron, E. Danfers, and Y. Vaadia, pp. 323–37. Springer-Verlag, Heidelberg and New York.

Raynor, Gilbert S., Eugene C. Ogden, and Janet V. Hayes. 1972. 'Dispersion and Deposition of Corn Pollen from Experimental Sources', *Agronomy Journal* **64**: 420–27.

Rea, Amadeo M. 1979. 'The Ecology of Pima Fields', *Environment Southwest* **484**: 1–6.

Redman, Charles L. 1993. *People of the Tonto Rim: Archaeological Discovery in Prehistoric Arizona*. Smithsonian Institution Press, Washington, D. C.

Reff, Daniel T. 1991. *Disease, Depopulation, and Culture Change in Northwestern New Spain, 1518–1764*. University of Utah Press, Salt Lake City.

Reher, Charles A. (editor). 1979. *Settlement and Subsistence Along the Lower Chaco River: The CGP Survey*. University of New Mexico Press, Albuquerque.

Ressler, John Quenton. 1966. 'Spanish Mission Water Systems, Northwest Frontier of New Spain'. Unpublished master's thesis, Department of Anthropology, University of Arizona, Tucson.

—— 1968. 'Indian and Spanish Water-Control on New Spain's Northwest Frontier', *Journal of the West* **7**: 10–17.

Reynolds, Edwin C. 1957. *The Seminoles*. University of Oklahoma Press, Norman.

Rhode, David. 1995. 'Estimating Agricultural Growing Capacity in the Zuni Region, West-Central New Mexico: A Water Allocation Model', in *Soil, Water, Biology, and Belief in Prehistoric and Traditional Southwestern Agriculture*, edited by H. Wolcott Toll, pp. 85–100. New Mexico Archaeological Council, Special Publication 2, Albuquerque.

Ribaut, Jean. 1927 [1563]. *The Whole True Discouerye of Terra Florida: A Facsimile*

Reprint of the London Edition of 1563. The Florida State Historical Society, Deland.

Rice, E. L. and R. E. Wilson. 1968. 'Alleopathy as Expressed by *Helianthus annus* and its Role in Old-field Succession', *Bulletin of the Torrey Botanical Club* **95**: 432–48.

Rice, Glen. 1978. 'An Archaeological Management Program for the Roosevelt Water Conservation District Floodway: Phase Two'. Unpublished report prepared for the U. S. Department of Agriculture, Soil Conservation Service, Phoenix.

——and Donald E. Simonis. 1977. 'An Archaeological Survey of a Portion of the Queen Creek Delta, Pinal County, Arizona', *Office of Cultural Resource Management Report* 31. Arizona State University, Tempe.

Riley, Carroll. L. 1976. 'Sixteenth Century Trade in the Greater Southwest', *Southern Illinois University Museum Research Records* 10. Southern Illinois University, Carbondale.

——1987. *The Frontier People: The Greater Southwest in the Protohistoric Period*. University of New Mexico Press, Albuquerque.

Riley, Thomas J. 1987. 'Ridged-Field Agriculture and the Mississippian Economic Pattern', in *Emergent Horticultural Economies of the Eastern Woodlands*, edited by William F. Keegan, pp. 295–304. Occasional Paper 7. Center for Archaeological Investigations, Southern Illinois University, Carbondale.

——1994. 'Ocmulgee and the Question of Mississippian Agronomic Practices', in *Ocmulgee Archaeology: 1936–1986,* edited by David J. Hally. pp. 96–104. University of Georgia Press, Athens.

——and Glen Freimuth. 1977. 'Prehistoric Agriculture in the Upper Midwest', *Field Museum of Natural History Bulletin* **48**: 4–8.

————1979. 'Field Systems and Frost Drainage in the Prehistoric Agriculture of the Upper Great Lakes', *American Antiquity* **44**: 271–85.

——Chaksana Said 1993. 'Fly Ash Analysis Supports Emergent Mississippian Agricultural Features at the Lunsford-Pulcher Site (11-S-40) in the American Bottoms, Illinois', *Plains Anthropologist* **38**: 177–86.

——Charles Moffat, and Glen Freimuth. 1981. 'Prehistoric Raised Fields in the Upper Midwestern United States: An Innovation in Response to Marginal Growing Conditions', *North American Archaeologist* **2**: 101–16.

——Richard Edging, and Jack Russen. 1990. 'Cultigens in Prehistoric Eastern North America', *Current Anthropology* **31**: 525–33.

Rindos, David. 1984. *The Origins of Agriculture: An Evolutionary Perspective*. Academic Press, Orlando.

Rinker, Jennifer Rebecca. 1998. 'The Bryce-Smith Project: Irrigated Agriculture and Habitation from AD 1000 to 1450, Lefthand Canyon, Safford Valley, Arizona'. Unpublished master's thesis, Department of Anthropology, University of Texas.

Ritchie, William A. 1969. *The Archaeology of New York State*. The Natural History Press, Garden City, New York.

Ritterbush, Lauren W. 1984. 'Prehistoric Water Collection and Conservation on Antelope Preserve, Wupatki National Monument, Arizona'. Unpublished master's thesis, Department of Anthropology, University of Kansas.

Robert, Joseph C. 1966. *The Story of Tobacco in America*. The University of North Carolina Press, Chapel Hill.

Robinson, William J. 1976. 'Mission Guevavi: Excavations in the Convento', *The Kiva* **42**: 135–75.

Rodgers, James B. n.d. [1970]. 'Prehistoric Agricultural Systems in the Vosberg Locality'. Unpublished paper presented at the 1970 Pecos Conference symposium on Prehistoric Southwestern Water Control Systems, Pecos, National Monument, New Mexico. On file at the Arizona State Museum, University of Arizona, Tucson.

——1974. *An Archaeological Survey of the Cave Buttes Dam Alternative Site and Reservoir, Arizona*. Anthropological Research Papers 8. Arizona State University, Tempe.

——1976. *An Archaeological Investigation of Buckeye Hills East, Maricopa County, Arizona*. Anthropological Research Papers 10. Arizona State University, Tempe.

——1977. *Archaeological Investigation of the Granite Reef Aqueduct, Cave Creek Archaeological District, Arizona*. Anthropological Research Papers 12. Arizona State University, Tempe.

——1978. 'The Fort Mountain Archaeological Complex: Cave Buttes, Arizona', in *Limited Activity and Occupation Sites: A Collection of Conference Papers*, compiled and edited by Albert E. Ward, pp. 147–64. Center for Anthropological Studies, Albuquerque.

——1985. 'Prehistoric Agricultural Variability in the Hohokam Northern Periphery', in *Hohokam Settlement and Economic Systems in the Central New River Drainage, Arizona*, edited by David E. Doyel and Mark D. Elson, pp. 249–96. Soil Systems Publications in Archaeology 4. Soil Systems Inc., Phoenix.

Rogel, Juan. 1946 [1570]. 'Pater Ioannes Rogel Petro Menendez', in *Monumenta Antiquae Floridae (1566–1572)*, pp. 471–79, edited by Felix Zubillaga. Apud Monumenta Historica Societatis IESU, Romae.

Rogers, Edward S. and Mary B. Black. 1976. 'Subsistence Strategy in the Fish and Hare Period, Northern Ontario: The Weagamow Ojibwa, 1880–1920', *Journal of Anthropological Research* **32**: 1–20.

Rogers, W. J. and H. L. Swift. 1970. *Frost and the Prevention of Frost Damage*. United States Department of Commerce, Silver Spring, Maryland.

Rohn, Arthur H. 1963*a*. 'An Ecological Approach to the Great Pueblo Occupation of the Mesa Verde, Colorado', *Plateau* **36**: 1–17.

——1963*b*. 'Prehistoric Soil and Water Conservation on Chapin Mesa, Southwestern Colorado', *American Antiquity* **28**: 441–55.

——1971. *Mug House: Wetherill Mesa Excavations, Mesa Verde National Park—Colorado*. National Park Service Archaeological Research Series Number 70, Washington, D. C.

——1972. 'Social Implications of Pueblo Water Management in the Northern San Juan', *Sonderdruck: Zeitschrift für Ethnologie*, **97**: 212–19.

——1977. *Cultural Change and Continuity on Chapin Mesa*. The Regents Press of Kansas, Lawrence.

Rojas Rabiela, Teresa (editora). 1983. *La Agricultura Chinampera: Compilacíon Histórica*. Universidad Autonoma Chapingo, Texcoco, Mexico.

Romans, Bernard. 1962 [1775]. *A Concise Natural History of East and West Florida*. University of Florida Press, Gainesville.

Rosier, James. 1905 [1605]. 'A True Relation of Most Prosperous Voyage Made this Present Yeere 1605, by Captaine George Waymouth in the Discoury of the Land of Virginia', in *Sailors' Narratives of Voyages along the New England Coast, 1524–1624*, edited by George Parker Winship, pp. 99–151. Burt Franklin, New York.

Rostlund, Erhard. 1957. 'The Evidence for the Use of Fish as Fertilizer in Aboriginal North America', *The Journal of Geography* **56**: 222–28.

Rountree, Helen C. 1989. *The Powhatan Indians of Virginia: Their Traditional Culture*. University of Oklahoma Press, Norman.

Russell, Emily W. 1981. 'Vegetation of Northern New Jersey Before European Settlement', *American Midland Naturalist* **105**: 1–12.

—— 1983. 'Indian-Set Fires in the Forests of the Northeastern United States', *Ecology* **64**: 78–88.

Russell, Frank. 1908. 'The Pima Indians', *Twenty-Sixth Annual Report of the Bureau of American Ethnology*, 1904–1905, pp. 3–389, Smithsonian Institution, Washington, D.C.

Russell, Howard S. 1961. 'New England Indian Agriculture', *Bulletin of the Massachusetts Archaeological Society* **22**: 58–61.

—— 1969. 'New England Agriculture from Champlain and Others', *Bulletin of the Massachusetts Archaeological Society* **31**: 11–18.

—— 1975. Letter, *Science* **189**: 944–6.

—— 1980. *Indian New England before the Mayflower*. University Press of New England, Hanover.

Russell, M. B. and R. E. Danielson. 1956. 'Time and Depth Patterns of Water Use by Corn', *Agronomy Journal* **48**: 163–5.

Russell, Scott C. 1978. 'The Agricultural Fieldhouse: A Navajo Limited Occupation and Special Use Site', in *Limited Activity and Occupation Sites: A Collection of Conference Papers*, compiled and edited by Albert E. Ward, pp. 35–40. Center for Anthropological Studies, Albuquerque.

—— 1983. 'Factors Affecting Agricultural Production in a Western Navajo Community'. Unpublished doctoral dissertation, Department of Anthropology, Arizona State University, Tempe.

Ruthenberg, Hans. 1976. *Farming Systems in the Tropics*. Clarendon Press, Oxford.

Sagard-Théodat, Gabriel. 1865 [1623]. *Le Grand Voyage du Pays des Hurons Situé en l'Amérique vers la Mer douce, ès derniers confins de la Nouvelle France dite Canada*. Librairie Tross, Paris.

—— 1939 [1623]. *The Long Journey to the Country of the Hurons*, translated by H. H. Langton and edited by George M. Wrong. The Champlain Society, Toronto.

—— 1964 [1623]. 'Sagard-Théodat's Description of Iroquoian Agriculture', in *The Valley of the Six Nations: A Collection of Documents on the Indian Lands of the Grand River*, edited by Charles M. Johnston, pp. 7–11. The Champlain Society for the Government of Ontario, University of Toronto Press, Ontario.

Salaman, Redcliffe N. 1940. 'Why "Jerusalem" Artichoke?' *Journal of the Royal Horticultural Society* **65**: 338–48.

Sale, Kirkpatrick. 1991. *The Conquest of Paradise: Christopher Columbus and the Columbian Legacy*. Alfred A. Knopf, New York.

Salisbury, Neal. 1982. *Manitou and Providence: Indians, Europeans, and the Making of New England, 1500–1643*. Oxford University Press, New York and Oxford.

Salwen, Bert. 1975. 'Post-Glacial Environments and Cultural Change in the Hudson River Basin', *Man in the Northeast* **10**: 43–70.

—— 1978. 'Indians of Southern New England and Long Island: Early Period', in

Handbook of North American Indians, Vol. 15, edited by Bruce G. Trigger, pp. 160–76. Smithsonian Institution, Washington, D. C.

Sanchez, Bartolomé. 1856 [1757]. 'Carta del Padre Bartolomé Sanchez, al Padre Prior y Rector Juan Antonio Baltasar en el año de 1757', in *Documentos para la Historia Mexico*, pp. 88–94. Cuarta Série, Tomo I. Imprenta de Vicente García Torres, Mexico.

Sandor, Jonathan A. 1987. 'Initial Investigation of Soils in Agricultural Terraces in the Colca Valley, Peru', *Pre-Hispanic Agricultural Fields in the Andean Region*, edited by William M. Denevan, Kent Mathewson, and Gregory Knapp, pp. 163–92. BAR International Services 359, Oxford.

—— P. L. Gersper, and J. W. Hawley. 1986. 'Soils at Prehistoric Agricultural Terracing Sites in New Mexico', *Agronomy Journal* **50**: 166–80.

—— 1990. 'Prehistoric Agricultural Terraces and Soils in the Mimbres Area, New Mexico', *World Archaeology* **22**: 70–86.

Sarmela, Matti. 1987. 'Swidden Cultivation in Finland as a Cultural System', *Suomen Antropodogi* **4**: 241–62.

Sasso, Robert F. n.d. 'The Issue of the Antiquity of Garden Beds and Corn Hills in Wisconsin', *Midcontinental Journal of Archeology*, forthcoming.

Sauer, Carl O. 1925. 'The Morphology of Landscape', *University of California Publications in Geography* **2**.2: 19–54.

—— 1971. *Sixteenth Century North America: The Land and the People as Seen by the Europeans*. University of California Press, Berkeley.

—— 1980. *Seventeenth Century North America*. Turtle Island Foundation, Berkeley, California.

—— and Donald D. Brand. 1930. 'Pueblo Sites in Southeastern Arizona', *University of California Publications in Geography* **3**: 415–59.

—— —— 1931. 'Prehistoric Settlements of Sonora, With Special Reference to Cerros de Trincheras', *University of California Publications in Geography* **5**: 67–148.

Sayles, Edwin B. 1936. 'An Archaeological Survey of Chihuahua, Mexico', *Medallion Paper* 17, Gila Pueblo, Globe, Arizona.

Scarry, C. Margaret (editor). 1993. *Foraging and Farming in the Eastern Woodlands*. University Press of Florida, Gainesville.

Schaber, Gerald G. and George J. Gumerman. 1969. 'Infrared Scanning Images: An Archeological Application', *Science* **164**: 712–13.

Schmidt, Louis B. and Earle D. Ross. 1925. *Readings in the Economic History of American Agriculture*. MacMillan, New York.

Schmidt, Robert H. and Rex E. Gerald. 1988. 'The Distribution of Conservation-Type Water Control Systems in the Northern Sierra Madre Occidental', *The Kiva* **53**: 165–79.

Schoolcraft, Henry R. 1851*a*. *Historical and Statistical Information Respecting the History, Condition and Prospects of the Indian Tribes of the United States*, Part I. Lippincott, Grambo and Co., Philadelphia.

—— 1851*b*. *Personal Memoirs of a Thirty Years Residence with the Indian Tribes on the American Frontier*. Lippincott, Grambo and Co. Philadelphia.

—— 1851*c*. *The American Indians: Their History, Conditions and Prospects, from Original Notes and Manuscripts*. Wanzer, Foot and Co., Rochester, New York.

Schrambling, Regina. 1988. 'Masachusetts Cranberry Ramble', *National Geographic Traveler* **5**.3: 58–66.

Schroeder, Albert H. 1943. 'Prehistoric Canals in the Salt River Valley, Arizona', *American Antiquity* **8**: 380–6.
——1960. 'The Hohokam, Sinagua and the Hakataya', *Archives of Archaeology* 5. Western Archaeological and Conservation Center, National Park Service, Tucson.
——1967–8. 'An Archeological Survey Adjacent to Hovenweep National Monument', *Southwestern Lore* **33**.3–4: 61–94.
——1979*a*. 'Pueblos Abandoned in Historic Times', in *Handbook of North American Indians*, Vol. 9, edited by Alfonso Ortiz, pp. 236–54. Smithsonian Institution, Washington, D. C.
——1979*b*. 'Pecos Pueblo', in *Handbook of North American Indians*, Vol. 9, edited by Alfonso Ortiz, pp. 430–7. Smithsonian Institution, Washington, D. C.
——and Homer F. Hastings. 1958. *Montezuma Castle National Monument, Arizona*. National Park Service Historical Handbook Series 27, Washington, D. C.
——and Dan S. Matson. 1965. *A Colony on the Move: Caspar Castaño de Sosa's Journal: 1590–1591*. School of American Research, Santa Fe.
Schwartz, Douglas W. 1960. 'Archaeological Investigations in the Shinumo Area of Grand Canyon, Arizona', *Plateau* **32**: 61–7.
——1967. *Conceptions of Kentucky Prehistory: A Case Study in the History of Archeology*. University of Kentucky Press, Lexington.
——1983. 'Havasupai', in *Handbook of North American Indians*, Vol. 10, edited by Alfonso Ortiz, pp. 13–24. Smithsonian Institution, Washington, D. C.
Scott, Aurelia C. 1993. 'Ancient Tech', *Invention and Technology* **9**: 34–44.
Sears, Elsie and William Sears. 1976. 'Preliminary Report on Prehistoric Corn Pollen from Fort Center, Florida', *Southeastern Archaeological Conference Bulletin* **19**: 53–6.
Sears, William H. 1971. 'Food Production and Village Life in Prehistoric Southeastern United States', *Archaeology* **24**: 322–9.
——1977. 'Seaborne Contacts between Early Cultures in Lower Southeastern United States and Middle through South America', in *The Sea in the Pre-Columbian World*, edited by Elizabeth P. Benson. pp. 1–13. Dumbarton Oaks Research Library and Collections, Washington, D. C.
——1982. *Fort Center: An Archaeological Site in the Lake Okeechobee Basin*. University Presses of Florida, Gainesville.
Sedelmayr, Jacobo. 1939 [1746]. 'Sedelmayr's Relación of 1746', translated and edited by Ronald L. Ives. Smithsonian Institution, Bureau of American Ethnology, Bulletin 123. *Anthropological Papers* **9**: 99–117. Washington D. C.
——1955 [1744–1751]. *Jacobo Sedmayr, Missionary, Frontiersman, Explorer in Arizona and Sonora, Four Original Manuscript Narratives*, translated and annotated by Peter Masten Dunne. Arizona Pioneers' Historical Society, Tucson.
Sedeño, Antonio. 1946 [1570]. 'Pater Antonius Sedeño Patri Francisco Borgiae', in *Monumenta Antiquae Floridae (1566–1572)*, pp. 421–28, edited by Felix Zubillaga. Apud Monumenta Historica Societatis IESU, Romae.
Serra, Junipero. 1967 [1769]. *Diario: The Journal of Padre Serra*, translated by Den F. Dixon. Don Diego's Liberial Associated Historians of San Diego. San Diego, California.
Setchell, William Albert. 1921. 'Aboriginal Tobaccos', *American Anthropologist* **23**: 397–427.
Seymour, Gregory R., David P. Doak, and Richard V. N. Ahlstrom. 1994. 'Prehistoric

Farming and Resource Use', in *The Sanchez Copper Project: Archaeological Investigations in the Safford Valley Graham Country, Arizona*, Vol. 1, edited by Gregory R. Seymour, Richard V. N. Ahlstrom, and David P. Doak, Chapter 8. SWCA Archaeological Report 94–82. SWCA Environmental Consultants, Tucson.

Shafer, Harry J. 1982. 'Prehistoric Agricultural Climax in Southwestern New Mexico: The Classic Mimbres Phase', in *Southwestern Agriculture: Pre-Columbian to Modern*, edited by Henry C. Dethloff and Irvin M. May, Jr., pp. 3–15. Texas A & M University Press, College Station.

Sharrock, Floyd W., David S. Dibble, and Keith M. Anderson. 1961. 'The Creeping Dune Irrigation Site in Glen Canyon, Utah', *American Antiquity* **27**: 188–202.

Shelley, Steven D. 1994. 'Cottonwood Boat Parking and Armer Area', in *The Roosevelt Rural Sites Study: Prehistoric Rural Settlements in the Tonto Basin*, Vol. 2, Part 1, edited by Richard Ciolek-Torrello, Steven D. Shelley, and Su Benaron, pp. 345–76. Technical Series 28. Statistical Research, Tucson.

—— and Richard Ciolek-Torrello. 1994. 'Grapevine Recreation and Stockpile Areas', in *The Roosevelt Rural Sites Study: Prehistoric Rural Settlements in the Tonto Basin*, Vol. 2, Part 1, edited by Richard Ciolek-Torrello, Steven D. Shelley, and Su Benaron, pp. 223–58. Technical Series 28. Statistical Research, Tucson.

Sheridan, Thomas E. and Thomas H. Naylor (editors). 1979. *Rarámuri: A Tarahumara Colonial Chronicle: 1607–1791*. Northland Press, Flagstaff.

Shetrone, Henry C. 1945. 'A Unique Prehistoric Irrigation Project', *Annual Report of the Smithsonian Institution*. Smithsonian Institution, Washington, D. C.

Shipek, Florence C. 1989. 'An Example of Intensive Plant Husbandry: The Kumeyaay of Southern California', in *Foraging and Farming: The Evolution of Plant Exploitation*, edited by David R. Harris and Gordon C. Hillman, pp. 159–70. Unwin Hyman, London.

—— 1993. 'Kumeyaay Plant Husbandry: Fire, Water, and Erosion Control Systems', in *Before the Wilderness: Environmental Management by Native Californians*, compiled and edited by Thomas C. Blackburn and Kat Anderson, pp. 379–88. Ballena Press. Anthropological Papers 40, Menlo Park, California.

Showalter, Pamela Sands. 1993. 'A Thematic Mapper Analysis of the Prehistoric Hohokam Canal System, Phoenix, Arizona', *Journal of Field Archaeology* **20**: 77–90.

Shreve, Forrest. 1934. 'Rainfall, Runoff, and Soil Moisture under Desert Conditions', *Annals of the Association of American Geographers* **24**: 131–56.

—— and Ira L. Wiggins. 1964. *Vegetation and Flora of the Sonoran Desert*. Stanford University Press, Stanford, California.

Shumacher, Paul. 1882. 'An Ancient Fortification in Sonora, Mexico', *The American Antiquarian and Oriental Journal* **4**: 227–9.

Siemens, Alfred H. 1983. 'Oriented Raised Fields in Central Veracruz', *American Antiquity* **48**: 85–102.

—— 1989. *Tierra Configurada: Investigaciones de los Vestigos de Agricultura Precolombina en Tierras Inundables Costeras desde el Norte de Veracruz hasta Belice*. Consejo Nacional para la Cultura y las Artes, Mexico City.

Sigüenza y Góngora, Carlos de. 1963 [1686]. *Documentos inéditos de Don Carlos de Sigüenza y Góngora*, edited by Irving A. Leonard. Mexico.

Silver, Shirley. 1978*a*. 'Chimariko', in *Handbook of North American Indians*, Vol. 8, edited by Robert F. Heizer, pp. 205–10. Smithsonian Institution, Washington, D. C.

——1978*b*. 'Shastan Peoples', in *Handbook of North American Indians*, edited by Robert F. Heizer, Vol. 8, pp. 211–24. Smithsonian Institution, Washington, D. C.

Silver, Timothy. 1990. *A New Face on the Countryside: Indians, Colonists, and Slaves in South Atlantic Forests, 1500–1800*. Cambridge University Press.

Simmons, Marc. 1972. 'Spanish Irrigation Practices in New Mexico', *New Mexico Historical Review* **47**: 135–50.

Simmons, William S. 1978. 'Narragansett', in *Handbook of North American Indians*, Vol. 15, edited by Bruce G. Trigger, pp. 190–7. Smithsonian Institution, Washington, D. C.

Skinner, S. Alan. 1965. 'A Survey of Field Houses at Sapawe, North Central New Mexico', *Southwestern Lore* **31**: 18–24.

Smith, Bruce D. 1984. '*Chenopodium* as a Prehistoric Domesticate in Eastern North America: Evidence from Russell Cave, Alabama', *Science* **226**: 165–7.

——1985. 'The Role of *Chenopodium* as a Domesticate in Pre-Maize Garden Systems of the Eastern United States', *Southeastern Archaeology* **4**: 51–72.

——1986. 'The Archaeology of the Southeastern United States: From Dalton to de Soto, 10,500–500 B. P.', in *Advances in World Archaeology*, Vol. 5, edited by Fred Wendorf and Angela E. Close, pp. 1–92. Academic Press, Orlando, Florida.

——1987. 'The Economic Potential of *Chenopodium Berlandieri* in Prehistoric Eastern North America', *Journal of Ethnobiology* **7**: 29–54.

——1989. 'Origins of Agriculture in Eastern America', *Science* **246**: 1566–71.

——1992. *Rivers of Change: Essays on Early Agriculture in Eastern North America*. Smithsonian Institution Press, Washington, D. C.

——1995. 'Seed Plant Domestication in Eastern North America', in *Last Hunters—First Farmers: New Perspectives on the Prehistoric Transition to Agriculture*, edited by T. Douglas Price and Anne Birgitte Gebauer, pp. 193–213. School of American Research Press, Santa Fe.

Smith, Charles, R. 1978. 'Tubatulabal', in *Handbook of North American Indians*, Vol. 8, edited by Robert F. Heizer, pp. 437–45. Smithsonian Institution, Washington, D. C.

Smith, John. 1907 [1612]. 'Description of Virginia and Proceedings of the Colonie', in *Narratives of Early Virginia*, edited by Lyon Gardiner Tyler, pp. 73–204, Charles Scribner's Sons, New York.

——1910*a* [1608]. 'A True Relation of Such Occurrences and Accidents of noate as Hapned in Virginia', in *Travels and Works of Captain John Smith*, edited by Edward Arber, pp. 1–40. Burt Franklin, New York.

——1910*b* [1616]. 'A Description of New England', in *Travels and Works of Captain John Smith*, edited by Edward Arber, pp. 175–234. Burt Franklin, New York.

——1910*c* [1606]. 'The Sixt Voyage: To Another Part of Virginia', in *Travels and Works of Captain John Smith*, edited by Edward Arber, pp. 343–82. Burt Franklin, New York.

——1910*d* [1630]. 'Advertisements For the unexperienced Planters of New-England, or anywhere', in *Travels and Works of Captain John Smith*, edited by Edward Arber, pp. 917–68. Burt Franklin, New York.

Smith, Landon D. 1978. 'Archaeological and Paleoenvironmental Investigations in the Cave Buttes Area North of Phoenix, Arizona', *The Arizona Archaeologist* **11**.

Smith, Philip E. L. 1976. *Food Production and Its Consequences*. Cummings Publishing Co., Menlo Park, California.

Snow, Dean R. 1978*a*. 'Late Prehistory of the East Coast', in *Handbook of North American Indians*, Vol. 15, edited by Bruce G. Trigger, pp. 58–69. Smithsonian Institution, Washington, D. C.

——1978*b*. 'Eastern Abenaki', in *Handbook of North American Indians*, Vol. 15, edited by Bruce G. Trigger, pp. 137–47. Smithsonian Institution, Washington, D. C.

——1980. *The Archaeology of New England*. Academic Press, New York.

Soleri, Daniela. 1989. 'Hopi Gardens', *Arid Lands Newsletter*, **29**: 11–14.

Speck, Frank G. 1940. *Penobscot Man: The Life History of a Forest Tribe in Maine*. University of Pennsylvania Press, Philadelphia.

Spellman, Charles W. 1948. 'The Agriculture of the Early North Florida Indians', *Florida Anthropologist* **48**: 37–48.

Spelman, Henry. 1910 [1613]. 'Relation of Virginea', in *Travels and Works of Captain John Smith*, edited by Edward Arber, pp. ci–cxxi, Burt Franklin, New York.

Spencer, Joseph E. 1966. *Shifting Cultivation in Southeastern Asia*. University of California Publications in Geography 19, Berkeley.

——1976. 'What's In a Name?—The Berkeley School', *Historical Geography Newsletter* **6**: 7–11.

——and Hale, G. A. 1961. 'The Origin, Nature, and Distribution of Agricultural Terracing', *Pacific Viewpoint* **2**: 1–40.

Sperka, Roger. 1964. 'The Senator Lake Site', *Wisconsin Archeologist* **43**: 94–104.

Spicer, Edward H. 1962. *Cycles of Conquest: The Impact of Spain, Mexico, and the United States on the Indians of the Southwest, 1533–1960*. University of Arizona Press, Tucson.

——1983. 'Yaqui', in *Handbook of North American Indians*, Vol. 10, edited by Alfonso Ortiz, pp. 250–63. Smithsonian Institution, Washington, D. C.

Spier, Leslie. 1917. 'An Outline for a Chronology of Zuni Ruins', *Anthropological Papers of the American Museum of Natural History* **18**: 207–331.

——1928. *Havasupai Ethnography*, Anthropological Papers of the American Museum of Natural History, Vol. 29, Part 3. The American Museum of Natural History, New York City.

——1933. *Yuman Tribes of the Gila River*. University of Chicago Press, Chicago.

Spinden, H. J. 1915. 'The Origin and Distribution of Agriculture in America', *Proceedings of the Nineteenth International Congress of Americanists*, edited by F. W. Hodge, pp. 269–72. Washington, D. C.

——1917. 'The Invention and Spread of Agriculture in America', *The American Museum Journal* **17**: 181–8.

——1950. *Tobacco is American: The Story of Tobacco before the Coming of the White Man*. The New York Public Library, New York.

Starbird, Charles M. 1928. *The Indians of the Androscoggin Valley*. Lewiston Journal Printshop, Lewiston, Maine.

Starna, William A., George R. Hamell, and William L. Butts. 1984. 'North Iroquoian Horticulture and Insect Infestation: A Cause for Village Removal', *Ethnohistory* **31**: 197–207.

——and Donald A. Kane, Jr. 1983. 'Phytoliths, Archaeology, and Caveats: A Case Study from New York State', *Man in the Northeast* **26**: 21–32.

St. Cosme, J. F. Buisson de. 1861 [1704]. 'Letter to the Bishop of Quebec', in *Early Voyages Up and Down the Mississippi*, edited by John Gilmary Shea, pp. 45–86. Joel Munsell, Albany.

Stevenson, Matilda Coxe. 1904. 'The Zuni Indians: Their Mythology, Esoteric Fraternities, and Ceremonies'. *Twenty-third Annual Report of the Bureau of American Ethnology, 1901–1902*, pp. 73–634. Smithsonian Institution, Washington, D. C.

——1915. 'Ethnobotany of the Zuni Indians', in *Thirtieth Annual Report of the Bureau of American Ethnology, 1908–1909*, pp. 31–103. Smithsonian Institution, Washington, D. C.

Steward, Julian H. 1929. 'Irrigation Without Agriculture', *Papers of the Michigan Academy of Science, Arts and Letters* **12**: 149–56.

——1933. 'Ethnography of the Owens Valley Paiute', *University of California Publications in American Archaeology and Ethnology* **33**: 233–350. University of California Press, Berkeley.

——1939. *Basin-Plateau Aboriginal Sociopolitical Groups*. Bureau of American Ethnology, Bulletin 120. Smithsonian Institution, Washington, D. C.

——1941. 'Cultural Element Distributions: XIII Nevada Shoshoni', *Anthropological Records, University of California* **4**: 209–59.

Stewart, Guy R. 1939. 'Conservation Practices in Primitive Agriculture of the Southwest', *Soil Conservation* **5**: 112–15.

——1940*a*. 'Conservation in Pueblo Agriculture: I. Primitive Practices', *Scientific Monthly* **51**: 201–20.

——1940*b*. 'Conservation in Pueblo Agriculture: II. Present-Day Flood Water Irrigation', *Scientific Monthly* **51**: 329–40.

——and Maurice Donnelly. 1943*a*. 'Soil and Water Economy in the Pueblo Southwest: I. Field Studies at Mesa Verde and Northern Arizona', *Scientific Monthly* **56**: 31–44.

————1943*b*. 'Soil and Water Economy in the Pueblo Southwest: II. Evaluation of Primitive Methods of Conservation', *Scientific Monthly* **56**: 134–44.

Stewart, Kenneth M. 1966. 'Mojave Indian Agriculture', *The Masterkey* **40**.1: 5–15.

——1977. 'Pueblo Indian Agriculture', in *The Native Americans*, edited by Robert F. Spencer and Jesse D. Jennings, p. 271. Harper and Row, New York.

——1983*a*. 'Yumans: Introduction', in *Handbook of North American Indians*, Vol. 10, edited by Alfonso Ortiz, pp. 1–3. Smithsonian Institution, Washington, D. C.

——1983*b*. 'Mohave', in Handbook of North American Indians, Vol. 10, edited by Alfonso Ortiz, pp. 55–70. Smithsonian Institution, Washington, D. C.

Stewart, Omer C. 1941. 'Cultural Element Distributions: XIV Northern Paiute', *Anthropological Records, University of California* **4**: 361–446.

Stewart, Tamara and Jim Walker. 1998. 'Shaping the Land', *American Archaeology* **2**.2: 32–3.

Stickney, Gardner P. 1896. 'Indian Use of Wild Rice', *American Anthropologist* **9**: 115–22.

Stiger, Mark A. 1979. 'Mesa Verde Subsistence Patterns from Basketmaker to Pueblo III', *The Kiva* **44**: 133–44.

Stites, Sara Henry. 1904. 'Economics of the Iroquois', *Bryn Mawr College Monographs*, Vol. 1. Lancaster, Pennsylvania.

Stout, A. B. 1914. 'Vegetable Foods of the American Indians', *Journal of the New York Botanical Gardens* **15**: 50–60.

Stowe, Gerald C. 1940. 'Plants Used by the Chippewa', *Wisconsin Archaeologist* **21**: 8–13.

Strachey, William. 1953 [1612]. *The Historie of Travell into Virginia Britania*. Hakluyt Society, London.

Stromgaard, Peter. 1989. 'Soil and Vegetation Changes under Shifting Cultivation in the Miombo of East Africa', *Geografiska Annaler* **71B**: 427–44.

Strong, Pauline Turner. 1979. 'Santa Ana Pueblo', in *Handbook of North American Indians*, Vol. 9, edited by Alfonso Ortiz, pp. 398–406. Smithsonian Institution, Washington, D. C.

Strong, W. D. 1933. 'The Plains Culture Area in the Light of Archaeology', *American Anthropologist* **35**: 271–87.

Struever, Stuart. 1964. 'The Hopewell Interaction Sphere in Riverine-Western Great Lakes Culture History', in *Hopewellian Studies*, edited by Joseph R. Caldwell and Robert L. Hall, pp. 87–106. Scientific Papers 12. Illinois State Museum, Springfield.

——1971. 'Implications of Vegetal Remains from an Illinois Hopewell Site', in *Prehistoric Agriculture*, edited by Stuart Struever, pp. 383–90. American Museum Sourcebooks in Anthropology. The American Museum of Natural History, The Natural History Press, Garden City, New York.

——and Kent D. Vickery. 1973. 'The Beginnings of Cultivation in the Midwest-Riverine Area of the United States', *American Anthropologist* **75**: 1197–1220.

Sturtevant, E. Lewis. 1885. 'Kitchen Garden Esculents of American Origin', *American Naturalist* **19**: 444–57, 542–53, 658–69.

Sturtevant, William C. 1965. 'Preliminary Annotated Bibliography on Eastern North American Indian Agriculture', *Proceedings of the 21st Southeastern Archaeological Conference*, Bulletin No. 3, pp. 1–24.

——1968. 'Lafitau's Hoes', *American Antiquity* **33**: 93–4.

Sullivan, Alan P. 1982. 'Mogollon Agrarian Ecology', *The Kiva* **48**: 1–16.

——and Christian E. Downum. 1991. 'Aridity, Activity, and Volcanic Ash Agriculture: A Study of Short-term Prehistoric Cultural-Ecological Dynamics', *World Archaeology* **22**: 271–87.

Sullivan, John. 1887 [1779]. 'Gen. Sullivan's Report: The Chronicle of his Expedition Against the Iroquois in 1779.—The Devastation of the Genesee County', in *Journals of the Military Expedition of Major General John Sullivan Against the Six Nations of Indians in 1779*, prepared by Frederick Cook, pp. 296–305. Knapp, Peck, and Thomson, Auburn, New York.

Sunseri, Alvin R. 1973. 'Agricultural Techniques in New Mexico at the Time of the Anglo-American Conquest', *Agricultural History* **47**: 329–37.

Suttles, Wayne. 1951. 'The Early Diffusion of the Potato among the Coast Salish', *Southwestern Journal of Anthropology* **7**: 281–8.

Swanton, John R. 1922. *Early History of the Creek Indians and their Neighbors.* Bureau of American Ethnology, Bulletin 73. Smithsonian Institution, Washington, D. C.

——1942. *Source Material on the History and Ethnology of the Caddo Indians.* Bureau of American Ethnology, Bulletin 132. Smithsonian Institution, Washington, D. C.

——1946. *The Indians of the Southeastern United States.* Bureau of American Ethnology, Bulletin 137. Smithsonian Institution, Washington, D. C.

Sykes, Clark M. 1980. 'Swidden Horticulture and Iroquoian Settlement', *Archaeology of Eastern North America* **8**: 45–52.

Tabeau, Pierre Antoine. 1939. *Tabeau's Narrative of Loisel's Expedition to the Upper Missouri*, edited by Annie Heloise Abel and translated by Rose Abel Wright. University of Oklahoma Press, Norman.

Tamarón y Romeral, Pedro. 1954 [1760]. *Bishop Tamarón's Visitation of New Mexico, 1760*, edited by Eleanor B. Adams. New Mexico Historical Society Publications in History 15, University of New Mexico Press, Albuquerque.

Temple, J. H. and George Sheldon. 1875. *A History of the Town of Northfield, Massachusetts, for 150 Years*. Joel Munsell, Albany, New York.

Thomas, Alfred B. 1932. *Forgotten Frontiers: A Study of the Spanish Indian Policy of Don Juan Bautista de Anza, Governor of New Mexico, 1777–1787*. University of Oklahoma Press, Norman.

Thomas, Cyrus. 1894. 'Report on the Mound Explorations of the Bureau of Ethnology', in *Twelfth Annual Report of the Bureau of Ethnology, 1890–91*. Smithsonian Institution, Washington, D. C.

——1903. *The Indians of North America in Historic Times*. George Barrie and Sons, Philadelphia.

Thomas, David Hurst, Lorann S. A. Pendleton, and Stephen C. Cappannari. 1986. 'Western Shoshone', in *Handbook of North American Indians*, Vol. 11, edited by Warren L. D'Azevedo, pp. 262–83. Smithsonian Institution, Washington, D. C.

Thomas, Peter A. 1976. 'Contrastive Subsistence Strategies and Land Use as Factors for Understanding Indian-White Relations in New England', *Ethnohistory* **23**: 1–18.

——1985. 'Cultural Change on the Southern New England Frontier, 1630–1665', in *Cultures in Contact: The Impact of European Contacts on Native American Cultural Institutions, A.D. 1000–1800*, edited by William W. Fitzhugh, pp. 131–61. Smithsonian Institution, Washington, D. C.

Thomas, Robert K. 1961. 'The Redbird Smith Movement', in *Symposium on Cherokee and Iroquois Culture*, edited by William N. Fenton and John Gulick, pp. 159–66. Bureau of American Ethnology, Bulletin 180. Smithsonian Institution, Washington, D. C.

Thompson, Daniel Q. and Ralph H. Smith. 1971. 'The Forest Primeval in the Northeast—A Great Myth?' in *Proceedings of the Annual Tall Timbers Fire Ecology Conference*, pp. 255–65. Tall Timbers Research Station, Tallahassee.

Thorne, D. Wynne. 1979. 'Soil Organic Matter, Microorganisms and Crop Production', in *Soil, Water and Crop Production*, edited by D. Wynne Thorne and Marlowe D. Thorne, pp. 82–95. AVI Publishing, Westport, Connecticut.

Thornton, Russell. 1987. *American Indian Holocaust and Survival: A Population History Since 1492*. University of Oklahoma Press, Norman.

Thruston, Gates P. 1890. *The Antiquities of Tennessee and the Adjacent States*. Robert Clarke, Cincinnati.

Thurston, H. David. 1990. 'Plant Disease Management Practices of Traditional Farmers', *Plant Diseases* **74**: 96–102.

Tiller, Veronica. 1983. 'Jicarilla Apache', in *Handbook of North American Indians*, Vol. 10, edited by Alfonso Ortiz, pp. 440–61. Smithsonian Institution, Washington, D. C.

Timbrook, John, John R. Johnson, and David E. Earle. 1993. 'Vegetation burning by the Chumash', in *Before the Wilderness: Environmental Management by Native*

Californians, compiled and edited by Thomas C. Blackburn and Kat Anderson, pp. 117–49. Ballena Press Anthropological Papers 40, Menlo Park, California.

Titiev, Mischa. 1944. *Old Oraibi: A Study of the Hopi Indians of Third Mesa*. Papers of the Peabody Museum of American Archaeology and Ethnology, Harvard University, Vol. 22, No. 1. Peabody Museum, Cambridge, Massachusetts.

Tjaden, Rex. L. 1979. 'Bordered Garden Plots and Field Houses Near Sapawe, North Central New Mexico: A Dry Farming Strategy'. Unpublished master's thesis, Department of Anthropology, Arizona State University, Tempe.

Toll, H. Wolcott, Mollie S. Toll, Marcia L. Newren, and William B. Gillespie. 1985. 'Experimental Corn Plots in Chaco Canyon: The Life and Hard Times of *Zea mays*'. L., in *Environment and Subsistence of Chaco Canyon, New Mexico*, edited by Frances Joan Mathien, pp. 79–133. Chaco Canyon Studies, Publications in Archeology 18E, Albuquerque.

Tonti, Henri de. 1844 (1693). 'Memoir Sent in 1693, on the Discovery of the Mississippi and the Neighboring Nations, by M. de La Salle, from the Year 1678 to the Time of his Death, and by the Sieur de Tonty to the Year 1691', in *On the Discovery of the Mississippi, and on the South-Western, Oregon, and North-Western Boundary of the United States*, translated by Thomas Falconer, pp. C47–96. Samuel Clarke, London.

Toulouse, Joseph H., Jr. 1945. 'Early Water Systems at Gran Quivira National Monument', *American Antiquity* **10**: 362–72.

Travis, Scott. C. 1990. 'The Prehistoric Agricultural Landscape of Wupatki National Monument', in *The Wupatki Archeological Inventory Survey Project: Final Report*, compiled by Bruce A. Anderson, pp. 411–54. Professional Paper 35. Southwest Cultural Resources Center, National Park Service, Santa Fe.

Treacy, John M. and William M. Denevan. 1994. 'The Creation of Cultivable Land Through Terracing', in *The Archeology of Garden and Field*, edited by Naomi F. Miller and Kathryn L. Gleeson, pp. 91–110. University of Pennsylvania Press, Philadelphia.

Treganza, Adan E. 1947. 'Possibilities of an Aboriginal Practice of Agriculture among the Southern Diegueno', *American Antiquity* **17**: 169–73.

Trigger, Bruce G. 1969. *The Huron: Farmers of the North*. Holt, Rinehart and Winston, New York.

—— 1976. *The Children of Aataentsic*. McGill–Queen's University Press, Montreal.

Trudeau, Jean Bautiste. 1952 [1796]. 'Trudeau's [Truteau's] Description of the Upper Missouri', in *Before Lewis and Clark: Documents Illustrating the History of the Missouri, 1785–1804*, edited by A. P. Nasatir, pp. 376–82. St. Louis Historical Documents Foundation, St. Louis.

Trumbull, Benjamin. 1818. *A Complete History of Connecticut*. Maltby, Goldsmith, and Wadsworth, New Haven.

Tsiang, T. C. 1948. *Soil Conservation: An International Study*. FAO Agricultural Studies 4. United Nations Food and Agriculture Organization, Washington, D. C.

Tuggle, H. David, J. Jefferson Reid, and Robert C. Cole, Jr. 1984. 'Fourteenth Century Mogollon Agriculture in the Grasshopper Region of Arizona', in *Prehistoric Agricultural Strategies in the Southwest*, edited by Suzanne K. Fish and Paul R. Fish, pp. 101–10. Anthropological Research Papers 33. Arizona State University, Tempe.

Turner II, B. L. 1992. 'Comments', in *Gardens of Prehistory: The Archaeology of Set-*

tlement Agriculture in Greater Mesoamerica, edited by Thomas W. Killian, pp. 263–73. University of Alabama Press, Tuscaloosa.
—— and Stephen B. Brush. 1987. 'Purpose, Classification, and Organization', in *Comparative Farming Systems*, edited by B. L. Turner II and Stephen B. Brush, pp. 1–10. The Guilford Press, New York.
—— William M. Denevan. 1985. 'Prehistoric Manipulation of Wetlands in the Americas: A Raised Field Perspective', in *Prehistoric Intensive Agriculture in the Tropics*, edited by I. S. Farrington, pp. 11–30. BAR International Series 232, Oxford.
—— William E. Doolittle. 1978. 'The Concept and Measure of Agricultural Intensity', *Professional Geographer* **30**: 297–301.
—— Peter D. Harrison. 1981. 'Prehistoric Raised-Field Agriculture in the Maya Lowlands', *Science* **213**: 399–405.
Turner, E. Randolph. 1985. 'Socio-Political Organization within the Powhatan Chiefdom and the Effects of European Contact, A.D. 1607–1646', in *Cultures in Contact: The Impact of European Contacts on Native American Cultural Institutions, A.D. 1000–1800*, edited by William W. Fitzhugh, pp. 193–224. Smithsonian Institution, Washington, D. C.
Turner, Nancy J. 1991. 'Burning Mountain Sides for Better Crops: Aboriginal Landscape Burning in British Columbia', *Journal of Ethnobiology* **11**: 286.
Turney, Omar A. 1924. *The Land of the Stone Hoe*. Arizona Republican Print Shop, Phoenix.
—— 1929. *Prehistoric Irrigation*. Arizona State Historian, Phoenix.
Tyler, Daniel. 1986. 'Dating the Caño Ditch: Detective Work in the Pojoaque Valley', *New Mexico Historical Review* **61**: 14–25.

USDA (United States Department of Agriculture) 1965. *Silvics of the Forest Trees of the United States*. Handbook No. 271. USDA, Washington, D. C.
Unger, P. W. 1971. 'Soil Profile Gravel Layers: (I) Effect on Water Storage, Distribution, and Evaporation', *Soil Science Society of America Proceedings* **35**: 631–4.

Vaillant, George C. 1941. *The Aztecs of Mexico*. Doubleday and Duran, Garden City.
Valverde, Antonio de. 1935 [1719]. 'Diary of the Campaign of Governor Antonio de Valverde Against the Ute and Comanche Indians, 1719', in *After Coronado: Spanish Exploration Northeast of New Mexico, 1696–1727*, translated and edited by Alfred Barnaby Thomas, pp. 110–33. University of Oklahoma Press, Norman.
Valverde, Francisco de. 1953*a* [1601]. 'Investigation of Conditions in New Mexico, 1601', in *Don Juan de Oñate: Colonizer of New Mexico, 1595–1628*, Vol. 2, translated and edited by George P. Hammond and Agapito Rey, pp. 623–62. The University of New Mexico Press, Albuquerque.
—— 1953*b* [1601]. 'Inquiry Concerning the New Provinces in the North, 1602', in *Don Juan de Oñate: Colonizer of New Mexico, 1595–1628*, Vol. 2, translated and edited by George P. Hammond and Agapito Rey, pp. 836–77. The University of New Mexico Press, Albuquerque.
Van der Donck, Adriaen. 1841 [1656]. 'A Description of the New Netherlands', in *Collections of the New York Historical Society*, 2nd series, Vol. 1, pp. 125–244. New York.
Van West, Carla. 1994. *Modeling Prehistoric Agricultural Productivity in Southwestern Colorado: A GIS Approach*. Reports of Investigations 67. Department of

Anthropology, Washington State University, Pullman, and Crow Canyon Archaeological Center, Cortez, Colorado.

Van West, Carla and Jeffrey H. Altschul. 1997. 'Environmental Variability and Agricultural Economics along the Lower Verde River, A.D. 750–1450', in *Vanishing River: Landscapes and Lives of the Lower Verde Valley, The Lower Verde Archaeological Project*, edited by Stephanie M. Whittlesey, Richard Ciolek-Torrello, and Jeffry H. Altschul, pp. 337–92. SRI Press, Tucson.

Vargas Rivero, Carlos A. 1983. El Ka'anche': Una Práctica Hortícola Maya. *Biotica* **8**: 151–73.

Vargas Zapata y Luxán Ponze de León, Diego de. 1940 [1692]. *First Expedition of Vargas into New Mexico, 1692*, translated by J. Manuel Espinosa. University of New Mexico Press, Albuquerque.

Vaughan, Alden T. 1965. *New England Frontier: Puritans and Indians 1620–1675*. Little, Brown and Co., Boston.

Vega, Garcilaso de la. 1951 [1605]. *The Florida of the Inca: A History of the Adelantado Hernando de Soto*, translated and edited by John Grier Varner and Jeannette Johnson Varner. University of Texas Press, Austin.

——1993 [1605]. 'La Florida', translated by Charmion Shelby and edited by David Bost, in *The De Soto Chronicles: The Expedition of Hernando De Soto to North America in 1539–1543*, Vol. 2, edited by Lawrence A. Clayton, Vernon James Knight, Jr., and Edward C. Moore, pp. 25–559. University of Alabama Press, Tuscaloosa.

Velarde, Juan. 1966 [1599]. 'Account Given by an Indian of the Flight of Leyva and Humaña from New Mexico', in *The Rediscovery of New Mexico, 1580–1594: The Explorations of Chamuscado, Espejo, Castaño de Sosa, Morlete, and Leyva de Bonilla and Humaña*, translated and edited by George P. Hammond and Agapito Rey, pp. 323–7. University of New Mexico Press, Albuquerque.

Velasco, Luis de. 1953 [1601]. 'Captain Velasco to the Viceroy, San Gabriel, March 22, 1601', in *Don Juan de Oñate: Colonizer of New Mexico, 1595–1628*, Vol. 2, translated and edited by George P. Hammond and Agapito Rey, pp. 608–18. University of New Mexico Press, Albuquerque.

Verano, John W. and Douglas H. Ubelaker (editors). 1993. *Disease and Demography in the Americas*. Smithsonian Institution Press, Washington, D. C.

Verrazano, Giovanni da. 1910 [1510]. 'The History of the Dauphine and its Voyage', in *Fifteenth Annual Report of the American Scenic and Historic Preservation Society*, pp. 179–202. J. B. Lyon Co., Albany.

——1970 [1524]. 'The Written Record of the Voyage of 1524: Translation of the Cellere Codex', by Susan Tarrow, in *The Voyages of Giovanni da Verrazzano: 1524–1528*, edited by Lawrence C. Wroth, pp. 133–43. The Pierpont Morgan Library and Yale University Press, New Haven.

Vivian, R. Gwinn. 1970. 'An Inquiry into Prehistoric Social Organization in Chaco Canyon, New Mexico', in *Reconstructing Prehistoric Pueblo Societies*, edited by William A. Longacre, pp. 69–83. University of New Mexico Press, Albuquerque.

——1972. 'Final Technical Letter Report for Prehistoric Water Conservation in Chaco Canyon, NSF Grant No. GS-3100'. Unpublished manuscript on file at the Arizona State Museum, University of Arizona, Tucson.

——1974. 'Conservation and Diversion: Water-Control Systems in the Anasazi Southwest', in *Irrigation's Impact on Society*, edited by Theodore E. Downing and

McGuire Gibson, pp. 95–112. Anthropological Papers of the University of Arizona, 25. University of Arizona Press, Tucson.
——1990. *The Chacoan Prehistory of the San Juan Basin*. Academic Press, San Diego.
——1991. 'Chacoan Subsistence', in *Chaco and Hohokam: Prehistoric Regional Systems in the American Southwest*, edited by Patricia L. Crown and W. James Judge, pp. 57–75. School of American Research Press, Santa Fe.
Voeglin, Erminie Wheeler. 1940. 'The Place of Agriculture in the Subsistence Economy of the Shawnee', *Papers of the Michigan Academy of Science and Letters* **26**: 513–20.
Vogel, J. C. and N. J. Van Der Merwe. 1977. 'Isotopic Evidence for Early Maize Cultivation in New York State', *American Antiquity* **42**: 238–42.

Waddell, Eric. 1972. *The Mound Builders: Agricultural Practices, Environment, and Society in the Central Highlands of New Guinea*. University of Washington Press, Seattle.
Wallace, Henry. 1983. 'The Mortars, Petroglyphs, and Trincheras on Rillito Peak', *The Kiva* **48**: 137–245.
Wallace, William J. 1978. 'Northern Valley Yokuts', in *Handbook of North American Indians*, Vol. 8, edited by Robert F. Heizer, pp. 462–70. Smithsonian Institution, Washington, D. C.
Wallis, Wilson D. and Ruth Sawtell Wallis. 1955. *The Micmac Indians of Eastern Canada*. University of Minnesota Press, Minneapolis.
Ward, Albert E. (compiler and editor). 1978*a*. *Limited Activity and Occupation Sites: A Collection of Conference Papers*. Center for Anthropological Studies, Albuquerque.
——1978*b*. 'Sinagua Farmers before the "Black Sand" Fell: An Example from Wupatki National Monument', in *Limited Activity and Occupation Sites: A Collection of Conference Papers*, compiled and edited by Albert E. Ward, pp. 135–46. Center for Anthropological Studies, Albuquerque.
Warner, Frederic W. 1972. 'The Foods of the Connecticut Indians', *Archaeological Society of Connecticut Bulletin* **37**: 27–47.
Waselkov, Gregory. 1977. 'Prehistoric Agriculture in the Central Mississippi Valley', *Agricultural History* **51**: 313–19.
——1997. 'Changing Strategies of Indian Field Location in the Early Historic Southeast', in *People, Plants, and Landscapes: Studies in Paleoethnobotany*, edited by Kristen J. Gremillion, pp. 179–94. University of Alabama Press, Tuscaloosa.
Wasley, William W. and Alfred E. Johnson. 1965. *Salvage Archaeology in Painted Rocks Reservoir, Western Arizona*. Anthropological Papers of the University of Arizona 9. University of Arizona Press, Tucson.
Waters, Michael R. 1987. 'Geoarchaeological Investigations of the Schuk Toak and San Xavier Study Areas', in *Archaeological Studies of the Avra Valley, Arizona for the Papago Water Supply Project*, edited by Allen Dart, pp. 207–20. Institute for American Research Anthropological Papers 9, Tucson.
——1988. 'Holocene Alluvial Geology and Geoarchaeology of the San Xavier Reach of the Santa Cruz River, Arizona', *Geological Society of America Bulletin* **100**: 479–91.

Waters, Michael R. and John J. Field. 1986. 'Geomorphic Analysis of Hohokam Settlement Patterns on Alluvial Fans Along the Western Flank of the Tortolita Mountains, Arizona', *Geoarchaeology* **1**: 329–45.

Watson, Patty Jo. 1974. 'Prehistoric Horticulturalists', in *Archaeology of the Mammoth Cave Area*, edited by Patty Jo Watson, pp. 233–8. Academic Press, New York.

——1985. 'The Impact of Early Horticulture in the Upper Drainages of the Midwest and the Midsouth', in *Prehistoric Food Production in North America*, edited by Richard I. Ford, pp. 99–147. Anthropological Papers 75, Museum of Anthropology, University of Michigan, Ann Arbor.

——1988. 'Prehistoric Gardening and Agriculture in the Midwest and Midsouth', in *Interpretations of Culture Change in the Eastern Woodlands During the Late Woodland Period*, edited by Richard W. Yerkes, pp. 39–67. Occasional Papers in Anthropology 3. The Ohio State University, Columbus.

——1995. 'Exploring the Transition to Agriculture', in *Lost Hunter—First Farmers: New Perspectives on the Prehistoric Transition to Agriculture*, edited by T. Douglas Price and Anne Birgitte Gebauer, pp. 21–37. School of American Research Press, Santa Fe.

Waugh, F. W. 1916. *Iroquois Foods and Food Preparation*. Anthropological Series 12, Memoir 86. Canada Department of Mines, Geological Survey, Ottawa.

Weatherford, Jack. 1988. *Indian Givers: How the Indians of the Americas Transformed the World*. Fawcett Columbine, New York.

Weatherwax, Paul. 1936. 'The Origin of the Maize Plant and Maize Agriculture in Ancient America', in *Symposium on Prehistoric Agriculture*. University of New Mexico Bulletin, pp. 11–18. Anthropological Series, Vol. 1, No. 5. University of New Mexico, Albuquerque.

——1954. *Indian Corn in Old America*. The Macmillan Co., New York.

Weaver, Donald E., Jr. 1972. 'A Cultural-Ecological Model for the Classic Hohokam Period in the Lower Salt River Valley, Arizona', *The Kiva* **38**: 43–52.

Webb, Clarence H. 1982. *The Poverty Point Culture* (2nd edn, revised), Geoscience and Man, Vol. 17. Louisiana State University School of Geosciences, Baton Rouge.

Weber, David J. 1992. *The Spanish Frontier in North America*. Yale University Press, New Haven.

Wedel, Waldo R. 1941. *Environment and Native Subsistence Economies in the Central Great Plains*. Smithsonian Miscellaneous Collections, Vol. 101, No. 3. Smithsonian Institution, Washington, D. C.

——1961. *Prehistoric Man on the Great Plains*. University of Oklahoma Press, Norman.

——1986. *Central Plains Prehistory: Holocene Environments and Culture Change in the Republican River Basin*. University of Nebraska Press, Lincoln.

Weigand, Phil C. 1993. 'Large-Scale Hydraulic Works in Prehistoric Western Mesoamerica', in *Economic Aspects of Water Management in the Prehispanic New World*, edited by Vernon L. Scarborough and Barry L. Issac, pp. 223–62. Research in Economic Anthropology, Supplement 7. JAI Press, Greenwich, Connecticut.

Weinstein, Laurie Lee. 1985. 'The Dynamics of Seventeenth Century Wampanoag Land Relations: The Ethnohistorical Evidence for Locational Change', *Bulletin of the Massachusetts Archaeological Society* **46**: 19–34.

Welch, John R. 1992. 'Irrigation Agriculture in the Tonto Basin', in *Proceedings of the*

Second Salado Conference, Globe, AZ, 1992, edited by Richard C. Lange and Stephen Germick, pp. 161–71. Arizona Archeological Society, Phoenix.

——1994. 'Archaeological Studies of Agricultural Context', in *The Roosevelt Rural Sites Study: Changing Land Use in the Tonto Basin*, Vol. 3, edited by Richard Ciolek-Torrello and John R. Welch, pp. 223–51. Technical Series 28. Statistical Research, Tucson.

Wellhausen, E. J., L. M. Roberts, E. Hernández X., and P. C. Mangelsdorf. 1951. *Razas de Maíz en México: su origen, características y distribución*. Folleto Tecnico 5. Oficina de Estudias Especiales, Secretaria de Agricultura y Ganaderia, Mexico.

Weltfish, Gene. 1937. *Caddoan Texts: Pawnee, South Band Dialects*. Publications of the American Ethnology Society, Vol. 17, edited by Franz Boas. G. E. Stechert and Co., New York.

——1965. *The Lost Universe*. Basic Books, New York.

Wenger, Gilbert R. 1980. *The Study of Mesa Verde National Park*. Mesa Verde Museum Association, Mesa Verde National Park, Colorado.

West, George A. 1904. 'Summary of the Archaeology of Racine County, Wisconsin', *Wisconsin Archaeologist*. **3**.1.

——1907. 'The Indian Authorship of Wisconsin Antiquities', *Wisconsin Archeologist* **6**.4.

Whitaker, Thomas W., and Hugh C. Cutler. 1965. 'Curcurbits and Cultures in America', *Economic Botany* **19**: 344–9.

White, Leslie A. 1941. 'The Cultivation of Cotton by Pueblo Indians of New Mexico', *Science* **94**: 162.

——1943. 'Punche: Tobacco in New Mexico History', *New Mexico Historical Review* **18**: 386–93.

White, Richard. 1983. *The Roots of Dependency: Subsistence, Environment, and Social Change among the Choctaws, Pawnees, and Navajos*. University of Nebraska Press, Lincoln.

Whiting, Alfred F. 1937. 'Hopi Indian Agriculture II: Seed Source and Distribution', *Museum Notes: Museum of Northern Arizona* **10**.5: 13–16.

——1939. *Ethnobotany of the Hopi*. Museum of Northern Arizona Bulletin 15, Flagstaff.

Whittlesey, Stephanie M. and Jeffrey A. Homburg. 1992. 'Excavation Results', in *Archaeological Investigations at Lee Canyon: Kayenta Anasazi Farmsteads in the Upper Basin, Coconino County, Arizona*, edited by Stephanie M. Whittlesey, pp. 47–74. Technical Series 38. Statistical Research, Tucson.

Wilbur, C. Keith. 1978. *The New England Indians*. The Globe Pequot Press, Chester, Connecticut.

Wilcox, David R. 1978. 'The Theoretical Significance of Fieldhouses', in *Limited Activity and Occupation Sites: A Collection of Conference Papers*, compiled and edited by Albert E. Ward, pp. 25–32. Center for Archaeological Studies, Albuquerque.

——1985. 'Preliminary Report on New Data on Hohokam Ballcourts', in *Proceedings of the 1983 Hohokam Symposium*, edited by Alfred E. Dittert, and Donald E. Dove, pp. 641–54. Arizona Archaeological Society Occasional Paper 2, Phoenix.

Wilhelm Jr., Gene. 1975. 'Dooryard Gardens and Gardening in the Black Community of Brushy, Texas', *Geographical Review* **65**: 73–92.

Wilken, Gene C. 1969. 'Drained-Field Agriculture: An Intensive Farming System in Tlaxcala, Mexico', *Geographical Review* **59**: 215–41.

Wilken, Gene C. 1987. *Good Farmers: Traditional Agricultural Resource Management in Mexico and Central America*. University of California Press, Berkeley, Los Angeles.

Will, George F. 1924. 'Indian Agriculture at its Northern Limits in the Great Plains Region of North America', *Annals do XX Congresso Internacional de Americanistas* **1**: 203–5. Rio de Janeiro.

—— and George E. Hyde. 1964 [1917]. *Corn among the Indians of the Upper Missouri*. University of Nebraska Press, Lincoln.

—— —— and Herbert J. Spinden. 1906. *The Mandans*. Papers of the Peabody Museum of American Archaeology and Ethnology, Vol. 3, No. 4. Harvard University, Cambridge, Massachusetts.

Williams, Anita Alvarez de. 1983. 'Cocopa', in *Handbook of North American Indians*, Vol. 10, edited by Alfonso Ortiz, pp. 99–112. Smithsonian Institution, Washington, D.C.

Williams, Jerry M. and Robert E. Lewis (editors). 1993. *Early Images of the Americas: Transfer and Invention*. University of Arizona Press, Tucson.

Williams, Lynden S. 1990. 'Agricultural Terrace Evolution in Latin America', *Yearbook, Conference of Latin Americanist Geographers* **16**: 82–93.

Williams, Michael. 1989. *Americans and Their Forests: A Historical Geography*. Cambridge University Press.

Williams, Roger. 1973 [1643]. *A Key into the Language of America*, edited by John J. Teunisser and Evelyn J. Hines. Wayne State University Press, Detroit.

Williams, Samuel Cole. 1930. *Adair's History of the American Indians*. Watauge Press, Tennessee.

Williams, Stephen. 1971. 'Round Table on Definition of Mississippian', *Newsletter of the Southeastern Archaeological Conference* **10**.2: 1–19.

—— and Jeffrey P. Brain. 1983. *Excavations at the Lake George Site, Yazoo County, Mississippi, 1958–1960*. Peabody Museum of Archaeology and Ethnology, Harvard University, Cambridge, Massachusetts.

Williamson, Frederick W. 1940. *Yesterday and Today in Louisiana Agriculture*. Hauser Printing, New Orleans.

Willoughby, Charles C. 1906. 'Houses and Gardens of the New England Indians', *American Anthropologist* **8**: 115–32.

—— 1907. 'The Virginia Indians in the Seventeenth Century', *American Anthropologist*, New Series **9**: 57–86.

—— 1927. 'The Wilderness and the Indian', in *Commonwealth History of Massachusetts*, Vol. 1, edited by Albert Bushnell Hart, pp. 127–58. The States History Company, New York.

—— 1935. *Antiquities of the New England Indians with Notes on the Ancient Cultures of the Adjacent Territories*. Peabody Museum of American Archaeology and Ethnology, Harvard University, Cambridge, Massachusetts.

Wills, W. H. 1988. 'Early Agriculture and Sedentism in the American Southwest: Evidence and Interpretations', *Journal of World Prehistory* **2**: 445–88.

—— 1995. 'Archaic Foraging and the Beginning of Food Production in the American Southwest', in *Lost Hunters—First Farmers: New Perspectives on The Prehistoric Transition to Agriculture*, edited by T. Douglas Price and Anne Birgitte Gebauer, pp. 215–42. School of American Research Press, Santa Fe.

Wilson, Gilbert Livingstone. 1915. 'Corn in Montana', *Montana Agricultural College Experiment Station Bulletin* 107. Montana State University, Boseman.

——1917. *Agriculture of the Hidatsa Indians: An Indian Interpretation.* Studies in the Social Sciences, No. 9. University of Minnesota, Minneapolis.

Wilson, H. Clyde. 1956. 'A New Interpretation of the Wild Rice District of Wisconsin', *American Anthropologist* **58**: 1059–64.

Wilson, John P. 1985. 'Early Piman Agriculture: A New Look', in *Southwestern Culture History: Collected Papers in Honor of Albert H. Schroeder*, edited by Charles H. Lange, pp. 129–38, Papers of the Archaeological Society of New Mexico, Vol. 10. Ancient City Press, Santa Fe.

Winchell, N. H. 1911. *The Aborigines of Minnesota.* The Minnesota Historical Society, St. Paul.

Windes, Thomas C. 1978. *Stone Circles of Chaco Canyon, Northwestern New Mexico.* Reports of the Chaco Center 5, Albuquerque.

——1987*a*. *Investigation at the Pueblo Alto Complex Chaco Canyon, New Mexico 1975–1979*: Vol. I, *Summary of Tests and Excavations at the Pueblo Alto Community.* Chaco Studies, Publications in Archeology 18F, Santa Fe.

——1987*b*. *Investigations at the Pueblo Alto Complex Chaco Canyon, New Mexico 1975–1979*: Vol. II, Part 2, *Architecture and Stratigraphy.* Chaco Studies, Publications in Archeology 18F. Santa Fe.

Winship, George Parker (translator and editor). 1904. *The Journey of Coronado.* A. S. Barnes and Co., New York.

Winter, Joseph C. 1974. 'Aboriginal Agriculture in the Southwest and Great Basin'. Unpublished doctoral dissertation, Department of Anthropology, University of Utah, Salt Lake City.

——1975. *Hovenweep 1974.* Archeological Report 1. Department of Anthropology, San Jose State University.

——1976. *Hovenweep 1975.* Archeological Report 2. Department of Anthropology, San Jose State University.

——1978. 'Anasazi Agriculture at Hovenweep, I: Field Systems', in *Limited Activity and Occupation Sites: A Collection of Conference Papers*, compiled and edited by Albert E. Ward, pp. 83–97. Center for Anthropological Studies, Albuquerque.

_____ and Patrick Hogan. 1986. 'Plant Husbandry in the Great Basin and Adjacent Northern Colorado Plateau', in *Anthropology of the Desert West: Essays in Honor of Jesse D. Jennings*, edited by Carol Cordie and Don Fowler. Anthropological Papers 110. University of Utah, Salt Lake City.

_____ and William J. Litzinger. 1976. 'Floral Indicators of Farm Fields', in *Hovenweep 1975*, Joseph C. Winter, pp. 123–68. Archaeological Report 2. Department of Anthropology, San Jose State University.

Winters, Howard D. 1981. 'Excavating in Museums: Notes on Mississippian Hoes and Middle Woodland Copper Gauges and Celts', in *The Research Potential of Anthropological Museum Collections*, edited by Anne-Marie E. Cantwell, James B. Griffin, and Nan A. Rothschild, pp. 17–34. Annals of the New York Academy of Sciences, Vol. 376. New York.

Winthrop Jr., John. 1863 [1636]. 'John Winthrop, Jr., to His Father', in *Collections of the Massachusetts Historical Society*, Vol. 6., pp. 514–515. Massachusetts Historical Society, Boston.

Wissler, Clark. 1922. *The American Indian: An Introduction to the Anthropology of the New World.* Oxford University Press, New York.

——1940. *Indians of the United States: Four Centuries of their History and Culture,*

The American Museum of Natural History Science Series. Doubleday, New York.

Wissler, Clark. 1967. 'Primitive Agriculture', *Encyclopaedia of the Social Sciences*, Vol. 1, pp. 572–4. Macmillan, New York.

Withers, Alexander Scott. 1908 [1780]. *Chronicles of Border Warfare or, a History of the Settlement by the Whites, of North-Western Virginia and of the Indian Wars and Massacres in that section of the State with Reflections, Anecdotes, & c.*, edited by Reuben Gold Thwaites. The Robert Clarke Co., Cincinnati.

Wittfogel, Karl. 1957. *Oriental Despotism: A Comparative Study of Total Power*. Yale University Press, New Haven.

Wolkomir, Richard. 1995. 'Bringing Ancient Ways to Our Farmers' Fields', *Smithsonian* **26**.8: 99–107.

Wood, J. Scott and Martin E. McAllister. 1984. 'Second Foundation: Settlement Patterns and Agriculture in the Northeastern Hohokam Periphery, Central Arizona', in *Prehistoric Agricultural Strategies in the Southwest*, edited by Suzanne K. Fish and Paul R. Fish, pp. 271–90. Anthropological Research Papers 33. Arizona State University, Tempe.

——— and Linda B. Kelley, and Michael A. Sullivan. 1987. *11,006 Years on the Tonto National Forest: Prehistory and History in Central Arizona.* Cultural Resources Inventory Report 87–68. Tonto National Forest, Phoenix.

Wood, William. 1967 [1634]. *Nevv Englands Prospect. A true, lively, and experimentall deſcription of that part of America.* A Library of American Puritan Writings: The Seventeenth Century, Vol. 9, edited by Sacvan Bercovitch. AMS Press, New York.

Woodbury, Richard B. 1960. 'The Hohokam Canals at Pueblo Grande, Arizona', *American Antiquity* **26**: 267–70.

—— 1961*a*. *Prehistoric Agriculture at Point of Pines, Arizona*. Memoirs of the Society for American Archaeology. 17, Salt Lake City.

—— 1961*b*. 'A Reappraisal of Hohokam Irrigation', *American Anthropologist* **63**: 550–60.

—— 1962. 'Systems of Irrigation and Water Control in Arid North America', *Akten des 34. Internationalen Amerikanistenkongresses 1960*. Verlag Ferdinand Berger, Horn-Wien.

—— 1963. 'Indian Adaptations to Arid Environments', in *Aridity and Man: The Challenge of the Arid Lands in the United States*, edited by Carle Hedge, pp. 55–85. American Association for the Advancement of Science, Washington, D. C.

—— n.d.*a* [1963]. 'Evidence of Prehistoric Farming in the Vicinity of Picuris, New Mexico'. Unpublished Paper on file at the Laboratory of Anthropology, Museum of New Mexico, Santa Fe.

—— n.d.*b* [1970]. 'The Zuni Area'. Unpublished paper presented at the 1970 Pecos Conference symposium on Prehistoric Southwestern Water Control Systems, Pecos, National Monument, New Mexico, on file at the Arizona State Museum, University of Arizona, Tucson.

Woods, William I. 1982. 'Analysis of Soils from the Carrier Mills Archaeological District', in *The Carrier Mills Archaeological Project: Human Adaptation in the Saline Valley, Illinois*, edited by Richard W. Jefferies and Brian M. Butler, pp. 1381–407. Center for Archaeological Investigations, Southern Illinois University, Carbondale.

——1986. 'Prehistoric Settlement and Subsistence in the Upland Cahokia Creek Drainage'. Unpublished doctoral dissertation, Department of Geography, University of Wisconsin–Milwaukee.

——1987. 'Maize Agriculture and the Late Prehistoric: A Characterization of Settlement Location Strategies', in *Emergent Horticultural Economies of the Eastern Woodlands*, edited by William F. Keegan, pp. 275–94. Occasional Paper 7. Center for Archaeological Investigations, Southern Illinois University, Carbondale.

——1993. 'A Study of Prehistoric Settlement-Subsistence Relationships in Southwestern Illinois', in *Proceedings of the First International Conference on Pedo-Archaeology*, edited by J. E. Foss, M. E. Thompson, and M. W. Morris, pp. 175–84. Soil Science Society of America Special Publications, 93-03.

_____ and Sidney G. Denny. 1980. *A Final Report of a Cultural Resource Shoreline Survey of the Rend Lake Reservoir, Franklin and Jefferson Counties, Illinois*. Report to the US Army Corps of Engineers, St. Louis District.

Woodward, Arthur. 1934. *Benjamin David Wilson's Observation of Early Days in California and New Mexico*. Annual Publication of the Historical Society of Southern California, Los Angeles.

Woosley, Anne I. 1980. 'Agricultural Diversity in the Prehistoric Southwest', *The Kiva* **45**: 317–35.

——n.d.*a* [1983]. 'Prehistoric Agriculture in the Northern Rio Grande'. Unpublished paper presented at the Annual Meeting of the Society for American Archaeology, Pittsburgh.

——n.d.*b* [1987]. 'Agricultural Competition and Cooperation on the Taos Plateau', Unpublished paper presented at the Annual Meeting of the Society for American Archaeology, Toronto.

——1986. 'Puebloan Prehistory of the Northern Rio Grande: Settlement, Population, Subsistence', *The Kiva* **51**: 143–64.

Wozniak, Frank. E. 1998. *Irrigation in the Rio Grande Valley, New Mexico: A Study and Annotated Bibliography of the Development of Irrigation Systems*. RMRS-P-2. United States Department of Agriculture, Forest Service, Rocky Mountain Research Station, Fort Collins, Colorado.

Wright, A. C. S. 1962. 'Some Terrace Systems of the Western Hemisphere and Pacific Islands', *Pacific Viewpoint* **3**: 97–101.

Wright, Ronald. 1992. *Stolen Continents: The 'New World' Through Indian Eyes*. Houghton Mifflin, Boston and New York.

Wright, Thomas E. 1993. *Keeping the Boys Busy: Archaeological and Documentary Investigations of AR-03-12-06-1391*. Archaeological Research Services Project Report 93:81, Tempe, Arizona.

Wyckoff, Don G. 1977. 'Secondary Forest Succession Following Abandonment of Mesa Verde', *The Kiva* **42**: 215–31.

Yarnell, Richard Asa. 1964. *Aboriginal Relationships between Culture and Plant Life in the Upper Great Lakes Region*. Anthropological Papers 23. The University of Michigan, Ann Arbor.

——1965*a*. 'Early Woodland Plant Remains and the Question of Cultivation', *Florida Anthropologist* **18**: 77–82.

——1965*b*. 'Implications of Distinctive Flora on Pueblo Ruins', *American Anthropologist* **67**: 662–74.

Yarnell, Richard Asa. 1969. 'Contents of Human Paleofeces', in *The Prehistory of Salts Cave, Kentucky*, edited by Patty Jo Watson, pp. 41–54. Reports of Investigations, Vol. 16. Illinois State Museum, Springfield.

——1972. 'Iva annua var. macropa: Extinct American Cultigen?', *American Anthropologist* **74**: 335–41.

——1974. 'Plant Food and Cultivation of the Salts Cavers,' in *Archeology of the Mammoth Cave Area*, edited by Patty Jo Watson, pp. 113–22. Academic Press, New York.

——1976. 'Early Plant Husbandry in Eastern North America', in *Cultural Change and Continuity: Essays in Honor of James Bennett Griffin*, edited by Charles E. Cleland, pp. 265–73. Academic Press, New York.

——1977. 'Native Plant Husbandry North of Mexico', in *Origins of Agriculture*, edited by Charles A. Reed, pp. 862–75. Mouton, The Hague.

——1978. 'Domestication of Sunflower and Sumpweed in Eastern North America', in *The Nature and Status of Ethnobotany*, edited by Richard I. Ford, pp. 289–99. Anthropological Papers 67. University of Michigan, Ann Arbor.

——1993. 'The Importance of Native Crops during the Late Archaic and Woodland Periods', in *Foraging in the Eastern Woodlands*, edited by C. Margaret Scarry, pp. 13–26. University Press of Florida, Gainesville.

Yen, D. 1989. 'The Domestication of Environment', in *Foraging and Farming: The Evolution of Plant Exploitation*, edited by David R. Harris, and Gordon C. Hillman, pp. 55–75. Unwin Hyman, London.

Zeisberger, David. 1910 [1776]. 'David Zeisberger's History of the Indians', edited by Archer Butler Hulbert and William Nathaniel Schwarze, in *Ohio Archaeological and Historical Quarterly* **19**.1–2.

Ziegler, Alan C. 1968. 'Quasi-Agriculture in North-Central California and Its Effects on Aboriginal Social Structure', *The Kroeber Anthropological Society Papers* **38**: 52–67.

Zigmond, Maurice L. 1986. 'Kawaiisu', in *Handbook of North American Indians*, Vol. 11, edited by Warren L. D'Azevedo, pp. 398–411. Smithsonian Institution, Washington, D. C.

Zimmerer, K. S. 1995. 'The Origins of Andean Irrigation', *Nature* **378**: 481–3.

INDEX